D0900395

MEANS REPAIR AND REMODELING ESTIMATING

Edward B. Wetherill

Illustrated by
Carl W. Linde

MEANS REPAIR AND REMODELING ESTIMATING

R.S. Means Company, Inc. Kingston, Massachusetts

R.S. MEANS COMPANY, INC.
CONSTRUCTION CONSULTANTS & PUBLISHERS
100 Construction Plaza
P.O. Box 800
Kingston, MA 02364-0800
(617) 585-7880

This book was edited by Mary Greene and Ernest Williams. Typesetting was supervised by Joan Marshman. The book and jacket were designed by Norman Forgit. Illustrations by Carl Linde.

Printed in the United States of America

10 9 8 7 6 5 4 3 2 1

Library of Congress Cataloging in Publication Data

ISBN 0-87629-144-2

TABLE OF CONTENTS

Foreword ix

PART I: THE ESTIMATING PROCESS 1

Chapter 1: Types of Estimates 5

Unit Price Estimates 7

Assemblies, or Systems, Estimates 8

Square Foot and Cubic Foot Estimates 13

Order of Magnitude Estimates 13

Chapter 2: The Site Visit and Evaluation 15

Foundations 16

Substructure 19

Superstructure 21

Exterior Closure 24

Roofing 27

Interior Construction 30

Conveying Systems 32

Mechanical 35

Electrical 38

General Conditions 42

Special Items 47

Site Work and Demolition 47

Summary 51

Chapter 3: The Quantity Takeoff 55

Using Pre-printed Forms 55

Summary of Takeoff Guidelines 62

Chapter 4: Pricing the Estimate 63

Sources of Cost Information 63

Obtaining and Recording Information 64

Types of Costs 64

Chapter 5: Direct Costs 67

Material 67

Labor 68

Equipment 73

Subcontractors 74

Project Overhead 74
Bonds 77
Sales Tax 77

Chapter 6: Indirect Costs 79
Taxes and Insurance 79
Office or Operating Overhead 79
Profit 81
Contingencies 81

Chapter 7: The Estimate Summary 85
Organizing the Estimate Data 85
Final Adjustments 86

Chapter 8: Pre-bid Scheduling 91
CPM Scheduling 91
Summary 94

PART II: ESTIMATING BY CSI DIVISION 95

Chapter 9: General Requirements 99
Personnel 100
Temporary Services 100
Temporary Construction 100
Job Clean-up 101
Bonds 101
Miscellaneous General Requirements 103
Other General Requirements Considerations 103

Chapter 10: Site Work/Demolition 105
Site Work 105
Demolition 105
Determining the Appropriate Equipment 109

Chapter 11: Concrete 111
Floor Slabs 111
Other Concrete Items 114

Chapter 12: Masonry 117
Brick 117
Concrete Block 119
Joint Reinforcing 119
Structural Facing Tile 126
Building Stone 126
Stone Floors 128
Glass Block 128
Tips for Estimating Masonry 131

Chapter 13: Metals 135
Structural Steel 135
Steel Joists and Deck 137
Miscellaneous and Ornamental Metals 137
Chapter 14: Wood and Plastics 145
Rough Carpentry 145
Laminated Construction 147
Finish Carpentry and Millwork 147
Chapter 15: Moisture and Thermal Protection 155
Waterproofing 155
Insulation 156
Shingles 157
Other Roofing Materials and Siding 161
Single-Ply Roofs 161
Sheet Metal 162
Roof Accessories 163
Chaper 16: Doors, Windows, and Glass 165
Wood Doors 167
Metal Doors and Frames 169
Special Doors 170
Glass and Glazing 173
Curtain Walls 178
Finish Hardware 178
Chapter 17: Finishes 183
Lath and Plaster 183
Drywall 185
Tile 189
Ceilings 191
Carpeting 195
Composition Flooring 198
Resilient Floors 198
Terrazzo 200
Wood Floors 201
Painting and Finishing 202
Wall Coverings 205
Chapter 18: Specialties 207
Partitions 208
Toilet Partitions, Dressing Compartments, and Screens 208
Chapter 19: Architectural Equipment 211
Chapter 20: Furnishings 213
Furniture 213
Furnishings Items 215
Packing, Crating, Insurance, Shipping, and Delivery 218

Chapter 21: Special Construction 219

Chapter 22: Conveying Systems 221

Chapter 23: Mechanical 227

Plumbing 227

Fire Protection 230

Heating, Ventilating, and Air Conditioning 234

Chapter 24: Electrical 239

Distribution Systems 239

Lighting and Power 243

Chapter 25: Using Means Repair & Remodeling Cost Data 253

Format and Data 253

Unit Price Section 256

Assemblies Section 263

Reference Section 263

City Cost Indexes 263

PART III: ESTIMATING EXAMPLES 271

Chapter 26: Unit Price Estimating Example 273

Chapter 27: Assemblies Estimating Example 375

Index 447

FOREWORD

In new construction, the developer, architect, and estimator have the advantage of being able to plan and analyze a project completely on paper before estimating and construction. There are relatively few unknown variables. Renovation, remodeling and repair, however, pose an entirely different set of problems, and the estimator must approach this type of work more carefully. Without proper analysis and estimating, such projects can easily involve cost overruns and time delays.

The object of this book is to establish proper methods to analyze and estimate repair and remodeling projects. Part I of the book, "The Estimating Process," focuses on types of estimates and the analysis of a project. Part II, "Estimating by CSI Division," provides a step-by-step explanation of the estimate from takeoff to pricing. The final chapter in Part II is a detailed explanation of how to use the annual cost reference, *Means Repair and Remodeling Cost Data*.

Part III contains two complete sample estimates. The first, a Unit Price estimate, is arranged according to the 16 major divisions of the Construction Specifications Institute's (CSI) MASTERFORMAT. The second estimating example of the book, a Systems, or Assemblies, estimate, uses a format that groups all the functional elements of a building into 12 Uniformat (sequential) construction divisions.

This book is written for contractors, architects, engineers, designers, developers, and building owners. A better understanding of the repair and remodeling process by all persons involved will establish a basis for a more efficient and cost effective project.

All prices and construction costs used in this book are found in *Means Repair and Remodeling Cost Data*, 1989 edition. Many pages of this annual publication have been reproduced to show the origin and development of cost data. The *Means Repair and Remodeling Cost Data* book is an up-to-date and thorough source of construction costs.

Part 1
THE ESTIMATING PROCESS

Part 1

THE ESTIMATING PROCESS

In new construction, the estimator is provided with a set of plans and specifications from which he can prepare a complete estimate. Repair and remodeling, however, requires more knowledge and experience of the estimator. Even the best set of plans and specifications for a renovation project cannot anticipate or include all restrictive, existing conditions or possible pitfalls that contractors and design professionals encounter when renovating an existing structure. Consequently, this type of project poses the greatest challenge to the estimator in the exercise of skill and professional judgment.

In this book, the term "estimator" is used as an all-inclusive title that may include the contractor, architect, owner, or any person who requires costs for a remodeling project. Before starting the estimate, the estimator must be able to analyze the structure. Existing conditions must be identified in terms of their effect – on the work to be performed, and, therefore, the estimate. The estimator must also decide what type of estimate is appropriate for the proposed renovation project. This decision should be based on:

- The amount of information supplied to the estimator
- The amount of time allowed to perform the estimate
- The purpose of the estimate

How many times have clients given estimators a small sketch on scrap paper or an oral description of the project, and said, "I need to know how much this will cost, right away"? Although this situation may be extreme, it does happen to everyone at some point. On the other hand, the client may give the estimator 40 sheets of plans, a specification book two inches thick, and a month to complete the estimate. Most remodeling projects fall somewhere between these two extremes. Depending on the above factors, the estimator must choose the type of estimate best suited for the project. This choice will affect how much time will be spent on the estimate as well as how accurate the result might be.

Chapter 1

TYPES OF ESTIMATES

Every cost estimate involves three basic steps. Without all three, the estimate cannot be completed. These steps are:

1. Designation of a "unit" of measure
2. Determination of quantity of units
3. Establishment of a reasonable cost per unit

The first step is designating an appropriate **unit** of measure. The units chosen will depend on the level of detail required (or known) at this stage of the project. In construction, such units could be as detailed as a square foot of drywall or as all-encompassing as a square foot of floor area. In auto repair, a unit could be an individual spark plug or a complete engine. Depending upon the estimator's intended use, the designation of the unit may imply only an isolated entity, or may describe the unit as *in place*. In building construction, these units are described as *material only* or *installed*, respectively. The *installed* unit includes both material and labor.

The second step of an estimate is the determination of the **quantity of units**. This is an actual counting process: how many square feet of drywall, how many spark plugs.

In construction, this process is called the "quantity takeoff." In order to perform this function successfully, the estimator should have a working knowledge of construction materials and methods. This knowledge helps to ensure that each quantity is correctly tabulated, and that items are not forgotten or omitted. The estimator with a sound construction knowledge is also more likely to account for all required work in the estimate. Experience is, therefore, invaluable.

The third step, and perhaps the most difficult to carry out, is the determination of a reasonable **cost for each unit.** If costs for units were to remain constant and there were no variables in construction, then there would be no need for estimates or estimators. Prices do fluctuate, however, and labor productivity varies; no two projects are exactly alike. It is this third step, determination of unit costs, that is most responsible for variations in estimating. For example, even though material costs for framing lumber may be the same for competing contractors, the labor costs for installing that material can vary because of a difference in productivity. One factor may be the use of specialized equipment, which can decrease installation time and, therefore, cost. Labor rates may also vary according to pay scales in different areas.

Generally, material prices fluctuate within the market. Cost differences occur from city to city and even from supplier to cross-town supplier. It is the experienced and well prepared estimator who can keep track of these variations and fluctuations and use them to best advantage when preparing accurate estimates.

What is the correct or accurate cost of a given construction project? Is it the total price that the owner pays to the contractor? Might not another reputable contractor perform the same work for a different cost, whether higher or lower? There is no one correct estimated cost for a given project. There are too many variables in construction. At best, the estimator can determine a very close approximation of what the final costs to the owner will be. The resulting accuracy of this approximation is directly affected by the amount of detail provided and the amount of time spent on the estimate.

Estimating for building construction, especially remodeling and renovation, is certainly not as simple as proceeding through three simple steps and arriving at a "magic figure." Detailed estimates for large projects require many weeks of hard work. The purpose of this text is to make the estimating process easier and more organized for the experienced estimator, and to provide those who are less experienced with a basis for sound estimating practice.

We begin with a discussion of the different types of estimates. All estimates break a construction project down into various stages of detail. By determining the quantities involved and the cost of each item, the estimator is able to complete an estimate. The units used to determine the quantities can be large in scale, like the number of apartments in a housing renovation, or very detailed, like the number of square feet of drywall for a repair job. The type of estimate used is based on the amount and detail of information supplied to the estimator, the amount of time available to complete the estimate, and the purpose of the estimate, such as for a preliminary feasibility study or a final construction bid. Depending on these criteria, there are four basic choices of estimate types:

1. **Unit Price** These estimates require working drawings and specifications. They are the most detailed and take the greatest amount of time to complete. The relative accuracy of a unit price estimate for repair and remodeling can be plus or minus (+ or −) 10%.
2. **Assemblies,** or **Systems** A Systems estimate is used when certain parameters of a renovation project are known, such as building size, general type of construction, type of heating system, etc. Relative accuracy can be + or − 15%.
3. **Square Foot and Cubic Foot** This type is used when only the size and proposed use for a renovation are known. These estimates are much faster to complete than unit price or systems estimates and can provide relative accuracy of + or − 20%.
4. **Order of Magnitude** These estimates are used for planning future renovation projects. Relative accuracy can be + or − 25%.

Variables in both the estimating process and the projects themselves are responsible for the differences in estimate accuracy. Figure 1.1 shows the relationship between the time spent preparing an estimate and the relative accuracy for each type of estimate. The accuracy described is based on an "average of reasonable bids," assuming that there are a number of competitive bids for a given project.

Unit Price Estimates

Unit Price estimates are the most accurate and detailed of the four estimate types, and take the most time to complete. Detailed working drawings and specifications should be available to the unit price estimator. Therefore, all decisions regarding materials and methods for the remodeling project should have been made. This effectively reduces the number of variables and "educated guesses" that can decrease the accuracy of an estimate.

The working drawings and specifications are needed to determine the quantities of material, equipment, and labor. Current and accurate costs for these items, in the form of unit prices, are also necessary to complete the estimate. These costs can come from different sources. Wherever possible, the estimator should use prices based on experience or costs from recent, similar projects. No two renovation projects are alike, however. Prices may, in some cases, be determined using an up-to-date industry source book, such as *Means Repair and Remodeling Cost Data*.

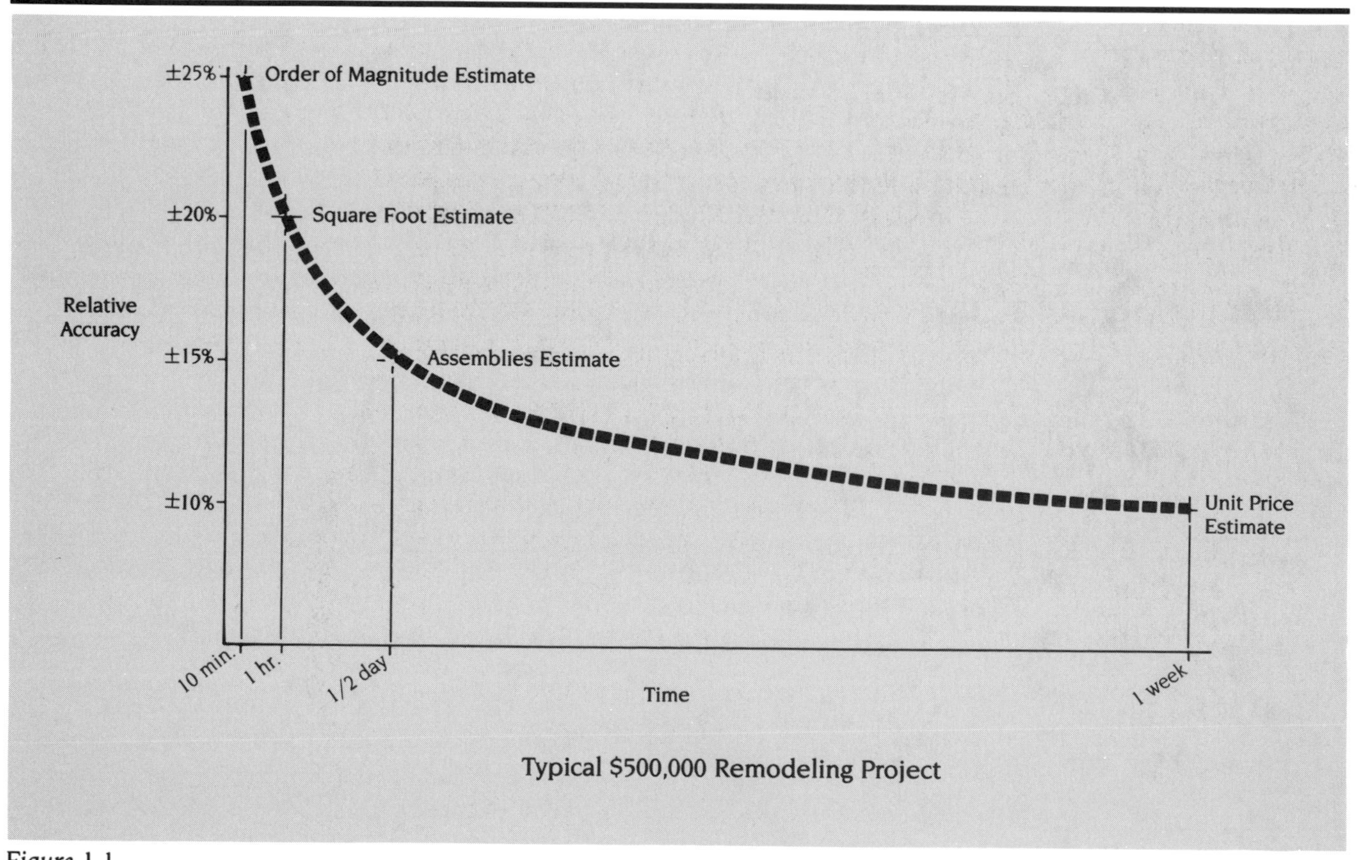

Figure 1.1

Because the preparation of unit price estimates requires a great deal of time and expense, this type of estimating is best suited for construction bids. It can also be effective for determining certain detailed costs in conceptual budgets or during design development.

Most construction specification manuals and cost reference books, including Means' annual cost data books, allocate all construction components into the 16 MASTERFORMAT divisions developed by the Construction Specifications Institute, Inc. and listed below.

CSI MASTERFORMAT Divisions

Division 1 – General Requirements
Division 2 – Site Work
Division 3 – Concrete
Division 4 – Masonry
Division 5 – Metals
Division 6 – Wood & Plastics
Division 7 – Thermal and Moisture Protection
Division 8 – Doors and Windows
Division 9 – Finishes
Division 10 – Specialties
Division 11 – Equipment
Division 12 – Furnishings
Division 13 – Special Construction
Division 14 – Conveying Systems
Division 15 – Mechanical
Division 16 – Electrical

This method of categorizing construction components provides a standard of uniformity that is widely used in the construction industry. Since a great number of architects use this system for specifications, it makes sense to base estimates and cost data on the same format. Figure 1.2 shows a page listing types of drywall from the Unit Price section of *Means Repair and Remodeling Cost Data*, 1989. Each unit price page contains a wealth of information that can be used for Unit Price estimates. The type of work to be performed is described in detail: what kind of crew is needed; how long it takes to perform the unit of work; and separate costs for material, labor, and equipment. Total costs are extended to include the installer's overhead and profit.

Figure 1.3 is a Means Condensed Estimate Summary form (from *Means Forms for Building Construction Professionals*). This form can be used to summarize the information that has been separately estimated by division. This form provides a checklist to ensure that estimators have included all divisions. It is also a concise means for determining the total costs of the Unit Price estimate. Please note at the top of the form the items: *Total Area*, *Volume*, and *Cost per* S.F./C.F. These are the basic "units" for the Order of Magnitude, Square Foot, and Cubic Foot estimates which may be used for budgeting or as a cross-check for similar projects in the future.

Assemblies, or Systems Estimates

Rising design and construction costs in recent years have made budgeting and cost efficiency increasingly important in the planning stages of remodeling and renovation projects. Never has it been so important to involve the estimator in the initial planning stages. Unit

092 | Lath, Plaster and Gypsum Board

092 600 | Gypsum Board Systems

		092 600 Gypsum Board Systems	CREW	DAILY OUTPUT	MAN-HOURS	UNIT	BARE COSTS MAT.	BARE COSTS LABOR	BARE COSTS EQUIP.	BARE COSTS TOTAL	TOTAL INCL O&P	
602	1400	On beams, columns, or soffits, standard, no finish included	2 Carp	675	.024	S.F.	.29	.51		.80	1.13	602
	1450	With thin coat plaster finish		475	.034		.38	.72		1.10	1.58	
	1600	Fire resistant, no finish included		675	.024		.33	.51		.84	1.18	
	1700	With thin coat plaster finish		475	.034		.42	.72		1.14	1.62	
	3000	½" thick, on walls or ceilings, standard, no finish included		1,900	.008		.20	.18		.38	.51	
	3100	With thin coat plaster finish		875	.018		.29	.39		.68	.95	
	3300	Fire resistant, no finish included		1,900	.008		.24	.18		.42	.55	
	3400	With thin coat plaster finish		875	.018		.33	.39		.72	.99	
	3450	On beams, columns, or soffits, standard, no finish included		675	.024		.30	.51		.81	1.14	
	3500	With thin coat plaster finish		475	.034		.39	.72		1.11	1.59	
	3700	Fire resistant, no finish included		675	.024		.34	.51		.85	1.19	
	3800	With thin coat plaster finish		475	.034		.43	.72		1.15	1.63	
	5000	⅝" thick, on walls or ceilings, fire resistant, no finish included		1,900	.008		.24	.18		.42	.55	
	5100	With thin coat plaster finish		875	.018		.33	.39		.72	.99	
	5500	On beams, columns, or soffits, no finish included		675	.024		.34	.51		.85	1.19	
	5600	With thin coat plaster finish		475	.034		.43	.72		1.15	1.63	
	6000	For high ceilings, over 8' high, add		3,060	.005		.10	.11		.21	.29	
	6500	For over 3 stories high, add per story	↓	6,100	.003	↓	.05	.06		.11	.15	
	9000	Minimum labor/equipment charge	1 Carp	2	4	Job		86		86	135	
608	0010	DRYWALL Gypsum plasterboard, nailed or screwed to studs,										608
	0100	unless otherwise noted										
	0150	⅜" thick, on walls, standard, no finish included	2 Carp	2,000	.008	S.F.	.19	.17		.36	.48	
	0200	On ceilings, standard, no finish included		1,800	.009		.19	.19		.38	.51	
	0250	On beams, columns, or soffits, no finish included	↓	675	.024	↓	.29	.51		.80	1.13	
	0270											
	0300	½" thick, on walls, standard, no finish included	2 Carp	2,000	.008	S.F.	.20	.17		.37	.49	
	0350	Taped and finished		965	.017		.25	.35		.60	.84	
	0400	Fire resistant, no finish included		2,000	.008		.24	.17		.41	.54	
	0450	Taped and finished		965	.017		.29	.35		.64	.89	
	0500	Water resistant, no finish included		2,000	.008		.30	.17		.47	.60	
	0550	Taped and finished		965	.017		.35	.35		.70	.95	
	0600	Prefinished, vinyl, clipped to studs	↓	900	.018	↓	.60	.38		.98	1.27	
	0650											
	1000	On ceilings, standard, no finish included	2 Carp	1,800	.009	S.F.	.20	.19		.39	.53	
	1050	Taped and finished		765	.021		.25	.45		.70	.99	
	1100	Fire resistant, no finish included		1,800	.009		.24	.19		.43	.57	
	1150	Taped and finished		765	.021		.29	.45		.74	1.04	
	1200	Water resistant, no finish included		1,800	.009		.30	.19		.49	.64	
	1250	Taped and finished		765	.021		.35	.45		.80	1.10	
	1500	On beams, columns, or soffits, standard, no finish included		675	.024		.30	.51		.81	1.14	
	1550	Taped and finished		475	.034		.35	.72		1.07	1.54	
	1600	Fire resistant, no finish included		675	.024		.34	.51		.85	1.19	
	1650	Taped and finished		475	.034		.39	.72		1.11	1.59	
	1700	Water resistant, no finish included		675	.024		.40	.51		.91	1.25	
	1750	Taped and finished		475	.034		.45	.72		1.17	1.65	
	2000	⅝" thick, on walls, standard, no finish included		2,000	.008		.24	.17		.41	.54	
	2050	Taped and finished		965	.017		.29	.35		.64	.89	
	2100	Fire resistant, no finish included		2,000	.008		.26	.17		.43	.56	
	2150	Taped and finished		965	.017		.31	.35		.66	.91	
	2200	Water resistant, no finish included		2,000	.008		.34	.17		.51	.65	
	2250	Taped and finished		965	.017		.39	.35		.74	1	
	2300	Prefinished, vinyl, clipped to studs	↓	900	.018	↓	.67	.38		1.05	1.35	
	2350											
	3000	On ceilings, standard, no finish included	2 Carp	1,800	.009	S.F.	.24	.19		.43	.57	
	3050	Taped and finished		765	.021		.29	.45		.74	1.04	
	3100	Fire resistant, no finish included		1,800	.009		.26	.19		.45	.59	
	3150	Taped and finished	↓	765	.021	↓	.31	.45		.76	1.06	

For expanded coverage of these items see *Means Interior Cost Data 1989*

141

Figure 1.2

Means® Forms

CONDENSED ESTIMATE SUMMARY

SHEET NO.

PROJECT | ESTIMATE NO.

LOCATION | TOTAL AREA/VOLUME | DATE

ARCHITECT | COST PER S.F./C.F. | NO. OF STORIES

PRICES BY: | EXTENSIONS BY: | CHECKED BY:

DIV.	DESCRIPTION	MATERIAL	LABOR	EQUIPMENT	SUBCONTRACT	TOTAL
1.0	**General Requirements**					
2.0	**Site Work**					
3.0	**Concrete**					
4.0	**Masonry**					
5.0	**Metals**					
6.0	**Carpentry**					
7.0	**Moisture & Thermal Protection**					
8.0	**Doors, Windows, Glass**					
9.0	**Finishes**					
10.0	**Specialties**					
11.0	**Equipment**					
12.0	**Furnishings**					
13.0	**Special Construction**					
14.0	**Conveying Systems**					
15.0	**Mechanical**					
16.0	**Electrical**					
	Subtotals					
	Sales Tax %					
	Overhead %					
	Subtotal					
	Profit %					
	Contingency %					
	Adjustments					
	TOTAL BID					

Figure 1.3

Price estimating, which requires more time and detailed information, is not an appropriate budgetary or planning tool. A faster and more cost effective method is needed at the planning stage of a renovation project. This is the **Assemblies** Estimate.

Instead of the 16 CSI divisions used for Unit Price estimating, the Assemblies estimate reorganizes separate trade items to reflect a logical, sequential approach to the construction of a project. The result is 12 "Uniformat" divisions that break the renovation project into assemblies, or systems. They are listed below:

Systems Estimating Divisions:

Division 1 – Foundations
Division 2 – Substructures
Division 3 – Superstructure
Division 4 – Exterior Closure
Division 5 – Roofing
Division 6 – Interior Construction
Division 7 – Conveying
Division 8 – Mechanical
Division 9 – Electrical
Division 10 – General Conditions
Division 11 – Special
Division 12 – Site Work and Demolition

Each of these Uniformat divisions may incorporate items from different unit pricing divisions. For example, when estimating an interior partition using the Assemblies approach, the estimator uses Division 6 – Interior Construction. (See Figure 1.4, a page from *Means Repair and Remodeling Cost Data*, 1989). When estimating the same interior partition using the Unit Price approach, the estimator would refer to CSI Masterformat Division 6 (for wood studs and baseboard), Division 7 (for insulation), and Division 9 (for drywall, taping, and paint). Conversely, a particular unit price item like cast-in-place concrete (Division 3) may be included in different Assemblies divisions: Division 1 – Foundations, Division 2– Substructures, Division 3 – Superstructure. The Assemblies method better reflects the way the contractor views the construction of a renovation project. Although it does not allow for the detail provided by the Unit Price approach, it is a faster way to develop costs.

A great advantage of the Assemblies estimate is that the estimator/designer is able to vary components of a renovation project and quickly determine the cost differential. The owner can then anticipate accurate budgetary requirements before final design, details, and dimensions are established.

In the Unit Price estimate, final details regarding the renovation project are available to the estimator. When using the Assemblies estimate, particularly for renovation projects, estimators must draw on their experience and knowledge of building code requirements, design options, and the ways in which existing conditions limit and restrict the proposed building renovations.

The Assemblies estimate should not be used as a substitute for the Unit Price estimate. While the "Assemblies" approach can be invaluable in the planning and budget stages of a renovation, the Unit Price method should be used when greater accuracy is required.

INTERIOR CONSTR. | 06.1-592 | Partitions, Wood Stud

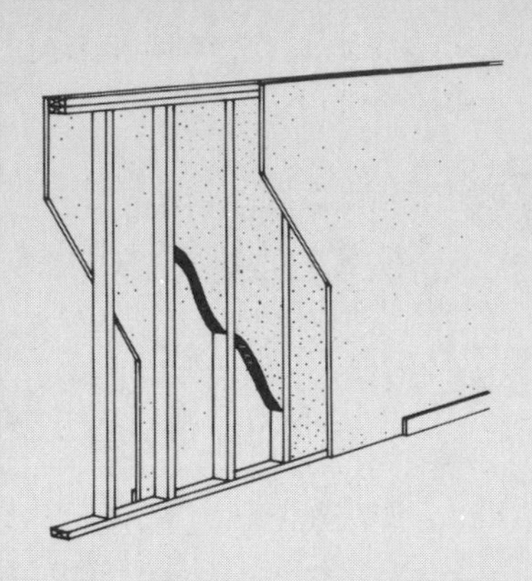

This page illustrates and describes a wood stud partition system including wood studs with plates, gypsum plasterboard - taped and finished, insulation, baseboard and painting. Lines 06.1-592-04 thru 10 give the unit price and total price per square foot for this system. Prices for alternate wood stud partition systems are on Line Items 06.1-592-13 thru 27. Both material quantities and labor costs have been adjusted for the system listed.

Factors: To adjust for job conditions other than normal working situations use Lines 06.1-592-29 thru 40.

Example: You are to install the system where material handling and storage present a serious problem. Go to Line 06.1-592-34 and apply these percentages to the appropriate MAT. and INST. costs.

LINE NO.	DESCRIPTION	QUANTITY	COST PER S.F.		
			MAT.	INST.	TOTAL
01	**Wood stud wall, 2"x4",16"O.C., dbl. top plate, sngl bot. plate, ⅝" dwl.**				
02	**Taped, finished and painted on 2 faces, insulation, baseboard, wall 8' high**				
03					
04	Wood studs, 2" x 4", 16" O.C., 8' high	1 S.F.	.35	.71	1.06
05	Gypsum drywall, ⅝" thick	2 S.F.	.53	.55	1.08
06	Taping and finishing	2 S.F.	.11	.55	.66
07	Insulation, 3-½" fiberglass batts	1 S.F.	.30	.17	.47
08	Baseboard, painted	.2 L.F.	.21	.43	.64
09	Painting, roller, 2 coats	2 S.F.	.20	.54	.74
10	TOTAL	S.F.	1.70	2.95	4.65
11					
12	For alternate wood stud systems:				
13	2" x 3" studs, 8' high, 16" O.C.	S.F.	1.68	2.91	4.59
14	24" O.C.		1.65	2.78	4.43
15	10' high, 16" O.C.		1.66	2.79	4.45
16	24" O.C.		1.61	2.68	4.29
17	2" x 4" studs, 8' high, 24" O.C.		1.67	2.81	4.48
18	10' high, 16" O.C.		1.68	2.81	4.49
19	24" O.C.		1.63	2.70	4.33
20	12' high, 16" O.C.		1.65	2.81	4.46
21	24" O.C.		1.60	2.70	4.30
22	2" x 6" studs, 8' high, 16" O.C.		1.88	3.03	4.91
23	24" O.C.		1.81	2.86	4.67
24	10' high, 16" O.C.		1.85	2.87	4.72
25	24" O.C.		1.77	2.74	4.51
26	12' high, 16" O.C.		1.80	2.90	4.70
27	24" O.C.	↓	1.74	2.74	4.48
28					
29	Cut & patch to match existing construction, add, minimum		2%	3%	
30	Maximum		5%	9%	
31	Dust protection, add, minimum		1%	2%	
32	Maximum		4%	11%	
33	Material handling & storage limitation, add, minimum		1%	1%	
34	Maximum		6%	7%	
35	Protection of existing work, add, minimum		2%	2%	
36	Maximum		5%	7%	
37	Shift work requirements, add, minimum			5%	
38	Maximum			30%	
39	Temporary shoring and bracing, add, minimum		2%	5%	
40	Maximum		5%	12%	
41					
42					

277

Figure 1.4

Square Foot and Cubic Foot Estimates

Square Foot and Cubic Foot estimates are appropriate when a building owner wants to know the cost of a renovation before the plans or even sketches are available. Often these costs are needed to determine whether it is economically feasible to proceed with a project, or to determine the best use (apartments, offices, etc.) for an existing structure.

Square foot costs for new construction can be found in the annually-updated *Means Square Foot Costs*. However, in remodeling and renovation, each existing building and each project is unique. Therefore, square foot estimating is not always effective for this kind of work. One building might have a two-year-old heating system requiring little work, while another may need all new equipment. One building might need a new roof, while another requires only patching. These variations challenge the estimator and make a site visit to determine and evaluate existing conditions critical to the square foot and cubic foot estimating process. The site visit and evaluation are covered in Chapter 2.

The best data available to the estimator is from past projects. (Please refer back to Figure 1.3, the Estimate Summary Form, which provides a convenient reference of such historical data, listing costs per S.F./C.F.) This data, together with the estimator's experience with the variables between projects, enables the estimator to use square foot and cubic foot estimates effectively for repair and remodeling work.

Order of Magnitude Estimates

Order of Magnitude estimates require the least amount of time to complete, and provide the lowest level of estimate accuracy. The information required is the **proposed use of the building** and the **number of units** (apartments, hospital beds, etc.). Figure 1.5, which shows an example of this level of information, is a page from *Means Building Construction Cost Data*, 1989. Note that under *Hospitals* and *Housing*, the final line items are assigned a cost per unit.

Order of Magnitude estimates are used primarily for planning purposes and primarily for new construction. The complexities of remodeling and renovation make the Order of Magnitude estimate ineffective unless costs from similar projects in similar existing buildings are available.

171 | S.F., C.F. and % of Total Costs

171 000 | S.F. & C.F. Costs

			UNIT	UNIT COSTS ¼	UNIT COSTS MEDIAN	UNIT COSTS ¾	% OF TOTAL ¼	% OF TOTAL MEDIAN	% OF TOTAL ¾	
410	9000	Per car, total cost	Car	5,375	7,325	10,300				410
	9500	Total: Mechanical & Electrical	"	385	580	710				
430	0010	**GYMNASIUMS**	S.F.	44.10	59.35	76				430
	0020	Total project costs	C.F.	2.20	3.06	3.90				
	1800	Equipment	S.F.	1.14	1.90	3.33	2%	3.20%	6.70%	
	2720	Plumbing		2.72	3.70	4.68	4.80%	7.20%	8.50%	
	2770	Heating, ventilating, air conditioning		2.89	5	8.20	7.40%	9.70%	14%	
	2900	Electrical		3.71	4.53	6.75	6.50%	8.90%	10.70%	
	3100	Total: Mechanical & Electrical	↓	8	12.45	16.35	16.70%	21.80%	27%	
	3500	See also division 114-801 & 114-805								
460	0010	**HOSPITALS**	S.F.	97.40	118	159				460
	0020	Total project costs	C.F.	7.15	8.60	11.50				
	1120	Roofing	S.F.	.71	1.77	2.88	.50%	1.20%	2.90%	
	1320	Finish hardware		.96	1.04	1.26	.60%	1%	1.20%	
	1540	Floor covering		.67	1.22	3.07	.50%	1.10%	1.60%	
	1800	Equipment		2.40	4.44	6.64	1.60%	3.80%	5.60%	
	2720	Plumbing		8.60	10.95	15	7.50%	9.10%	10.80%	
	2770	Heating, ventilating, air conditioning		9.30	15.50	21.75	8.40%	13%	16.70%	
	2900	Electrical		10.05	13.45	20.45	10%	12.30%	15.10%	
	3100	Total: Mechanical & Electrical	↓	29.45	40.15	59.35	28.20%	36.60%	40.30%	
	9000	Per bed or person, total cost	Bed	26,400	51,400	67,700				
	9900	See also division 117-001								
480	0010	**HOUSING** For the Elderly	S.F.	47.35	59.40	74.50				480
	0020	Total project costs	C.F.	3.34	4.63	6.05				
	0100	Sitework	S.F.	3.38	5.10	7.35	6%	8.20%	12.10%	
	0500	Masonry		1.42	4.98	7.60	2.10%	6.50%	10.40%	
	0730	Miscellaneous metals		1.39	1.85	4.88	1.40%	2.10%	3.10%	
	1120	Roofing		.92	1.59	2.81	1.20%	2.10%	3%	
	1140	Dampproofing		.27	.37	.85	.20%	.40%	.70%	
	1340	Windows		.64	1.09	2.35	1.10%	1.50%	2.40%	
	1350	Glass & glazing		.17	.49	.91	.20%	.40%	.90%	
	1530	Drywall		2.81	3.69	5.91	3.70%	4.10%	4.60%	
	1540	Floor covering		.76	1.13	1.63	.90%	1.40%	1.90%	
	1570	Tile & marble		.36	.52	.78	.50%	.60%	.80%	
	1580	Painting		1.52	2.14	3.32	2%	2.60%	3.10%	
	1800	Equipment		1.09	1.51	2.39	1.80%	3.20%	4.40%	
	2510	Conveying systems		1.11	1.52	2.02	1.70%	2.30%	2.80%	
	2720	Plumbing		3.58	4.96	7.24	8.30%	9.70%	10.90%	
	2730	Heating, ventilating, air conditioning		1.60	2.48	3.50	3.20%	5.60%	7.10%	
	2900	Electrical		3.48	4.87	6.85	7.50%	9%	10.60%	
	2910	Electrical incl. electric heat		3.98	7.50	8.90	9.60%	11%	13.30%	
	3100	Total: Mechanical & Electrical	↓	8.65	12.20	15.90	18.40%	21.90%	24.70%	
	9000	Per rental unit, total cost	Unit	41,800	49,500	54,400				
	9500	Total: Mechanical & Electrical	"	8,300	10,400	12,300				
500	0010	**HOUSING** Public (low-rise)	S.F.	36.45	49.90	67.90				500
	0020	Total project costs	C.F.	3.01	3.92	4.95				
	0100	Sitework	S.F.	4.77	6.70	10.35	8.30%	11.70%	16.10%	
	1800	Equipment		1.04	1.69	2.76	2.20%	3.10%	4.60%	
	2720	Plumbing		2.60	3.59	4.61	7.10%	9%	11.50%	
	2730	Heating, ventilating, air conditioning		1.39	2.60	2.95	4.40%	6%	6.40%	
	2900	Electrical		2.29	3.29	4.62	5%	6.50%	8.10%	
	3100	Total: Mechanical & Electrical	↓	6.80	9.70	13.55	15.60%	19.20%	22.20%	
	9000	Per apartment, total cost	Apt.	39,700	45,000	56,200				
	9500	Total: Mechanical & Electrical	"	6,675	9,150	11,500				
510	0010	**ICE SKATING RINKS**	S.F.	33.90	47.35	77.70				510
	0020	Total project costs	C.F.	1.92	2.41	2.84				
	2720	Plumbing	S.F.	1.02	1.50	2.29	3.10%	3.20%	4.60%	
	2900	Electrical	"	2.67	3.51	4.87	5.70%	7%	10.10%	

362

For expanded coverage of these items see *Means Square Foot Cost Data 1989*

Figure 1.5

Chapter 2
THE SITE VISIT AND EVALUATION

In repair and remodeling, it is of the utmost importance for the estimator to conduct a site visit and evaluation before performing the estimate. To create a reliable estimate, the estimator must become familiar with each project, and this requires an inspection of the existing conditions.

The extent and detail of the plans and specifications supplied to the estimator have an important influence on the site visit and the proper completion of the estimate. As stated previously, this information may range from minimal to extensive. Before conducting the site visit, the estimator should examine all information provided. As important as it is to perform the site evaluation before making the estimate, it is just as important to know what the project entails before visiting the site. For example, the specifications may state:

> *"All existing mortar joints shall be scraped to a depth of at least 1/2", using hand tools only, before tuck pointing individual joints. Care shall be taken not to allow new mortar to stain existing porous bricks."*
>
> or:
>
> *"Clean and patch existing masonry joints."*

Clearly, the two statements could have entirely different implications.

Generally, the architect involved in the planning and supervision of the project will answer most questions about methods and materials. If an architect is not involved in the supervision of the project, then the estimator is responsible for the interpretation of the plans and specifications. However, the estimator should not make assumptions or interpretations in areas where he lacks experience or knowledge of specific conditions.

When there is no architect involved in the project, it is often the estimator's responsibility to make sure that all items are included in the estimate and that the roles of the subcontractors do not overlap. This is especially important in renovation, where omissions or overlaps are more likely to occur. For example, wood backing for plumbing fixtures could be included in both carpentry and plumbing subcontracts. At the same time, providing holes through existing walls for sprinkler piping could be omitted, as each trade assumes that this item is the responsibility of another. This type of problem is less likely to occur in new construction where all parties are familiar with established precedents. It cannot be stressed enough, however, that each remodeling project is different and requires special attention and diligence.

If possible, subcontractors and engineers should accompany the estimator on the site visit. Undoubtedly, questions will arise specific to certain trades. The estimator should know how existing conditions will affect the work to be performed.

Knowing the applicable Building, Fire, and Energy Codes also helps the estimator to anticipate any requirements that may not be included in the plans and specifications. It should be noted that Building, Fire, and Energy Codes for existing structures are often different from those that apply to new buildings.

The estimator for remodeling projects should also be familiar with older types of construction, such as mill-type wood construction built before extensive use of steel and concrete, or the terra cotta-encased structural steel method introduced with the advent of fire-resistant construction. Each of these older types of construction requires a different approach to remodeling.

This chapter concerns the site visit and evaluation. While one text cannot possibly describe all the variables in commercial renovation, we will present an effective approach for evaluating an existing building: what to look for, how to look, and what to anticipate.

The best way to evaluate and analyze an existing structure is from the bottom up. By beginning the inspection of existing conditions at the bottom, the estimator can observe the structure as it was built. Structural elements can be identified and followed up through the building. The mechanical and electrical systems can also be traced from below to better understand the distribution above.

Throughout the site visit and evaluation, the estimator should keep in mind the renovation plans and the building's ultimate use, noting any conditions or potential problems that may not have been incorporated into the plans and specifications. The architect or owner should be consulted on any questionable conditions.

The following evaluation is based on the twelve "Uniformat" divisions used in Assemblies estimating and explained and listed in Chapter 1. The Uniformat divisions represent the chronological order of construction and, thereby, provide a good format for the site visit and evaluation, no matter what estimating method is used.

Throughout the discussion of the site visit and evaluation, the person performing the inspection is referred to as the *estimator*. This reference is used primarily for consistency. The inspector could, in fact, be the building owner or any professional involved in a renovation project, such as the architect, engineer, or contractor. The project is enhanced if all involved parties perform a site visit to gain a better understanding of the renovation process.

Foundations

Materials

The foundation of an existing building may consist of stone rubble, concrete, wood piles, or another material. The foundation material and, if possible, its thickness, should be noted. This information is important in planning remodeling work. For example, if new utilities are to be provided, the estimator should know if the subcontractor has to drill through 12" of concrete block or hand chisel through 54" of stone and mortar.

Moisture Problems

In cases of water or moisture problems, the plans and specifications may call for foundation waterproofing. Depending on the amount of information provided for the work, the estimator may have to determine if the method chosen is the best solution to the problem. The solution can be as simple as repairing an undiscovered buried water pipe leak or as extensive as total exterior excavation and sealing and coating of the entire foundation. If possible, the source of the problem should be determined, and the solution estimated before work begins.

The inside surface of the foundation should be inspected. The condition of the masonry, mortar, or concrete can reveal subsurface conditions. For example, spalling, or deteriorated mortar, in only a small area suggests a localized problem.

The exposed portion of the foundation's exterior should also be inspected. Downspout locations should be noted and examined for adequate run-off, away from the foundation. Potential site drainage problems, low spots, standing water, and clogged catch basins and sewer drains should be noted. If the source of the water problem cannot be found readily, the estimator should consult the architect or owner to discuss the problem and possible solutions before starting the work.

Settling

The estimator should note cracks or any signs of settling at the foundation, interior bearing walls, or columns. If conditions suggest unusual amounts of settling, or if there are any questions about the foundation's structural integrity, an engineer should be consulted.

Excavation and Access Problems

If specified work is to be performed on the foundation, the estimator must determine how to do the work and what equipment to use. If workers must excavate or place concrete, is there adequate access for machinery or will the work be performed by hand? Workers must often create access for equipment and required materials. When exterior foundation work is to be done, an adjoining property may be involved, particularly if the building is close to lot lines. In such cases, space in lots adjacent to renovation projects may be rented, whether for actual work access or material storage. This arrangement can, however, become an expensive item, one which may or may not have been included in the original estimate.

Another item to consider for excavation is sheet piling, which may be necessary to protect adjacent areas and/or to prevent the collapse of trenches. See Figure 2.1 for an illustration.

For interior foundation work, where much of the work and delivery of materials must be performed manually, workers sometimes gain access by cutting holes in floors or walls. Expenses include not only the cost of opening the holes, but also the cost of labor and materials to patch the opening.

The estimator should try to envision the work to be done on the project, including all intermediate steps, and those items which may not be directly specified or obvious. Such an approach will help to prevent omissions, reducing the chance of unforeseen costs and change orders.

Different Construction Methods

When inspecting the foundation, the estimator is well advised to take nothing for granted. Older methods of construction can be very different from those used today. Above grade, the differences can be seen; below, they can be a mystery. Figure 2.2 is an example of a situation in which looks can be deceiving. Note how the basement slab, obviously added after the building was constructed, is actually below the base of the foundation (with no footing). Depending on the proposed use of the building, this condition could be a potential problem and a costly "extra" to rectify. Again, the estimator should consult an engineer if there is any question about structural integrity and how it affects the work or the proposed building use.

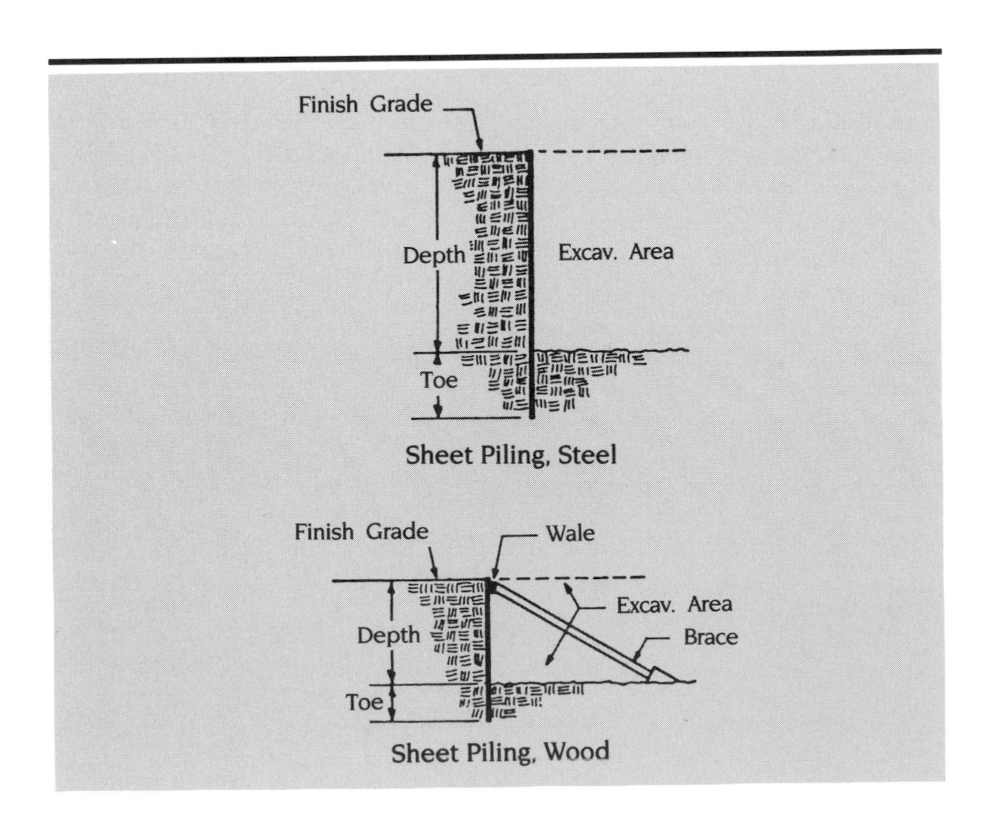

Figure 2.1

Substructure

The substructure consists of the basement floor or the slab on grade. Older buildings may lack concrete slabs, having only soil floors. In many older buildings, basement floors were constructed of brick or wood placed directly on the soil. Where concrete slabs do exist in older buildings, they may have been installed at some point after the building was constructed (as in Figure 2.2). In such cases, there may be no reinforcing.

Modern compaction methods and some of the materials used in subgrading today were not as widely used in older buildings. This is the reason why settling and cracking are more evident in older buildings. Any cracks should be checked for excessive movement, and the general condition of the slab should be checked for level and for

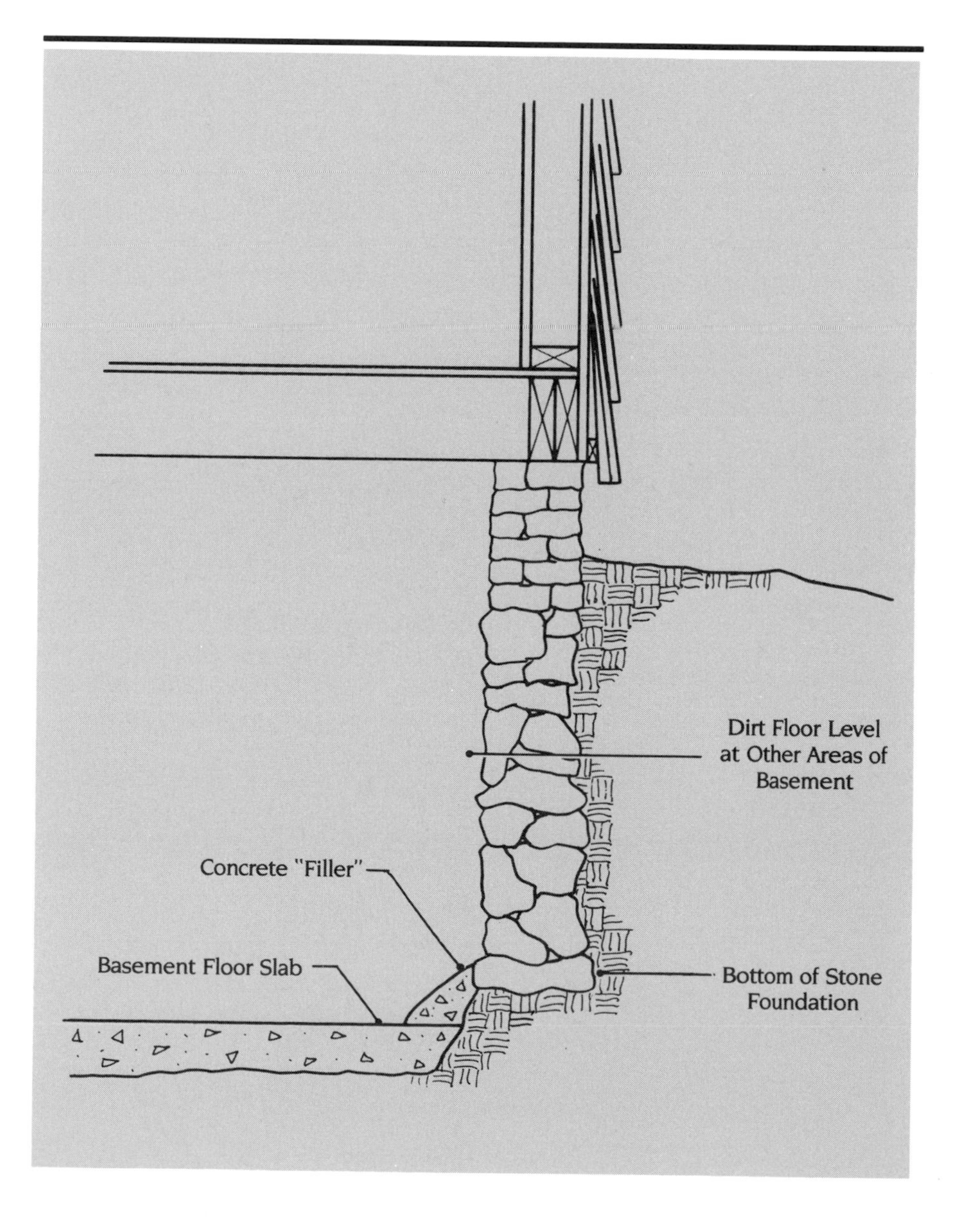

Figure 2.2

signs of deterioration. Any existing penetrations through the slab should be examined as an indication of the thickness. Remember, especially in older buildings, slab thickness may not be constant.

As is the case with foundations, water seepage can be a problem at floor slabs. Placing a new reinforced slab with a vapor barrier is an effective, but expensive and not always feasible solution. If the instructions provided are not specific, the estimator may have to determine the source of the problem and how best to repair the existing slab for less cost.

The estimator should note the time of year when the site visit is conducted and the amount of precipitation over recent weeks. If there has been a dry spell, potential water problems may not be evident. Any water marks should be noted. Cracks are another indication of a moisture problem.

Drainage

Any existing interior drainage should be checked for clogs or broken pipes. If a new drainage system is to be installed, the estimator should check the proposed path and ultimate outlet to ensure that existing conditions will not inhibit the installation. Particular drainage requirements may be specified by Building Codes for boilers and hot water heaters.

Access

If the plans and specifications call for a new basement floor slab, or concrete pads for equipment, builders face the same problems of access for equipment and materials as they do when performing foundation work. Hand work is often the only way to excavate and compact the subgrade.

Utilities and Mechanical Work

The estimator should be especially aware of the work involved in new utility and mechanical installations. Clearances for the installation of new equipment should be verified. Presumably, a structural engineer will have checked the bearing capacities of existing slabs. If not, and there is any question regarding heavy equipment, an analysis should be done.

A specified new electrical service can pose many problems. In urban areas, space for exterior transformers may not be available. Consequently, transformer vaults are required in basements. The contractor must provide the utility with access for the transformers. Utility companies may not permit installation of utility equipment and service switch-overs during normal working hours, in which case this work must be performed at night or on weekends. This is especially likely when the building remains partially occupied during the renovation. For these reasons, overtime work must be anticipated.

Other Considerations

Before beginning the estimate, the estimator must determine from his evaluation of the substructure how the specified work is to be performed. Particular attention should be given to access for materials and equipment.

If an architect or engineer has not thoroughly analyzed the building, the estimator should examine and check the location of existing structural elements as seen in the basement: columns, beams, interior bearing walls. These are important factors in the analysis of the structure above. On upper floors, the structural members may not be exposed or readily evident. Columns above may be buried in interior partitions. The plans may call for a large opening to be cut in a bearing wall that was thought to be non-bearing. In some cases, columns could have been removed in the past with no provisions for the structural consequences. Conversely, there exist structures on upper floors that appear to be columns, but are not. These elements might possibly be eliminated. Even if this information is included in the plans and specifications, the estimator should have first-hand knowledge of the location of structural elements in the building.

Superstructure

The superstructure, floor systems, ceilings, and stairs should be thoroughly inspected and evaluated as part of the site visit. The estimator should consult a structural engineer on any questions regarding structural integrity or bearing capacity if these issues are not addressed in the plans and specifications. The plans and specifications may generally state that the contractor shall patch and repair floors to match existing conditions. This may be the only information provided. It is then the responsibility of the estimator to determine the type and extent of work involved.

For example, in an old mill building, steam pipes are to be removed and the holes patched. It may be a simple operation to attach a plate to the underside of the floor and fill the hole with concrete. But if the ceiling below is to be sandblasted as the finished ceiling, or if the floor is to be sanded and refinished, much more work is involved. A wood plug may have to be custom-fitted, or, possibly, a whole flooring plank may have to be replaced with an original board that is to be removed from an inconspicuous location in the building. Clearly, there is a wide range of circumstances and costs depending on the work that must actually be done. These facts must be taken into account in preparing the estimate.

At this point in the site visit, a hammer and wrecking bar may be the estimator's most useful tools. For example, the materials in the floor systems must be identified as mill-type wood planking, wood joists, steel joist with concrete slab, or another material. It is often necessary to open holes in ceilings and to inspect under floor finishes. Different flooring systems may occur in the same building.

Floors and Ceilings

Floors in older buildings are often out of level, sometimes extremely. Settling is common in buildings built before modern engineering standards and Building Codes were established. The plans and specifications may call for the floors to be leveled with a lightweight "self-leveling" concrete. Not only is the placement of the concrete often difficult, but the determination of quantities involved can also be a problem. It often takes more concrete than originally thought, and if a large amount is required, a structural engineer may have to be consulted to determine whether or not the existing structure can support it.

When new floor systems are to be added in existing buildings, the estimator must be able to visualize the complete sequence of construction. Depending on the floor system specified in the plans, placement of new structural members can be difficult at best. Figure 2.3 shows three typical new floor systems that may be used in remodeling and renovation projects. Instead of installing larger pieces (such as steel beams) directly from the truck, with large equipment, workers may have to handle materials manually, moving them two or three times, in order to position the pieces before erecting the system. If the job calls for very long structural members, some fabrication may be needed on-site in order to accommodate restricted access. Depending on the bearing capacity of existing structural members, the existing structure may require structural reinforcement or a new structural support system, beginning with new footings.

Adding new floors in commercial renovation can be very costly, and may represent a significant percentage of the total cost of a renovation project. Consequently, the owner or architect's final decision to add a new floor is often not final until the estimator has determined the costs. Thus, it is the estimator's responsibility to be thorough and to include all costs that may result from such work.

Stairs

Particularly under today's Building Codes, stairways are an extremely important focus of commercial renovation. Often in older buildings, only one *unenclosed* stairway exists where two *enclosed* stairways will be required. If no architect is involved in the project, the estimator should consult local building inspectors as early as possible to make sure that the renovation plans comply with access and egress requirements.

When new stairways are specified as part of remodeling work, installation becomes a major factor. For example, it may not be possible to preassemble large sections to be lowered into place. In such cases, workers must carry smaller pieces to the site and assemble them by hand. Railings often must be built in place. These factors require more labor, and obviously greater expense than does new construction.

Existing stairways, as well as new ones, will most likely have to be enclosed in firewalls. Achieving the required continuity of firewalls can be very difficult in existing buildings. Where the existing ceilings are suspended, made of plaster or another material, they must be cut. The firewall must then be constructed on the underside of the deck above, and the existing structure patched to meet the new work. The firewalls also must be extended through attic spaces to the underside of the roof deck. Again, the estimator must anticipate the extra labor involved.

Other Considerations

The estimator should look for patched or repaired areas in the flooring and ceilings, as these may be signs of significant damage. Fire or water damage may have been covered up. Where possible, existing steel should be inspected for corrosion. Throughout the site evaluation, the estimator should recall the proposed renovations and how the existing conditions will affect the work.

Much commercial renovation that was performed before the relatively recent general acceptance of uniform Building Codes would not be

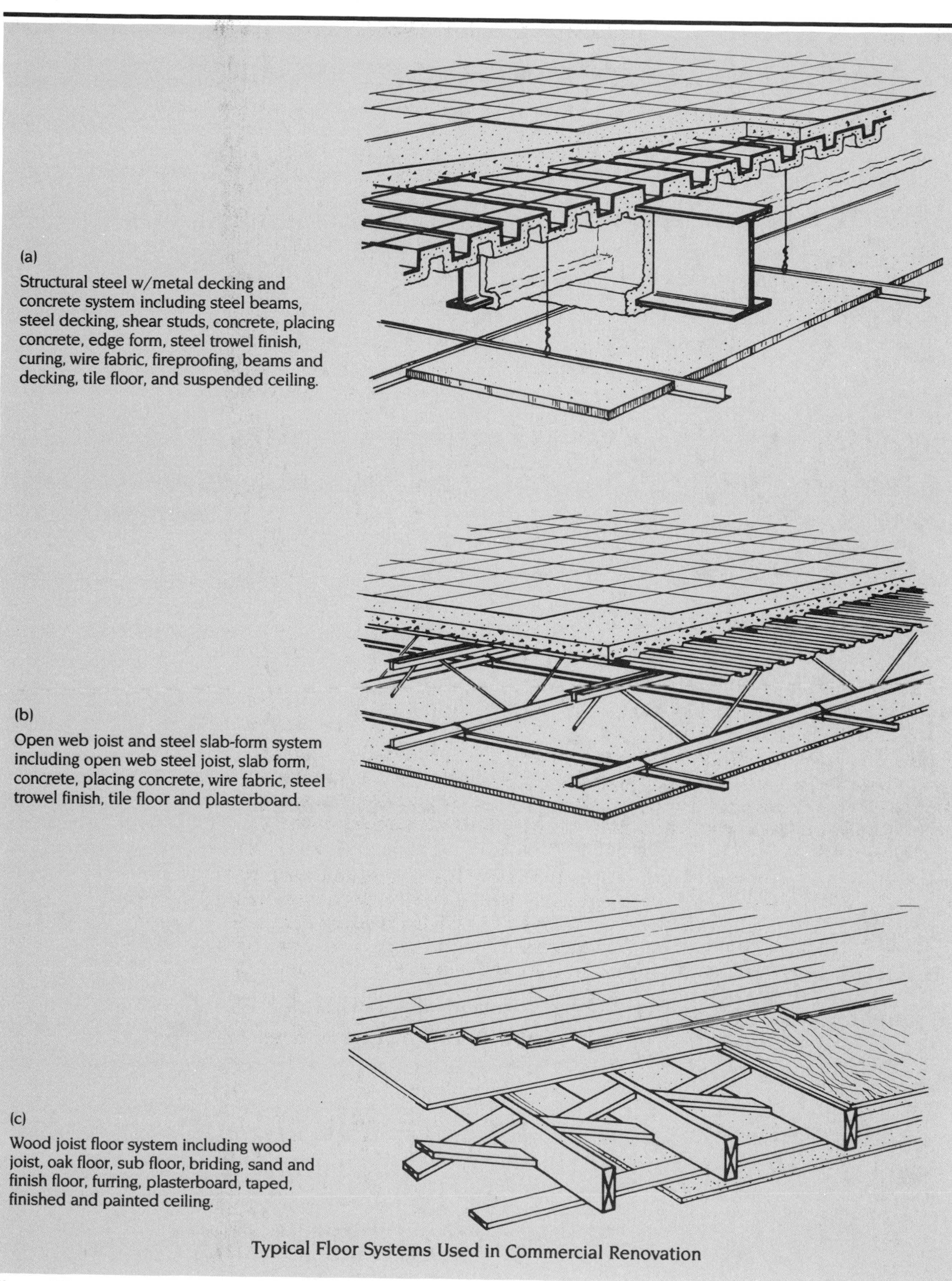

Typical Floor Systems Used in Commercial Renovation

Figure 2.3

acceptable under today's standards. Such poor quality work must be uncovered and dealt with properly in buildings that are to undergo remodeling. Openings in the floors of these older buildings have often been filled with materials of weaker structural capacity than the original surrounding floor system. These factors all have an effect on the renovation to be performed, and the estimator should be aware of the consequences. The owner or architect should be notified immediately of any such conditions.

Exterior Closure

Having determined the work to be performed, based on the plans and specifications, the estimator must thoroughly inspect the entire exterior of a building. If the building is tall, the exterior should be checked from upper story windows. Much can be overlooked from a ground level inspection. For example, when the project calls for repointing of any deteriorated or loose masonry joints, the estimator must determine the area of wall surface where repointing is required. Without close inspection, areas of deteriorated mortar may go undetected until the work has begun. All sides of the building should be checked. Different weather exposures will exhibit varied amounts of deterioration. Conditions will also vary over a particular wall surface.

Materials

A great number of older commercial buildings have some form of masonry exterior, whether stone or brick. In addition, cast iron, copper, and bronze were often used for storefronts, cornices, and facades. The properties of these materials differ greatly from the concrete, manufactured brick, and aluminum that are so common in new construction. Therefore, a knowledge of these different materials and installation methods is crucial to both estimator and architect when renovating older buildings.

When inspecting the masonry, the estimator should note the type of brick or stone, and if possible, the type of mortar used. Also, note the joint size and conditions, paying particular attention to the condition of joints between dissimilar materials (e.g., brick-stone, brick-wood). Due to differences in thermal expansion and contraction of different materials, this type of joint is most susceptible to deterioration. The information gathered from such an inspection will be used to determine the best way to perform the specified renovation work, and when calculating costs.

Cleaning

When the plans call for the exterior of a brick building to be cleaned (often before repair), the estimator must carefully read the specifications before deciding the appropriate cleaning method. Older brick is usually very porous and softer than modern brick. Sandblasting and harsh chemical washing can cause extensive damage to such exteriors, and architects and preservation officials often reject these methods. High pressure water with a mild chemical cleaner can be effective, but this approach is much more time consuming. Hand scrubbing, the most labor-intensive choice, may be necessary. Therefore, the estimator and architect alike must be aware of the materials involved and the appropriate working methods.

When exterior repairs or replacement are specified, materials to match the existing surrounding conditions are often required. These materials may have to be taken from the existing exterior. If existing materials must be reused elsewhere, the estimator should include extra labor expense for special handling to prevent damage to these materials.

Windows and Doors

When the project calls for new windows, the estimator should determine if the windows can be easily installed from the interior, or if scaffolding is required. Some types of replacement windows require only the removal of the existing sash. This type is installed around or within the existing frames. If the existing frames must be removed, more labor is involved in both demolition and installation.

In most renovation projects, doors and windows must be custom-made to fit existing openings, or the openings must be altered to accept standard sizes. Both options generally take much time, work, and money. Figures 2.4 and 2.5 show two buildings with very different types of windows. The building in Figure 2.4 has a different size and shape of window at each floor. The windows in the building in Figure 2.5 are all rectangular and are of only two sizes. The estimator must remember that like these examples, every job is different and must be evaluated carefully and independently.

Historic Preservation Guidelines

Especially on older buildings, the exterior closure can be a significant factor in commercial renovation. Today, a heightened appreciation for fine old architecture and a rising interest in historic preservation present new considerations in dealing with older buildings. Not only should the exterior be thoroughly inspected, but if an architect is not involved, the estimator should determine whether local preservation officials have any jurisdiction over renovations to the exterior of the building. Renovation projects have been legally stopped because historic guidelines were not being met.

The building in Figure 2.4 was renovated according to required historic preservation guidelines. Under these guidelines the existing windows could have been replaced only by custom-made wood windows exactly replicating the existing windows. The millwork subcontractor's estimate for this work was so great that the owner chose to have the existing windows repaired. The estimator bid the repair work at 30% of the replacement cost. However, when the repair work was completed, the eventual cost was over 50% of the cost of replacement. A more careful and thorough evaluation would have prevented this expensive error.

Other Considerations

When working on the exterior of an existing building, estimators must take into account factors other than actual labor and materials. For example, scaffolding often must be erected along the entire exterior perimeter. This, in turn, involves pedestrian protection, sidewalk permits, and, depending on the weather, tarpaulins to enclose the scaffolding (see Figure 2.6). If the exterior work is lengthy, the rental expense for these items can be significant. When possible, the estimator should determine whether swing staging might be an adequate, less costly alternative.

Figure 2.4

Figure 2.5

The estimator should carefully read and understand the drawings and specifications for every remodeling project. Required methods of work and materials will vary from project to project and will be greatly influenced by the age of the building and the role of historic preservation in the project.

Roofing

Materials

In commercial renovation, the estimator will encounter many different types and combinations of roofing structures and materials. For example, older houses may have two layers of asphalt shingles over cedar shingles. A flat built-up roof on an old building may have patches over patches. If the project includes roof work, the estimator must know what the type and condition of the existing roofing is in order to make a proper estimate of the work. If access is possible, the estimator should closely examine the existing roof structure, materials, flashing, and penetrations.

Figure 2.6

Insulation and Venting

When examining the roof structure, the estimator should note any existing insulation. The specifications may call for an "R" value of 19, leaving the methods and materials up to the estimator. As seen in Figure 2.7, 6" of glass fiber insulation and 3" of paper-backed urethane have approximately the same "R" value, but differ greatly in cost. By evaluating related work to be performed, the estimator must choose which materials are appropriate.

For example, many older commercial buildings have heavy timber trussed roof systems. Owners and architects today like to expose these trusses and wood plank roof decks for aesthetic reasons. At the same time, there may be no attic space or access to spaces under the roof deck. Under these conditions (if the roofing materials are to be replaced), the more expensive foam insulation may be required.

A wet sprinkler system in an attic space makes it necessary to install insulation above the piping to prevent freezing. If the architect has made no provisions for the ventilation that should be included in this newly-insulated space, then the estimator must address this issue. Without proper ventilation, the addition of insulation can cause condensation problems.

Assessing Damage

The roof structural system is often hidden from view by a ceiling. In such cases, inspection holes should be made or access gained to examine the structure. The structural members, and especially the roof decking, should be checked for water marks, damage, or deterioration. When older roofing materials are removed for replacement, portions of existing decking or sheathing, sometimes over the whole roof, must also be replaced in many instances. If possible, such conditions should be uncovered before work starts.

By examining the ceiling under the roof structure and under any eaves, the estimator can also detect signs of water damage or leaks. Knowing the locations of problems at the underside of the roof will be helpful when inspecting above.

Inspecting the Roof Surface

The estimator should try to gain access to the roof surface for close examination of the roofing materials. Built-up roofing should be inspected for visible cracks and bubbles. If the specifications call only for patching of the existing material, the estimator must determine the limits of this work. The amount of required patching may be so extensive that the estimator concludes that a new roof will cost little more than the repair. In this case, the client may benefit from the estimator's experienced opinion and suggested alternatives to the proposed work. Not only might the client get a better job, but the estimator may get more work for providing better service.

Flashing, Gutters, and Downspouts

Flashing, gutters, and downspouts should be examined to determine materials and evaluate their condition. If the roofing, but not flashing materials, are to be replaced, extra care and labor expense will be required to protect the flashing. If the materials are very old, then disturbing the system may cause problems. When copper or lead gutters and flashing must be repaired by soldering, the estimator must account for the additional cost of adequate fire protection.

05.9-300	Roofing & Ceiling Finish Selective Price Sheet			
LINE NO.	DESCRIPTION	COST PER S.F.		
		MAT.	INST.	TOTAL
01	Roofing, built-up, asphalt roll roof, 3 ply organic/mineral surface	.39	.71	1.10
02	3 plies glass fiber felt type iv, 1 ply mineral surface	.56	.79	1.35
03	Cold applied, 3 ply		.27	.27
04	Coal tar pitch, 4 ply asbestos felt	.64	.91	1.55
05	Mopped, 3 ply glass fiber	.56	.99	1.55
06	4 ply organic felt	.70	.90	1.60
07	Elastomeric, hypalon, neoprene unreinforced	1.60	1.68	3.28
08	Polyester reinforced	1.76	1.98	3.74
09	Neoprene, 5 coats 60 mils	4.79	5.86	10.65
10	Over 10,000 S.F.	4.46	3.04	7.50
11	PVC, traffic deck sprayed	1.27	3.05	4.32
12	With neoprene	1.38	1.23	2.61
13	Shingles, asbestos, strip, 14" x 30", 325#/sq.	1.41	.69	2.10
14	12' x 24", 167#/sq.	.74	.76	1.50
15	Shake, 9.35" x 16" 500#/sq.	2.37	1.23	3.60
16				
17	Asphalt, strip, 210-235#/sq.	.33	.50	.83
18	235-240#/sq.	.33	.55	.88
19	Class A laminated	.57	.63	1.20
20	Class C laminated	.52	.68	1.20
21	Slate, buckingham, 3/16" thick	3.69	1.56	5.25
22	Black, 1/4" thick	3.69	1.56	5.25
23	Wood, shingles, 16" no. 1, 5" exp.	1.21	1.09	2.30
24	Red cedar, 18" perfections	1.27	.98	2.25
25	Shakes, 24", 10" exposure	.94	1.11	2.05
26	18", 8-½" exposure	.88	1.37	2.25
27	Insulation, ceiling batts, fiberglass, 3-½" thick, R11	.24	.17	.41
28	6" thick, R19	.40	.20	.60
29	9" thick, R30	.55	.24	.79
30	12" thick, R38	.79	.28	1.07
31	Mineral, 3-½" thick, R13	.31	.17	.48
32	Fiber, 6" thick, R19	.48	.18	.66
33	Roof deck, fiberboard, 1" thick, R2.78	.30	.34	.64
34	Mineral, 2" thick, R5.26	.67	.34	1.01
35	Perlite boards, ¾" thick, R2.08	.33	.34	.67
36	2" thick, R5.26	.80	.39	1.19
37	Polystyrene extruded, R5.26, 1" thick,	.36	.18	.54
38	2" thick	.70	.22	.92
39	Urethane paperbacked, 1" thick, R6.7	.53	.27	.80
40	3" thick, R25	1.03	.34	1.37
41	Foam glass sheets, 1-½" thick, R3.95	1.79	.34	2.13
42	2" thick, R5.26	2.27	.34	2.61
43	Ceiling, plaster, gypsum, 2 coats	.34	1.56	1.90
44	3 coats	.47	1.86	2.33
45	Perlite or vermiculite, 2 coats	.39	1.81	2.20
46	3 coats	.62	2.27	2.89
47	Gypsum lath, plain ⅜" thick	.39	.35	.74
48	½" thick	.42	.37	.79
49	Firestop, ½" thick	.42	.42	.84
50	⅝" thick	.44	.46	.90
51	Metal lath, rib, 2.75 lb.	.19	.40	.59
52	3.40 lb.	.22	.43	.65
53	Diamond, 2.50 lb.	.13	.41	.54
54	3.40 lb.	.14	.51	.65
55	Drywall, taped and finished standard, ½" thick	.28	.71	.99
56	⅝" thick	.32	.72	1.04
57	Fire resistant, ½" thick	.32	.72	1.04
58	⅝" thick	.34	.72	1.06

272

Figure 2.7

Caution is needed for this work, even to the point of hiring personnel to remain at the end of the day to ensure that there is no smoldering material.

Penetrations

Even if no general roofing work is specified, the estimator should examine the plans to see if any existing plumbing vents or other penetrations are to be removed or relocated, thereby requiring incidental patching. It is also possible that new penetrations, vents, equipment curbs, smoke hatches, and skylights will require flashing and roofing work which are not specifically shown on the drawings. In this case, the estimator must know what conditions exist and, therefore, what work is required.

The Importance of an Inspection

Roofing repair and patching must be planned and performed with extra care. Just one or two call-backs to repair leaks can severely erode the contractor's profit margin. Knowing what is involved and what to expect before beginning the work will help to eliminate future problems.

Interior Construction

Restoration

It cannot be emphasized enough that every existing building is unique in some way and should be given thorough and independent examination. The restoration of existing interior features in older buildings, such as brick walls, plaster cornices, and wood floors, is becoming more popular as a design concept in commercial renovation. Restoration of these features requires expertise, in estimating as well as in construction, whether the work involves repair, patching, or even duplication of existing features. If existing doors or moldings are to be matched, they may have to be custom milled. Plaster cornice work, almost unheard of in new construction, may have to be repaired or recreated. Clearly, this is demanding work, for estimator and contractor alike.

Complications

Interior construction in commercial renovation can be more difficult than new construction, from the initial delivery of materials to final installation. For example, existing windows may have to be removed to allow for delivery of new materials by lift truck. When the building is more than three stories high, workers must place material, either by crane or by hand, carrying it up stairs or in an elevator. When an existing elevator is used, materials, such as sheets of drywall, may have to be cut to fit into the elevator. These restrictions entail not only the extra costs for material handling, but also the extra labor involved, such as taping more joints upon installation. Only in very large renovation projects is it cost-effective to install a temporary exterior construction elevator.

During the site visit, the estimator should anticipate material handling problems and check clearances for large items which cannot be broken down, like large sheets of glass. The timing of deliveries is also critical. Large items must be on-site before construction restricts access.

Variations in Construction Standards

Every building to be renovated is different, and the new interior construction must be built around existing conditions. Ceiling heights in older buildings are often higher than those found in new construction and can vary from floor to floor. The two foot module of modern materials was not a standard in the past. A 10'-3" high ceiling necessitates buying 12' studs and drywall for partitions. The estimator should check all existing ceiling heights if they are not shown clearly on the drawings.

Labor Considerations

The labor involved in constructing a relatively simple stud wall can almost double when working around existing conditions. For example, the upper plate may have to be secured with mastic and toggle bolts to an existing hollow terra-cotta ceiling. Or, the new wall may have to be fitted around ceiling beams. If the floor is extremely out of level, each stud may have to be cut to a different length. Extra inside and outside corners are necessary when workers have to box existing piping.

The estimator cannot assume that a wall is just a simple straight run without evaluating the existing conditions. Similarly, installation of ceilings, whether acoustical grid type, drywall, or plaster, may require soffits to accommodate existing conditions or to enclose new mechanical and electrical systems.

Analysis Based on Experience

Typical instructions in plans and specifications for commercial renovation are: "Prepare existing surfaces to receive new finishes" or "Refinish existing surfaces." When the existing surface is a plaster wall, the work may be as simple as light sanding, or as extensive as complete replacement of the wall by hand-scraping and replastering. In this example, the estimator must determine the soundness of the existing conditions in order to estimate the amount of work involved. When scraping loose plaster, what was thought to be a minor patching job can easily spread to include large areas that were not anticipated. If the condition of the existing surfaces is questionable or very deteriorated, the estimator may suggest alternatives to the architect or owner, such as covering an unsound plaster wall with drywall. Especially if the project is being competitively bid, it is in the estimator's best interest to use experience and ingenuity to suggest comparably good, but less costly alternatives.

Protection of Adjacent Work and Surroundings

When performing remodeling and renovation work, the estimator must take into account many indirect factors. For example, when sandblasting interior surfaces, extensive precautions must be taken to control the vast amounts of dust created. Large, high-volume window fans are required to exhaust the dust, yet pedestrian protection from the exhaust must also be addressed. Consequently, sandblasting may have to be performed at night or on weekends, and may require special permits.

If the building is partially occupied, floors and ceilings must also be covered or sealed to prevent dust dispersion. Adjacent surfaces must be masked or otherwise protected. Often, the costs involved in preparation can exceed the actual cost of the sandblasting.

Preparation costs can also become a major factor when refinishing wood floors. Repair of damaged areas or replacement of random individual boards can involve much handwork in the removal of the existing boards and custom fitting of new pieces. If the original floor was installed with exposed nails, each nail must be hand set before sanding.

Firewalls

The installation of required firewalls at stairways and tenant separations may involve additional expense due to the layout of the existing building. Most Building Codes require vertical continuity of firewalls and fireproofing of structural members that support the firewalls. Vertical continuity is not always possible and may involve fireproofing ceilings, beams, or columns outside of the stairway enclosure. While these conditions are almost always addressed in the plans and specifications, the estimator should pay particular attention to these elements during the site visit, and calculate the labor involved in constructing firewalls.

The elements covered in this section for interior construction demonstrate the importance of thoroughly understanding the existing conditions and the work to be performed. The estimator must use every applicable resource, including experience and knowledge, to prepare a complete and proper estimate.

Conveying Systems

For the installation of an elevator in commercial renovation, the best estimates are clearly competitive bids from elevator subcontractors. In other than budget pricing, such bids are necessary. The estimator should closely review the elevator subcontractors' bids, which may involve more exclusions than inclusions. It is the estimator's responsibility to make sure that such items as construction and preparation of the shaft, structural supports for rails and equipment, access for drilling machinery if required, and construction of the machine room are included in the estimate. In other words, the estimator may have to include—and price—all work other than the actual, direct installation of the elevator equipment.

Often in older buildings there is no shaft, or the existing shaft for an old freight or passenger elevator does not conform to requirements for the proposed new elevator. Pre-engineered standard size elevators often must be altered or custom-built to conform to existing conditions. In either case, the installing subcontractor should be consulted, and the costs included in the estimate. Where new floor openings are to be cut or existing openings altered, structural details in the plans and specifications may be general. The estimator should determine if the structural configuration is similar at all levels. As stated previously, floor systems and materials can vary in existing buildings. If there are questions regarding the structure, the estimator should consult an engineer.

Hydraulic Elevators

Unless there is adequate overhead structure and space for pulleys and machinery at the top of the shaft, hydraulic elevators are usually specified in commercial renovation. (See Figure 2.8.) Machinery for hydraulic elevators is usually placed adjacent to and at the base of the shaft. While standard hydraulic elevators eliminate the

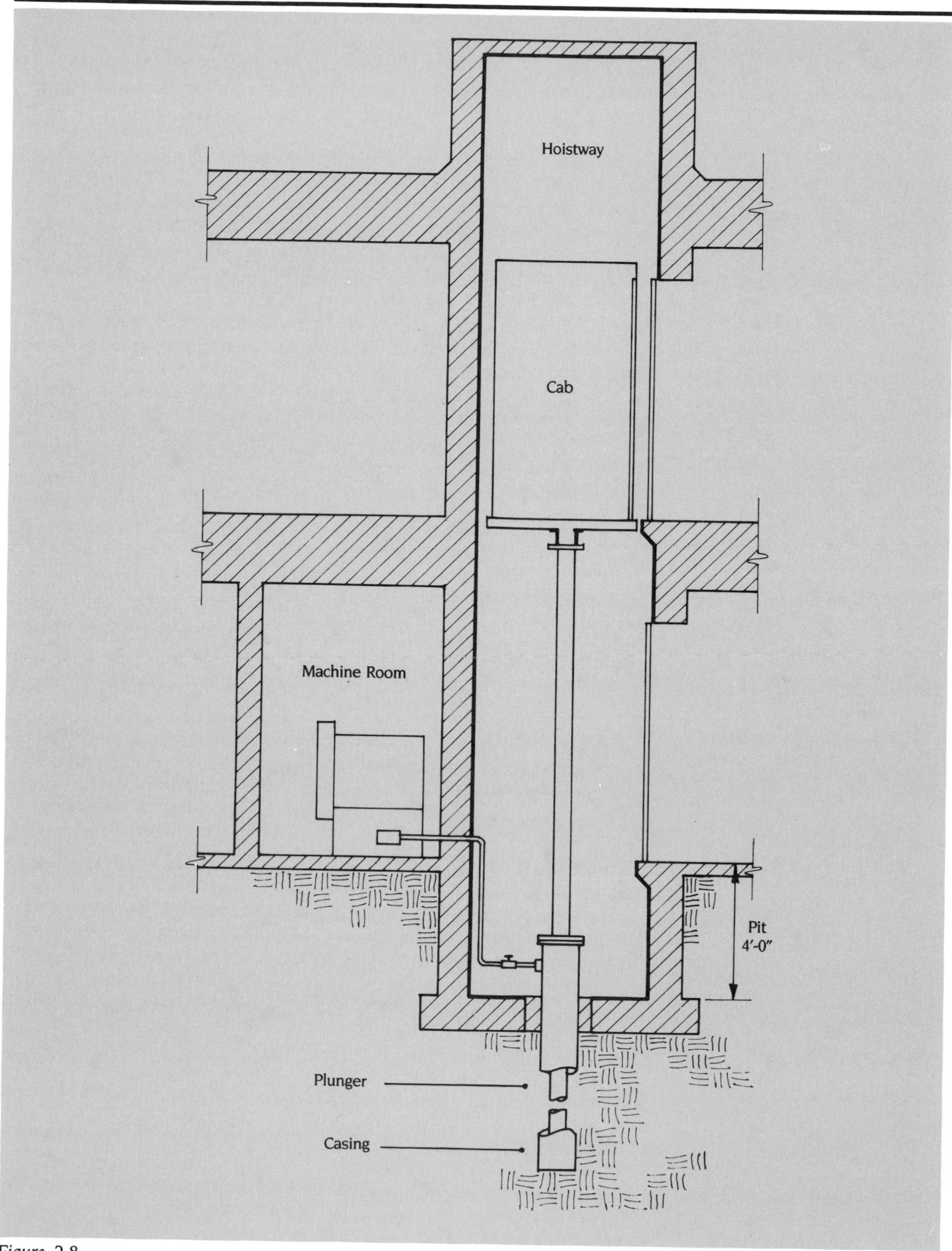

Figure 2.8

requirements at the top of the shaft, they do require a piston shaft to be drilled, often to a depth equal to the height of elevator travel. In new construction, drilling is a relatively easy operation because it is usually performed before erection of the structure. The tailings, waste, and mud from drilling can often be disposed of at the site. In renovation, however, the drilling can become a difficult and expensive proposition. The drilling rig often must be dismantled and reassembled by hand within the existing structure, and the debris created by this operation can be a problem. While this work is usually performed as part of the elevator subcontract, the estimator must be aware of the associated costs and include them in the estimate.

The following example illustrates the importance of a thorough site visit and analysis, followed by proper planning and provisions. In the building shown in Figure 2.5, an interesting problem occurred during the installation of a hydraulic elevator. An existing freight elevator shaft was to be used for the new passenger elevator. At the base of the shaft was a metal pan. The estimator had included the cost of cutting a hole in the pan for drilling as required by the elevator subcontract. After two hours of attempting to cut the metal with an acetylene torch, workers determined that the metal was cast iron and would not melt at the temperature of the torch. Attempts to break the brittle metal with a sledgehammer also failed. After many telephone calls, project managers located a welding subcontractor with a high temperature heli-arc welder capable of cutting what was found to be 2-1/2" thick cast iron. The cost for the work of the welding subcontractor alone was $1,600.00. This amount does not include the cost of labor for the five hours of previous attempts to cut the hole. Determination of responsibility for these extra costs – the owner or the contractor – lead to costly litigation.

This is obviously an isolated and unique example, but it is appropriate to show how important the site evaluation can be. While every conceivable problem may not be discovered, the estimator's thorough inspection and evaluation can reduce the probability that such hidden problems will occur.

Disposal of Drilling Waste

The costs of collecting, draining, and removing drilling waste (usually 90% water, 10% solids) must be included in the estimate. Especially in urban settings, the contractor is not allowed to dispose of this material in public sewers. The waste often must be somewhat filtered and hauled to an appropriate dump site. Provisions must be made for the costs and logistics of these operations.

Other Considerations

There are many other factors involved in the preparation of an existing building for elevators. Shaft enclosures are usually of fire-rated drywall or "shaftwall" masonry, or a combination. For example, elevator pits and roof-top enclosures with smoke hatches are often required. When contractors dig pits, water may be encountered, calling for the installation of pumps. The estimator should carefully examine the plans and specifications for such requirements, (including the elevator subcontractor's own requirements) and evaluate existing conditions in order to prepare a proper estimate.

Other conveying systems often encountered in commercial renovation are wheelchair lifts and pneumatic tube systems. While these are specialty items, usually installed by the supplier, the estimator still must anticipate the work involved in preparing the existing structure for such installations.

Until recent years, accessibility for the handicapped had not been addressed in building design and construction. Almost all Building Codes today require modifications for the handicapped when a commercial building is renovated. When there is no room for exterior or sidewalk ramps, especially in urban locations, electric wheelchair lifts are often specified. The estimator should carefully evaluate the installation, which usually requires extensive cutting and patching.

The existing conditions at locations for pneumatic tube installation should also be closely inspected. Invariably, structural members will occur in places preventing straight run installations of the tubes, requiring offsets. Unless these problems are accounted for in the estimate, the contractor may have to bear any extra costs for irregular installations.

When conveying systems have been installed, the finished work must be protected during the remainder of the construction period. In commercial renovation, where construction elevators are usually not provided, the brand new passenger elevator can be damaged and abused if workers use it to carry materials, equipment, and tools. The estimator must account for this needed protection in estimating the total project costs.

Safety

During construction of an elevator shaft, and throughout the entire renovation project, the estimator *must* plan for the requirements and costs of job safety. Temporary railings and toe boards must be securely erected and maintained at floor openings. When welding equipment or cutting torches are being used, an adequate number of fire extinguishers must be supplied and maintained. Time should be allotted to monitor fire safety during hazardous operations. These are just some of the arrangements that must be made, and costs that must be included for this essential element of the project.

The best estimates for conveying systems are competitive bids by installing subcontractors. Nevertheless, the estimator must be aware of and anticipate the work and costs involved.

Mechanical

When estimating mechanical systems, the estimator should read and study the architectural, structural, and mechanical plans and specifications before visiting the site. In this way, the estimator is sure to include all items and can examine the site to determine how the proposed work will be affected by existing conditions. The estimator should make as few assumptions as possible, and direct any questions to the architect or engineer. Mechanical plans are often prepared by consulting engineers hired by the architect. Plans prepared by different parties should be compared for possible variation of scale or inconsistencies regarding proposed locations for installation. Symbols used on the plans should be checked to ensure conformity with the estimator's interpretations.

In new construction, the mechanical systems are incorporated and designed into the planning of the structure. Architectural consideration is given to make mechanical installations as practical and efficient as possible. In renovation, however, mechanical systems must be designed to suit existing systems.

Utilities

The basement, if there is one, is the best place to begin the inspection of the mechanical systems. The existing utility connections should be noted and located on a plan. The estimator should verify sizes of sewer connections and of water and gas services. If the building has sprinklers, check for two water services: domestic and fire protection.

The estimator should also inspect the exterior of the building for shut-offs, man-holes, and any other indications of utility connections. Where new utilities are specified, closely examine the proposed locations and surrounding conditions. For example, a building with an individual sewage disposal system (septic tank and leaching field) is to be tied into a public sewer. The estimator should be aware of the location of the proposed connection, requirements of the local public works department, and the contractor responsible for trenching and backfilling. Many municipalities require that the existing septic or holding tanks be removed and backfilled. If no engineer is involved, the estimator should investigate all of these possibilities.

Piping

The estimator should inspect the general condition of existing piping that is to remain in place. Corrosion or pitting of the exposed piping may indicate similar conditions at those areas where piping is concealed in walls or ceilings. Where old piping is to remain in place, the estimator should, if possible, try to inspect the inside of the pipes for scaling. Often old pipes, especially galvanized steel, may be encrusted so that flow is severely restricted. When such conditions are found, the owner or architect should be notified to effect a solution before work begins.

Plumbing Installations

Where existing plumbing fixtures are to remain, gaskets and seals should be checked for general condition. If existing fixtures are to be replaced, rough-in dimensions should be measured to ensure compliance with the new fixtures.

The proposed plumbing installation should be visually followed in the building inspection. In renovation, the typical efficient "stacked bathrooms" of new construction are not always feasible. Consequently, piping may require many jogs and extra fittings, and there may be substantially more labor involved than in new construction. If a tenant is occupying the space below a proposed plumbing installation, the estimator must take into account the work that will be required below the floor. If the ceiling below is concealed spline or plaster, large sections may have to be replaced. This work may have to be done at off-hours on an overtime basis, and existing finished space may require protection.

Sprinkler Systems

Proposed sprinkler installations require the same thorough preparation and evaluation as the other plumbing systems. Existing sprinkler valves should be checked. If no regular maintenance has been performed, the valves may not be operable. If the existing sprinkler system is connected to a fire alarm system, the estimator may have to provide for extra labor to keep the system operational during renovations. Each morning, the fire alarm monitoring agency must be notified and the system drained. Similarly, at the end of each day, workers must notify the monitoring agency again to refill and activate the system. These necessary operations may require up to two hours or 25% of each working day.

If there is no engineer involved in the project, it may become the estimator's responsibility to bring an old sprinkler system into compliance with current codes and regulations. Along with local Fire and Building Code officials, the owner's insurance company should be consulted on sprinkler system requirements. The following items should be noted.

- the age of the sprinkler heads (usually engraved on the fused link)
- the location of the heads
- sizes and condition of piping
- existing water pressure

A possible extreme example: If the existing water pressure is not sufficient to meet regulation, a supplementary electric fire pump may be required. The installation of a fire pump has many ramifications. For example, the existing electrical service must be adequate to handle what is usually a large horsepower motor. Codes may also call for an emergency power generator, which, in turn, includes many indirect costs. Clearly, when a job becomes this involved, a mechanical engineer should design the system. Building owners, however, often wish to avoid such costs and request that the contractor do the planning.

When an engineer designs and specifies a new sprinkler system, the estimator must compare the plans with the existing conditions. New plans are often prepared from old as-built plans which may not be accurate or current. What are shown as proposed straight pipe runs may, in fact, require many offsets, fittings, and more labor to conform to existing conditions. The estimator should walk through the space to be renovated, keeping the proposed system in mind, and anticipating any installation problems.

Heating and Cooling Systems

Similar design and installation problems can occur with heating and cooling systems. The engineer may provide information on the type of system and its capacity, and diffuser or register locations. However, only line diagrams may be provided for piping or ductwork. In these cases, the contractor is usually required to furnish shop drawings showing the actual configuration of the installation. Especially with ductwork, the estimator should examine accompanying architectural and structure drawings to ensure that there is adequate space for the proposed work. Proposed equipment locations should also be inspected. Any questions about bearing capacities for equipment, such as roof-top compressors, should be directed to the structural engineer.

Other Considerations

The estimator should visualize all of the proposed mechanical installations as a whole. Even if one engineering firm has provided complete mechanical drawings and specifications, it is likely that different individuals designed each system, and conflicts may occur. For example, sprinkler pipes have been inadvertently run through planned ductwork. Such conflicts should be discovered and rectified as early as possible. When following the paths of the proposed work, the estimator must be aware of any obstacles that may restrict or complicate installation. Penetrations through masonry walls require much hand work. Penetrations of ductwork through firewalls is sometimes not allowed by local codes. If they are allowed, fire dampers will be required.

Even if the specifications do not require shop drawings, the estimator may find that sketches or drawings of the proposed work, including all mechanical systems, will be helpful in preventing problems and in preparing a complete estimate.

Mechanical work in commercial renovation can be challenging and even fun, requiring ingenuity, innovation, and the resources of the experienced estimator.

Electrical

Reviewing the Plans and Specifications

As with mechanical systems, the electrical plans and specifications should be thoroughly reviewed so that the estimator understands the full extent of the work involved before the site visit.

The electrical estimator should have adequate knowledge of both national and local Electric Codes. Even when an engineer prepares the plans and specifications, the following clause is often included.

> *"Perform all electrical work in compliance with applicable codes and ordinances, even when in conflict with the drawings and specifications."*

If inspectors reject work for non-compliance with codes, it may become the contractor's responsibility to rectify the condition. When analyzing the plans and specifications, the estimator should make notes on questionable installations and advise the architect or engineer as soon as possible.

The electrical drawings will show wiring layouts and circuiting, but often do not show wire types. Different localities have varying requirements regarding the uses of non-metallic sheathed cable ("Romex"), BX cable (flexible metallic), EMT (conduit) and galvanized steel conduit.

Figure 2.9 shows the differences in the cost of wiring similar devices using different materials. Note the cost of a duplex receptacle for each type of wiring. The lowest to highest costs vary by over 100%. This variation shows why it is so important to know what type of wiring is required in particular localities. Similarly, aluminum wiring was once widely used as a less expensive alternative to copper, but is now often disallowed as wiring material. Since copper prices have been fluctuating greatly in recent years, the estimator must keep abreast of suppliers' current prices for this type of wiring.

Service Connections

The site evaluation should begin at the existing incoming electric service. The estimator should attempt to determine the size and condition of wiring, switches, meter trim, and other components. If the service is to be revamped or increased, the estimator may be required to determine what materials to reuse and what to replace. In older buildings, all electrical wiring and equipment often must be replaced. If the existing wiring is not very old, some equipment may be rebuilt or reused, thereby saving costs. These determinations require experience and ingenuity. Local electrical inspectors should also be consulted.

For a new electric service, the estimator should contact the electric utility company as soon as possible. Scheduling for new connections should be established. Often, adjacent buildings may be affected by an electric service switchover, and the work must therefore be performed at night or on weekends. If the existing service must be removed before connecting the new, the estimator may have to include the costs of a temporary generator in the estimate.

The Distribution System

The entire electrical installation should be visualized and kept in mind by the estimator conducting the site visit and analysis. If there is no engineer involved in the project, the estimator must determine the best distribution system. If the building is to have many separately-metered areas, there may be several options available. Some local electric utilities require that all meters be installed in one location. This approach requires expensive, and possibly long, individual feeders to each tenant. On the other hand, if remote meter locations are allowed, only one main feeder is required to each remote multiple meter location. Distribution feeders can be much shorter in this case, and the system can be less expensive. (See Figure 2.10). The estimator should be able to analyze the options to determine the best system for each renovation project.

Wiring

At some point in new construction, all spaces are established, yet still open and accessible for the electrical installation. This can be before concrete slabs are poured or before walls and ceilings are closed. In the renovation of existing buildings, however, gaining access to spaces for wiring can be difficult, costly, and time-consuming.

If existing walls and ceilings are to remain, access holes must be made to snake wiring, and then the holes must be patched. Not only can cutting and patching become a major expense in renovation electrical work, but wiring often must be installed in circuitous routes to avoid obstacles. This situation involves both more labor and more material. In renovation wiring, the shortest route from point A to point B is not always a straight line. An example is when duplex receptacles are being installed in existing plaster walls. To wire horizontally from one outlet box to the next would involve removing large amounts of plaster, drilling each stud, and then patching. Instead, each box must be individually fed from a junction box,

09.9-500	Wiring Devices Selective Price Sheet			
LINE NO.	DESCRIPTION	COST EACH		
		MAT.	INST.	TOTAL
01	Using non-metallic sheathed, cable, air conditioning receptacle	9.15	29.85	39
02	Disposal wiring	8.01	32.99	41
03	Dryer circuit	22.88	53.12	76
04	Duplex receptacle	9.15	22.85	32
05	Fire alarm or smoke detector	44	29	73
06	Furnace circuit & switch	12.58	49.42	62
07	Ground fault receptacle	40.70	37.30	78
08	Heater circuit	9.15	36.85	46
09	Lighting wiring	9.15	18.85	28
10	Range circuit	34.10	75.90	110
11	Switches single pole	9.15	18.85	28
12	3-way	11.44	24.56	36
13	Water heater circuit	11.44	58.56	70
14	Weatherproof receptacle	66	49	115
15	Using BX cable, air conditioning receptacle	13.73	35.27	49
16	Disposal wiring	11.44	39.56	51
17	Dryer circuit	28.60	64.40	93
18	Duplex receptacle	13.73	27.27	41
19	Fire alarm or smoke detector	44	29	73
20	Furnace circuit & switch	16.02	58.98	75
21	Ground fault receptacle	46.20	44.80	91
22	Heater circuit	12.58	44.42	57
23	Lighting wiring	13.73	22.27	36
24	Range circuit	47.30	87.70	135
25	Switches, single pole	13.73	22.27	36
26	3-way	17.16	29.84	47
27	Water heater circuit	16.02	69.98	86
28	Weatherproof receptacle	70.40	59.60	130
29	Using EMT conduit, air conditioning receptacle	16.02	43.98	60
30	Disposal wiring	14.87	49.13	64
31	Dryer circuit	31.90	78.10	110
32	Duplex receptacle	16.02	33.98	50
33	Fire alarm or smoke detector	49.50	43.50	93
34	Furnace circuit & switch	20.59	73.41	94
35	Ground fault receptacle	49.50	55.50	105
36	Heater circuit	16.02	54.98	71
37	Lighting wiring	17.16	27.84	45
38	Range circuit	53.90	111.10	165
39	Switches, single pole	16.02	27.98	44
40	3-way	19.45	36.55	56
41	Water heater circuit	19.45	85.55	105
42	Weatherproof receptacle	74.80	75.20	150
43	Using aluminum conduit, air conditioning receptacle	19.69	58.95	87
44	Disposal wiring	18.15	65.50	84
45	Dryer circuit	34.10	105.30	140
46	Duplex receptacle	19.25	45.35	65
47	Fire alarm or smoke detector	58.30	58.95	115
48	Furnace circuit & switch	26.40	98.25	125
49	Ground fault receptacle	53.90	73.70	130
50	Heater circuit	20.40	73.70	94
51	Lighting wiring	21.50	36.85	58
52	Range circuit	61.60	147.40	210
53	Switches, single pole	25.30	36.85	62
54	3-way	27.50	49.15	77
55	Water heater circuit	27.50	117.90	145
56	Weatherproof receptacle	83.60	98.25	180
57	Using galvanized steel conduit	18.81	62.70	82
58	Disposal wiring	17.00	70.20	87

322

Figure 2.9

perhaps at the ceiling. When rigid conduit is needed for fire alarms or other purposes, many more offsets and bends may be necessary in order to adapt the wiring to existing conditions. Similarly, when wiring is to be exposed, rigid conduit or surface-mounted raceways are often specified and must be run at right angles to the existing structure.

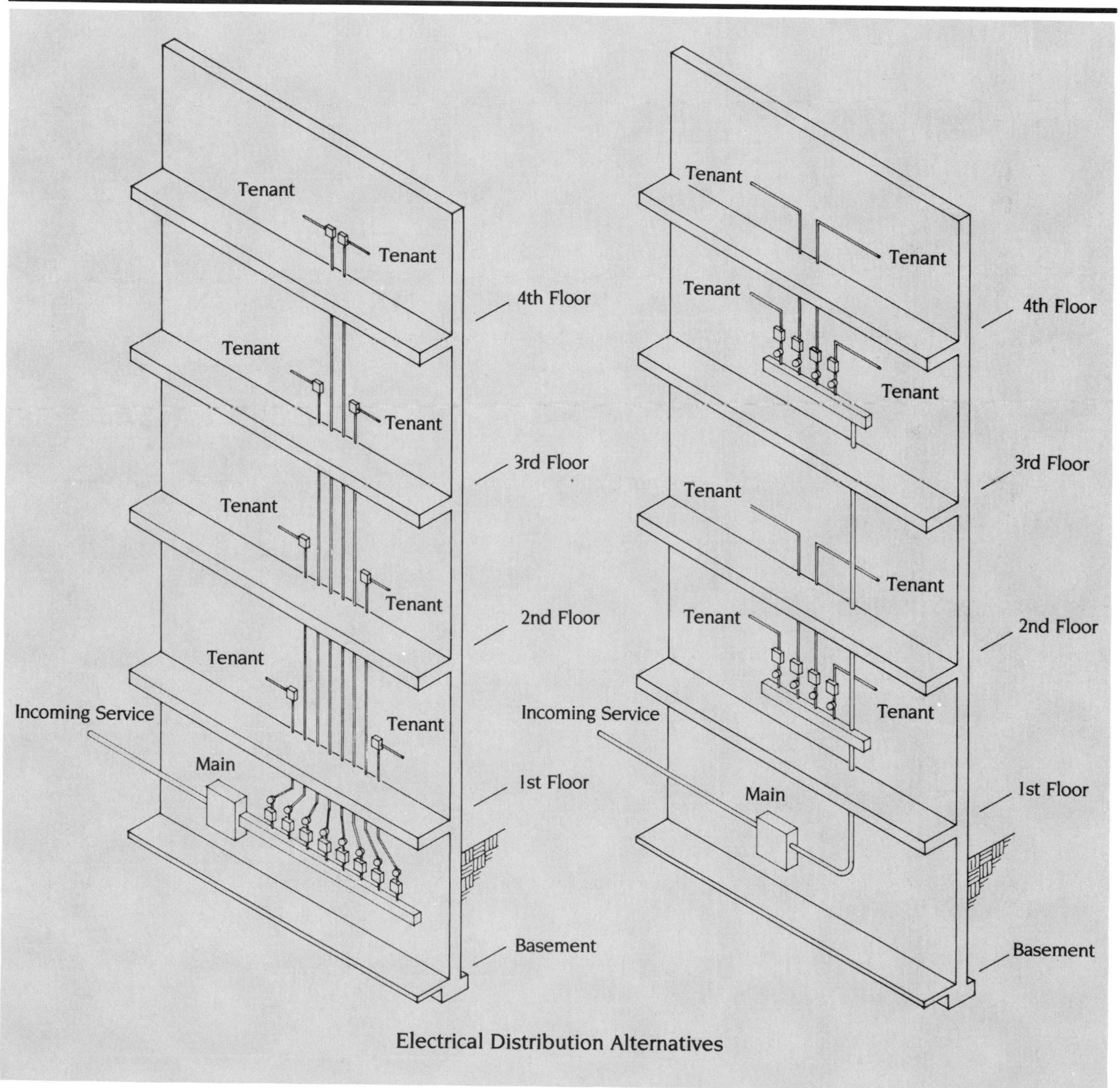

Figure 2.10

Temporary Lighting and Power

The estimate should also include temporary lighting and power. The estimator must thoroughly evaluate the plans, specifications, and examine the site to ensure that adequate provisions are made for this item. For example, if scaffolding is to be erected at the exterior, weatherproof temporary lighting is necessary for pedestrian protection. Certain construction equipment, like mortar mixers and welders, may have special temporary electrical needs.

General Conditions

The estimator must not only include all the "hard costs," or materials and labor, to produce the actual finished project, but also the "soft costs," those indispensible items that are required for the performance of the work. The following is a partial list of items that are included in the General Conditions, or General Requirements:

- Overhead and main office expenses
- Profit
- Sales and employer's taxes
- Bonds
- Insurance
- Architectural and engineering fees
- Testing & borings
- General working conditions
- Temporary requirements & construction
- Contractor equipment

Added together, these items can represent a large percentage of the cost of the proposed renovation project. Incorporating these items into the estimating process will be discussed in Chapter 9. The following section deals with how the estimator anticipates the cost of General Conditions while evaluating the site and analyzing the existing conditions.

The first seven items mentioned above are determined and included while preparing the actual estimate. Items 8 through 10, however, must be determined in part while inspecting the building to be renovated.

Figure 2.11 shows a page from *Means Repair and Remodeling Cost Data*, 1989. Please note section 110-032, "FACTORS". The seven factors listed are those that most commonly increase the costs of a renovation project. The estimator must visualize the proposed work and keep in mind how existing conditions will affect the work.

Cutting and Patching

Almost every renovation or remodeling job involves much cutting and patching to match existing conditions. Workers must open access holes in existing walls to install wire and pipe. Often, original components, moldings, and trim must be carefully removed and then replaced. Damaged pieces must be recreated or repaired at some expense.

For example, a job calls for a new plumbing vent to be installed through an old slate shingle roof. A copper sleeve and flashing must be custom made to match the existing work. Cutting the hole for the pipe can be difficult. Chances are, shingles at the hole and elsewhere near the work will break and need replacement. The scaffolding or roof jacks must be installed with extreme care to prevent such

010 | Overhead

		010 000 Overhead	CREW	DAILY OUTPUT	MAN-HOURS	UNIT	BARE COSTS MAT.	BARE COSTS LABOR	BARE COSTS EQUIP.	BARE COSTS TOTAL	TOTAL INCL O&P	
004	0011	**ARCHITECTURAL FEES** (10)										004
	0020	For work to $10,000				Project					15%	
	0040	To $25,000									13%	
	0060	To $100,000									10%	
	0080	To $500,000									8%	
	0090	To $1,000,000				↓					7%	
016	0011	**CONSTRUCTION MANAGEMENT FEES**										016
	0060	For work to $10,000				Project					10%	
	0070	To $25,000									9%	
	0090	To $100,000									6%	
	0100	To $500,000									5%	
	0110	To $1,000,000									4%	
020	0010	**CONTINGENCIES** Allowance to add at conceptual stage									20%	020
	0050	Schematic stage									15%	
	0100	Preliminary working drawing stage									10%	
	0150	Final working drawing stage				↓					2%	
028	0010	**ENGINEERING FEES** Educational planning consultant, minimum				Contrct					4.10%	028
	0100	Maximum									10.10%	
	0400	Elevator & conveying systems, minimum (11)									2.50%	
	0500	Maximum									5%	
	1000	Mechanical (plumbing & HVAC), minimum									4.10%	
	1100	Maximum				↓					10.10%	
	1200	Structural, minimum				Project					1%	
	1300	Maximum				"					2.50%	
032	0010	**FACTORS** To be added to construction costs for particular job (143)										032
	0200											
	0500	Cut & patch to match existing construction, add, minimum				Costs	2%	3%				
	0550	Maximum					5%	9%				
	0800	Dust protection, add, minimum					1%	2%				
	0850	Maximum					4%	11%				
	1100	Equipment usage curtailment, add, minimum					1%	1%				
	1150	Maximum					3%	10%				
	1400	Material handling & storage limitation, add, minimum					1%	1%				
	1450	Maximum					6%	7%				
	1700	Protection of existing work, add, minimum					2%	2%				
	1750	Maximum					5%	7%				
	2000	Shift work requirements, add, minimum						5%				
	2050	Maximum						30%				
	2300	Temporary shoring and bracing, add, minimum					2%	5%				
	2350	Maximum				↓	5%	12%				
036	0011	**FIELD PERSONNEL**										036
	0020											
	0180	Project manager, minimum				Week		925		925	1,405	
	0200	Average						1,030		1,030	1,555	
	0220	Maximum						1,170		1,170	1,790	
	0240	Superintendent, minimum						875		875	1,325	
	0260	Average						975		975	1,480	
	0280	Maximum				↓		1,095		1,095	1,655	
040	0010	**INSURANCE** Builders risk, standard, minimum				Job					.22%	040
	0050	Maximum									.59%	
	0200	All-risk type, minimum (2)									.25%	
	0250	Maximum				↓					.62%	
	0400	Contractor's equipment floater, minimum				Value					.90%	
	0450	Maximum				"					1.60%	
	0600	Public liability, average				Job					1.55%	
	0810	Workers compensation & employer's liability										
	2000	Range of 36 trades in 50 states, excl. wrecking, minimum				Payroll		1.80%				
	2100	Average				"		12.50%				

For expanded coverage of these items see *Means Building Construction Cost Data 1989* 1

Figure 2.11

breakage. The cost of performing such work could be many times greater than that for similar work on an asphalt shingle roof. The estimator must be familiar with the materials in order to determine how the work is to be performed and what it will cost.

Dust Protection

When a renovation project is to be performed in a partially occupied building, the estimator is often required to protect adjacent areas not involved in the project from dust, noise, and general disturbance. Workers must employ certain methods to prevent the production and dispersal of dust while performing the work. These precautions can slow production.

When the main lobby and only elevator are in constant use by tenants and must also be used for materials and equipment, the estimator has to provide for constant cleaning of these areas. Door and window openings at the work site must be sealed to prevent dust dispersion. If a dust problem occurs in adjacent areas, the contractor may be required to hire an outside cleaning contractor.

Especially in urban areas, dumpsters have to be covered, and laborers must use trash chutes (see Figure 2.12) instead of throwing materials out of a window. These requirements are not only for pedestrian protection, but also for the reduction of dust problems on the site. Covering a dumpster has the added benefit of saving the contractor from having to haul the entire neighborhood's garbage, which has a tendency to mysteriously appear in dumpsters each morning.

Equipment Usage

In new construction, work can be planned and scheduled to take full advantage of the most efficient and productive construction equipment. For example, hydraulic elevator piston shafts are drilled before erection of the superstructure to allow the mobile drilling rig free access. In commercial renovation, however, such access is often severely curtailed. Work must be performed with small, less effective equipment, or even by hand. This approach can add substantially to the labor costs of the project.

The use of equipment may also be restricted because of noise. In a partially occupied building, the specifications may state that the contractor must cease a noisy operation when a tenant disturbance results in a complaint. The options then available are to forego the use of the equipment for more time-consuming, quieter methods, or to perform the work during overtime hours when the building is not occupied. Both result in higher construction costs. The estimator must anticipate such restrictions and include the appropriate costs in the estimate.

Material Handling and Storage

Limitations regarding the delivery, storage, and handling of construction materials can also add to the cost of renovation projects. The specifications may impose a restriction that allows delivery and placement of materials only at certain off-peak times of the day. Local authorities may further restrict the use of the street, or the building may limit the use of loading areas or elevators. All parties having an influence on such restrictions should be consulted before the estimate is prepared.

If delivery and handling are restricted to certain times, it may be necessary to have all personnel stop other work to assist with unloading and placement of materials during those allowed times. This stopping and starting work during the course of the day and, possibly, adding to the number of required deliveries—can significantly reduce productivity.

Inside the building, there may be insufficient bearing capacities or lack of area for material storage. Materials stored outside must be

Figure 2.12

adequately protected from the elements. Architects usually specify that materials should be stored in the environment where they are to be used for a stated period of time before placement. This is a requirement for proper installation. If interior storage areas are limited, only small amounts of material can be placed at a time. Often, materials must be carried up stairs or cut to fit inside elevators.

Protection of Adjacent Areas

When materials are delivered or work is performed in a building where certain areas are finished or occupied, contractors must protect the finished work. For example, the contractor may be responsible for providing temporary coverings, moving pads, or plywood for elevators, and floor runners for carpeted areas. The estimator must determine appropriate protection measures, and must be sure to include the costs in the estimate.

Shift Work

Shift work or overtime may be required in order to provide some kinds of protection for adjacent work. In such cases, the estimator should discuss the conditions involved with the architect and owner to be sure that all requirements are met and that appropriate costs are included in the estimate. For smaller projects, even a minor amount of unanticipated overtime work can severely erode the profit margin.

Temporary Shoring and Bracing

When contractors perform structural work in an existing building, temporary shoring or bracing may be necessary. If bracing is not included in the plans and specifications, the estimator should consult the engineer on the structural requirements of such work.

Temporary Structures

Temporary requirements for construction can add substantially to the costs of renovation. For example, a field office, either inside the building or in a trailer, may be required. The field office may include a telephone, heat, and lighting for the duration of the project.

Barricades, scaffolding, temporary railings, and fences are other temporary items that may not be directly specified. In new construction, these items may be included in the estimate as established costs or percentages determined from previous similar jobs. Each commercial renovation project is unique, however, and has particular requirements.

Temporary Non-Construction Personnel

Hiring non-construction personnel on a temporary basis may also be necessary. For example, local authorities in cities may require a police officer to direct traffic or to protect pedestrians when work or material deliveries are performed at the street. Watchmen, sometimes with guard dogs, may be necessary to prevent theft and vandalism. It is helpful if estimators are familiar with the particular conditions of the neighborhood in which the renovation project is to take place. With this knowledge, they can anticipate and include such indirect costs in the estimate.

Coordination of Subcontractors

Another intangible item that affects the cost of a commercial renovation is the coordination of subcontractors. The division of labor and responsibility is often unclear in this kind of construction work, since existing conditions might make it impossible to use the standard new construction approach. The estimator must anticipate potential misunderstandings by visualizing the complete renovation process, including the appropriate supervision for each particular project. With information gathered from the site visit, the estimator must determine how many workers will be needed and what equipment can and cannot be used.

Other appropriate items are shown in Figure 2.13 from *Means Repair and Remodeling Cost Data*. Section 010-042, "Job Conditions," lists more general items that will affect the estimate. Please note that the items discussed in Figures 2.11 and 2.13 all involve percentages of project costs to be added or deducted. The estimator must use his own judgment based on experience to incorporate these figures into the estimate, without losing accuracy.

Special Items

Special items that are usually included in renovations are kitchen, cabinets, toilet accessories, toilet partitions, wood or coal stoves, and fireplaces. These items are often clearly specified, and materials prices can be easily obtained. During the site visit, the estimator should inspect the proposed locations for these installations. If the installation is to be performed on existing surfaces, the estimator must be sure that there is adequate backing or space for specified recessed equipment. Labor costs can increase quickly when workers have to remove existing wall finishes, install backing, and patch the area.

When installing wood-burning stoves or metal or masonry fireplaces, contractors should consult a structural engineer to determine whether there is adequate bearing capacity at the proposed location for these items, or whether the structure is to be modified. The estimator should also examine the existing framing that is to be penetrated by the flue or chimney to determine the extent and type of alterations needed. An inspection of these elements may show that the proposed location is at a bearing structural member that cannot be removed.

The estimator should be aware of local fire safety codes and ordinances regulating the installation of stoves and fireplaces. These codes are constantly being updated because of the increasing popularity of wood stove use. If there are any questions, an engineer or local fire official should be consulted. Again, the estimator must know exactly what is specified and how the existing conditions will affect the work.

Site Work and Demolition

Excavation

Usually, site work in commercial renovation is relatively limited. The economy of scale that can be applied to site work for new construction is often not applicable for renovation work. Trenching and excavation around an existing building usually involves cutting and patching concrete or asphalt. If the particular job is small or if

010 | Overhead

	010 000 Overhead		CREW	DAILY OUTPUT	MAN-HOURS	UNIT	BARE COSTS MAT.	LABOR	EQUIP.	TOTAL	TOTAL INCL O&P	
040	2200	Maximum (7)				Payroll		105%				040
042	0010	**JOB CONDITIONS** Modifications to total										042
	0020	project cost summaries										
	0100	Economic conditions, favorable, deduct				Project					2%	
	0200	Unfavorable, add									5%	
	0300	Hoisting conditions, favorable, deduct									2%	
	0400	Unfavorable, add									5%	
	0500	General Contractor management, experienced, deduct									2%	
	0600	Inexperienced, add									10%	
	0700	Labor availability, surplus, deduct									1%	
	0800	Shortage, add									10%	
	0900	Material storage area, available, deduct									1%	
	1000	Not available, add									2%	
	1100	Subcontractor availability, surplus, deduct									5%	
	1200	Shortage, add									12%	
	1300	Work space, available, deduct									2%	
	1400	Not available, add				↓					5%	
048	0010	**MAIN OFFICE EXPENSE** Average for General Contractors										048
	0020	As a percentage of their annual volume										
	0030	Annual volume to $50,000, minimum				% Vol.				20%		
	0040	Maximum								30%		
	0060	To $100,000, minimum								17%		
	0070	Maximum								22%		
	0080	To $250,000, minimum								16%		
	0090	Maximum								19%		
	0110	To $500,000, minimum								14%		
	0120	Maximum								16%		
	0130	To $1,000,000, minimum								10%		
	0140	Maximum				↓				8%		
052	0010	**MARK-UP** For General Contractors for change										052
	0100	of scope of job as bid										
	0200	Extra work, by subcontractors, add				%					10%	
	0250	By General Contractor, add									15%	
	0400	Omitted work, by subcontractors, deduct									5%	
	0450	By General Contractor, deduct									7.50%	
	0600	Overtime work, by subcontractors, add									15%	
	0650	By General Contractor, add									10%	
	1000	Installing contractors, on their own labor, minimum						43%				
	1100	Maximum						68%				
058	0010	**OVERHEAD** As percent of direct costs, minimum (4)								5%		058
	0050	Average								15%		
	0100	Maximum (6)				↓				30%		
062	0010	**OVERHEAD & PROFIT** Allowance to add to items in this										062
	0020	book that do not include Subs O&P, average				%				30%		
	0100	Allowance to add to items in this book that (4)										
	0110	do include Subs O&P, minimum				%					5%	
	0150	Average									10%	
	0200	Maximum									15%	
	0290	Typical, by size of project, under $50,000								40%		
	0310	Under $100,000								35%		
	0320	Under $500,000								25%		
	0330	$500,000 to $1,000,000				↓				20%		
070	0010	**PERMITS** Rule of thumb, most cities, minimum				Job					.50%	070
	0100	Maximum				"					2%	
082	0010	**SMALL TOOLS** As % of contractor's work, minimum (4)				Total					.50%	082
	0100	Maximum				"					2%	

2

For expanded coverage of these items see *Means Building Construction Cost Data 1989*

Figure 2.13

there is inadequate space for the use of large equipment, the work may have to be performed by hand. To prevent damage, workers must take extra care when digging near existing buried utilities or piping.

The estimator must try to gather as much information as possible when inspecting for proposed site work. Any holes or test pits should be examined for soil type and stability. Shut-offs and manholes for utilities should be located or verified on a site plan. If site work is to be performed on public sidewalks or in the street, the estimator should consult local authorities to determine what requirements must be met. For example, a renovation project may call for a new sewer connection in the street. After a visual examination, the estimator concludes that the job involves only cutting and patching the concrete sidewalk and asphalt pavement, trenching, and backfilling. Upon further investigation, however, the estimator learns that the existing sidewalk contains metal reinforcing, that the street has a six-inch reinforced concrete base that must be cut and patched to match the existing work, and that under local regulations, a policeman must be hired for traffic control. Furthermore, low overhead wires severely restrict the use of large equipment. As can be seen, the project can become much more involved than originally planned if it is not properly investigated and evaluated at the start.

Demolition

Demolition can be the most challenging aspect of estimating for commercial renovation. When the job calls for extensive demolition, it may be best to obtain prices from local subcontractors who are familiar with local regulations, hauling requirements, and dump site locations. Local demolition subcontractors are also the most familiar with the values of and procedures for disposing of salvageable materials, a factor that may reduce costs of demolition.

When the specifications call for the removal of asbestos or other hazardous materials, the estimator should always consult with local authorities and a licensed subcontractor. Not only does asbestos present a health hazard, but EPA and OSHA regulations impose stiff fines and penalties if hazardous materials are improperly handled and discarded.

Demolition in commercial renovation can be divided into three phases:

1. The actual dismantling of the existing structures, including labor and equipment.
2. Handling the debris. This includes the transport of material to an on-site container or truck, and may include the installation and rental of a trash chute and/or dumpsters.
3. Hauling the rubbish to an approved dump site.

Identifying Materials

It is important to be accurate when identifying the materials and determining the limits of demolition. Figure 2.14 shows a page from *Means Repair and Remodeling Cost Data*, 1989. Note Section 020-732, "Wall & Partition Demolition." Removal of gypsum plaster on metal lath ($.72 per square foot) is three times more costly than the removal of nailed drywall ($.22 per square foot). If the estimator does not take the time to identify the material, expensive mistakes can occur.

020 | Subsurface Investigation and Demolition

020 700 | Selective Demolition

		020 700 Selective Demolition	CREW	DAILY OUTPUT	MAN-HOURS	UNIT	BARE COSTS MAT.	BARE COSTS LABOR	BARE COSTS EQUIP.	BARE COSTS TOTAL	TOTAL INCL O&P	
728	0010	**SAW CUTTING** Asphalt over 1000 L.F., 3" deep	B-89	775	.021	L.F.	.21	.39	.26	.86	1.14	728
	0020	Each additional inch of depth		1,250	.013		.05	.24	.16	.45	.62	
	0400	Concrete slabs, mesh reinforcing, per inch of depth		960	.017		.25	.32	.21	.78	1.01	
	0420	Rod reinforcing, per inch of depth		550	.029		.33	.55	.37	1.25	1.64	
	0800	Concrete walls, plain, per inch of depth	A-1A	100	.080		.23	1.35	.69	2.27	3.18	
	0820	Rod reinforcing, per inch of depth		60	.133		.33	2.25	1.16	3.74	5.25	
	1200	Masonry walls, brick, per inch of depth		146	.055		.23	.92	.48	1.63	2.26	
	1220	Block walls, solid, per inch of depth		122	.066		.23	1.10	.57	1.90	2.65	
	5000	Wood sheathing to 1" thick, on walls	1 Carp	200	.040			.86		.86	1.37	
	5020	On roof	"	250	.032			.68		.68	1.10	
	9000	Minimum labor/equipment charge	A-1	2	4	Job		67	26	93	135	
	9950	See also Div. 020-125 core drilling										
730	0010	**TORCH CUTTING** Steel, 1" thick plate	A-1A	95	.084	L.F.		1.42	.73	2.15	3.08	730
	0040	1" diameter bar	"	210	.038	Ea.		.64	.33	.97	1.39	
	1000	Oxygen lance cutting, reinforced concrete walls										
	1040	12" to 16" thick walls	A-1A	10	.800	L.F.		13.50	6.95	20.45	29	
	1080	24" thick walls	"	6	1.330	"		22	11.55	33.55	49	
	1090	Minimum labor/equipment charge	A-1	1	8	Job		135	51	186	275	
	1100	See also division 051-240										
732	0010	**WALLS AND PARTITIONS DEMOLITION**										732
	0100	Brick, 4" to 12" thick	B-9	220	.182	C.F.		3.14	.53	3.67	5.60	
	0200	Concrete block, 4" thick		1,000	.040	S.F.		.69	.12	.81	1.24	
	0280	8" thick		810	.049			.85	.14	.99	1.53	
	1000	Drywall, nailed	1 Clab	1,000	.008			.13		.13	.22	
	1020	Glued and nailed		900	.009			.15		.15	.24	
	1500	Fiberboard, nailed		900	.009			.15		.15	.24	
	1520	Glued and nailed		800	.010			.17		.17	.27	
	2000	Movable walls, metal, 5' high		300	.027			.45		.45	.72	
	2020	8' high		400	.020			.34		.34	.54	
	2200	Metal or wood studs, finish 2 sides, fiberboard	B-1	520	.046			.81		.81	1.30	
	2250	Lath and plaster		260	.092			1.62		1.62	2.60	
	2300	Plasterboard (drywall)		520	.046			.81		.81	1.30	
	2350	Plywood		450	.053			.93		.93	1.50	
	3000	Plaster, lime and horsehair, on wood lath	1 Clab	400	.020			.34		.34	.54	
	3020	On metal lath		335	.024			.40		.40	.65	
	3400	Gypsum or perlite, on gypsum lath		410	.020			.33		.33	.53	
	3420	On metal lath		300	.027			.45		.45	.72	
	3800	Toilet partitions, slate or marble		5	1.600	Ea.		27		27	43	
	3820	Hollow metal		8	1	"		16.85		16.85	27	
	9000	Minimum labor/equipment charge		4	2	Job		34		34	54	
734	0010	**WINDOW DEMOLITION**										734
	0020											
	0200	Aluminum, including trim, to 12 S.F.	A-1A	12	.667	Ea.		11.25	5.80	17.05	24	
	0240	To 25 S.F.		8	1			16.85	8.70	25.55	37	
	0280	To 50 S.F.		4	2			34	17.35	51.35	73	
	0320	Storm windows, to 12 S.F.		20	.400			6.75	3.47	10.22	14.65	
	0360	To 25 S.F.		16	.500			8.45	4.34	12.79	18.30	
	0400	To 50 S.F.		12	.667			11.25	5.80	17.05	24	
	0600	Glass, minimum	1 Clab	200	.040	S.F.		.67		.67	1.08	
	0620	Maximum	"	150	.053	"		.90		.90	1.44	
	1000	Steel, including trim, to 12 S.F.	A-1A	10	.800	Ea.		13.50	6.95	20.45	29	
	1020	To 25 S.F.		7	1.140			19.25	9.90	29.15	42	
	1040	To 50 S.F.		3	2.670			45	23	68	98	
	2000	Wood, including trim, to 12 S.F.	1 Clab	16	.500			8.45		8.45	13.55	
	2020	To 25 S.F.		12	.667			11.25		11.25	18.05	
	2060	To 50 S.F.		6	1.330			22		22	36	

For expanded coverage of these items see *Means Site Work Cost Data 1989*

Figure 2.14

Establishing the Limits

Determining the limits of demolition is often the estimator's responsibility. Even if a demolition drawing is included in the plans and specifications (and this is often not the case), it may have been based on obsolete plans. Or, the specifications may state: "Contractor is to remove and dispose of all materials not to remain as part of the work." This is a vague statement that leaves all responsibility to the estimator.

If there is a demolition plan, the estimator should walk through the renovation site to verify the location and dimensions and to identify the materials to be removed. If there is no plan of the existing layout, the estimator should make a sketch, using the *proposed* floor plan as a reference. To help determine quantities, the estimator should measure ceiling heights. Measurements should be written on the sketch at the site. The estimator should leave as little as possible to memory when preparing the estimate.

Protecting Adjacent Materials

When certain existing materials are to remain, the estimator must choose the method of work that offers effective protection of these components. Sometimes skilled labor may be necessary to remove certain items, while retaining adjacent materials. The estimator may also have to determine when it is more economical to remove existing material completely and build new, or when it is cheaper to cut, patch, and alter existing work to conform to new specifications.

Removal of Debris

Once the materials marked for demolition have been dismantled, the estimator must decide on the best method for moving the debris to a dumpster or truck for removal from the site. If the building is not tall, a covered slide may be constructed with relative ease. If the building is many stories above grade, however, alternatives must be examined. An existing freight elevator might be available, but if it can only transport small loads, the process can be very time-consuming. A trash chute is another possibility. Figure 2.15 (from *Means Repair and Remodeling Cost Data*, 1989) shows the costs involved in erecting this device. It should be noted that in addition to these costs, the estimator should include the costs of support scaffolding and the possibility that workers will need a crane to erect it.

The estimator must also draw on experience with past projects in determining the number of dumpsters or truckloads required. Most disposal contractors include landfill costs in the dumpster rental fees, but the estimator should verify what this service includes. If the estimator decides that hauling by truck is the best alternative, he should contact local landfills to determine the cost of dumping and to make sure that they will accept construction materials.

Demolition and the removal of materials is rarely well-defined in plans and specifications. It is a tricky facet of commercial renovation, and one where the estimator could easily underestimate the costs involved. The whole process of the work must be well thought out and planned.

Summary

The most important point to be made regarding the site evaluation is that the estimator must have a thorough understanding of the

020 | Subsurface Investigation and Demolition

020 550 | Site Demolition

			CREW	DAILY OUTPUT	MAN-HOURS	UNIT	BARE COSTS MAT.	BARE COSTS LABOR	BARE COSTS EQUIP.	BARE COSTS TOTAL	TOTAL INCL O&P	
554	2930	15" diameter	B-6	150	.160	L.F.		2.90	1.16	4.06	5.85	554
	2960	24" diameter		120	.200			3.62	1.45	5.07	7.35	
	3000	36" diameter		90	.267			4.83	1.93	6.76	9.80	
	3200	Steel, welded connections, 4" diameter		160	.150			2.72	1.09	3.81	5.50	
	3300	10" diameter		80	.300			5.45	2.18	7.63	11	
	3390	Minimum labor/equipment charge		3	8	Job		145	58	203	295	
	3500	Railroad track removal, ties and track	B-14	110	.436	L.F.		7.75	1.58	9.33	14.15	
	3600	Ballast		500	.096	C.Y.		1.71	.35	2.06	3.11	
	3700	Remove and re-install ties & track using new bolts & spikes		50	.960	L.F.		17.10	3.48	20.58	31	
	3800	Turnouts using new bolts and spikes		1	48	Ea.		855	175	1,030	1,550	
	3890	Minimum labor/equipment charge		5	9.600	Job		170	35	205	310	
	4000	Sidewalk removal, bituminous, 2-½" thick	B-6	325	.074	S.Y.		1.34	.54	1.88	2.71	
	4050	Brick, set in mortar		185	.130			2.35	.94	3.29	4.76	
	4100	Concrete, plain		160	.150			2.72	1.09	3.81	5.50	
	4200	Mesh reinforced		150	.160			2.90	1.16	4.06	5.85	
	4290	Minimum labor/equipment charge	B-39	12	4	Job		71	9.75	80.75	125	

020 600 | Building Demolition

			CREW	DAILY OUTPUT	MAN-HOURS	UNIT	BARE COSTS MAT.	BARE COSTS LABOR	BARE COSTS EQUIP.	BARE COSTS TOTAL	TOTAL INCL O&P	
604	0010	**BUILDING DEMOLITION** Large urban projects, incl. disposal, steel	B-8	21,500	.003	C.F.		.06	.08	.14	.18	604
	0050	Concrete		15,300	.004			.08	.11	.19	.25	
	0080	Masonry		20,100	.003			.06	.09	.15	.19	
	0100	Mixture of types, average		20,100	.003			.06	.09	.15	.19	
	0500	Small bldgs, or single bldgs, no salvage included, steel	B-3	14,800	.003			.06	.09	.15	.19	
	0600	Concrete		11,300	.004			.08	.11	.19	.25	
	0650	Masonry		14,800	.003			.06	.09	.15	.19	
	0700	Wood		14,800	.003			.06	.09	.15	.19	
	1000	Single family, one story house, wood, minimum				Ea.					1,900	
	1020	Maximum									3,250	
	1200	Two family, two story house, wood, minimum									2,450	
	1220	Maximum									4,875	
	1300	Three family, three story house, wood, minimum									3,150	
	1320	Maximum									6,250	
	1400	Gutting building, see division 020-716										
608	0010	**DISPOSAL ONLY** Urban buildings with salvage value allowed										608
	0020	Including loading and 5 mile haul to dump										
	0200	Steel frame	B-3	430	.112	C.Y.		2.04	2.99	5.03	6.55	
	0300	Concrete frame		365	.132			2.40	3.53	5.93	7.70	
	0400	Masonry construction		445	.108			1.97	2.89	4.86	6.30	
	0500	Wood frame		247	.194			3.54	5.20	8.74	11.35	
612	0010	**DUMP CHARGES** Typical urban city, fees only										612
	0100	Building construction materials				C.Y.					10.50	
	0200	Demolition lumber, trees, brush									12.50	
	0300	Rubbish only									10.50	
	0500	Reclamation station, usual charge				Ton					30	
620	0010	**RUBBISH HANDLING** The following are to be added to the										620
	0020	demolition prices										
	0400	Chute, circular, prefabricated steel, 18" diameter	B-1	40	.600	L.F.	9.65	10.50		20.15	27	
	0440	30" diameter	"	30	.800	"	18.70	14		32.70	43	
	0600	Dumpster, (debris box container), 5 C.Y., rent per week				Ea.					155	
	0700	10 C.Y. capacity									190	
	0800	30 C.Y. capacity									265	
	0840	40 C.Y. capacity									315	
	1000	Dust partition, 6 mil polyethylene, 4' x 8' panels, 1" x 3" frame	2 Carp	2,000	.008	S.F.	.16	.17		.33	.45	
	1080	2" x 4" frame	"	2,000	.008	"	.27	.17		.44	.57	
	2000	Load, haul to chute & dumping into chute, 50' haul	2 Clab	21.50	.744	C.Y.		12.55		12.55	20	
	2040	100' haul	"	16.50	.970	"		16.35		16.35	26	

For expanded coverage of these items see *Means Site Work Cost Data 1989*

12

Figure 2.15

proposed work and must perform a detailed inspection of the site to determine how the work will be affected by the existing conditions. The size or age of an existing building should not determine how thorough the site visit should be. In fact, with a smaller project, a particular error or oversight can have a larger percentage effect on the accuracy of the estimate than with a larger job.

The estimator should always verify dimensions and feasibility, and should be familiar with the types of construction and materials in the existing building. Applying past experience, knowledge, and all available resources is key to the successful preparation of complete and accurate estimates for commercial renovation.

Chapter 3

THE QUANTITY TAKEOFF

The quantity takeoff–the counting of units–is the basis of estimating. This phase of the estimate is regarded as so important in England that those who perform the takeoff are registered and licensed as *Quantity Surveyors.* Not only is the takeoff essential for the application of prices, but it can also be used effectively in other aspects of a construction project.

The quantity takeoff should be organized so that the information gathered can be used to future advantage. For example, scheduling can be made easier if items are taken off and listed by construction phase, or by floor. Material purchasing will similarly benefit from consistent organization. Units for each item should be used consistently throughout the whole project–from takeoff to cost control. In this way, the original estimate can be equitably compared to any progress reports and final cost reports. The result is better record keeping and monitoring.

When working with the plans during the quantity takeoff, consistency is very important. If each job is approached in the same manner, a pattern will develop, such as moving from the lower floors to the top, clockwise or counterclockwise. The particular method is not important, as long as it is used consistently. By maintaining a regular pattern, duplications as well as omissions and errors can be minimized.

Using Pre-printed Forms

Pre-printed forms provide an excellent means for developing consistent patterns. Figures 3.1 to 3.3 are examples of such forms. The *Quantity Sheet* (Figure 3.1) is designed purely for quantity accumulation. Note that one list of materials and dimensions can be used for up to four different areas or segments of work. Figure 3.2 shows a *Cost Analysis Sheet,* which can be used in conjunction with a quantity sheet. Totals of quantities are transferred to this sheet for pricing and extensions. Figure 3.3, a *Consolidated Estimate Sheet,* is designed to be used for both quantity takeoff and pricing on one form. Part III of this book (the sample estimates) shows how such forms can be used effectively.

Custom-designed forms are another option. The important thing is that if employees use the same types of forms, then communication and coordination of the estimating process will proceed more smoothly. One estimator will be able to more easily understand the work of another. R. S. Means has published a book completely

Means® Forms

QUANTITY SHEET

SHEET NO.

ESTIMATE NO.

PROJECT

LOCATION

ARCHITECT

DATE

TAKE OFF BY

EXTENSIONS BY:

CHECKED BY:

DESCRIPTION	NO.	DIMENSIONS			UNIT	UNIT	UNIT	UNIT

Figure 3.1

Means® Forms

COST ANALYSIS

SHEET NO.

PROJECT

ESTIMATE NO.

ARCHITECT

DATE

TAKE OFF BY: QUANTITIES BY: PRICES BY: EXTENSIONS BY: CHECKED BY:

DESCRIPTION	SOURCE/DIMENSIONS			QUANTITY	UNIT	MATERIAL		LABOR		EQ./TOTAL	
						UNIT COST	TOTAL	UNIT COST	TOTAL	UNIT COST	TOTAL

Figure 3.2

Means® Forms

CONSOLIDATED ESTIMATE

SHEET NO.

PROJECT

CLASSIFICATION

ESTIMATE NO.

LOCATION

ARCHITECT

DATE

TAKE OFF BY

QUANTITIES BY

PRICES BY

EXTENSIONS BY

CHECKED BY

DESCRIPTION	NO.	DIMENSIONS			QUANTITIES UNIT		MATERIAL UNIT COST	MATERIAL TOTAL	LABOR UNIT COST	LABOR TOTAL	EQUIPMENT UNIT COST	EQUIPMENT TOTAL	TOTAL UNIT COST	TOTAL TOTAL
					UNIT	UNIT								

Figure 3.3

devoted to forms and their use, entitled *Means Forms for Building Construction Professionals*. Scores of forms are provided, along with filled-in examples, and instructions for use.

Another important type of form is the schedule. The two most important forms the interior estimator can use, *Door and Frame Schedule* and a *Room Finish Schedule*, are shown in Figures 3.4 and 3.5, respectively. These schedules help to efficiently organize much data involved in a project, in the following ways:

- The designer can be sure that all work is included in the drawings and specifications.
- The estimator can be sure that all items are included in the estimate.
- The contractor can be sure of proper installation instructions.

Appropriate and easy-to-use forms are the first and most important of the "tools of the trade" for estimators. Other tools useful to the estimator may include scales, rotometers, mechanical counters, and colored pencils.

Time-saving Measures

A number of shortcuts can be used for the quantity takeoff. If approached logically and systematically, these techniques help to save time without sacrificing accuracy. Consistent use of accepted abbreviations saves the time of writing things out. An abbreviations list might be posted in a conspicuous place to provide a consistent pattern of definitions for use within an office.

All dimensions–whether printed, measured, or calculated–that can be used for determining quantities of more than one item should be listed on a separate sheet and posted for easy reference. Posted overall dimensions can also be used to quickly check for order of magnitude errors.

Rounding off, or decreasing the number of significant digits, should be done only when it will not statistically affect the resulting product. The estimator must use good judgment to determine instances when rounding is appropriate. An overall two or three percent variation in a competitive market can often be the difference between winning a contract or losing a job, or between profit or no profit. The estimator should establish rules for rounding to achieve a consistent level of precision. In general, it is best not to round numbers until the final summary of quantities.

The final summary is also the time to convert units of measure into standards for practical use (square feet of wallcovering to number of rolls, for example). This is done to keep the quantities in the same units as they are purchased, handled, and recorded in cost reports.

Be sure to quantify (count) and include "labor only" items that are not shown on the plans. Such items may or may not be indicated in the specifications and might include clean-up, special labor for handling materials, and furniture set-up.

Means® Forms

DOOR
AND FRAME SCHEDULE

PAGE ______ OF ______

PROJECT ______ ARCHITECT ______ DATE ______

LOCATION ______ OWNER ______ BY ______

	DOOR NO.	DOOR							FRAME					FIRE RATING		HARDWARE		REMARKS
		SIZE			MAT.	TYPE	GLASS	LOUVER	MAT.	TYPE	DETAILS					SET NO.	KEYSIDE ROOM NO.	
		W	H	T							JAMB	HEAD	SILL	LAB	CON			

Figure 3.4

Means® Forms

ROOM
FINISH SCHEDULE

PAGE ____ OF ____

PROJECT ____ ARCHITECT ____ BY ____ DATE ____

ROOMS			FLOORS							BASES						WAINSCOTS						WALLS							CEILING										REMARKS
		ROOM NAME	MATERIALS							MATERIALS						MATERIALS					HT.	MATERIALS							MATERIALS					MTG					
			1	2	3	4	5	6	P	1	2	3	4	5	P	1	2	3	4	P		1	2	3	4	5	W	P	1	2	3	4	5	P	F	S			

Figure 3.5

Summary of Takeoff Guidelines

The following list is a summary of the previous discussion, together with some additional guidelines for the quantity takeoff.

- Use pre-printed forms.
- Transfer carefully when copying numbers from one sheet to the next.
- List dimensions (width, length, height) in consistent order.
- Verify the scale of drawings before using them as a basis for measurement.
- Mark drawings neatly and consistently as quantities are counted.
- Be alert to changes in scale, or notes such as "N.T.S." (not to scale). Sometimes drawings have been photographically reduced.
- Include required items which may not appear on the drawings or in the specifications.

The four most important points:

- Print legibly.
- Be organized.
- Use common sense.
- Be consistent.

Chapter 4

PRICING THE ESTIMATE

When the quantities have been counted, values, in the form of unit costs, must be applied and mark-ups (e.g., overhead and profit) added in order to determine the total "selling" price (the quote). Depending upon the chosen estimating method and the level of detail required, these unit costs may be direct or "bare", costs or may include overhead, profit, or contingencies. In unit price estimating, the unit costs most commonly used are "bare", or "unburdened." Items such as overhead and profit are usually added to the total bare costs at the estimate summary. (Chapter 1 includes a discussion of the different types of estimates.)

Sources of Cost Information

One of the most difficult aspects of the estimator's job is determining accurate and reliable bare cost data. Sources for such data are varied, but can be categorized in terms of their relative reliability.

In-house Historical Cost Data

The most reliable of any cost information is accurate, up-to-date, well-kept records of completed work by the estimator's own company. There is no better cost for a particular construction item than the *actual* cost to the contractor of that item from another recent job, modified (if necessary) to meet the requirements of the project being estimated.

Subcontractor Bids

Bids from responsible subcontractors are the next most reliable source of cost data. Any estimating inaccuracies are essentially absorbed by the subcontractor. A subcontract bid is a known, fixed cost, prior to the project. Whether the price is "right" or "wrong" does not matter (as long as it is a responsible competitive bid with no apparent errors). The bid price is what the appropriate portion of work will cost. No estimating is required, except for possible verification of the quote and comparison with other subcontractors' quotes.

Vendor Quotes

Quotations by vendors for material costs are, for the same reasons, as reliable as subcontract bids. In this case, however, the estimator must apply estimated labor costs. Thus the "installed" price for a particular item may be more variable.

Published Cost Data

If the estimator has no cost records for a particular item and is unable to obtain a quotation, then another reliable source of price information is current construction cost books, such as *Means Repair and Remodeling Cost Data*. Means presents all such data in the form of national averages. These figures can be adjusted to local conditions, a procedure that will be explained later in this book. In addition to being a source of primary costs, current annual construction cost data books can also be useful as reference tools (for productivity data, estimating methods, and design criteria) or as cross-checks for verifying costs obtained from unfamiliar sources.

Lacking cost information from any of the above-mentioned sources, the estimator may have to rely on experience and personal knowledge of the field to develop costs.

No matter which source of cost information is used, the system and sequence of pricing should be the same as that used for the quantity takeoff. This consistent approach should continue through both accounting and cost control during work on the project.

Obtaining and Recording Information

Whenever possible, all price quotations from vendors or subcontractors should be confirmed in writing. Qualifications and exclusions should be clearly stated. The requirements and times quoted should be checked to be sure that they are complete and as specified. One way to check these items is to prepare a standard form to be used by all subcontractors and vendors submitting quotations. This form, generally called a "request for quote," should include all of the appropriate questions, so that the estimator can obtain all needed information. The form also provides a standard format for organizing the information.

The above procedures are ideal, but in the realistic haste of estimating and bidding, quotations are often received verbally, either in person or by telephone. In such circumstances, gathering all pertinent information becomes even more crucial, because omissions are more likely. Using a pre-printed form, such as the one shown in Figure 4.1, is essential to ensure that all required information and qualifications are obtained and understood. How often has the subcontractor or vendor said, "I didn't know that I was supposed to include that?" With the help of such forms, the appropriate questions will be asked and bids will be better understood and complete.

Types of Costs

All costs included in a unit price estimate can be divided into two types, *direct* and *indirect*. Direct costs are those dedicated solely to the physical construction of a specific project. Material, labor, equipment, and subcontract costs, as well as project overhead costs, are all direct.

Indirect costs are usually added to the estimate at the summary stage and are most often calculated as percentages of the direct costs. They include such items as taxes, insurance, overhead, profit, and contingencies. It is the indirect costs that account for the greatest variation in estimates among different bidders.

Means® Forms

TELEPHONE QUOTATION

PROJECT	DATE
FIRM QUOTING	TIME
ADDRESS	PHONE ( )
ITEM QUOTED	BY
	RECEIVED BY

WORK INCLUDED	AMOUNT OF QUOTATION
DELIVERY TIME **TOTAL BID**	
DOES QUOTATION INCLUDE THE FOLLOWING: If ☐ NO is checked, determine the following:	
STATE & LOCAL SALES TAXES ☐ YES ☐ NO MATERIAL VALUE	
DELIVERY TO THE JOB SITE ☐ YES ☐ NO WEIGHT	
COMPLETE INSTALLATION ☐ YES ☐ NO QUANTITY	
COMPLETE SECTION AS PER PLANS & SPECIFICATIONS ☐ YES ☐ NO DESCRIBE BELOW	
EXCLUSIONS AND QUALIFICATIONS	
ADDENDA ACKNOWLEDGEMENT **TOTAL ADJUSTMENTS**	
ADJUSTED TOTAL BID	
ALTERNATES	
ALTERNATE NO.	
ALTERNATE NO.	
ALTERNATE NO.	
ALTERNATE NO.	
ALTERNATE NO.	
ALTERNATE NO.	
ALTERNATE NO.	

Figure 4.1

Types of Costs in a Construction Estimate

Direct Costs	Indirect Costs
Material	Taxes
Labor	Insurance
Equipment	Office Overhead
Subcontractors	Profit
Project Overhead	Contingencies
Sales Tax	
Bonds	

A clear understanding of direct and indirect cost factors is a fundamental part of pricing the estimate. The following chapters address the components of direct and indirect costs in detail.

Chapter 5

DIRECT COSTS

Direct costs can be defined as those necessary for the completion of the project, in other words, the "hard," or unburdened costs. Material, labor, and equipment are among the more obvious items in this category. While subcontract costs include the overhead and profit (indirect costs) of the subcontractor, they are considered to be direct costs to the prime contractor. Also included are certain project overhead costs for items that may be necessary for construction. Examples are rolling scaffolding, tools, and temporary power and lighting. Sales tax and bonds are additional direct costs, since they are essential for the performance of the project.

Material

When careful attention is given to quantity takeoff and pricing, estimates of material cost should be quite accurate. For a high level of accuracy, the material unit prices (especially for expensive finish materials) must be reliable and current. The most reliable source of material costs is a quotation from a vendor for the particular job in question. Ideally, the vendor should have access to the plans and specifications for verification of quantities and specified products. Material quotes and specified submittals for approval are services that most suppliers readily provide.

Analyzing Material Quotations

While material pricing seems straightforward, there are certain considerations that the estimator must address when analyzing material quotations. For example, the reputation of the vendor is a significant factor. Can the vendor "deliver," both figuratively and literally? Estimators may choose not to rely on a "competitive" lower price from an unknown vendor, but will instead use a slightly higher price from a known, reliable vendor. Experience is the best judge for such decisions.

There are many other questions that the estimator should ask. How long is the price guaranteed? When does the price guarantee begin? Is there an escalation clause? Does the price include delivery charges or sales tax, if required? Where is the point of FOB? This can be an extremely important factor. Do product guarantees and warranties comply with the specification requirements? Will there be adequate and appropriate storage space available? If not, can staggered shipments be made? Note that most of these questions are addressed in the form in Chapter 4, Figure 4.1. More information should be obtained, however, to ensure that a quoted price is accurate and competitive. A written quotation should always follow a verbal one.

The estimator must be sure that the quotation or obtained price is for the materials according to the plans and specifications. Architects, engineers, and designers may write into the specifications that:

- A particular type or brand of product must be used, with no substitution allowed. (This is known as a *proprietary specification*.)
- A particular type of brand of product is specified, but alternate brands of equal quality and performance may be accepted *upon approval*.
- No particular type or brand is specified. (This is a *generic specification*.)

Depending on the specifications, the estimator may be able to find an acceptable, less expensive alternative. In some cases, these substitutions can substantially lower the cost of a project. Note also that many specification packages require that "catalogue cuts," or published product data, be submitted for approval for certain materials as part of the bid proposal. In this case, there is pressure on the estimator to obtain the lowest possible price on materials that will meet the specified criteria.

There are still other considerations that should have a bearing on the final choice of a vendor. For example, lead time–the amount of time between order and delivery–must be determined and considered. It does not matter how competitive or low a quote is if the material cannot be delivered to the job site in time to support the schedule. If a delivery date is promised, is there a guarantee or a penalty clause for late delivery?

The estimator should also determine if there are any unusual payment requirements. Cash flow for a company can be severely affected if a large material purchase, thought to be payable in 30 days, is delivered C.O.D. In such cases, truck drivers may not allow unloading until payment has been received. The requirements for possible financing must be determined at the estimating phase so that the cost of borrowing money, if necessary, can be included.

If unable to obtain the quotation of a vendor from whom the material would be purchased, the estimator has other sources for obtaining material prices. These include, in order of reliability:

1. *Current* price lists from manufacturers' catalogs. Be sure to check that the list is for "contractor prices."
2. Cost records from previous jobs. Historical costs must be updated to reflect present market conditions.
3. Reputable and current annual unit price cost books, such as *Means Repair and Remodeling Cost Data*. Such books usually represent national averages and must be factored to local markets.

No matter which price source is used, the estimator must be sure to include an allowance for any burdens, such as delivery costs, taxes, or finance charges, over the actual cost of the material.

Labor

In order to determine the installation cost for each item of construction, the estimator must know two pieces of information: first, the *labor rate* (hourly wage or salary) of the worker, and second, how much time a worker will need to complete a given unit of the installation–in other words, the *productivity* or output. Wage rates are usually known going into a project, but productivity may be very difficult to predict before work gets under way.

Labor Rates

To estimators working for contractors, the construction labor rates that the contractor pays will be known, well-documented, and constantly updated. Projected labor rate increases for construction jobs of long duration should also be anticipated. Estimators for owners, architects, or engineers must determine labor rates from outside sources. Annual unit price data books, such as *Means Repair and Remodeling Cost Data*, provide national average labor wage rates. The unit costs for labor are based on these averages. Figure 5.1 shows national average *union* rates for the construction industry based on January 1, 1989. Figure 5.2 lists national average *non-union* rates, again based on January 1, 1989 (from *Means Open Shop Building Construction Cost Data*, 1989).

If more localized union labor rates are required, the following sources are available to the estimator. Union locals can provide rates (as well as negotiated increases) for a particular location. Employer bargaining groups can usually provide labor cost data as well. R. S. Means Co., Inc. publishes *Means Labor Rates for the Construction Industry* on an annual basis. This book lists the union labor rates by trade for over 300 U.S. and Canadian cities.

Determination of non-union, or "open shop" rates is much more difficult. In larger cities, there are often employer organizations that represent non-union contractors. These organizations may have records of local pay scales, but ultimately, wage rates are determined by individual contractors.

Productivity

Productivity is the least predictable of all factors for repair, remodeling, and renovation. Nevertheless, it is important to determine as accurately as possible the prevailing productivity. The best source of labor productivity or labor units (and therefore labor costs) is the estimator's well-kept records from previous projects. If there are no company records for productivity, cost data books, such as *Means Repair & Remodeling Cost Data*, and productivity reference books, such as *Means Man-Hour Standards for Construction*, can be invaluable.

Included with the Means listing for each individual construction item is the designation of a suggested crew make-up. The crew is the minimum grouping of workers that can be expected to accomplish the task efficiently. Figure 5.3, a typical page from *Means Repair & Remodeling Cost Data*, includes this crew information, and indicates the man-hours as well as the average daily output of the designated crew for each construction item.

The estimator who has neither company records nor the sources described above must put together the appropriate crews and determine the expected output or productivity. This type of estimating should only be attempted based on strong experience and considerable exposure to construction methods and practices. There are rare occasions when this approach is necessary to estimate a particular item or a new technique. Even then, the new labor units are often extrapolated from existing figures for similar work, rather than being created from scratch.

Installing Contractor's Overhead & Profit

Below are the **average** installing contractor's percentage mark-ups applied to base labor rates to arrive at typical billing rates.

Column A: Labor rates are based on union wages averaged for 30 major U.S. cities. Base rates including fringe benefits are listed hourly and daily. These figures are the sum of the wage rate and employer-paid fringe benefits such as vacation pay, employer-paid health and welfare costs, pension costs, plus appropriate training and industry advancement funds costs.

Column B: Workers' Compensation rates are the national average of state rates established for each trade.

Column C: Column C lists average fixed overhead figures for all trades. Included are Federal and State Unemployment costs set at 6.2%; Social Security Taxes (FICA) set at 7.65%; Builder's Risk Insurance costs set at 0.34%; and Public Liability costs set at 1.55%. All the percentages except those for Social Security Taxes vary from state to state as well as from company to company.

Column D and E: Percentages in Columns D and E are based on the presumption that the installing contractor has annual billing of $500,000 and up. Overhead percentages may increase with smaller annual billing. The overhead percentages for any given contractor may vary greatly and depend on a number of factors, such as the contractor's annual volume, engineering and logistical support costs, and staff requirements. The figures for overhead and profit will also vary depending on the type of job, the job location, and the prevailing economic conditions. All factors should be examined very carefully for each job.

Column F: Column F lists the total of columns B, C, D, and E.

Column G: Column G is Column A (hourly base labor rate) multiplied by the percentage in Column F (O&P percentage).

Column H: Column H is the total of Column A (hourly base labor rate) plus Column G (Total O&P).

Column I: Column I is Column H multiplied by eight hours.

		A		B	C	D	E	F	G	H	I
		Base Rate Incl. Fringes		Work-ers' Comp. Ins.	Average Fixed Over-head	Over-head	Profit	Total Overhead & Profit		Rate with O & P	
Abbr.	Trade	Hourly	Daily					%	Amount	Hourly	Daily
Skwk	Skilled Workers Average (35 trades)	$21.90	$175.20	12.5%	15.7%	16.0%	15.0%	59.2%	$12.95	$34.85	$278.80
	Helpers Average (5 trades)	16.60	132.80	13.8				60.5	10.05	26.65	213.20
	Foremen Average, Inside (50¢ over trade)	22.40	179.20	12.5				59.2	13.25	35.65	285.20
	Foremen Average, Outside ($2.00 over trade)	23.90	191.20	12.5				59.2	14.15	38.05	304.40
Clab	Common Building Laborers	16.85	134.80	13.9				60.6	10.20	27.05	216.40
Asbe	Asbestos Workers	23.85	190.80	11.6				58.3	13.90	37.75	302.00
Boil	Boilermakers	23.90	191.20	7.9				54.6	13.05	36.95	295.60
Bric	Bricklayers	22.10	176.80	11.3				58.0	12.80	34.90	279.20
Brhe	Bricklayer Helpers	17.10	136.80	11.3				58.0	9.90	27.00	216.00
Carp	Carpenters	21.40	171.20	13.9				60.6	12.95	34.35	274.80
Cefi	Cement Finishers	20.80	166.40	7.9				54.6	11.35	32.15	257.20
Elec	Electricians	24.25	194.00	5.2				51.9	12.60	36.85	294.80
Elev	Elevator Constructors	24.15	193.20	6.8				53.5	12.90	37.05	296.40
Eqhv	Equipment Operators, Crane or Shovel	22.40	179.20	8.8				55.5	12.45	34.85	278.80
Eqmd	Equipment Operators, Medium Equipment	21.65	173.20	8.8				55.5	12.00	33.65	269.20
Eqlt	Equipment Operators, Light Equipment	20.60	164.80	8.8				55.5	11.45	32.05	256.40
Eqol	Equipment Operators, Oilers	18.45	147.60	8.8				55.5	10.25	28.70	229.60
Eqmm	Equipment Operators, Master Mechanics	23.10	184.80	8.8				55.5	12.80	35.90	287.20
Glaz	Glaziers	22.05	176.40	10.4				57.1	12.60	34.65	277.20
Lath	Lathers	21.60	172.80	8.4				55.1	11.90	33.50	268.00
Marb	Marble Setters	21.95	175.60	11.3				58.0	12.75	34.70	277.60
Mill	Millwrights	22.15	177.20	8.4				55.1	12.20	34.35	274.80
Mstz	Mosaic and Terrazzo Workers	21.45	171.60	7.0				53.7	11.50	32.95	263.60
Pord	Painters, Ordinary	20.10	160.80	10.4				57.1	11.50	31.60	252.80
Psst	Painters, Structural Steel	20.85	166.80	34.9				81.6	17.00	37.85	302.80
Pape	Paper Hangers	20.25	162.00	10.4				57.1	11.55	31.80	254.40
Pile	Pile Drivers	21.50	172.00	21.9				68.6	14.75	36.25	290.00
Plas	Plasterers	21.10	168.80	11.0				57.7	12.15	33.25	266.00
Plah	Plasterer Helpers	17.35	138.80	11.0				57.7	10.00	27.35	218.80
Plum	Plumbers	24.45	195.60	6.2				52.9	12.95	37.40	299.20
Rodm	Rodmen (Reinforcing)	23.15	185.20	22.0				68.7	15.90	39.05	312.40
Rofc	Roofers, Composition	19.75	158.00	25.7				72.4	14.30	34.05	272.40
Rots	Roofers, Tile & Slate	19.85	158.80	25.7				72.4	14.35	34.20	273.60
Rohe	Roofer Helpers (Composition)	14.65	117.20	25.7				72.4	10.60	25.25	202.00
Shee	Sheet Metal Workers	24.05	192.40	8.8				55.5	13.35	37.40	299.20
Spri	Sprinkler Installers	25.25	202.00	6.7				53.4	13.50	38.75	310.00
Stpi	Steamfitters or Pipefitters	24.55	196.40	6.2				52.9	13.00	37.55	300.40
Ston	Stone Masons	22.00	176.00	11.3				58.0	12.75	34.75	278.00
Sswk	Structural Steel Workers	23.25	186.00	26.8				73.5	17.10	40.35	322.80
Tilf	Tile Layers (Floor)	21.50	172.00	7.0				53.7	11.55	33.05	264.40
Tilh	Tile Layer Helpers	17.05	136.40	7.0				53.7	9.15	26.20	209.60
Trlt	Truck Drivers, Light	17.45	139.60	11.8				58.5	10.20	27.65	221.20
Trhv	Truck Drivers, Heavy	17.60	140.80	11.8				58.5	10.30	27.90	223.20
Sswl	Welders, Structural Steel	23.25	186.00	26.8				73.5	17.10	40.35	322.80
Wrck	*Wrecking	16.80	134.40	28.1				74.8	12.55	29.35	234.80

*Not included in Averages.

Figure 5.1

Installing Contractor's Overhead & Profit

Below are the **average** installing contractor's percentage mark-ups applied to base labor rates to arrive at typical billing rates.

Column A: Labor rates are based on average open shop wages for 7 major U.S. regions. Base rates including fringe benefits are listed hourly and daily. These figures are the sum of the wage rate and employer-paid fringe benefits such as vacation pay and employer-paid health costs.

Column B: Workers' Compensation rates are the national average of state rates established for each trade.

Column C: Column C lists average fixed overhead figures for all trades. Included are Federal and State Unemployment costs set at 6.2%; Social Security Taxes (FICA) set at 7.65%; Builder's Risk Insurance costs set at 0.34%; and Public Liability costs set at 1.55%. All the percentages except those for Social Security Taxes vary from state to state as well as from company to company.

Column D and E: Percentages in Columns D and E are based on the presumption that the installing contractor has annual billing of $500,000 and up. Overhead percentages may increase with smaller annual billing. The overhead percentages for any given contractor may vary greatly and depend on a number of factors, such as the contractor's annual volume, engineering and logistical support costs, and staff requirements. The figures for overhead and profit will also vary depending on the type of job, the job location, and the prevailing economic conditions. All factors should be examined very carefully for each job.

Column F: Column F lists the total of columns B, C, D, and E.

Column G: Column G is Column A (hourly base labor rate) multiplied by the percentage in Column F (O&P percentage).

Column H: Column H is the total of Column A (hourly base labor rate) plus Column G (Total O&P).

Column I: Column I is Column H multiplied by eight hours.

		A		B	C	D	E	F	G	H	I
Abbr.	Trade	Base Rate Incl. Fringes		Workers' Comp. Ins.	Average Fixed Overhead	Overhead	Profit	Total Overhead & Profit		Rate with O & P	
		Hourly	Daily					%	Amount	Hourly	Daily
Skwk	Skilled Workers Average (35 trades)	$14.25	$114.00	12.5%	15.7%	26.8%	10%	65.0%	$ 9.25	$23.50	$188.00
	Helpers Average (5 trades)	10.80	86.40	13.8		27.0		66.5	7.20	18.00	144.00
	Foremen Average, Inside (50¢ over trade)	14.75	118.00	12.5		26.8		65.0	9.60	24.35	194.80
	Foremen Average, Outside ($2.00 over trade)	16.25	130.00	12.5		26.8		65.0	10.55	26.80	214.40
Clab	Common Building Laborers	10.95	87.60	13.9		25.0		64.6	7.05	18.00	144.00
Asbe	Asbestos Workers	15.50	124.00	11.6		30.0		67.3	10.45	25.95	207.60
Boil	Boilermakers	15.55	124.40	7.9		30.0		63.6	9.90	25.45	203.60
Bric	Bricklayers	14.35	114.80	11.3		25.0		62.0	8.90	23.25	186.00
Brhe	Bricklayer Helpers	11.10	88.80	11.3		25.0		62.0	6.90	18.00	144.00
Carp	Carpenters	13.90	111.20	13.9		25.0		64.6	9.00	22.90	183.20
Cefi	Cement Finishers	13.50	108.00	7.9		25.0		58.6	7.90	21.40	171.20
Elec	Electricians	15.75	126.00	5.2		30.0		60.9	9.60	25.35	202.80
Elev	Elevator Constructors	15.70	125.60	6.8		30.0		62.5	9.80	25.50	204.00
Eqhv	Equipment Operators, Crane or Shovel	14.55	116.40	8.8		28.0		62.5	9.10	23.65	189.20
Eqmd	Equipment Operators, Medium Equipment	14.05	112.40	8.8		28.0		62.5	8.80	22.85	182.80
Eqlt	Equipment Operators, Light Equipment	13.40	107.20	8.8		28.0		62.5	8.40	21.80	174.40
Eqol	Equipment Operators, Oilers	12.00	96.00	8.8		28.0		62.5	7.50	19.50	156.00
Eqmm	Equipment Operators, Master Mechanics	15.00	120.00	8.8		28.0		62.5	9.40	24.40	195.20
Glaz	Glaziers	14.35	114.80	10.4		25.0		61.1	8.75	23.10	184.80
Lath	Lathers	14.05	112.40	8.4		25.0		59.1	8.30	22.35	178.80
Marb	Marble Setters	14.25	114.00	11.3		25.0		62.0	8.85	23.10	184.80
Mill	Millwrights	14.40	115.20	8.4		25.0		59.1	8.50	22.90	183.20
Mstz	Mosaic and Terrazzo Workers	13.95	111.60	7.0		25.0		57.7	8.05	22.00	176.00
Pord	Painters, Ordinary	13.05	104.40	10.4		25.0		61.1	7.95	21.00	168.00
Psst	Painters, Structural Steel	13.55	108.40	34.9		25.0		85.6	11.60	25.15	201.20
Pape	Paper Hangers	13.15	105.20	10.4		25.0		61.1	8.05	21.20	169.60
Pile	Pile Drivers	14.00	112.00	21.9		30.0		77.6	10.85	24.85	198.80
Plas	Plasterers	13.70	109.60	11.0		25.0		61.7	8.45	22.15	177.20
Plah	Plasterer Helpers	11.30	90.40	11.0		25.0		61.7	6.95	18.25	146.00
Plum	Plumbers	15.90	127.20	6.2		30.0		61.9	9.85	25.75	206.00
Rodm	Rodmen (Reinforcing)	15.05	120.40	22.0		28.0		75.7	11.40	26.45	211.60
Rofc	Roofers, Composition	12.85	102.80	25.7		25.0		76.4	9.80	22.65	181.20
Rots	Roofers, Tile & Slate	12.90	103.20	25.7		25.0		76.4	9.85	22.75	182.00
Rohe	Roofer Helpers (Composition)	9.55	76.40	25.7		25.0		76.4	7.30	16.85	134.80
Shee	Sheet Metal Workers	15.65	125.20	8.8		30.0		64.5	10.10	25.75	206.00
Spri	Sprinkler Installers	16.40	131.20	6.7		30.0		62.4	10.25	26.65	213.20
Stpi	Steamfitters or Pipefitters	15.95	127.60	6.2		30.0		61.9	9.85	25.80	206.40
Ston	Stone Masons	14.30	114.40	11.3		25.0		62.0	8.85	23.15	185.20
Sswk	Structural Steel Workers	15.10	120.80	26.8		28.0		80.5	12.15	27.25	218.00
Tilf	Tile Layers (Floor)	14.00	112.00	7.0		25.0		57.7	8.10	22.10	176.80
Tilh	Tile Layer Helpers	11.10	88.80	7.0		25.0		57.7	6.40	17.50	140.00
Trlt	Truck Drivers, Light	11.35	90.80	11.8		25.0		62.5	7.10	18.45	147.60
Trhv	Truck Drivers, Heavy	11.45	91.60	11.8		25.0		62.5	7.15	18.60	148.80
Sswl	Welders, Structural Steel	15.10	120.80	26.8		28.0		80.5	12.15	27.25	218.00
Wrck	*Wrecking	10.95	87.60	28.1		25.0		78.8	8.65	19.60	156.80

*Not included in Averages.

Figure 5.2

062 | Finish Carpentry

				DAILY	MAN-		BARE COSTS				TOTAL	
		062 500 | Prefin. Wood Paneling	CREW	OUTPUT	HOURS	UNIT	MAT.	LABOR	EQUIP.	TOTAL	INCL O&P	
504	4600	Plywood, "A" face, birch, V.C., ½" thick, natural	F-2	450	.036	S.F.	1.40	.76	.03	2.19	2.80	504
	4700	Select		450	.036		1.50	.76	.03	2.29	2.91	
	4900	Veneer core, ¾" thick, natural		320	.050		1.40	1.07	.05	2.52	3.31	
	5000	Select		320	.050		1.60	1.07	.05	2.72	3.53	
	5200	Lumber core, ¾" thick, natural	↓	320	.050	↓	2	1.07	.05	3.12	3.97	
	5250											
	5500	Plywood, knotty pine, ¼" thick, A2 grade	F-2	450	.036	S.F.	1.50	.76	.03	2.29	2.91	
	5600	A3 grade		450	.036		1.60	.76	.03	2.39	3.02	
	5800	¾" thick, veneer core, A2 grade		320	.050		1.80	1.07	.05	2.92	3.75	
	5900	A3 grade		320	.050		1.50	1.07	.05	2.62	3.42	
	6100	Aromatic cedar, ¼" thick, plywood		400	.040		1.40	.86	.04	2.30	2.96	
	6200	¼" thick, particle board	↓	400	.040	↓	.65	.86	.04	1.55	2.14	
	9000	Minimum labor/equipment charge	1 Carp	2	4	Job		86		86	135	
		062 550 | Prefin. Hardboard Panel										
554	0010	**PANELING, HARDBOARD**										554
	0020											
	0050	Not incl. furring or trim, hardboard, tempered, ⅛" thick	F-2	500	.032	S.F.	.26	.68	.03	.97	1.42	
	0100	¼" thick		500	.032		.32	.68	.03	1.03	1.48	
	0300	Tempered pegboard, ⅛" thick		500	.032		.29	.68	.03	1	1.45	
	0400	¼" thick		500	.032		.36	.68	.03	1.07	1.53	
	0600	Untempered hardboard, natural finish, ⅛" thick		500	.032		.23	.68	.03	.94	1.38	
	0700	¼" thick		500	.032		.30	.68	.03	1.01	1.46	
	0900	Untempered pegboard, ⅛" thick		500	.032		.26	.68	.03	.97	1.42	
	1000	¼" thick		500	.032		.32	.68	.03	1.03	1.48	
	1200	Plastic faced hardboard, ⅛" thick		500	.032		.42	.68	.03	1.13	1.59	
	1300	¼" thick		500	.032		.57	.68	.03	1.28	1.76	
	1500	Plastic faced pegboard, ⅛" thick		500	.032		.41	.68	.03	1.12	1.58	
	1600	¼" thick		500	.032		.50	.68	.03	1.21	1.68	
	1800	Wood grained, plain or grooved, ¼" thick, minimum		500	.032		.37	.68	.03	1.08	1.54	
	1900	Maximum		425	.038	↓	.67	.81	.04	1.52	2.07	
	2100	Moldings for hardboard, wood or aluminum, minimum		500	.032	L.F.	.27	.68	.03	.98	1.43	
	2200	Maximum	↓	425	.038	"	.68	.81	.04	1.53	2.08	
	9000	Minimum labor/equipment charge	1 Carp	2	4	Job		86		86	135	
		062 600 | Board Paneling										
604	0010	**PANELING, BOARDS**										604
	0020											
	6400	Wood board paneling, ¾" thick, knotty pine	F-2	300	.053	S.F.	.77	1.14	.05	1.96	2.74	
	6500	Rough sawn cedar		300	.053		1.19	1.14	.05	2.38	3.20	
	6700	Redwood, clear, 1" x 4" boards		300	.053		3.10	1.14	.05	4.29	5.30	
	6900	Aromatic cedar, closet lining, boards	↓	275	.058	↓	1.89	1.25	.05	3.19	4.14	
	9000	Minimum labor/equipment charge	1 Carp	2	4	Job		86		86	135	
		062 700 | Misc Finish Carpentry										
704	0010	**BEAMS, DECORATIVE** Rough sawn cedar, non-load bearing, 4" x 4"	2 Carp	180	.089	L.F.	1.50	1.90		3.40	4.70	704
	0100	4" x 6"		170	.094		2.50	2.01		4.51	6	
	0200	4" x 8"		160	.100		3.20	2.14		5.34	6.95	
	0300	4" x 10"		150	.107		3.75	2.28		6.03	7.80	
	0400	4" x 12"		140	.114		4.75	2.45		7.20	9.15	
	0500	8" x 8"	↓	130	.123	↓	6.75	2.63		9.38	11.65	
	1100	Beam connector plates see div 060-512										
	9000	Minimum labor/equipment charge	1 Carp	3	2.670	Job		57		57	92	
720	0010	**FIREPLACE MANTEL BEAMS** Rough texture wood, 4" x 8"		36	.222	L.F.	3.45	4.76		8.21	11.45	720
	0100	4" x 10"	↓	35	.229	"	4	4.89		8.89	12.25	

For expanded coverage of these items see *Means Interior Cost Data 1989*

79

Figure 5.3

Equipment

Construction equipment used for interior projects can be a very expensive item. For example, for a sixth floor job, the elevator may be too small to transport the drywall. Consequently, a crane may have to be used, a window removed, and other work essentially halted during delivery. These procedures can cost thousands of dollars. Estimators must carefully address the issue of equipment and related expenses, when required. Equipment costs can be divided into the two following categories:

- **Rental, lease, or ownership costs.** These costs may be determined based on hourly, daily, weekly, monthly, or annual increments. These fees or payments only buy the "right" to use the equipment and do not include operating costs.
- **Operating Costs.** Once the "right" of use is obtained, costs are incurred for actual use or operation. These costs may include fuel, lubrication, maintenance, and parts.

Equipment costs, as described above, do not include the labor expense of operators. However, some cost books and suppliers may include the operator in the quoted price for equipment as an "operated" rental cost. In other words, the equipment is priced as if it were a subcontract cost.

Equipment ownership costs apply to both leased and owned equipment: the operating costs of equipment, whether rented, leased or owned, are available from the following sources (listed in order of reliability):

1. The company's own records
2. Annual cost books containing equipment operating costs, such as *Means Repair and Remodeling Cost Data*
3. Manufacturers' estimates
4. Reference books dealing with equipment operating costs

Operating costs include fuel, lubrication, expendable parts replacement, minor maintenance, transportation, and mobilizing costs. For estimating purposes, the equipment ownership and operating costs should be listed separately. In this way, the decision to rent, subcontract, or purchase can be decided on a project-by-project basis.

Allocating Equipment Costs

There are two commonly used methods for including equipment costs in a construction estimate. The first is to include the equipment as a part of the construction task for which it is used. In this case, costs are included in each line item as a separate unit price. The advantage of this method is that costs are allocated to the division or task that actually incurs the expense. As a result, more accurate records can be kept for each installed component. A disadvantage of this method occurs in the pricing of equipment that may be used for many different tasks. Duplication of costs can occur in this instance. Another disadvantage is that the budget may be left short for the following reason: the estimate may only reflect two hours for a crane truck, when the minimum cost of such a crane is usually a daily (8-hour) rental charge.

The second method for including equipment costs in the estimate is to keep all such costs separate and to include them in Division 1 as a part of Project Overhead. The advantage of this method is that all equipment costs are grouped together, and that machines that may be used for several tasks are included (without duplication). One disadvantage is that for future estimating purposes, equipment costs will be known only on a job basis and not per installed unit.

Whichever method is used, the estimator must be consistent, and must be sure that all equipment costs are included, but not duplicated. The estimating method should be the same as that chosen for cost monitoring and accounting. In this way, the data will be available both for monitoring the project costs and for bidding future projects.

Subcontractors

Subcontractors may account for a large percentage of a remodeling or renovation project. When subcontractors are used, quotations should be solicited and analyzed in the same way as material quotes. A primary concern is that the bid covers the work according to the plans and specifications, and that all appropriate work alternates and allowances, if any, are included. Any inclusions should be clearly stated and explained.

If the bid is received orally, a form such as that shown in Chapter 4, Figure 4.1, will help to ensure that it is documented accurately. Any unique scheduling or payment requirements must be noted and evaluated. Such requirements could affect (restrict or enhance) the normal progress of the project, and should, therefore, be known in advance.

The estimator should know how long the subcontract bid will be honored. This time period usually varies from 30 to 90 days and is often included as a condition in complete bids.

The estimator should also know or verify the bonding capability and capacity of unfamiliar subcontractors, when required. This may be necessary when bidding in a new location. Other than word of mouth, these inquiries may be the only way to confirm subcontractor reliability.

Project Overhead

Project overhead represents those construction costs that are usually included in Division One–General Requirements. Typical project overhead items are supervisory personnel, clean-up labor, and temporary heat and power. While these items may not be directly part of the physical structure, they are a part of the project. Project overhead, like all other direct costs, can be separated into material, labor, and equipment components.

Figures 5.4a and 5.4b are examples of a form which can help ensure that all appropriate costs are included. This form is for general construction, but could easily be adapted to the specific requirements of repair and remodeling projects.

Some may not agree that certain items (such as equipment or scaffolding) should be included as Project Overhead, and might prefer to list such items in another division. Ultimately, it is not important *where* each item is incorporated into the estimate, but that *every item is included somewhere.*

Project Overhead often includes time-dependent items; equipment rental, supervisory labor, and temporary utilities are examples. The

Means® Forms

PROJECT OVERHEAD SUMMARY

SHEET NO.

PROJECT

ESTIMATE NO.

LOCATION

ARCHITECT

DATE

QUANTITIES BY:

PRICES BY:

EXTENSIONS BY:

CHECKED BY:

DESCRIPTION	QUANTITY	UNIT	MATERIAL/EQUIPMENT		LABOR		TOTAL COST	
			UNIT	TOTAL	UNIT	TOTAL	UNIT	TOTAL
Job Organization: Superintendent								
Project Manager								
Timekeeper & Material Clerk								
Clerical								
Safety, Watchman & First Aid								
Travel Expense: Superintendent								
Project Manager								
Engineering: Layout								
Inspection/Quantities								
Drawings								
CPM Schedule								
Testing: Soil								
Materials								
Structural								
Equipment: Cranes								
Concrete Pump, Conveyor, Etc.								
Elevators, Hoists								
Freight & Hauling								
Loading, Unloading, Erecting, Etc.								
Maintenance								
Pumping								
Scaffolding								
Small Power Equipment/Tools								
Field Offices: Job Office								
Architect/Owner's Office								
Temporary Telephones								
Utilities								
Temporary Toilets								
Storage Areas & Sheds								
Temporary Utilities: Heat								
Light & Power								
Water								
PAGE TOTALS								

Page 1 of 2

Figure 5.4a

Means® Forms

DESCRIPTION	QUANTITY	UNIT	MATERIAL/EQUIPMENT		LABOR		TOTAL COST	
			UNIT	TOTAL	UNIT	TOTAL	UNIT	TOTAL
Total Brought Forward								
Winter Protection: Temp. Heat/Protection								
Snow Plowing								
Thawing Materials								
Temporary Roads								
Signs & Barricades: Site Sign								
Temporary Fences								
Temporary Stairs, Ladders & Floors								
Photographs								
Clean Up								
Dumpster								
Final Clean Up								
Punch List								
Permits: Building								
Misc.								
Insurance: Builders Risk								
Owner's Protective Liability								
Umbrella								
Unemployment Ins. & Social Security								
Taxes								
City Sales Tax								
State Sales Tax								
Bonds								
Performance								
Material & Equipment								
Main Office Expense								
Special Items								
TOTALS:								

Page 2 of 2

Figure 5.4b

cost for these items depends on the duration of the project. A preliminary schedule should, therefore, be developed *prior* to completion of the estimate, so that time items can be properly counted. This topic is covered in more detail in Chapter 8, "Pre-Bid Scheduling."

Bonds

Although bonds are really a type of "direct cost," they are priced and based on the total "bid" or "selling price." For that reason, they are generally figured after indirect costs have been added. Bonding requirements for a project will be specified in Division 1 – General Requirements – of the specifications, and will be included in the construction contract.

Sales Tax

Sales tax varies from state to state, and often from city to city within a state. Larger cities may have a sales tax in addition to the state sales tax. Some localities also impose separate sales taxes on labor and equipment.

When bidding takes place in an unfamiliar location, the estimator should check with local agencies regarding the amount and the method of payment of sales tax. Local authorities may require owners to withhold payments to out-of-state contractors until payment of all required sales tax has been verified. Sales tax, often taken for granted or even omitted, can be as much as 7.5% of material costs. Indeed, this can represent a significant portion of the project's total cost. Conversely, some clients and/or their projects may be tax exempt. If this fact is unknown to the estimator, a large dollar amount for sales tax might be needlessly included in a bid.

Chapter 6
INDIRECT COSTS

Indirect costs are the "costs of doing business." These expenses are sometimes referred to as "burden" to the project. Indirect costs may include certain fixed, or known, expenses and percentages, as well as costs which can be variable and subjectively determined. For example, government authorities require payment of certain taxes and insurance, usually based on labor costs and determined by trade. This expense is a type of *fixed* indirect cost. Office overhead, if well understood and established, can also be considered as a relatively fixed percentage. Profit and contingencies, however, are more variable and subjective. These figures are often determined based on the judgment and discretion of the person responsible for the company's growth and success.

If the direct costs for the same project have been carefully determined, they should not vary significantly from one estimator to another. It is the indirect costs that are often most responsible for variations among bids.

The direct costs of a project must be itemized, tabulated, and totalled before the indirect costs can be applied to the estimate. The most prevalent indirect costs include:

- Taxes and Insurance
- Office or Operating Overhead
- Profit
- Contingencies

Taxes and Insurance

The taxes and insurance included as indirect costs are most often related to the costs of labor and/or the type of work. This category may include Workers' Compensation, Builder's Risk, and Public Liability Insurance, as well as employer-paid social security tax and federal and state unemployment insurance. By law, the employer must pay these expenses. Rates are based on the type and salary of the employees, as well as the location and/or type of business.

Office or Operating Overhead

Office overhead, or the cost of doing business, is perhaps one of the main reasons why so many contractors are unable to realize a profit, or even to stay in business. The problem shows up in two ways: either a company does not know its true overhead cost and, therefore, fails to mark up its costs enough to recover them; or management does not restrain or control overhead costs effectively and fails to remain competitive.

If a contractor does not know the costs of operating the business, then, more than likely, these costs will not be recovered. Many companies survive, and even turn a profit, by simply adding an arbitrary percentage for overhead to each job, without knowing how the percentage is derived or what is included. When annual volume changes significantly, whether by increase or decrease, or when office staff and/or expenses increase, the previously used percentage for overhead may no longer be valid. When such a change occurs, the owner often finds that the company is not doing as well as before and cannot determine the reasons. Chances are, overhead costs are not being fully recovered. The following is a list of items that may be included when determining office overhead costs.

Owner: This should include only a reasonable base salary and does not include profits. An owner's salary is *not* a company's profit.

Engineer/Estimator/Project Manager: Since the owner may be primarily on the road getting business, the project manager runs the daily operation of the company and is responsible for estimating. In some operations, the estimator who successfully wins a bid becomes the "project manager" and is responsible to the owner for that project's profitability.

Secretary/Receptionist/Bookkeeper: This person manages office operations and handles paperwork. A talented individual in this position can be a tremendous asset.

Office Staff Insurance & Taxes: These costs are for main office personnel and should include, but are not limited to, the following:

- Worker's compensation insurance
- FICA
- Unemployment insurance
- Medical & other insurance
- Profit sharing, pension, etc.

Physical Plant Expenses: Whether the office, warehouse and/or yard are rented or owned, roughly the same costs are incurred. Telephone and utility costs will vary depending on the size of the building and the type of business. Office equipment includes items such as a copy machine, typewriters, computer, and furniture.

Professional Services: Accountant fees may be primarily for quarterly audits and annual tax return preparation. Legal fees go towards collecting receivables and contract disputes. In addition, a prudent contractor will have *every* contract read by his lawyer prior to signing.

Miscellaneous: There are many expenses that could be placed in this category. Association dues, seminars, travel, and entertainment may be included. Advertising includes the Yellow Pages, promotional materials, etc.

Uncollected Receivables: This amount can vary greatly, and is often affected by the overall economic climate. Depending on the timing of "uncollectibles," cash flow can be severely restricted and can cause serious financial problems, even for large companies. Sound cash planning and anticipation of such possibilities are essential.

In order for a company to stay in business *without losses* (profit is not yet a factor), not only must all direct construction costs be paid, but all additional overhead costs must be recovered during the year in order to operate the office. The most common method for recovering these costs is to apply this percentage to each job over the course of the year. The percentage may be calculated and applied in two ways:

- Office overhead applied as a percentage of *labor costs only*. This method requires that labor and material costs be estimated separately.
- Office overhead applied as a percentage of *total project costs*. This is appropriate where material and labor costs are not estimated separately.

The estimator must also remember that, if volume changes significantly, then the percentage for office overhead should be recalculated to reflect current conditions. Since salaries are the major portion of office overhead costs, any changes in office staff should also be figured into an accurate percentage for overhead.

It should be noted that an additional percentage is commonly applied to material costs, for handling, regardless of the method of recovering overhead costs. This percentage is more easily calculated if material costs are estimated and listed separately.

Profit

Determining a fair and reasonable percentage to be included for profit is not an easy task. This responsibility is usually left to the owner or chief estimator. Experience is crucial in anticipating what profit the market will bear. The economic climate, competition, knowledge of the project, and familiarity with the architect, designer, or owner all affect the way in which profit is determined.

Contingencies

Like profit, contingencies can also be difficult to quantify. Especially appropriate in preliminary budgets, the addition of a contingency is meant to protect the contractor as well as to give the owner a realistic estimate of potential project costs.

A contingency percentage should be based on the number of "unknowns" in a project, or the level of risk involved. This percentage should be inversely proportional to the amount of planning that has been done for the project and the detail of available information. If complete plans and specifications are supplied and the estimate is thorough and precise, then there is little need for a contingency. Figure 6.1 from *Means Repair and Remodeling Cost Data*, 1989, lists suggested contingency percentages that may be added to an estimate based on the stage of planning and development.

If an estimate is priced and each individual item is rounded up, or "padded," this is, in essence, adding a contingency. This method can cause problems, however, because the estimator can never be quite sure of what is the actual cost and what is the "padding," or safety margin, for each item. At the summary, the estimator cannot determine exactly how much has been included as a contingency factor for the project as a whole. A much more accurate and controllable approach is to price the estimate precisely and then add one contingency amount at the bottom line.

Also shown in Figure 6.1 is a section entitled "Factors." In planning and estimating repair and remodeling projects, there are many factors that can affect the project cost beyond the basic material and labor. The economy of scale usually associated with new construction

010 | Overhead

	010 000	Overhead	CREW	DAILY OUTPUT	MAN-HOURS	UNIT	BARE COSTS MAT.	LABOR	EQUIP.	TOTAL	TOTAL INCL O&P	
004	0011	**ARCHITECTURAL FEES** (10)										004
	0020	For work to $10,000				Project					15%	
	0040	To $25,000									13%	
	0060	To $100,000									10%	
	0080	To $500,000									8%	
	0090	To $1,000,000									7%	
016	0011	**CONSTRUCTION MANAGEMENT FEES**										016
	0060	For work to $10,000				Project					10%	
	0070	To $25,000									9%	
	0090	To $100,000									6%	
	0100	To $500,000									5%	
	0110	To $1,000,000									4%	
020	0010	**CONTINGENCIES** Allowance to add at conceptual stage									20%	020
	0050	Schematic stage									15%	
	0100	Preliminary working drawing stage									10%	
	0150	Final working drawing stage									2%	
028	0010	**ENGINEERING FEES** Educational planning consultant, minimum				Contrct					4.10%	028
	0100	Maximum									10.10%	
	0400	Elevator & conveying systems, minimum (11)									2.50%	
	0500	Maximum									5%	
	1000	Mechanical (plumbing & HVAC), minimum									4.10%	
	1100	Maximum									10.10%	
	1200	Structural, minimum				Project					1%	
	1300	Maximum				"					2.50%	
032	0010	**FACTORS** To be added to construction costs for particular job (143)										032
	0200											
	0500	Cut & patch to match existing construction, add, minimum				Costs	2%	3%				
	0550	Maximum					5%	9%				
	0800	Dust protection, add, minimum					1%	2%				
	0850	Maximum					4%	11%				
	1100	Equipment usage curtailment, add, minimum					1%	1%				
	1150	Maximum					3%	10%				
	1400	Material handling & storage limitation, add, minimum					1%	1%				
	1450	Maximum					6%	7%				
	1700	Protection of existing work, add, minimum					2%	2%				
	1750	Maximum					5%	7%				
	2000	Shift work requirements, add, minimum						5%				
	2050	Maximum						30%				
	2300	Temporary shoring and bracing, add, minimum					2%	5%				
	2350	Maximum					5%	12%				
036	0011	**FIELD PERSONNEL**										036
	0020											
	0180	Project manager, minimum				Week		925		925	1,405	
	0200	Average						1,030		1,030	1,555	
	0220	Maximum						1,170		1,170	1,790	
	0240	Superintendent, minimum						875		875	1,325	
	0260	Average						975		975	1,480	
	0280	Maximum						1,095		1,095	1,655	
040	0010	**INSURANCE** Builders risk, standard, minimum				Job					.22%	040
	0050	Maximum									.59%	
	0200	All-risk type, minimum (2)									.25%	
	0250	Maximum									.62%	
	0400	Contractor's equipment floater, minimum				Value					.90%	
	0450	Maximum				"					1.60%	
	0600	Public liability, average				Job					1.55%	
	0810	Workers compensation & employer's liability										
	2000	Range of 36 trades in 50 states, excl. wrecking, minimum				Payroll		1.80%				
	2100	Average				"		12.50%				

For expanded coverage of these items see *Means Building Construction Cost Data 1989*

1

Figure 6.1

often has no influence on the cost of repair and remodeling. Small quantities of components may have to be custom-fabricated at great expense. Work schedule coordination between trades frequently becomes difficult and work area restrictions can lead to subcontractor quotations with start-up and shut-down costs exceeding the cost of the specified work. The factors explained below and shown in Figure 6.1 are those normally associated with the loss of productivity encountered in repair and remodeling projects.

1. Cutting and patching to match the existing construction can often lead to an economical trade-off of, for example, removing entire walls rather than creating new door and window openings. Substitutions for materials that are no longer manufactured can be expensive. Piping and ductwork runs may not be as straight as in new construction and wiring may have to be snaked through walls and floors.
2. Dust and noise protection of adjoining non-construction areas can alter usual construction methods.
3. Equipment usage curtailment resulting from the physical limitations of the project may force workmen to use slow hand-operated equipment instead of power tools.
4. The confines of an enclosed building can have a costly influence on movement and material handling. Low capacity elevators and stairwells may be the only access to the upper floors of a multi-story building.
5. On some repair or remodeling projects, existing construction and completed work must be secured or otherwise protected from possible damage during ongoing construction. In certain areas, completed work must be guarded to prevent theft and vandalism.
6. Work may have to be done during other than normal shifts and/or around an existing production facility that is in operation during the repair and remodeling project.
7. Shoring and bracing may be required to support the building while structural changes are being made. Another expense is the allowance for temporary storage of construction materials on above grade floors.

The exact percentages used to modify the estimate are left to the estimator's judgment based upon the job conditions. The figures shown are for guidance only and should be used accordingly.

Chapter 7

THE ESTIMATE SUMMARY

At the pricing stage of the estimate, there is typically a large amount of paperwork that must be assembled, analyzed, and tabulated. This documentation generally consists of the following items.

- Quantity takeoff sheets for all work items (Figure 3.1)
- Material suppliers' written quotations
- Equipment or material suppliers' or subcontractors' telephone quotations (Figure 4.1)
- Subcontractors' written quotations
- Equipment suppliers' quotations
- Pricing sheets (Figures 3.2 and 3.3)
- Estimate summary sheets

Organizing the Estimate Data

In the "real world" of estimating, many quotations, especially for large material purchases and for subcontracts, are not received until the last minute before the bidding deadline. Therefore, a system is needed to efficiently handle the paperwork and to ensure that all relevant information will be transferred once (and only once) from the quantity takeoff to the cost analysis sheets. Some general rules for this process are as follows:

- Write on only one side of any document, where possible.
- Use Telephone Quotation forms for uniformity in recording prices received orally.
- Document the source of every quantity and price.
- Keep the entire estimate in one or more compartmentalized folders.
- If you are pricing your own materials, number and code each takeoff sheet and each pricing extension sheet as it is created. At the same time, keep an index list of each sheet by number. If a sheet is to be abandoned, write "VOID" on it, but do not discard it. Keep it until the bid is accepted; it will serve as a reference in accounting for all pages and sheets.

All subcontract costs should be properly noted and listed separately. These costs contain the subcontractors' mark-ups and may be treated differently from other direct costs when the estimator calculates the prime contractor's overhead and profit.

After all the unit prices and allowances have been entered on the pricing sheets, the costs are extended. In making the extensions, ignore the cents column and round all totals to the nearest dollar. In a column of figures, the cents will average out and will not be of consequence. Finally, each subdivision is added and the results checked, preferably by someone other than the person doing the extensions.

It is important to check the larger items for order of magnitude errors. If the total costs are divided by the floor area, the resulting square foot cost figures can be used to quickly pinpoint areas that are out of line with expected square foot costs. These cost figures should be recorded for comparison to past projects and as a resource for future estimating.

The takeoff and pricing method, as discussed, has been to utilize a Quantity Sheet for the material takeoff, and to transfer the data to a Cost Analysis form for pricing the material, labor, and equipment items.

An alternative to this method is a consolidation of the takeoff task and pricing on a single form. This approach works well for smaller bids and for change orders. An example, the Consolidated Estimate form, is shown in Figure 7.1. The same sequences and recommendations used to complete the Quantity Sheet and Cost Analysis form should be followed when using the Consolidated Estimate form to price the estimate.

Final Adjustments

When the pricing of all direct costs is complete, the estimator has two choices: to make all further price changes and adjustments on the Cost Analysis or Consolidated Estimate sheets, or to transfer the total costs for each subdivision to an Estimate Summary sheet so that all further price changes, until bid time, will be done on one sheet. Any indirect cost mark-ups and burdens will be figured on this sheet. An example of an Estimate Summary is shown in Figure 7.2.

Unless the estimate has a limited number of items, it is recommended that direct costs be transferred to an Estimate Summary sheet. This step should be double-checked since an error of transposition may easily occur. Pre-printed forms can be useful, although a plain columnar form may suffice. This summary, with the page numbers referencing from each extension sheet, can also serve as an index.

A company that repeatedly uses certain standard listings can save valuable time by having a custom Estimate Summary sheet printed. The printed CSI division and subdivision headings (shown in Figure 7.2) serve as another type of checklist, which can be used to ensure that all required costs are included. Appropriate column headings or categories for any estimate summary form could be as follows:

- Material
- Labor
- Equipment
- Subcontractor
- Total

Means® Forms

CONSOLIDATED ESTIMATE

SHEET NO.

PROJECT | CLASSIFICATION | ESTIMATE NO.

LOCATION | ARCHITECT | DATE

TAKE OFF BY | QUANTITIES BY | PRICES BY | EXTENSIONS BY | CHECKED BY

DESCRIPTION	NO.	DIMENSIONS			QUANTITIES				MATERIAL		LABOR		EQUIPMENT		TOTAL	
						UNIT		UNIT	UNIT COST	TOTAL	UNIT COST	TOTAL	UNIT COST	TOTAL	UNIT COST	TOTAL

Figure 7.1

Means® Forms

CONDENSED ESTIMATE SUMMARY

SHEET NO.

PROJECT | ESTIMATE NO.

LOCATION | TOTAL AREA/VOLUME | DATE

ARCHITECT | COST PER S.F./C.F. | NO. OF STORIES

PRICES BY: | EXTENSIONS BY: | CHECKED BY:

DIV.	DESCRIPTION		MATERIAL	LABOR	EQUIPMENT	SUBCONTRACT	TOTAL
1.0	**General Requirements**						
2.0	**Site Work**						
3.0	**Concrete**						
4.0	**Masonry**						
5.0	**Metals**						
6.0	**Carpentry**						
7.0	**Moisture & Thermal Protection**						
8.0	**Doors, Windows, Glass**						
9.0	**Finishes**						
10.0	**Specialties**						
11.0	**Equipment**						
12.0	**Furnishings**						
13.0	**Special Construction**						
14.0	**Conveying Systems**						
15.0	**Mechanical**						
16.0	**Electrical**						
	Subtotals						
	Sales Tax	%					
	Overhead	%					
	Subtotal						
	Profit	%					
	Contingency	%					
	Adjustments						
	TOTAL BID						

Figure 7.2

When the items have been listed in the proper columns, the categories are added and appropriate mark-ups applied to the total dollar values. Different percentages may be added to the sum of each column at the estimate summary. These percentages may include the following items, as discussed in Chapter 6.

- Taxes and Insurance
- Overhead
- Profit
- Contingencies

Chapter 8

PRE-BID SCHEDULING

The need for planning and scheduling is clear once the contract is signed and work commences on the project. However, some scheduling is also important during the bidding stage for the following reasons:

- To determine whether or not the project can be completed in the allotted or specified time using normal crew sizes
- To identify potential overtime requirements
- To determine the time requirements for supervision
- To anticipate possible temporary heat and power requirements
- To price certain general requirement items and overhead costs
- To budget for equipment usage
- To anticipate and justify material and equipment delivery requirements

The schedule produced prior to bidding may be a simple bar chart or network diagram that includes overall quantities, probable delivery times, and available manpower. Network scheduling methods, such as the Critical Path Method (CPM) and the Precedence Chart simplify pre-bid scheduling because they do not require time-scaled line diagrams.

CPM Scheduling

In the CPM diagram, each activity is represented by an arrow. Nodes indicate start/stop between activities. The Precedence Diagram, on the other hand, shows the activity as a node, with arrows used to denote precedence relationships between the activities. The precedence arrows may be used in different configurations to represent the sequential relationships between activities. Simple examples of CPM and Precedence diagrams are shown in Figures 8.1 and 8.2, respectively. In both systems, duration times are indicated along each path. The sequence (path) of activities requiring the most total time represents the shortest possible time (critical path) in which those activities may be completed.

For example, in both Figures 8.1 and 8.2, activities A, B, and C require 20 successive days for completion before activity G can begin. Activity paths for D and E (15 days) are shorter and can easily be completed during the 20-day sequence. Therefore, this 20-day sequence is the shortest possible time (i.e., the "critical path") for the completion of these activities – before activity G can begin.

Past experience or a prepared rough schedule may suggest that the time specified in the bidding documents is insufficient to complete the required work. In such cases, a more comprehensive schedule should be produced prior to bidding; this schedule will help to determine the added overtime or premium time work costs required to meet the completion date.

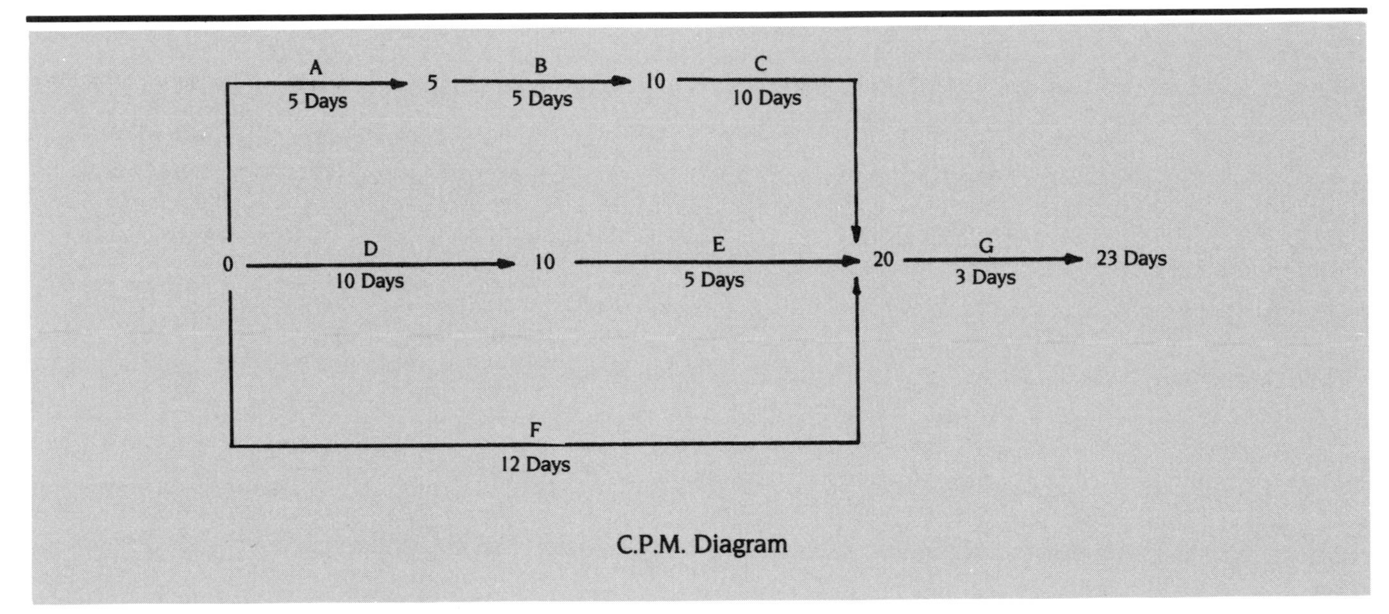

Figure 8.1

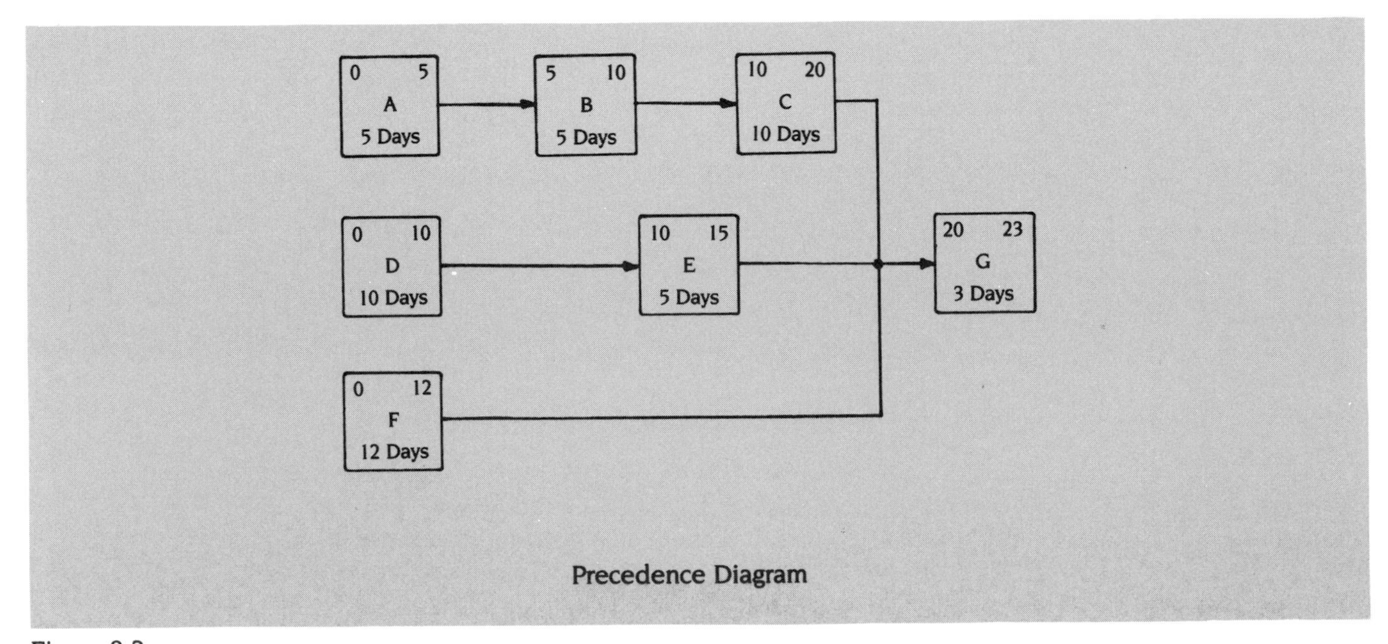

Figure 8.2

For example, a preliminary schedule is needed to determine the supervision and manning requirements of a small office renovation. A rough schedule for the interior work might be produced as shown in Figure 8.3. The man-days used to develop this schedule are derived from output figures determined in the estimate. Output can be determined based on the figures in *Means Repair and Remodeling Cost Data*, 1989. Man-days can also be figured by dividing the total labor cost shown on the estimate by the cost per man-day for each appropriate tradesperson.

As shown, the preliminary schedule can be used to determine supervision requirements, to develop appropriate crew sizes, and as a basis for ordering materials. All of these factors must be considered at this preliminary stage in order to determine how to meet the required completion date.

Summary

A pre-bid schedule can provide much more information than simple job duration. It can also be used to refine the estimate by introducing realistic manpower projections. Furthermore, the schedule may help the contractor to adjust the structure and size of the company based on projected requirements for months, even years ahead. It should also be noted that a schedule can become an effective tool for negotiating contracts.

For more detailed information on construction scheduling methods, see *Means Scheduling Manual* by F. William Horsley.

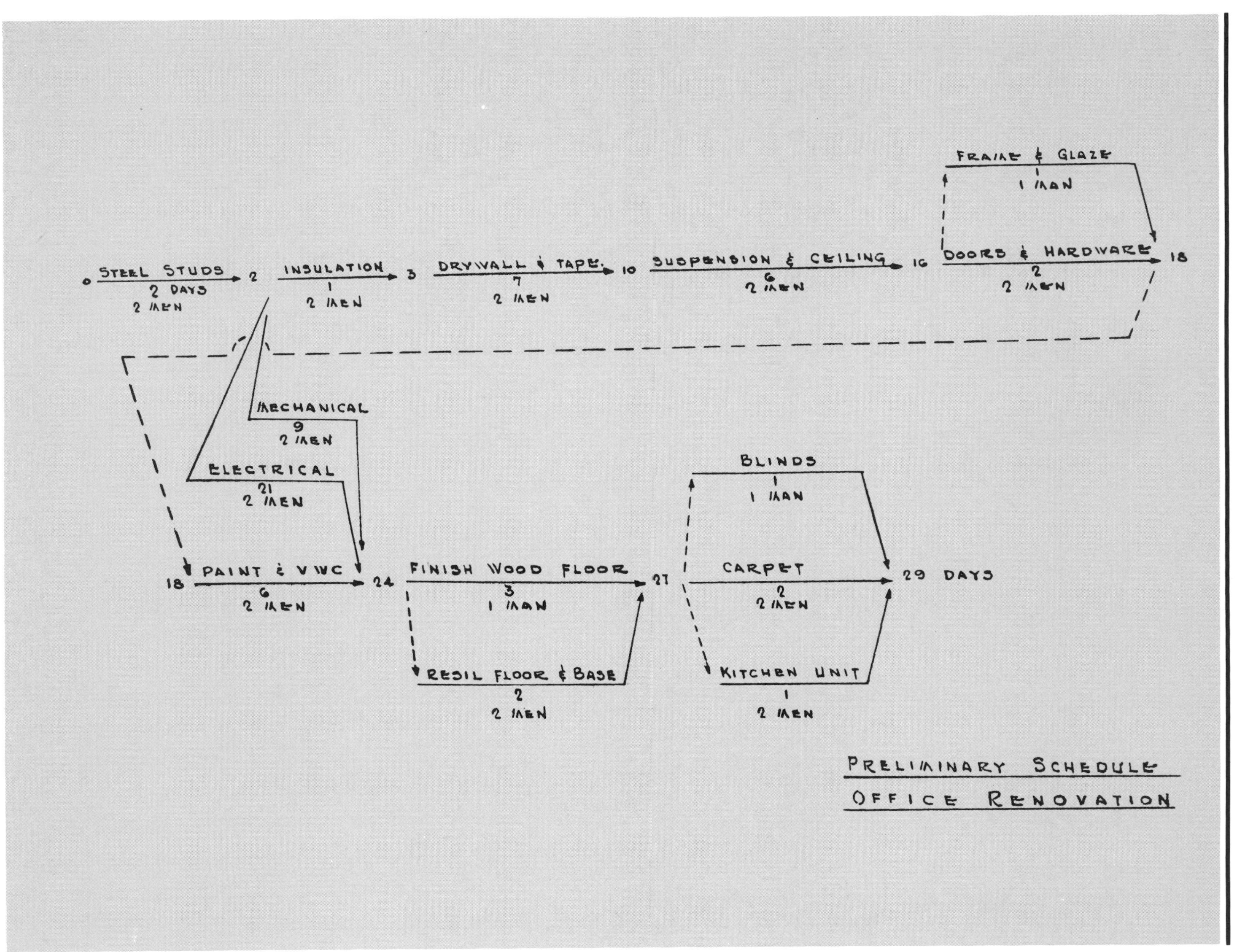
0
STEEL STUDS
2 DAYS
2 MEN
2
INSULATION
1
2 MEN
3
DRYWALL & TAPE
7
2 MEN
10
SUSPENSION & CEILING
6
2 MEN
16
DOORS & HARDWARE
2
2 MEN
18
FRAME & GLAZE
1
1 MAN
MECHANICAL
9
2 MEN
ELECTRICAL
21
2 MEN
18
PAINT & VWC
6
2 MEN
24
FINISH WOOD FLOOR
3
1 MAN
RESIL FLOOR & BASE
2
2 MEN
27
BLINDS
1
1 MAN
CARPET
2
2 MEN
KITCHEN UNIT
1
2 MEN
29 DAYS
PRELIMINARY SCHEDULE
OFFICE RENOVATION

Figure 8.3

Part II:
ESTIMATING BY CSI DIVISION

Part II:

ESTIMATING BY CSI DIVISION

There are two basic approaches to creating an estimate. One is to proceed with the quantity takeoff and pricing in a sequence similar to the order in which a building is constructed. This is the approach taken for an Assemblies estimate. The advantage of this method is that it enables experienced estimators to visualize the construction process. As a result, omissions are less likely. The basic disadvantage of the Assemblies method is that it is not easy to determine and track costs for each item by division (or subcontract). Costs for a particular trade may be spread throughout the estimate.

The second approach to estimating is by CSI MASTERFORMAT division (see Chapter 1). Most architectural specifications today are written according to this format, using the 16 divisions. Trades and subcontracts are generally limited to work within one division (e.g., Mechanical, or Masonry). It makes sense, therefore, that the estimate should also be organized in the CSI format. Using this method, the estimator may have to be more careful to include all required work. Nevertheless, the advantages outweigh potential drawbacks.

Most specifications contain references to related work for all divisions. These references serve as an aid to ensure completeness, but the estimator for the general contractor still has the responsibility of making sure that subcontractors include all that is specified and required for the appropriate portions of the project. The estimator must also decide what work will be subcontracted and what will be performed by the work force of the general contractor. These decisions can have a significant effect on the final cost of a project. Traditionally, work has been done at a lower cost using the general contractor's own labor force. With the specialization of trades, however, subcontractors can often perform work faster, and thus at a lower cost.

Cost accounting and control is another area affected by the choice of estimating method. Whichever method is chosen should be used consistently, from estimating–to field reporting–to final analysis. The first approach, Assemblies estimating in the sequence of construction, brings the same challenge to cost accounting as it does to estimating–the difficulty of keeping similar items grouped together as the work of single trades.

The second method of estimating and cost accounting, by CSI MASTERFORMAT division, may take longer when it comes to compiling all costs for each division (e.g., until all drywall or electrical work is complete). However, each trade will be separated and the records will be in accordance with the specifications. Since material purchases and scheduling of manpower and subcontractors are based on the construction sequence, these items should be calculated based on the project schedule. This schedule is established after and formulated from the unit price estimate, and is the basis of project coordination.

Project specifications may contain, or owners may request, a list of alternates which must be included with a submitted bid. These alternates become a series of mini-estimates within the total project estimate. Each alternate may include deductions of some items from the project, originally specified, as well as the addition of other items. Mistakenly regarded as incidentals, alternates are often not addressed until the project estimate is complete. In order to efficiently determine alternate costs, without performing a completely new estimate for each, the project estimate must be organized with the alternates in mind. For example, if items are to be deducted as part of an alternate, they must be separated in the project estimate. Similarly, when measuring for quantities, pertinent dimensions and quantities for the "adds" of an alternate should be listed separately. If alternates are considered in this way, they can be estimated quickly and accurately.

Part II (Chapters 9 through 24) presents specific guidelines for estimating each of the CSI MASTERFORMAT divisions. Two complete sample estimates follow in Part III of this book. The first is performed using the unit price CSI MASTERFORMAT division method. The second is an Assemblies estimate, carried out in the sequence of the actual construction.

Chapter 9

GENERAL REQUIREMENTS

When estimating by CSI division, the estimator must be careful to include all items which, while not attributable to one trade only, or directly to the physical construction of the building, are nevertheless required to successfully complete the project. These items are included in Division 1 – General Requirements.

Often referred to as the "General Conditions" or "Project Overhead," such items are usually set forth in the first part of the specifications. Some requirements may not be directly specified, even though they are required to perform the work. Standardized sets of General Conditions have been developed by various segments of the construction industry, such as those by the American Institute of Architects, the Consulting Engineers Council/U.S., National Society of Professional Engineers, and others. These standardized documents usually include:

- Definitions
- Contract document descriptions
- Contractor's rights and responsibilities
- Architect/Engineer's authority and responsibilities
- Owner's rights and responsibilities
- Variation from contract provisions
- Payment requirements and restrictions
- Requirements for the performance of the work
- Insurance and bond requirements
- Job conditions and operation

Since these documents are generic, additions, deletions and modifications unique to specific projects are usually included in Supplementary General Conditions.

Estimated costs for Division 1 are often recorded on a standardized form or checklist. Pre-printed forms or checklists are helpful to ensure that all requirements are included and priced. Many of the costs are dependent upon work in other divisions, or on the total cost and/or duration of the job. Project overhead costs should be determined throughout the estimating process and finalized when all other divisions have been estimated and a preliminary schedule established.

The following are brief discussions of various items that may be included as project overhead. Depending on the size and type of project, costs for these items may easily represent a significant portion of a repair or remodeling project's total costs (20% to 50%, or more). The goal is to develop an approach to estimating that will ensure that all project requirements are included.

Personnel

Job site personnel may be included as either *project overhead* or *office overhead*. This distinction often depends on the size of the project and the contractor's accounting methods. For example, if a project is large enough to require a full-time superintendent, then all time-related costs for that person may be considered project overhead. If, on the other hand, the superintendent is responsible for a number of smaller jobs, the expense may be either included in office overhead, or proportioned for each job.

Depending on the size of the project, a carpenter and/or laborer may be assigned to the job on a full-time basis for miscellaneous work. This individual would be directly responsible to the job superintendent for various tasks. The costs of this work would, therefore, not be attributable to any specific division, and might appropriately be included as project overhead.

Temporary Services

Required temporary services may or may not be included in the specifications. The following is a typical statement in the specifications or scope of work.

> *"Contractor shall supply all material, labor, equipment, tools, utilities, and other items and services required for the proper and timely performance of the work and completion of the project."*

The owner and designer may feel that such a statement eliminates ambiguity. To the estimator, this statement means that many items must be estimated. Temporary utilities, such as heat, light, power, water, and fire protection are major considerations. The estimator must not only account for anticipated monthly (or time-related) costs, but should also be sure that installation and removal costs are included, whether by the appropriate subcontractor or the general contractor.

Storage

If there is inadequate space inside the building, an office trailer and/or storage trailers or containers may be required. Even if these facilities are owned by the contractor, depreciation and other related costs should still be allocated to the job as part of project overhead. Telephone, utility, and temporary toilet facilities are other costs that must be included when required.

Security

Depending on the location and type of project, some security services may be required. In addition to security personnel or guard dogs, fences, gates, special lighting, and alarms may also be needed.

Temporary Construction

Temporary construction may also involve many items which are not specified in the construction documents. Partitions, doors, fences, and barricades may be required to delineate or isolate portions of the building or site during construction.

Worker Protection

In addition to the above items, some temporary construction may be required solely for the protection of workers. Depending on the project size, an OSHA representative may visit the site to ensure that all safety precautions are being observed.

Protection of Finished Work

Workers will almost always use a new or existing permanent elevator for access throughout the building. Even though this use is almost always restricted, precautionary measures should always be taken to protect the doors and cabs. Invariably, some damage occurs. Protection of any and all finished surfaces (throughout the project) should be priced and included in the estimate.

Types of temporary services and construction, with costs, from *Means Repair and Remodeling Cost Data*, 1989, are shown in Figure 9.1.

Job Clean-up

An amount should always be carried in the estimate for clean-up of the construction area, both during the construction process and upon completion. The clean-up can be broken down into three basic categories, which can be estimated separately:

- Continuous (daily or otherwise) cleaning of the project area
- Rubbish handling and removal
- Final clean-up

Costs for continuous cleaning can be included as an allowance, or estimated by required man-hours (in some cases, a full-time laborer is appropriate). Rubbish handling should include barrels; wheeled carts; a trash chute, if necessary; dumpster rental; and disposal fees. Disposal fees vary depending upon the project and location. A permit may also be required.

Costs for final clean-up should be based on past projects and may include subcontract costs for items such as the cleaning of windows and waxing of floors. Included in the costs for final clean-up may be an allowance for repair of minor damage to finished work.

Bonds

Bonding requirements for a project are specified in the Division 1 – General Requirements section of the specifications. They are also included in the construction contract. Various types of bonds may be required for a given project. Some common types are listed below.

Bid Bond

A form of bid security executed by the bidder or principal, and by a surety (bonding company) to guarantee that the bidder will enter into a contract within a specified time and furnish any required Performance of Labor and Material Payment bonds.

Completion Bond

Also known as "Construction" or "Contract" bond, it is a guarantee by a surety to the owner that the contractor will pay for all labor and materials used in the performance of the contract as per the construction documents. The claimants under the bond have direct contracts with the contractor or any subcontractor.

Performance Bond

(1) A guarantee that a contractor will perform a job according to the terms of the contracts. (2) A bond of the contractor in which a surety guarantees to the owner that the work will be performed in

014 | Quality Control

		014 100 Testing Services	CREW	DAILY OUTPUT	MAN-HOURS	UNIT	BARE COSTS MAT.	LABOR	EQUIP.	TOTAL	TOTAL INCL O&P	
108	4950	6" modified mold				Ea.					145	108
	5100	Shear tests, triaxial, minimum									145	
	5150	Maximum									475	
	5550	Technician for inspection, per day, earthwork									210	
	5650	Bolting									250	
	5750	Roofing									210	
	5790	Welding									250	
	5820	Non-destructive testing, dye penetrant				Day					295	
	5840	Magnetic particle									295	
	5860	Radiography									460	
	5880	Ultrasonic									315	
	6000	Welding certification, minimum				Ea.					74	
	6100	Maximum				"					260	

015 | Construction Facilities and Temporary Controls

		015 100 Temporary Utilities	CREW	DAILY OUTPUT	MAN-HOURS	UNIT	BARE COSTS MAT.	LABOR	EQUIP.	TOTAL	TOTAL INCL O&P	
104	0010	**TEMPORARY UTILITIES**										104
	0100	Heat, incl. fuel and operation, per week, 12 hrs. per day	1 Skwk	8.75	.914	CSF Flr	15.20	20		35.20	49	
	0200	24 hrs. per day	"	4.50	1.780		20	39		59	84	
	0350	Lighting, incl. service lamps, wiring & outlets, minimum	1 Elec	34	.235		2.06	5.70		7.76	10.95	
	0360	Maximum	"	17	.471		4.64	11.40		16.04	22	
	0400	Power for temporary lighting only, per month, minimum/month								.93	1.02	
	0450	Maximum/month								2.37	2.61	
	0600	Power for job duration incl. elevator, etc., minimum								45	50	
	0650	Maximum								93	100	
		015 200 Temporary Construction										
204	0010	**PROTECTION** Stair tread, 2" x 12" planks, 1 use	1 Carp	75	.107	Tread	.98	2.28		3.26	4.74	204
	0100	Exterior plywood, ½" thick, 1 use		65	.123		.45	2.63		3.08	4.72	
	0200	¾" thick, 1 use		60	.133		.82	2.85		3.67	5.50	
208	0010	**TEMPORARY CONSTRUCTION** See division 010-094 & 015-300										208
		015 250 Construction Aids										
254	0014	**SCAFFOLDING, STEEL TUBULAR** Rent, 1 use per mo., no plank										254
	0090	Building exterior, 1 to 5 stories	3 Carp	16.80	1.430	C.S.F.	13	31		44	63	
	0200	To 12 stories (14)	4 Carp	15	2.130		12.20	46		58.20	87	
	0310	13 to 20 stories	5 Carp	16.75	2.390		11.40	51		62.40	95	
	0460	Building interior walls, (area) up to 16' high	3 Carp	22.70	1.060		12.50	23		35.50	50	
	0560	16' to 40' high		18.70	1.280		12.75	27		39.75	58	
	0800	Building interior floor area, up to 30' high		90	.267	C.C.F.	3.90	5.70		9.60	13.45	
	0900	Over 30' high	4 Carp	100	.320	"	4.40	6.85		11.25	15.85	
255	0011	**SCAFFOLDING SPECIALTIES**										255
	0050											
	1200	Sidewalk bridge, heavy duty steel posts & beams, including										
	1210	parapet protection & waterproofing										
	1220	8' to 10' wide, 2 posts	3 Carp	15	1.600	L.F.	10.80	34		44.80	67	
	1230	3 posts	"	10	2.400	"	16.50	51		67.50	100	

4

For expanded coverage of these items see *Means Building Construction Cost Data 1989*

Figure 9.1

accordance with the contract documents. Except where prohibited by statute, the performance bond is frequently combined with the labor and material payment bond.

Surety Bond

A legal instrument under which one party agrees to answer to another party for the debt, default, or failure to perform of a third party.

Miscellaneous General Requirements

Many other items must be taken into account when costs are being determined for project overhead. The following items are among the major considerations.

Scaffolding or Rolling Platforms

It is important to determine who is responsible for the rental, erection, and dismantling of scaffolding, since it will often be used by more than one trade. If a subcontractor is responsible, it may be necessary to leave the scaffolding in place long enough for use by other trades. Different types of scaffolding are illustrated in Figure 9.2.

Small Tools

An allowance, based on past experience, should be carried for small tools. This sum should cover the cost of hand tools as well as small power tools supplied by the contractor. Small tools often break down or "walk," and a certain amount of replacement is inevitable.

Permits

Various types of permits may be required depending on local codes and regulations. Both the necessity of the permit and the responsibility for acquiring it must be determined before bidding. If the work is being done in an unfamiliar location, local building officials should be consulted regarding unusual or unknown requirements.

Insurance

Insurance coverage for each project and locality–above and beyond the normal required operating insurance–should be reviewed to ensure that coverage is adequate. The contract documents will often specify certain required policy limits. The estimator should anticipate the need for specific policies or riders.

Other General Requirements Considerations

Other items commonly included in project overhead are: photographs, models, job signs, costs for sample panels and materials for owner/architect approval, and an allowance for replacement of broken glass. For some materials, such as imported goods or custom-fabricated items, both shipping costs and off-site storage fees can be expected. An allowance should also be included for anticipated costs pertaining to punch list items. These costs are likely to be based on past experience.

Some project overhead costs can be calculated at the beginning of the estimate. Others are included when the estimating process is under way. Still other costs are estimated last since they depend on the total cost and duration of the project. Because many of the overhead items are not directly specified, the estimator must use experience and visualize the construction process to ensure that all requirements are met. *It is not important when or where these items are included, but that they are included.* One contract may list certain costs as *project overhead*, while another contractor allocates the same costs (and responsibility) to a subcontractor. Either way, the costs must be and are recorded in the estimate.

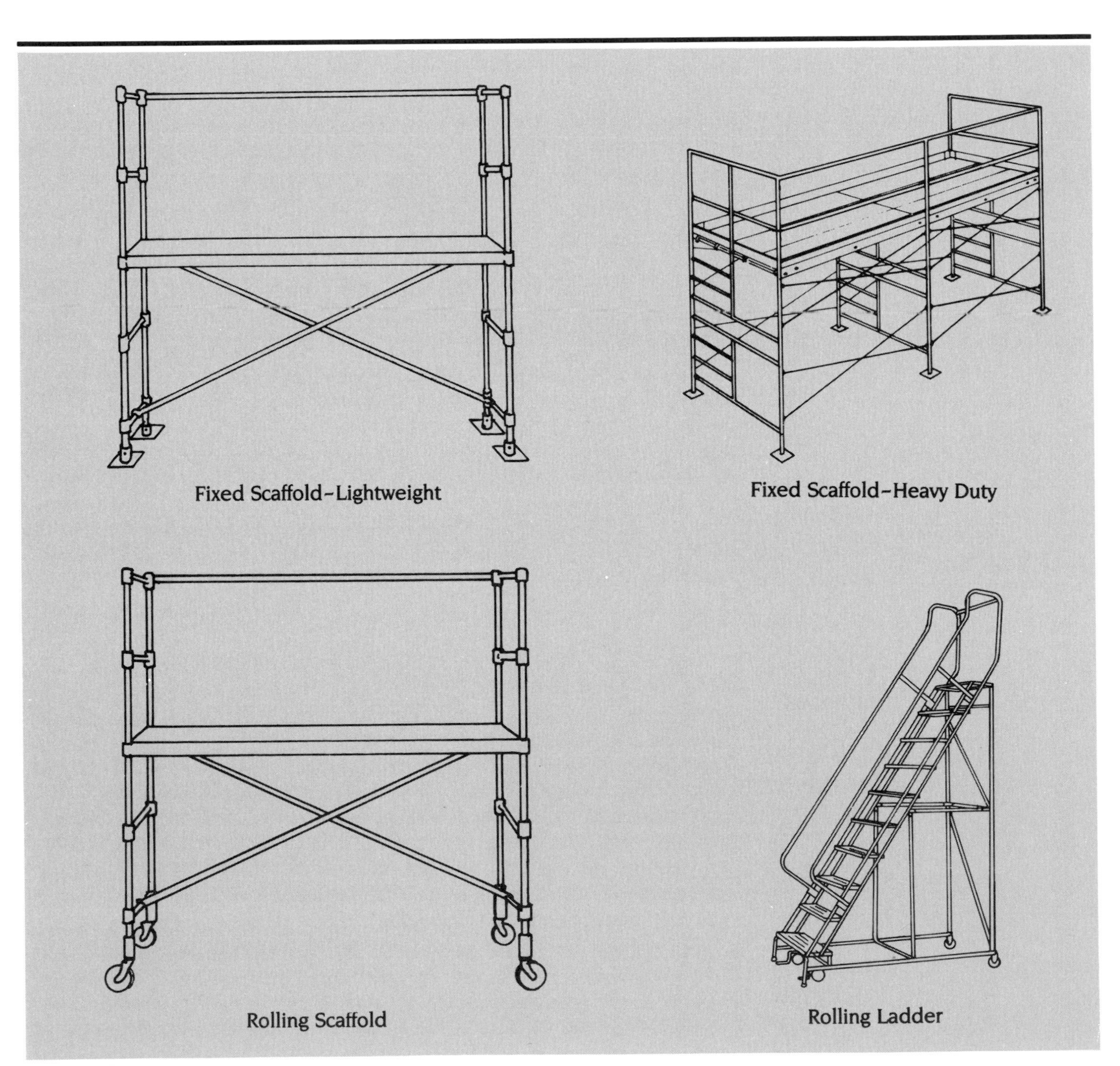

Figure 9.2

Chapter 10

SITE WORK/ DEMOLITION

Site Work

In most remodeling and renovation projects, any site work involved is usually small relative to the amount of interior work. However small, the analysis of site work to be done is an important part of the quantity takeoff. Because most site work requires special tools or heavy equipment, its role in commercial renovation requires special attention. The estimator must determine not only the appropriate equipment options, but also the cost effectiveness.

Minimum equipment charges may be in effect if the quantities of work are small. For example, Figure 10.1, from *Means Repair and Remodeling Cost Data*, 1989 shows costs for core drilling. The estimator has determined that only two six-inch holes are required through a four-inch concrete slab. Line 020-125-0700 would suggest the cost to be 2 × $28, or $56. However, line 020-125-2050 in Figure 10.2 shows the *minimum charge* to be $99. Clearly, the minimum charge must be carried in the estimate. Not only must the estimator determine the exact quantities involved, but he must also employ careful judgment and experience to anticipate the consequences of actual conditions.

Demolition

In contrast to conventional site work (such as excavation, site utilities, and paving), demolition is, in most cases, a major component of remodeling work. For takeoff and estimating purposes, demolition should be broken down into individual components. Often, demolition is not directly included in the plans and specifications. A review of Division 020, Subsurface Investigation and Demolition, in *Means Repair and Remodeling Cost Data*, 1989, may help the estimator to be sure that all required work is included in the estimate. Unless the estimator has had extensive experience with selective demolition, or unless a local subcontractor can provide a bid, each item to be demolished should be listed separately. When performing the quantity takeoff, requirements for handling, hauling, and dumping the debris must also be included as separate items.

When the contractor needs only a preliminary or budget cost for demolition, the estimator may have to determine only the size of the job (square feet or cubic feet) of the proposed demolition area, and then base costs on previous jobs, or use cost data such as that shown in Figure 10.3. The estimator, however, must observe caution when using this technique. As emphasized, every remodeling project

017 | Contract Closeout

		017 100 | Final Cleaning	CREW	DAILY OUTPUT	MAN-HOURS	UNIT	BARE COSTS MAT.	BARE COSTS LABOR	BARE COSTS EQUIP.	BARE COSTS TOTAL	TOTAL INCL O&P	
104	0050	Cleanup of floor area, continuous, per day	A-5	8	2.250	M.S.F.	1.50	38	3.41	42.91	66	104
	0100	Final	"	11.50	1.570	"	1.60	26	2.37	29.97	47	

020 | Subsurface Investigation and Demolition

		020 120 | Std Penetration Tests	CREW	DAILY OUTPUT	MAN-HOURS	UNIT	BARE COSTS MAT.	BARE COSTS LABOR	BARE COSTS EQUIP.	BARE COSTS TOTAL	TOTAL INCL O&P	
123	0010	**BORINGS** Initial field stake out and determination of elevations	A-6	1	16	Day		310		310	485	123
	0100	Drawings showing boring details				Total		165		165	255	
	0200	Report and recommendations from P.E.						365		365	560	
	0300	Mobilization and demobilization, minimum	B-55	4	6			100	120	220	295	
	0350	For over 100 miles, per added mile		450	.053	Mile		.91	1.06	1.97	2.62	
	0600	Auger holes in earth, no samples, 2-½" diameter		78.60	.305	L.F.		5.20	6.05	11.25	15	
	0650	4" diameter		67.50	.356			6.05	7.05	13.10	17.45	
	0800	Cased borings in earth, with samples, 2-½" diameter		55.50	.432			7.35	8.60	15.95	21	
	0850	4" diameter		32.60	.736			12.55	14.60	27.15	36	
	1000	Drilling in rock, "BX" core, no sampling	B-56	34.90	.458			8.60	13.20	21.80	28	
	1050	With casing & sampling		31.70	.505			9.45	14.50	23.95	31	
	1200	"NX" core, no sampling		25.92	.617			11.55	17.75	29.30	38	
	1250	With casing and sampling		25	.640			12	18.40	30.40	39	
	1400	Drill rig and crew with light duty rig	B-55	1	24	Day		410	475	885	1,175	
	1450	With heavy duty rig	B-56	1	16	"		300	460	760	980	
	1500	For inner city borings add, minimum									10%	
	1510	Maximum									20%	
125	0010	**DRILLING, CORE** Reinforced concrete slab, up to 6" thick slab										125
	0020	Including layout and set up										
	0100	1" diameter core	B-89A	43	.372	Ea.	2.02	7.20	1.06	10.28	14.90	
	0150	Each added inch thick, add		350	.046		.34	.89	.13	1.36	1.93	
	0300	3" diameter core		37	.432		4.49	8.40	1.23	14.12	19.65	
	0350	Each added inch thick, add		222	.072		.75	1.40	.20	2.35	3.29	
	0500	4" diameter core		34	.471		6	9.10	1.34	16.44	23	
	0550	Each added inch thick, add		212	.075		1	1.46	.21	2.67	3.67	
	0700	6" diameter core		27	.593		7.30	11.50	1.68	20.48	28	
	0750	Each added inch thick, add		180	.089		1.21	1.72	.25	3.18	4.36	
	0900	8" diameter core		19	.842		10	16.30	2.39	28.69	40	
	0950	Each added inch thick, add		118	.136		1.67	2.63	.38	4.68	6.45	
	1100	10" diameter core	A-1	17	.471		13.50	7.95	3.02	24.47	31	
	1150	Each added inch thick, add	"	106	.075		2.25	1.27	.48	4	5.05	
	1300	12" diameter core					16.30	9	3.43	28.73	36	
	1350	Each added inch thick, add	A-1	89	.090		2.72	1.51	.58	4.81	6.05	
	1500	14" diameter core		13	.615		20	10.35	3.95	34.30	43	
	1550	Each added inch thick, add		74	.108		3.37	1.82	.69	5.88	7.40	
	1700	18" diameter core		6.30	1.270		26	21	8.15	55.15	72	
	1750	Each added inch thick, add		36	.222		4.50	3.74	1.43	9.67	12.55	
	1760	For horizontal holes, add to above								30%	30%	
	1770	Prestressed hollow core plank, 6" thick										
	1780	1" diameter core	B-89A	65	.246	Ea.	1.31	4.77	.70	6.78	9.85	
	1790	Each added inch thick, add		375	.043		.21	.83	.12	1.16	1.68	
	1800	3" diameter core		55	.291		2.95	5.65	.83	9.43	13.15	
	1810	Each added inch thick, add		225	.071		.50	1.38	.20	2.08	2.97	
	1820	4" diameter core		54	.296		3.94	5.75	.84	10.53	14.45	
	1830	Each added inch thick, add		210	.076		.66	1.48	.22	2.36	3.33	

10

For expanded coverage of these items see *Means Site Work Cost Data 1989*

Figure 10.1

020 | Subsurface Investigation and Demolition

		020 120 Std Penetration Tests	CREW	DAILY OUTPUT	MAN-HOURS	UNIT	BARE COSTS MAT.	LABOR	EQUIP.	TOTAL	TOTAL INCL O&P	
125	1840	6″ diameter core	B-89A	46	.348	Ea.	4.79	6.75	.99	12.53	17.10	125
	1850	Each added inch thick, add		175	.091		.80	1.77	.26	2.83	4	
	1860	8″ diameter core		31	.516		6.55	10	1.46	18.01	25	
	1870	Each added inch thick, add		135	.119		1.08	2.30	.34	3.72	5.25	
	1880	10″ diameter core	A-1	28	.286		8.80	4.81	1.84	15.45	19.45	
	1890	Each added inch thick, add		115	.070		1.08	1.17	.45	2.70	3.56	
	1900	12″ diameter core		25	.320		10.65	5.40	2.06	18.11	23	
	1910	Each added inch thick, add		95	.084		1.78	1.42	.54	3.74	4.83	
	1950	Minimum charge for above, 3″ diameter core	B-89A	6.35	2.520	Total		49	7.15	56.15	86	
	2000	4″ diameter core		6.15	2.600			50	7.40	57.40	89	
	2050	6″ diameter core		5.50	2.910			56	8.25	64.25	99	
	2100	8″ diameter core		5.10	3.140			61	8.90	69.90	105	
	2150	10″ diameter core		4.50	3.560			69	10.10	79.10	120	
	2200	12″ diameter core		3.70	4.320			84	12.25	96.25	145	
	2250	14″ diameter core		3.20	5			97	14.20	111.20	170	
	2300	18″ diameter core		3	5.330			105	15.15	120.15	180	

		020 550 Site Demolition	CREW	DAILY OUTPUT	MAN-HOURS	UNIT	BARE COSTS MAT.	LABOR	EQUIP.	TOTAL	TOTAL INCL O&P	
554	0010	**SITE DEMOLITION** No hauling, abandon catch basin or manhole	B-6	7	3.430	Ea.		62	25	87	125	554
	0020	Remove existing catch basin or manhole		4	6			110	44	154	220	
	0030	Catch basin or manhole frames and covers stored		13	1.850			33	13.40	46.40	68	
	0040	Remove and reset		7	3.430			62	25	87	125	
	0045	Minimum labor/equipment charge		4	6	Job		110	44	154	220	
	0600	Fencing, barbed wire, 3 strand	2 Clab	430	.037	L.F.		.63		.63	1.01	
	0650	5 strand		280	.057			.96		.96	1.55	
	0700	Chain link, remove only, 8′ to 10′ high		310	.052			.87		.87	1.40	
	0800	Guide rail, remove only		85	.188			3.17		3.17	5.10	
	0850	Remove and reset		35	.457			7.70		7.70	12.35	
	0890	Minimum labor/equipment charge		4	4	Job		67		67	110	
	0900	Hydrants, fire, remove only	2 Plum	4.70	3.400	Ea.		83		83	125	
	0950	Remove and reset		1.40	11.430	"		280		280	425	
	0990	Minimum labor/equipment charge		2	8	Job		195		195	300	
	1000	Remove masonry walls, block or tile, solid	B-5	1,800	.036	C.F.		.68	.41	1.09	1.52	
	1100	Cavity		2,200	.029			.56	.33	.89	1.24	
	1200	Brick, solid		900	.071			1.36	.81	2.17	3.04	
	1300	With block		1,130	.057			1.08	.65	1.73	2.42	
	1400	Stone, with mortar		900	.071			1.36	.81	2.17	3.04	
	1500	Dry set		1,500	.043			.81	.49	1.30	1.82	
	1590	Minimum labor/equipment charge	A-1	4	2	Job		34	12.85	46.85	68	
	1710	Pavement removal, bituminous, 3″ thick	B-38	690	.058	S.Y.		1.10	1.50	2.60	3.39	
	1750	4″ to 6″ thick		420	.095			1.81	2.46	4.27	5.55	
	1800	Bituminous driveways		680	.059			1.12	1.52	2.64	3.44	
	1900	Concrete to 6″ thick, mesh reinforced		255	.157			2.97	4.06	7.03	9.15	
	2000	Rod reinforced		200	.200			3.79	5.15	8.94	11.70	
	2100	Concrete 7″ to 24″ thick, plain		13.10	3.050	C.Y.		58	79	137	180	
	2200	Reinforced		9.50	4.210	"		80	110	190	245	
	2250	Minimum labor/equipment charge		6	6.670	Job		125	170	295	390	
	2300	With hand held air equipment, bituminous	B-39	1,900	.025	S.F.		.45	.06	.51	.79	
	2320	Concrete to 6″ thick, no reinforcing		1,200	.040			.71	.10	.81	1.24	
	2340	Mesh reinforced		830	.058			1.03	.14	1.17	1.80	
	2360	Rod reinforced		765	.063			1.12	.15	1.27	1.95	
	2390	Minimum labor/equipment charge	B-38	6	6.670	Job		125	170	295	390	
	2400	Curbs, concrete, plain	B-6	325	.074	L.F.		1.34	.54	1.88	2.71	
	2500	Reinforced		220	.109			1.97	.79	2.76	4	
	2600	Granite curbs		355	.068			1.22	.49	1.71	2.48	
	2700	Bituminous curbs		830	.029			.52	.21	.73	1.06	
	2790	Minimum labor/equipment charge		6	4	Job		72	29	101	145	
	2900	Pipe removal, concrete, no excavation, 12″ diameter		175	.137	L.F.		2.48	.99	3.47	5.05	

For expanded coverage of these items see *Means Site Work Cost Data 1989* 11

Figure 10.2

020 | Subsurface Investigation and Demolition

		020 550 Site Demolition	CREW	DAILY OUTPUT	MAN-HOURS	UNIT	BARE COSTS MAT.	BARE COSTS LABOR	BARE COSTS EQUIP.	BARE COSTS TOTAL	TOTAL INCL O&P	
554	2930	15″ diameter	B-6	150	.160	L.F.		2.90	1.16	4.06	5.85	554
	2960	24″ diameter		120	.200			3.62	1.45	5.07	7.35	
	3000	36″ diameter		90	.267			4.83	1.93	6.76	9.80	
	3200	Steel, welded connections, 4″ diameter		160	.150			2.72	1.09	3.81	5.50	
	3300	10″ diameter		80	.300			5.45	2.18	7.63	11	
	3390	Minimum labor/equipment charge		3	8	Job		145	58	203	295	
	3500	Railroad track removal, ties and track	B-14	110	.436	L.F.		7.75	1.58	9.33	14.15	
	3600	Ballast		500	.096	C.Y.		1.71	.35	2.06	3.11	
	3700	Remove and re-install ties & track using new bolts & spikes		50	.960	L.F.		17.10	3.48	20.58	31	
	3800	Turnouts using new bolts and spikes		1	48	Ea.		855	175	1,030	1,550	
	3890	Minimum labor/equipment charge		5	9.600	Job		170	35	205	310	
	4000	Sidewalk removal, bituminous, 2-½″ thick	B-6	325	.074	S.Y.		1.34	.54	1.88	2.71	
	4050	Brick, set in mortar		185	.130			2.35	.94	3.29	4.76	
	4100	Concrete, plain		160	.150			2.72	1.09	3.81	5.50	
	4200	Mesh reinforced		150	.160			2.90	1.16	4.06	5.85	
	4290	Minimum labor/equipment charge	B-39	12	4	Job		71	9.75	80.75	125	
		020 600 Building Demolition										
604	0010	**BUILDING DEMOLITION** Large urban projects, incl. disposal, steel	B-8	21,500	.003	C.F.		.06	.08	.14	.18	604
	0050	Concrete		15,300	.004			.08	.11	.19	.25	
	0080	Masonry		20,100	.003			.06	.09	.15	.19	
	0100	Mixture of types, average		20,100	.003			.06	.09	.15	.19	
	0500	Small bldgs, or single bldgs, no salvage included, steel	B-3	14,800	.003			.06	.09	.15	.19	
	0600	Concrete		11,300	.004			.08	.11	.19	.25	
	0650	Masonry		14,800	.003			.06	.09	.15	.19	
	0700	Wood		14,800	.003			.06	.09	.15	.19	
	1000	Single family, one story house, wood, minimum				Ea.					1,900	
	1020	Maximum									3,250	
	1200	Two family, two story house, wood, minimum									2,450	
	1220	Maximum									4,875	
	1300	Three family, three story house, wood, minimum									3,150	
	1320	Maximum									6,250	
	1400	Gutting building, see division 020-716										
608	0010	**DISPOSAL ONLY** Urban buildings with salvage value allowed										608
	0020	Including loading and 5 mile haul to dump										
	0200	Steel frame	B-3	430	.112	C.Y.		2.04	2.99	5.03	6.55	
	0300	Concrete frame		365	.132			2.40	3.53	5.93	7.70	
	0400	Masonry construction		445	.108			1.97	2.89	4.86	6.30	
	0500	Wood frame		247	.194			3.54	5.20	8.74	11.35	
612	0010	**DUMP CHARGES** Typical urban city, fees only										612
	0100	Building construction materials				C.Y.					10.50	
	0200	Demolition lumber, trees, brush									12.50	
	0300	Rubbish only									10.50	
	0500	Reclamation station, usual charge				Ton					30	
620	0010	**RUBBISH HANDLING** The following are to be added to the										620
	0020	demolition prices										
	0400	Chute, circular, prefabricated steel, 18″ diameter	B-1	40	.600	L.F.	9.65	10.50		20.15	27	
	0440	30″ diameter	"	30	.800	"	18.70	14		32.70	43	
	0600	Dumpster, (debris box container), 5 C.Y., rent per week				Ea.					155	
	0700	10 C.Y. capacity									190	
	0800	30 C.Y. capacity									265	
	0840	40 C.Y. capacity									315	
	1000	Dust partition, 6 mil polyethylene, 4′ x 8′ panels, 1″ x 3″ frame	2 Carp	2,000	.008	S.F.	.16	.17		.33	.45	
	1080	2″ x 4″ frame	"	2,000	.008	"	.27	.17		.44	.57	
	2000	Load, haul to chute & dumping into chute, 50′ haul	2 Clab	21.50	.744	C.Y.		12.55		12.55	20	
	2040	100′ haul	"	16.50	.970	"		16.35		16.35	26	

12

For expanded coverage of these items see *Means Site Work Cost Data 1989*

Figure 10.3

is different and the amount and difficulty of demolition – whether selective or complete – will vary from job to job.

Determining the Appropriate Equipment

When tabulating the quantities involved in all aspects of site work, the estimator must keep in mind the methods that will be used and the restrictions involved in order to determine the appropriate equipment. For example, the plans may specify a trench for a buried tank. The estimator, having determined the size of the trench, must decide upon the best and least expensive method of work. See Figure 10.4, derived from costs in *Means Repair and Remodeling Cost Data*, 1989.

When the estimator looks only at the actual cost of excavating and backfilling, digging by hand is almost ten times more expensive than using a backhoe. The daily rental, however (usually the minimum charge), makes the backhoe the more costly alternative. In certain cases when quantities are small, it can be more economical to work by hand. However, if other work on that job can be done by the same piece of equipment during the same day, then the equipment may become the less expensive alternative. When determining quantities involving expensive equipment, the estimator must be aware of potential savings due to economy of scale.

Trench Excavation

Line Item	Unit Price	Cost
022-254-1400 Excavation by hand	$27.00/CY	$ 95.85
022-204-0010 Backfill by hand	$15.45/CY	$ 54.84
Actual cost by hand		$150.69
022-254-0090 Backhoe excavation	$ 3.79/CY	$ 13.45
022-204-1300 Backhoe backfill	$ 0.98/CY	$ 3.47
Actual cost by backhoe		$ 16.92
016-408-0400 Minimum daily rental of backhoe		$175.00

Trench: 8′ length x 3′ width x 4′ depth = 96 cubic feet
3.55 cubic yards
(to be excavated)

Figure 10.4

Chapter 11
CONCRETE

Concrete work in remodeling may entail placing a floor slab, placing a new topping on an existing floor, constructing new column foundations, or providing equipment foundations. Figure 11.1 illustrates some different types of concrete work commonly used in remodeling projects.

Estimating concrete work can be performed using two basic methods. The first is to estimate all of the individual components — formwork, reinforcing, concrete, placement, and finish — individually. This is the most accurate, but also the most time-consuming method, and is not always practical when the concrete work is only a small portion of the total project.

The second method is to estimate using costs that incorporate all of the components into a *system*, or *assembly*. This type of inclusive pricing is shown in Figure 11.2. When using this method, the estimator must be sure that the system to be estimated is, in fact, the same as that for which the assembly has been developed. Slight variations in the design of a system can significantly affect costs.

Regardless of the scope of the work required, all concrete construction involves the following basic components:

- Formwork
- Reinforcing steel or welded wire fabric
- Concrete
- Placement
- Finishing

The following sections describe various concrete design details frequently found in remodeling projects. Reinforcing, strength, placement, and finish requirements are included where applicable.

Floor Slabs

Floor slabs (on grade) should be placed on compacted granular fill, such as gravel or crushed stone, which has been covered by a polyethylene vapor barrier. Slabs may be reinforced or unreinforced. Reinforcing is usually provided by welded wire fabric, which prevents any cracks from expanding.

Large concrete floors are usually placed in strips which extend across the building at column lines or at 20' to 30' widths. Construction joints may be keyed or straight and may contain smooth or deformed reinforcing which, in turn, can be wrapped or greased on one side to allow horizontal movement and to control cracking.

Control joints are intended to limit cracking to designated lines. They may be established by saw-cutting the partially cured concrete slab to a specified depth or by applying a pre-formed metal strip to create a crack line. Many specifications require a boxed out section around columns, with that area to be concreted after the slab has been placed and cured. This helps to prevent future cracks due to settling.

Expansion joints are generally used against confining walls, foundations, and existing construction. They are commonly constructed from a pre-formed expansion material. The finish of the slabs is usually dictated by their use and varies between a screed (rough) finish (associated with two-course floors) to a steel trowel-

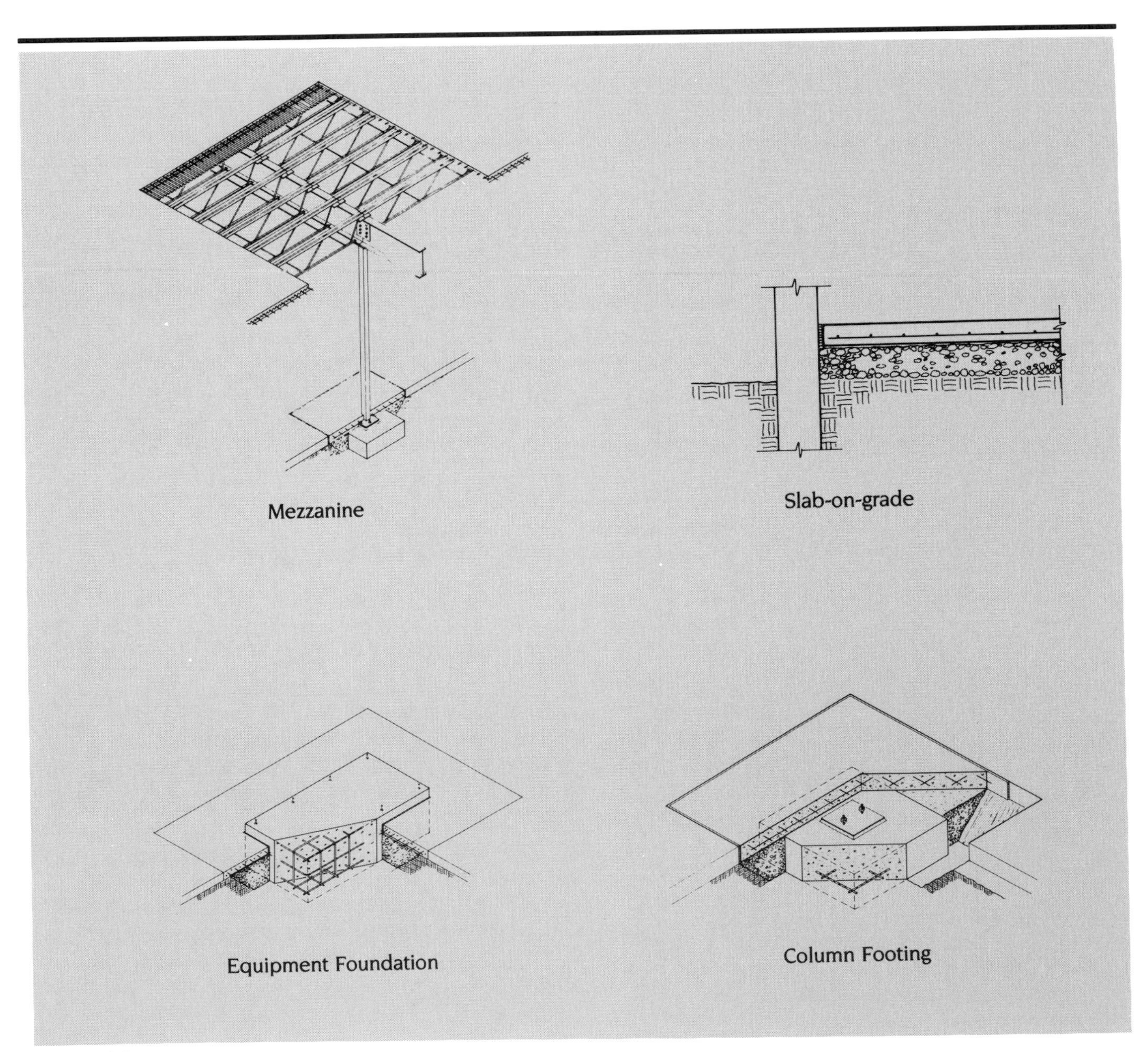

Figure 11.1

033 | Cast-In-Place Concrete

	033 100 | Structural Concrete		Crew	Daily Output	Man-Hours	Unit	Bare Costs Mat.	Bare Costs Labor	Bare Costs Equip.	Bare Costs Total	Total Incl O&P	
118	2120	Maximum				Lb.	1.50			1.50	1.65	118
	2200	Water reducing admixture, average				Gal.	7.75			7.75	8.55	
126	0010	**CONCRETE, READY MIX** Regular weight, 2000 psi (42)				C.Y.	52.30			52.30	58	126
	0100	2500 psi					54			54	59	
	0150	3000 psi (43)					55.80			55.80	61	
	0200	3500 psi					57.50			57.50	63	
	0250	3750 psi					57.90			57.90	64	
	0300	4000 psi					58.25			58.25	64	
	0350	4500 psi					60			60	66	
	0400	5000 psi					62.35			62.35	69	
	1000	For high early strength cement, add					10%					
	2000	For all lightweight aggregate, add				↓	45%					
	3000	For integral colors, 2500 psi, 5 bag mix										
	3100	Red, yellow or brown, 1.8 lb. per bag, add				C.Y.	12.60			12.60	13.85	
	3200	9.4 lb. per bag, add					66			66	73	
	3400	Black, 1.8 lb. per bag, add					13.10			13.10	14.40	
	3500	7.5 lb. per bag, add (41)					55			55	61	
	3700	Green, 1.8 lb. per bag, add					27			27	30	
	3800	7.5 lb. per bag, add				↓	115			115	125	
130	0010	**CONCRETE IN PLACE** Including forms (4 uses), reinforcing										130
	0050	steel, including finishing unless otherwise indicated										
	0100	Average for concrete framed building, (35)										
	0110	including finishing	C-17B	15.75	5.210	C.Y.	105	115	16.70	236.70	320	
	0130	Average for substructure only, simple design, incl. finishing		29.07	2.820		75	63	9.05	147.05	195	
	0150	Average for superstructure only, including finishing	↓	13.42	6.110	↓	113	135	19.60	267.60	365	
	0200	Base, granolithic, 1" x 5" high, straight (50)	C-10	175	.137	L.F.	.13	2.67	.35	3.15	4.71	
	0220	Cove	"	140	.171	"	.13	3.34	.44	3.91	5.85	
	0300	Beams, 5 kip per L.F., 10' span	C-17A	6.28	12.900	C.Y.	185	290	22	497	685	
	0350	25' span		7.40	10.950		145	245	18.50	408.50	570	
	0500	Chimney foundations, minimum		26.70	3.030		98	68	5.10	171.10	220	
	0510	Maximum		19.70	4.110		113	92	6.95	211.95	280	
	0700	Columns, square, 12" x 12", minimum reinforcing (49)		4.60	17.610		200	395	30	625	880	
	0720	Average reinforcing	↓	4.10	19.760		285	440	33	758	1,050	
	0740	Maximum reinforcing	C-17B	3.84	21.350		420	475	69	964	1,300	
	0800	16" x 16", minimum reinforcing	C-17A	6	13.500		175	300	23	498	695	
	0820	Average reinforcing (47)	"	4.97	16.300		280	365	28	673	915	
	0840	Maximum reinforcing	C-17B	8.34	9.830		460	220	32	712	890	
	1200	Columns, round, tied, 16" diameter, minimum reinforcing		13.02	6.300		245	140	20	405	515	
	1220	Average reinforcing		8.30	9.880		350	220	32	602	770	
	1240	Maximum reinforcing (49)		6.05	13.550		560	300	43	903	1,150	
	1300	20" diameter, minimum reinforcing		17.35	4.730		230	105	15.15	350.15	435	
	1320	Average reinforcing		10.43	7.860		370	175	25	570	715	
	1340	Maximum reinforcing	↓	7.47	10.980	↓	500	245	35	780	980	
	1700	Curbs, formed in place, 6" x 18", straight,	C-15	400	.180	L.F.	3.20	3.63	.12	6.95	9.45	
	1750	Curb and gutter	"	170	.424	"	5.05	8.55	.27	13.87	19.55	
	3800	Footings, spread under 1 C.Y.	C-17B	35.95	2.280	C.Y.	80	51	7.30	138.30	175	
	3850	Over 5 C.Y.	C-17C	73.91	1.120		76	25	5.40	106.40	130	
	3900	Footings, strip, 18" x 9", plain	C-17B	29.24	2.800		69	63	9	141	185	
	3950	36" x 12", reinforced		51.42	1.590		75	36	5.10	116.10	145	
	4000	Foundation mat, under 10 C.Y.		32.32	2.540		124	57	8.15	189.15	235	
	4050	Over 20 C.Y.	↓	47.37	1.730		113	39	5.55	157.55	190	
	4200	Grade walls, 8" thick, 8' high	C-17A	9.50	8.530		115	190	14.40	319.40	445	
	4250	14' high	C-20	7.30	8.770		167	160	71	398	515	
	4260	12" thick, 8' high	C-17A	13.50	6		134	135	10.15	279.15	370	
	4270	14' high	C-20	11.60	5.520		124	100	45	269	345	
	4300	15" thick, 8' high	C-17B	18.40	4.460		91	99	14.30	204.30	275	
	4350	12' high	C-20	14.80	4.320	↓	106	79	35	220	280	

For expanded coverage of these items see *Means Concrete Cost Data 1989*

43

Figure 11.2

treated finish (common for exposed concrete floors). Dropped areas or deeper slabs may be used under concentrated loads, such as masonry walls. Formed depressions to receive other floor materials, such as mud-set ceramic tile or terrazzo, may also be necessary, at an additional cost to the slab-on-grade system.

Concrete floors on slab form, or centering, is a widely used structural system because of its lightweight, fast erection time, and flexibility in bay sizes. Figure 11.3 illustrates typical concrete floor slab construction details.

Concrete toppings may include new concrete slabs placed over existing floors, with colors, hardeners, integral toppings, or abrasive finishes. Granolithic topping 1/2" to 2" in thickness may be placed over existing slabs with the proper surface preparation.

Other Concrete Items

Other required concrete should be estimated separately. Such work may include locker bases, closures around pipes in pipe chases, and grouting around door sills, frames, and recessed hardware.

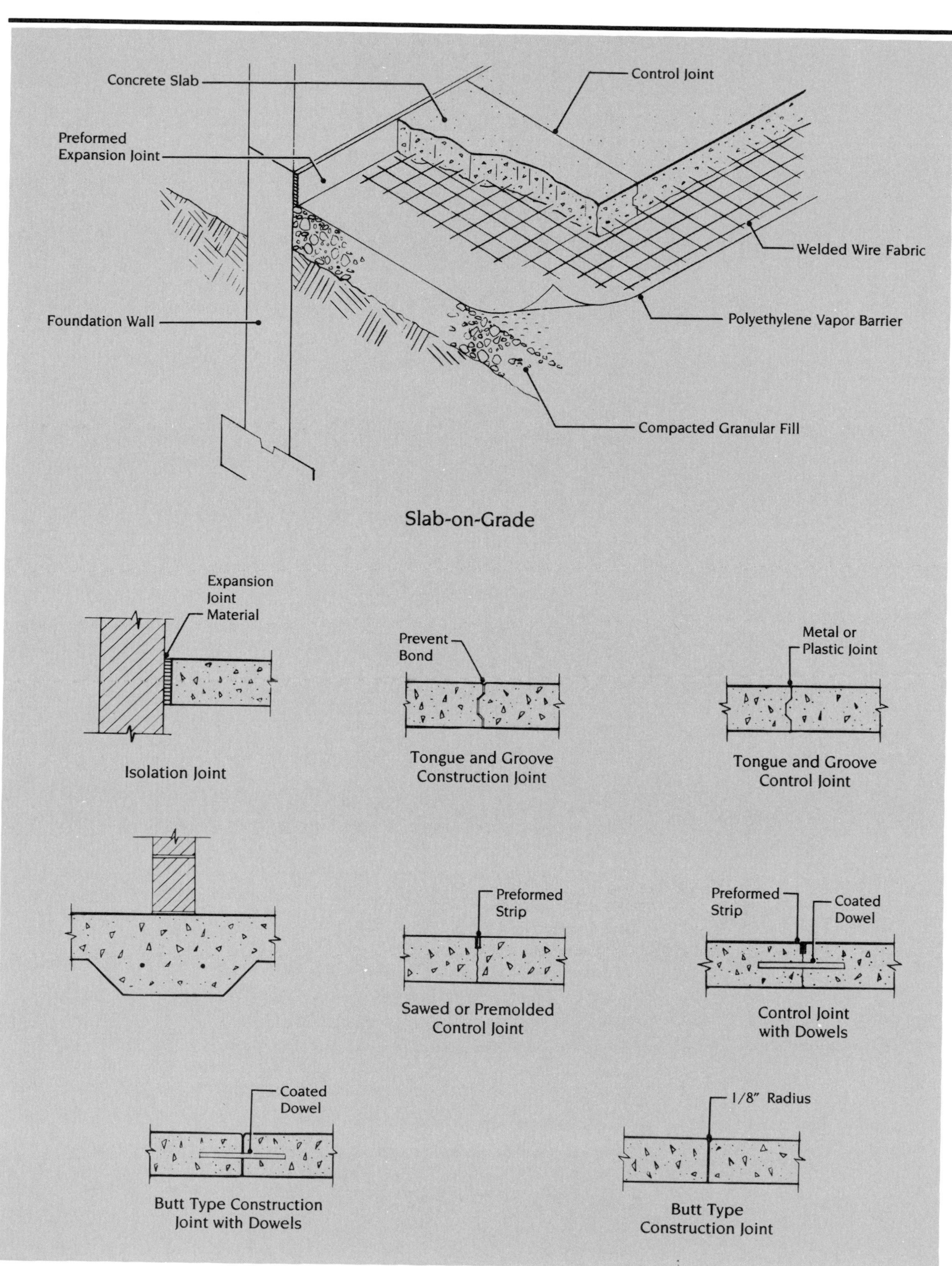

Concrete Slab
Control Joint
Preformed
Expansion Joint
Welded Wire Fabric
Foundation Wall
Polyethylene Vapor Barrier
Compacted Granular Fill
Slab-on-Grade
Expansion
Joint
Material
Isolation Joint
Prevent
Bond
Tongue and Groove
Construction Joint
Metal or
Plastic Joint
Tongue and Groove
Control Joint
Preformed
Strip
Sawed or Premolded
Control Joint
Preformed
Strip
Coated
Dowel
Control Joint
with Dowels
Coated
Dowel
Butt Type Construction
Joint with Dowels
1/8" Radius
Butt Type
Construction Joint

Figure 11.3

Chapter 12
MASONRY

A variety of masonry materials and methods are used in remodeling and renovation. The methods of installation and types of work also vary, even within the same project. The estimator can categorize masonry work according to exterior walls, interior partitions, masonry restoration, and repair. Depending on the structural design of masonry units, the reinforcement and mortar requirements also vary. Other items to be identified and listed are insulation, embedded items, special finishes, scaffolding-related work accessories, and shoring requirements.

After the estimator has identified and analyzed the walls and partitions, each kind or combination of masonry unit is then measured and entered on the takeoff sheets. The items might be grouped and listed in the order in which they would be constructed on the job:

1. Walls Below Grade
2. External Walls
3. Chimney
4. Interior Partitions
5. Cleaning and Repair
6. Miscellaneous Brick-work

This system of organization allows the estimator to tackle one part of the work at a time. Once a consistent takeoff pattern is established and followed, there is less chance that an item will be overlooked or duplicated.

The number of units per square foot is multiplied by the area of the wall to determine the total quantity involved. For items installed in a course, the quantity per linear foot is calculated and multiplied by the length involved (such as a cove base).

Before starting the quantity takeoff, the estimator should make notes based on the masonry specifications of all items that must be estimated. These notes will include the kind of masonry units, the bonds, the mortar type, joint reinforcement, grout, miscellaneous installed items, scaffolding, and cleaning.

Brick

Masonry, especially brick, is generally priced by the unit, by the piece, or per thousand units. The quantity of units is determined from area measurements of walls and partitions. The areas are converted from square feet into number of units by appropriate multipliers. The multipliers are a function of the size of the masonry unit, the pattern or coursing of the masonry unit, the thickness of mortar joints, and the thickness of the wall. Figure 12.1 illustrates various types and sizes of contemporary brick, and provides a chart

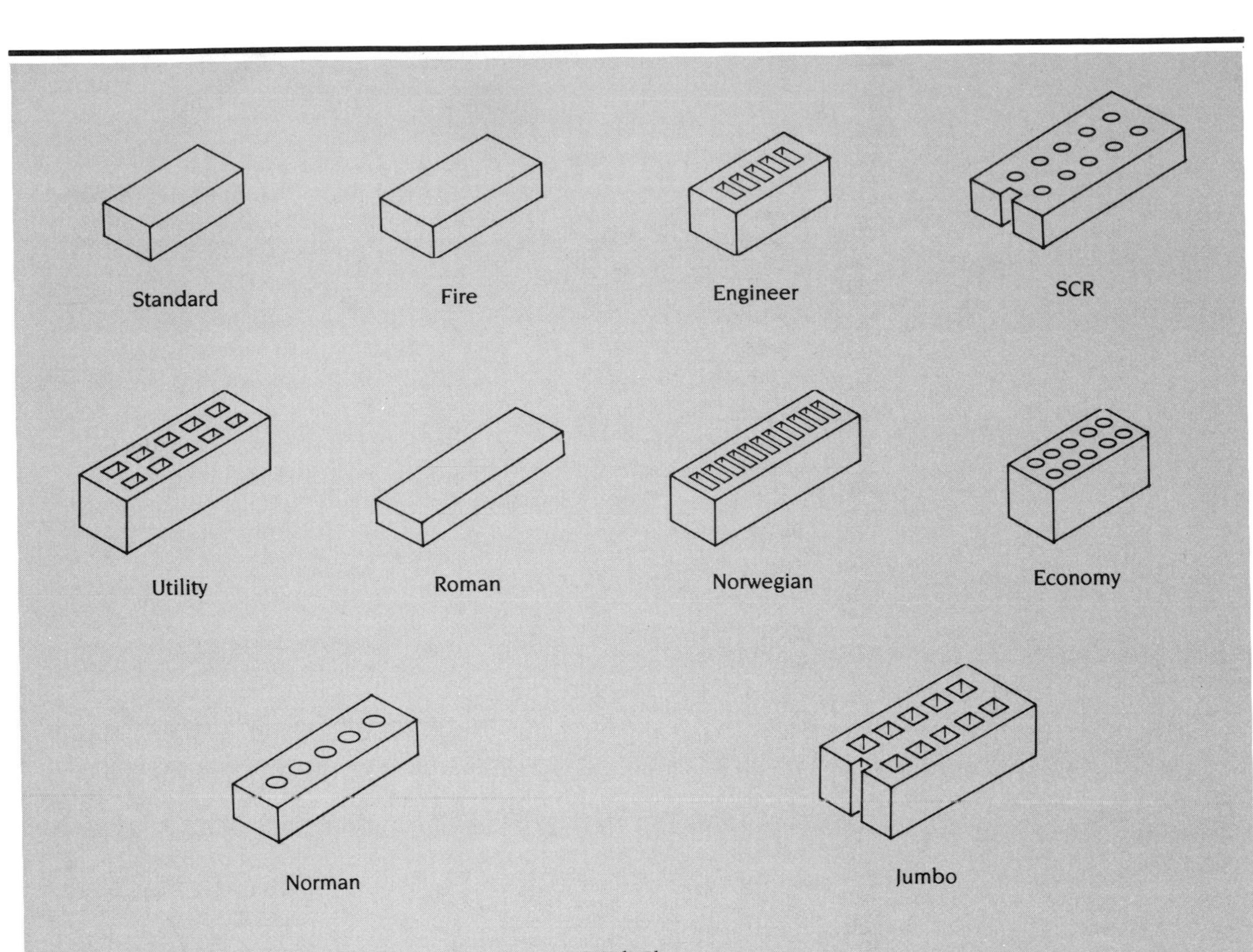

Brick Shapes

Brick & Mortar Quantities

Running Bond					
Number of Brick per SF of Wall - Single Wythe with 1/2" Joints				CF of Mortar per M Bricks, Waste Included	
Type Brick	Nominal Size (Incl. Mortar) L H W	Modular Coursing	Number of Brick per SF	1 Wythe	2 Wythe
Standard	$8 \times 2\frac{2}{3} \times 4$	3C x 8"	6.75	12.9	16.5
Economy	8 x 4 x 4	1C = 4"	4.50	14.6	19.6
Engineer	$8 \times 3\frac{1}{5} \times 4$	5C = 16"	5.63	13.6	17.6
Fire	$9 \times 2\frac{1}{2} \times 4\frac{1}{2}$	2C = 5"	6.40	550# fireclay	—
Jumbo	12 x 4 x 6 or 8	1C = 4"	3.00	34.0	41.4
Norman	$12 \times 2\frac{2}{3} \times 4$	3C = 8"	4.50	17.8	22.8
Norwegian	$12 \times 3\frac{1}{5} \times 4$	5C = 16"	3.75	18.5	24.4
Roman	12 x 2 x 4	2C = 4"	6.00	17.0	20.7
SCR	$12 \times 2\frac{2}{3} \times 6$	3C = 8"	4.50	26.7	31.7
Utility	12 x 4 x 4	1C = 4"	3.00	19.4	26.8

For Other Bonds Standard Size Add to S F Quantities in Table to Left		
Bond Type	Description	Factor
Common	full header every fifth course	+20%
	full header every sixth course	+16.7%
English	full header every second course	+50%
Flemish	alternate headers every course	+33.3%
	every sixth course	+5.6%
Header = W x H exposed		+100%
Rowlock = H x W exposed		+100%
Rowlock stretcher = L x W exposed		+33.3%
Soldier = H x L exposed		—
Sailor = W x L exposed		−33.3%

Figure 12.1

of brick and mortar quantities. Cement bonding and coursing patterns for brick are shown in Figures 12.2 and 12.3.

Concrete Block

For reasons of strength, versatility, and economy, concrete blocks are frequently used for constructing masonry walls and partitions in remodeling projects. They may be used for interior bearing walls, infill panels, interior partitions, back-up for brick, shaft enclosures, and fire walls.

Concrete blocks are manufactured in two basic types, solid and hollow, and in various strength ratings. If the cross-sectional area, exclusive of voids, is 75% or greater than the gross area of the block, then it is classified as "solid block." If the specified area is below the 75% figure, then the block is classified as "hollow block." The strength of concrete block is determined by the compressive strength of the type of concrete used in its manufacture, or by the equivalent compressive strength, which is based on the gross area of the block, including voids. Applications of various types of concrete block are shown in Figures 12.4a and 12.4b. Typical block shapes are shown in Figure 12.5.

There are several special aggregates that can be used to manufacture lightweight blocks. These blocks can be identified by the weight of the concrete mixture used in their manufacture. Regular weight block is made from 125 lb. per cubic foot (PCF) concrete, and lightweight block from 105 to 85 PCF concrete. Costs for lightweight concrete as compared to regular weight range from 20% more for 105 PCF block to 50% more for 85 PCF block. Blocks of various strengths, grades, finishes, and weights should be taken off separately, as costs may vary considerably.

Concrete block partitions may be erected of nominal 4", 6", 8", 10", or 12" thick concrete masonry units. The units may be regular weight or lightweight and solid or 75% solid in horizontal profile. Normal units are nominal 16" long, but some manufacturers produce a nominal 24" unit. Partitions may be left exposed, painted, epoxy-coated, or furred and then covered with gypsum board or paneling. Partitions may also be plastered (directly on the block) or covered with self-furring lath and then plastered. Figure 12.6 shows various types of concrete block specialties: lintels, corners, bond beam, and joint reinforcing.

Joint Reinforcing

Joint reinforcing and individual ties serve as important components of the various types of brick and concrete block walls. These elements must be included in the estimate. Two types of joint reinforcing are available for block walls: the *truss type* and the *ladder type* (shown in Figure 12.6). Because the truss type provides better load distribution, it is normally used in bearing walls. The ladder type is usually installed in light-duty walls that serve non-bearing functions. Both types of joint reinforcing may also be used to tie together the inner and outer wythes of composite or cavity-design walls. Corrugated strips, as well as Z-type, rectangular, and adjustable wall ties, may also be used for this purpose. Generally, one metal wall tie should be installed for each 4-1/2 square feet of brick or stone veneer. Although both types of joint reinforcing may be used as ties, individual ties should not be used as joint reinforcing to control cracking.

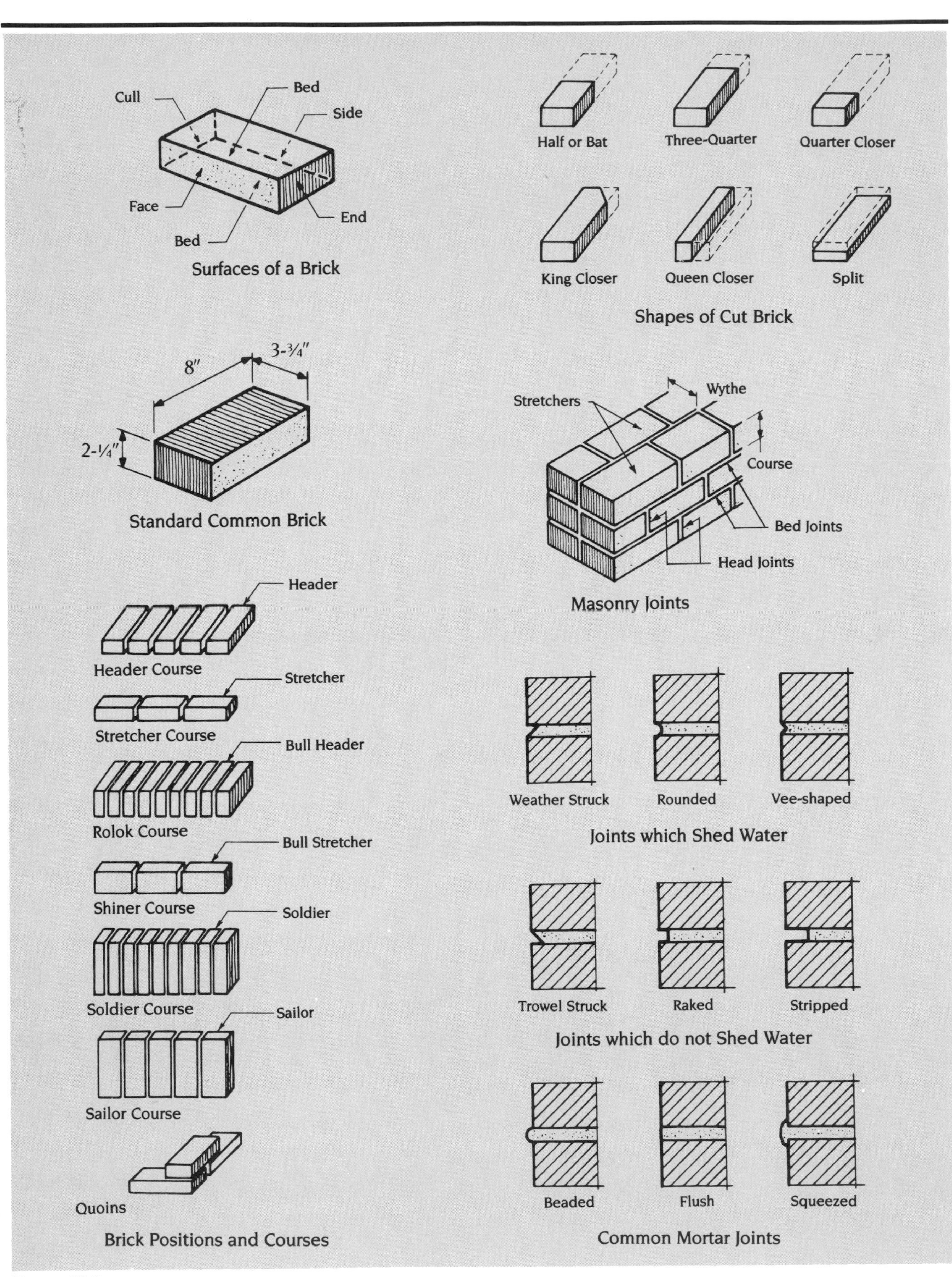
Cull
Bed
Side
Face
End
Bed
Surfaces of a Brick
Half or Bat
Three-Quarter
Quarter Closer
King Closer
Queen Closer
Split
Shapes of Cut Brick
8″
3-¾″
2-¼″
Standard Common Brick
Stretchers
Wythe
Course
Bed Joints
Head Joints
Masonry Joints
Header
Header Course
Stretcher
Stretcher Course
Bull Header
Rolok Course
Bull Stretcher
Shiner Course
Soldier
Soldier Course
Sailor
Sailor Course
Quoins
Brick Positions and Courses
Weather Struck
Rounded
Vee-shaped
Joints which Shed Water
Trowel Struck
Raked
Stripped
Joints which do not Shed Water
Beaded
Flush
Squeezed
Common Mortar Joints

Figure 12.2

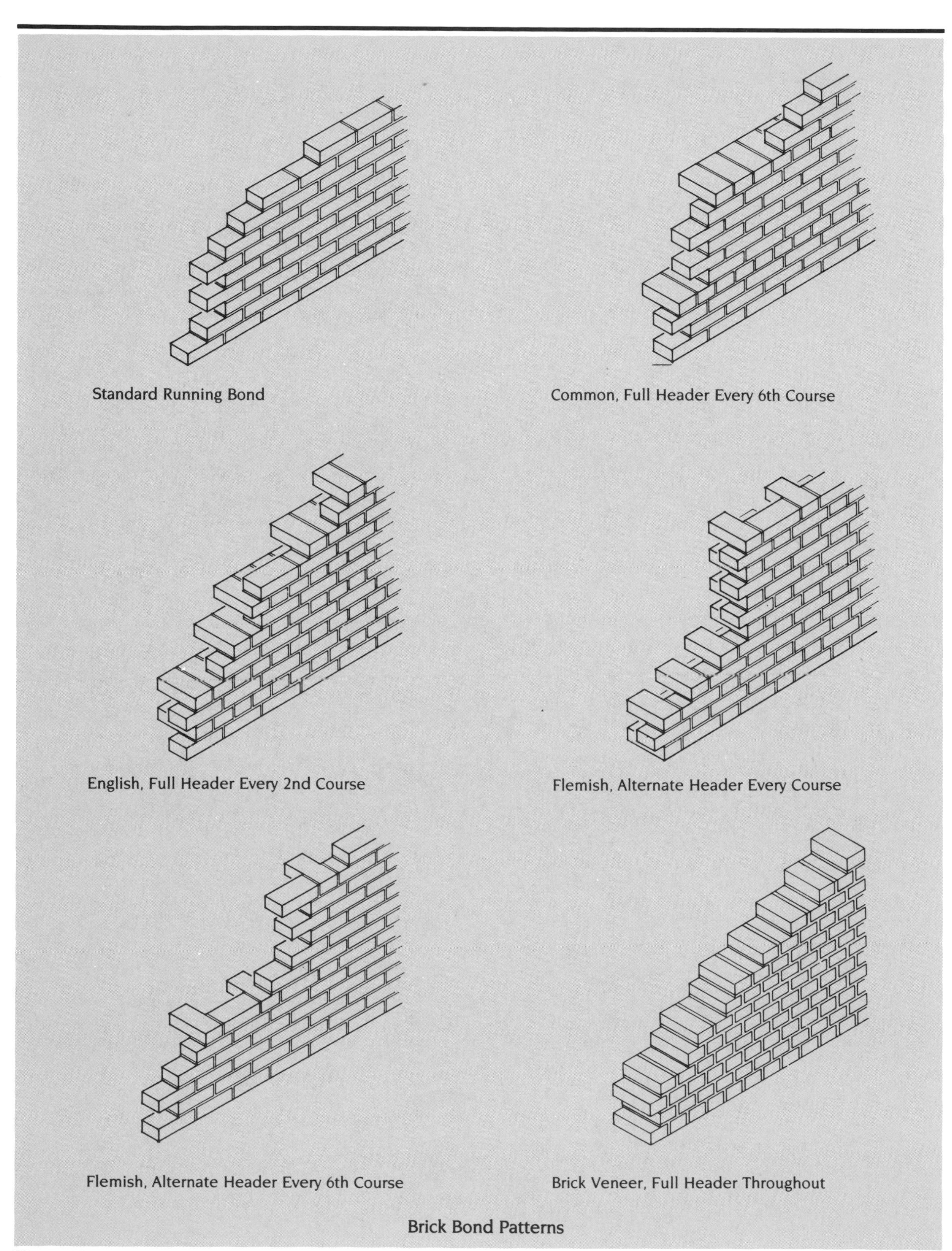

Brick Bond Patterns

Figure 12.3

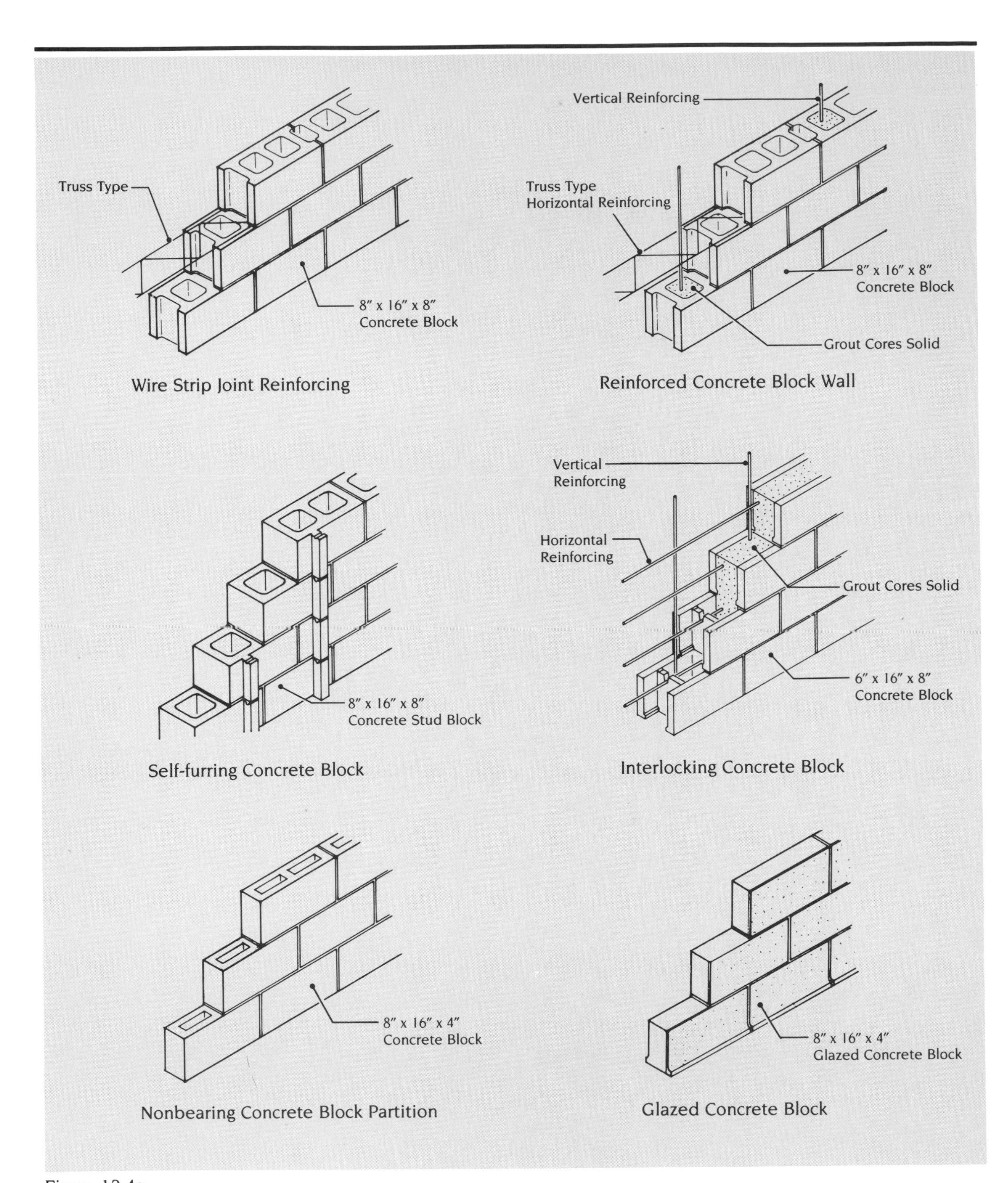

Truss Type
8″ x 16″ x 8″
Concrete Block
Wire Strip Joint Reinforcing
Vertical Reinforcing
Truss Type
Horizontal Reinforcing
8″ x 16″ x 8″
Concrete Block
Grout Cores Solid
Reinforced Concrete Block Wall
8″ x 16″ x 8″
Concrete Stud Block
Self-furring Concrete Block
Vertical
Reinforcing
Horizontal
Reinforcing
Grout Cores Solid
6″ x 16″ x 8″
Concrete Block
Interlocking Concrete Block
8″ x 16″ x 4″
Concrete Block
Nonbearing Concrete Block Partition
8″ x 16″ x 4″
Glazed Concrete Block
Glazed Concrete Block

Figure 12.4a

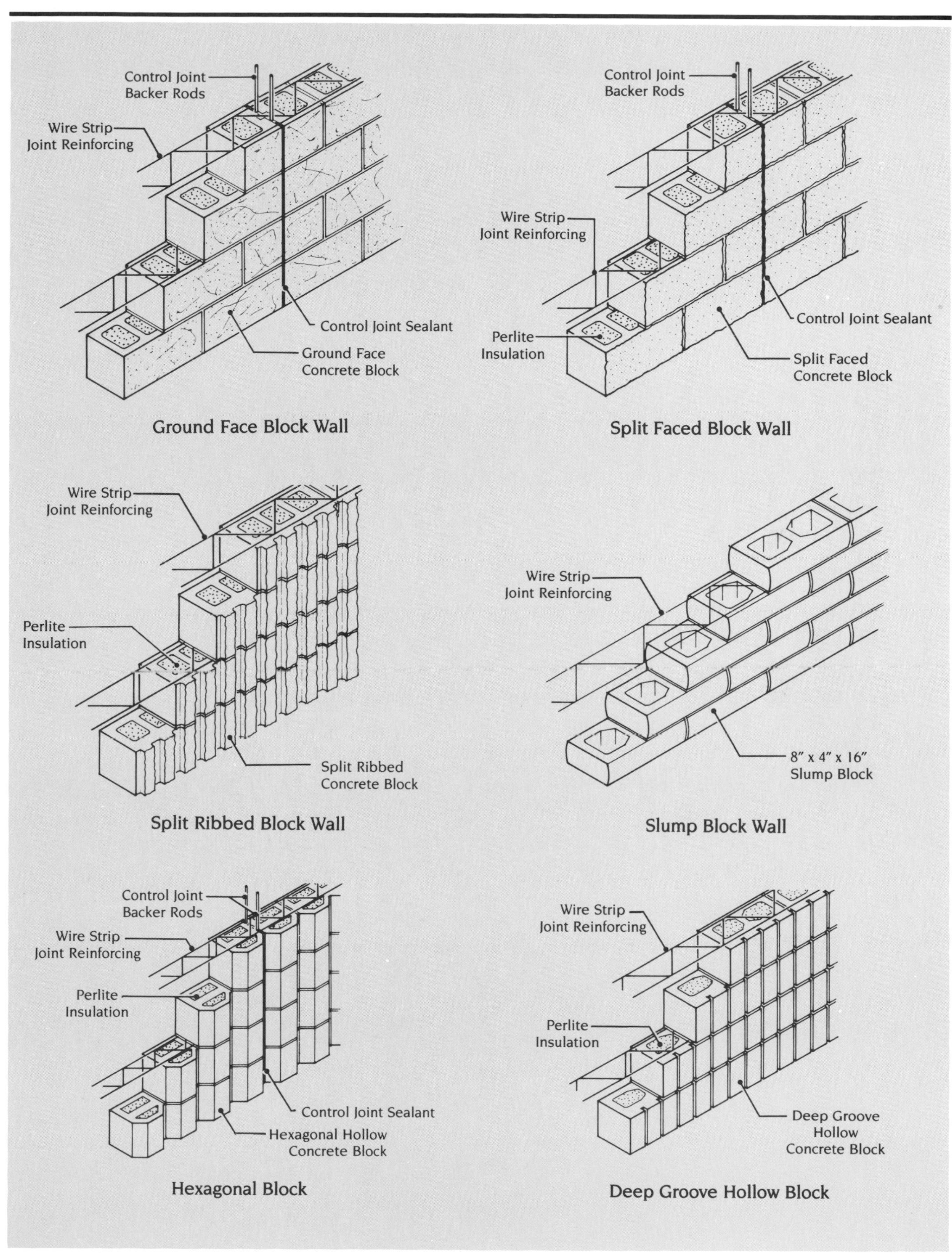

Control Joint
Backer Rods
Wire Strip
Joint Reinforcing
Control Joint Sealant
Ground Face
Concrete Block
Ground Face Block Wall
Control Joint
Backer Rods
Wire Strip
Joint Reinforcing
Perlite
Insulation
Control Joint Sealant
Split Faced
Concrete Block
Split Faced Block Wall
Wire Strip
Joint Reinforcing
Perlite
Insulation
Split Ribbed
Concrete Block
Split Ribbed Block Wall
Wire Strip
Joint Reinforcing
8" x 4" x 16"
Slump Block
Slump Block Wall
Control Joint
Backer Rods
Wire Strip
Joint Reinforcing
Perlite
Insulation
Control Joint Sealant
Hexagonal Hollow
Concrete Block
Hexagonal Block
Wire Strip
Joint Reinforcing
Perlite
Insulation
Deep Groove
Hollow
Concrete Block
Deep Groove Hollow Block

Figure 12.4b

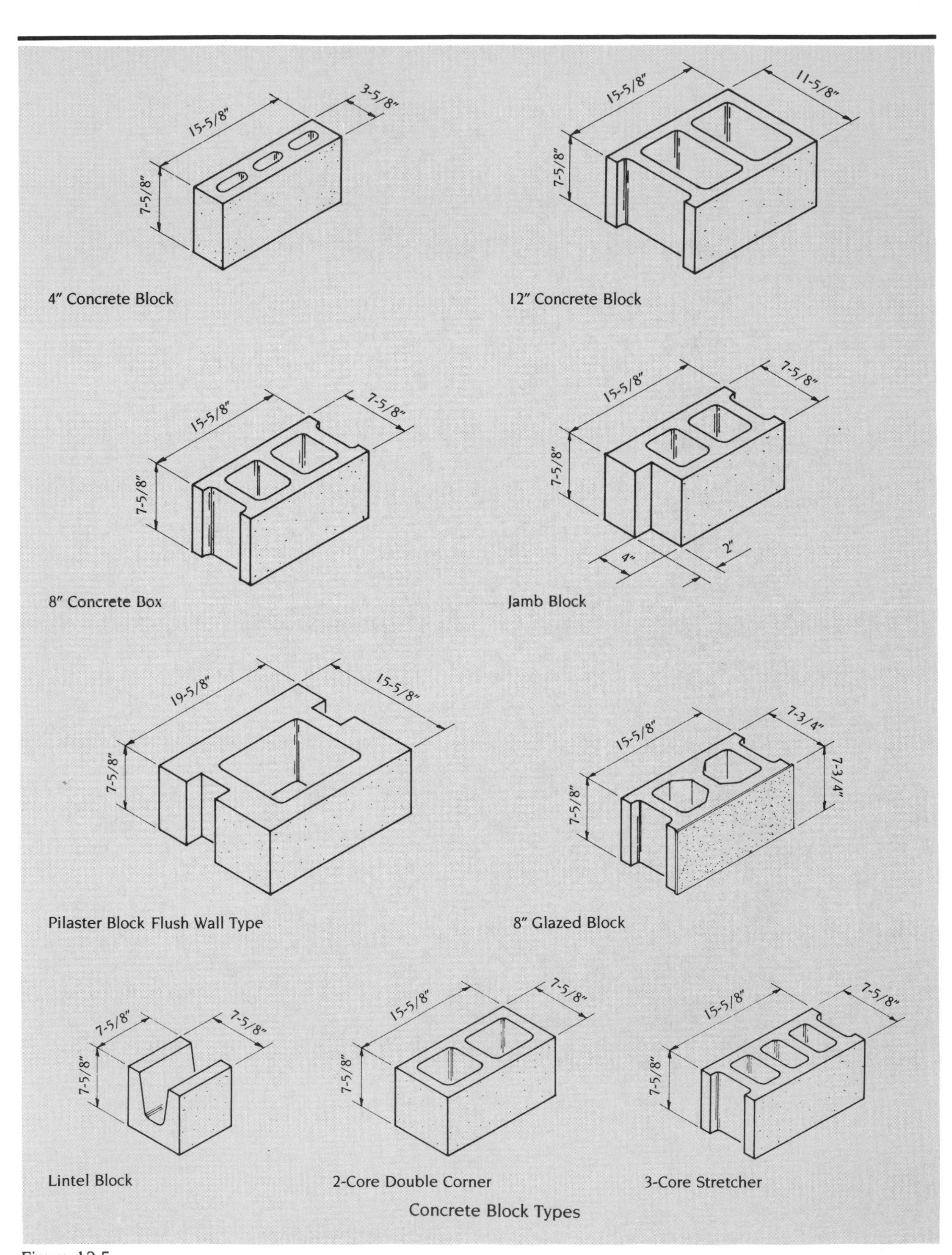

Concrete Block Types

Figure 12.5

Structural reinforcement is commonly required in concrete block walls, especially in those that are load-bearing. Deformed steel bars

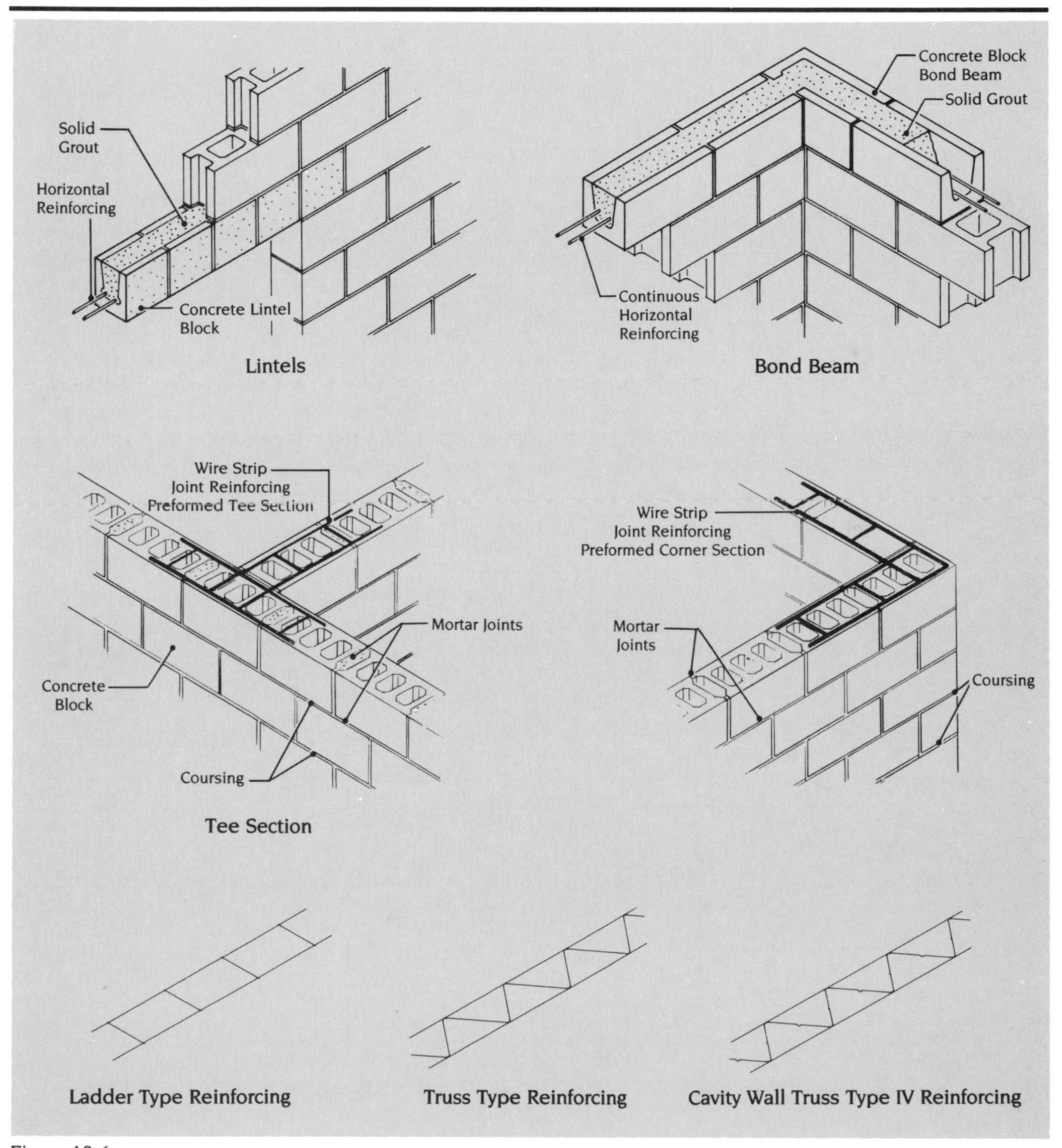

Figure 12.6

may be used as vertical reinforcement when grouted into the block voids, and as horizontal reinforcement when installed above openings and in bond beams. Horizontal and vertical bars may be grouted into the void that is normally used as the collar joint in a composite wall. Lintels should be installed to carry the weight of the wall above openings. Steel angles, built-up steel members, bond beams filled with steel bars and grout, and precast shapes may function as lintels.

Structural Facing Tile

Structural glazed facing tile (SGFT) is kiln-fired structural clay with an integral impervious ceramic face. It may also be manufactured with an acoustical perforated face. Because glazed tile resists stains, marks, impact, abrasion, fading, and cracking, it is ideally suited for use in school corridors, locker rooms, rest rooms, kitchens, and other places where cleanliness and indestructibility are primary considerations.

Structural glazed facing tile is commonly available in a large selection of colors and color combinations in the 6T series, with 5-1/3" x 12" nominal face and in the 8W series, with 8" x 16" nominal face, both in 2", 4", 6", and 8" widths. Some manufacturers produce a 4W series with 8" x 8" nominal face in 2" and 4" widths. Some available tile shapes include stretchers, bullnose jamb or corner, square jamb or corner, covered internal corner, recessed cove base, non-recessed cove base, bullnose sill, square sill, and universal miter. Walls with openings and returns usually require partition layout drawings to establish quantities of special shapes. The different shapes of structural glazed facing tile are shown in Figure 12.7.

Building Stone

For reasons of durability and unique appearance, stone is well suited to a wide range of decorative applications as a building material. It can be installed in small units, referred to as "building stone," which can be assembled in many different systems, with or without mortar, to create walls and veneers of all sorts. Stone can also be employed as a material in larger units, such as stone panels, which are installed with elaborate anchor and framing systems and used as decorative wall facings in high-rise office and other commercial buildings. The cost of stone building materials varies considerably with location and depends on the available supply and the distance that it must be transported. Other significant cost factors include the extent of quarrying and subsequent processing required to extract and produce the finished product.

Building stone is available in random or pre-cut sizes and shapes. Small, irregular building stone that has been quarried in random sizes is called "rubble" or "fieldstone." This material, which is sold by the ton, is commonly installed with mortar. Fieldstone may be split by hand on site to provide a flat exterior surface for a patterned wall or fireplace.

Small stone units may also be quarry-split and processed to meet aesthetic requirements. For example, decorative building stone can be purchased by the ton in 4" thick slabs, available in lengths ranging from 6" to 14" and in heights ranging from 2" to 16". These pieces are commonly installed with mortar to create veneer walls of

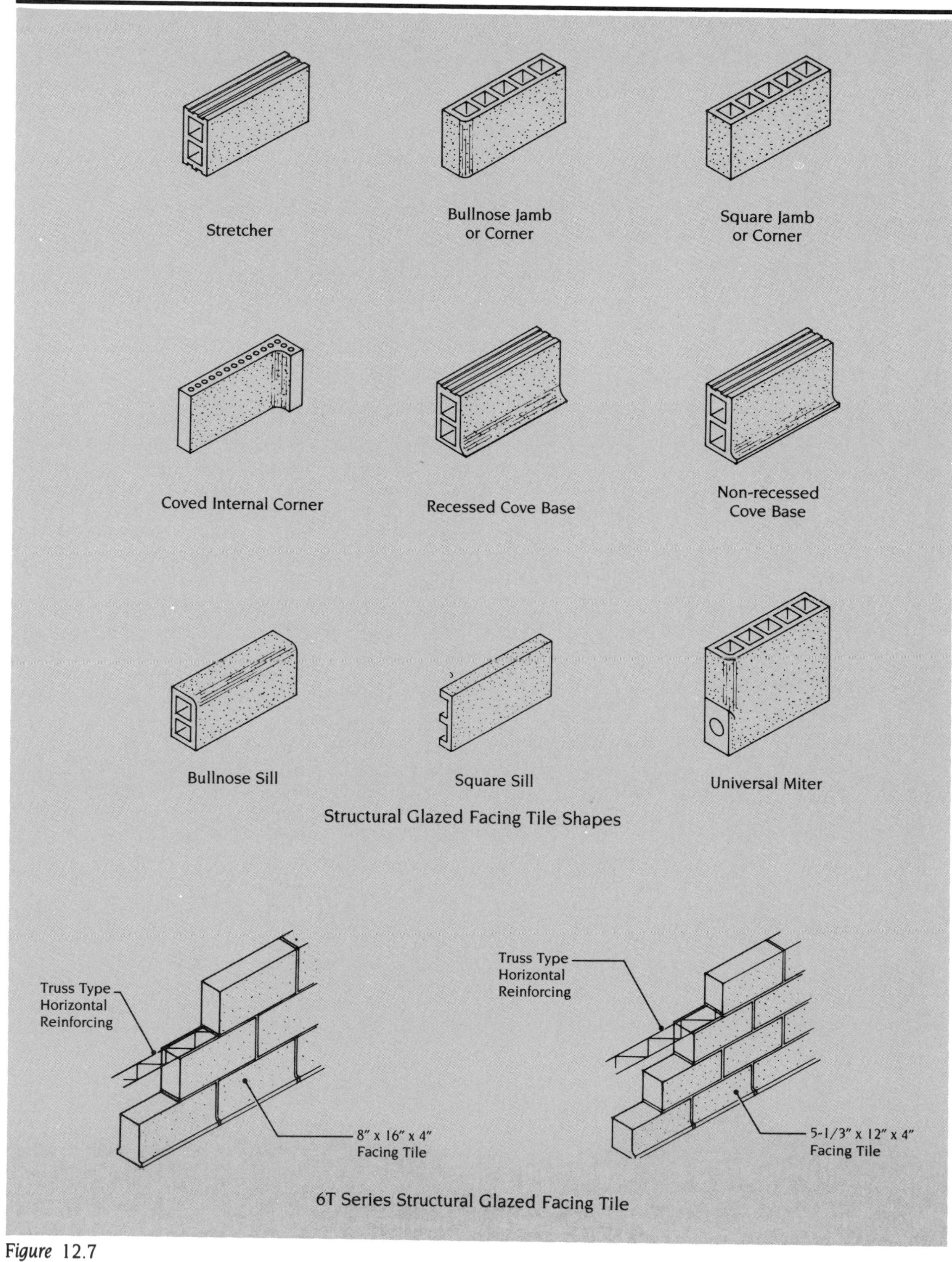
Stretcher
Bullnose Jamb
or Corner
Square Jamb
or Corner
Coved Internal Corner
Recessed Cove Base
Non-recessed
Cove Base
Bullnose Sill
Square Sill
Universal Miter
Structural Glazed Facing Tile Shapes
Truss Type
Horizontal
Reinforcing
8" x 16" x 4"
Facing Tile
Truss Type
Horizontal
Reinforcing
5-1/3" x 12" x 4"
Facing Tile
6T Series Structural Glazed Facing Tile

Figure 12.7

varying patterns, such as *ledge stone, spider web, uncoursed rectangular,* and *squared*. Ashlar stone, also priced by the ton, is building stone that has been sawn on the edges so as to produce a rectangular face. This shape makes ashlar stone another possible veneer material because the pieces can be arranged in either a regular or random-coursed pattern within the face of a wall. Typical patterns are shown in Figure 12.8.

Stone veneer can be tied to the back-up wall with galvanized ties or 8″ stone headers in a method similar to that used in brick veneer walls. The coverage of stone veneer ranges from 35 to 50 square feet per ton for 4″ wide veneer, with correspondingly reduced coverages per ton for veneers of 6″ and 8″ in width.

Stone Floors

Stone floors may be constructed from any type of stone that meets the durability standards and aesthetic requirements of the proposed location. Some commonly used stone flooring materials include slate, flagstone, granite, and marble. Stone may be laid in a patterned or random design. Some popular stone patterns are shown in Figure 12.9. Stone may be randomly cut, or uniform in size and shape. It is also available in patterned sets. The stones may feature neat sawn edges or the irregular shapes of field-cut edges. The various possible exposed face finishes for the stones include: *natural cleft, sawn, sawn and polished,* and any specialized finish available for the type of stone that is functionally and aesthetically appropriate.

Stone floors may be placed on mortar beds or applied on mastic adhesive. When the floor is installed on a mortar bed, the stone may vary slightly in thickness, with the total floor usually measuring 1-1/2″ to 2″ thick from subfloor to finish. When the stone floor is thin set, or laid in mastic, gauged stones with a constant thickness must be used. After the stone flooring material has been placed, mortar, or pre-mixed grouting material (available in various colors), is used to fill the joints between the stones. Some stone flooring materials with consistent sizes and regular edges may not require grouted joints. Applying the sealant is the next step. There are many different types of sealant available for use with the various stone types.

Glass Block

Glass block, a popular building material in the 1930's , has attracted new interest in the past few years in the construction industry. It is used predominantly for the construction of interior partitions and provides the dual advantage of admitting light while providing privacy.

Glass block can be placed on a raised base, plate, or sill, provided that the surfaces to be mortared are primed with asphalt emulsion. Wall recesses and channel track that receive the glass block should be lined with expansion strips before oakum filler and caulking are applied. Horizontal joint reinforcing is specified for flexural as well as shrinkage control and is laid in the glass block joints along with the mortar. End blocks are anchored to the adjacent construction with metal anchors, if there are no other provisions for attachment. If intermediate support is required, vertical I-shaped stiffeners can be either installed in the plane of the wall or adjacent to it. In any case, the stiffeners should be tied to the wall with wire anchors. The top of

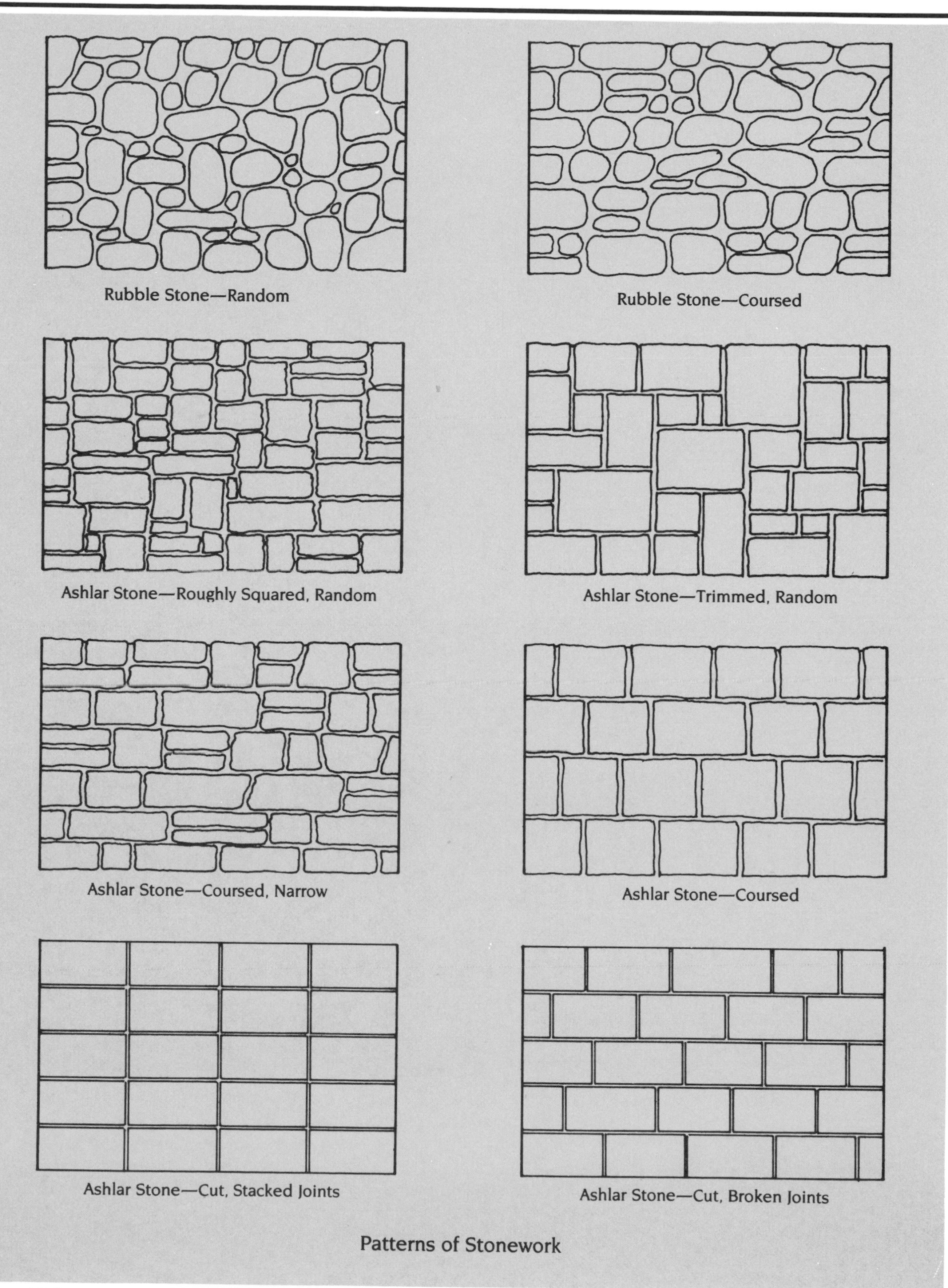

Patterns of Stonework

Figure 12.8

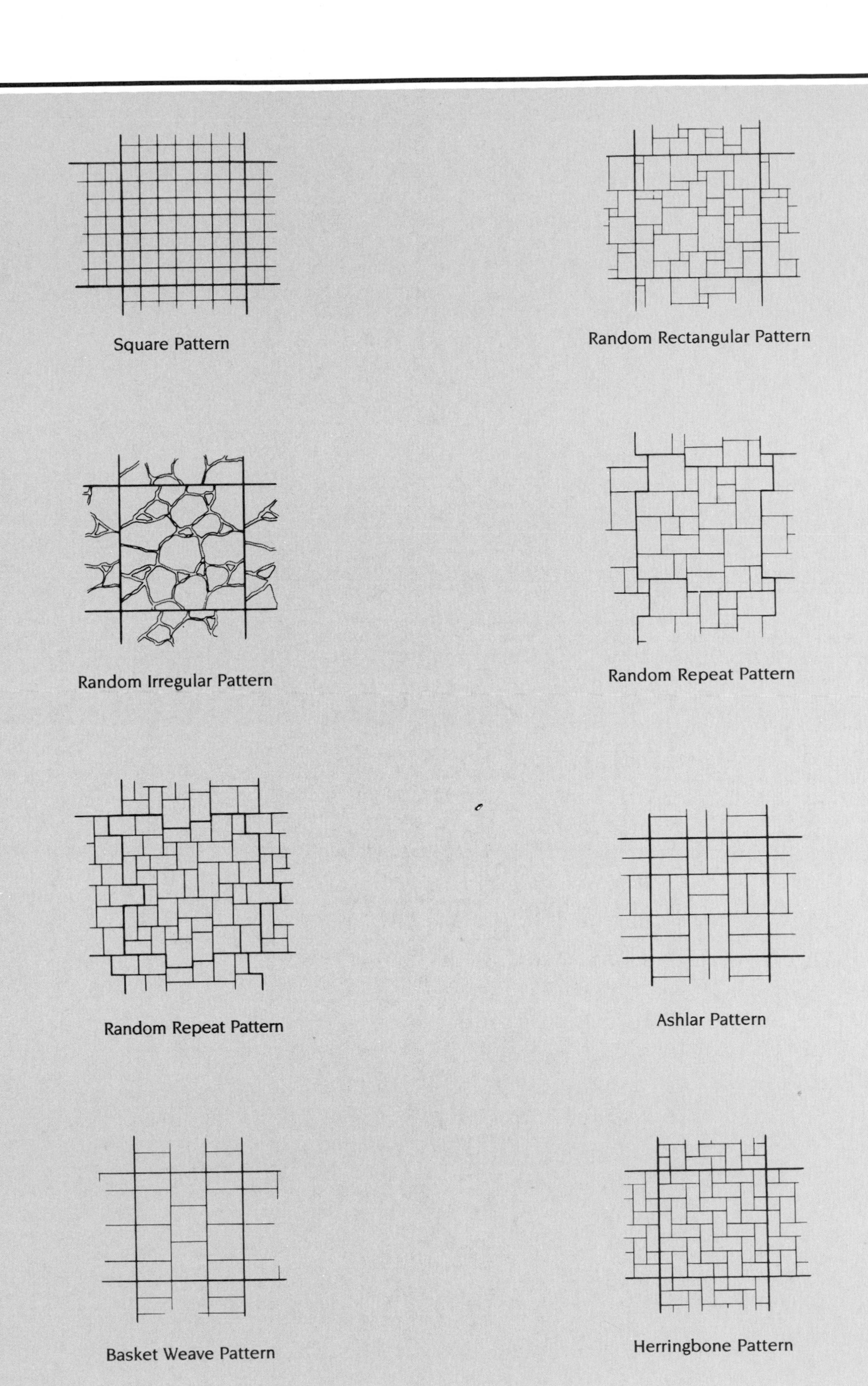
Square Pattern
Random Rectangular Pattern
Random Irregular Pattern
Random Repeat Pattern
Random Repeat Pattern
Ashlar Pattern
Basket Weave Pattern
Herringbone Pattern

Figure 12.9

the wall is supported between angles or in a channel track similar to the jambs. Typical construction details for glass block are shown in Figure 12.10.

Glass block is manufactured in sizes from 6″ by 6″ to 12″ by 12″, and in thicknesses from 3″ to 4″. The block may be hollow or fused brick; the latter allows a clear view through the block. Inserts can be manufactured into the block to reduce solar transmission. Typical quantities used to estimate glass block are shown in Figure 12.11.

Tips for Estimating Masonry

In remodeling and renovation, the quantities of masonry work may be small and might involve localized areas of new work and varying amounts of repair and restoration. When the amount of new work is small, the estimator must remember that factors for economy of scale may not be applicable. The cost of mobilization for the work may exceed actual cost of installation.

Masonry *restoration*, however, can often be a large project. The estimator must be very careful during the quantity takeoff to ensure that all areas requiring work are included. Photographs, notes, and sketches from the site visit are particularly useful. Stringent controls are often placed on masonry restoration and must be considered in the estimate.

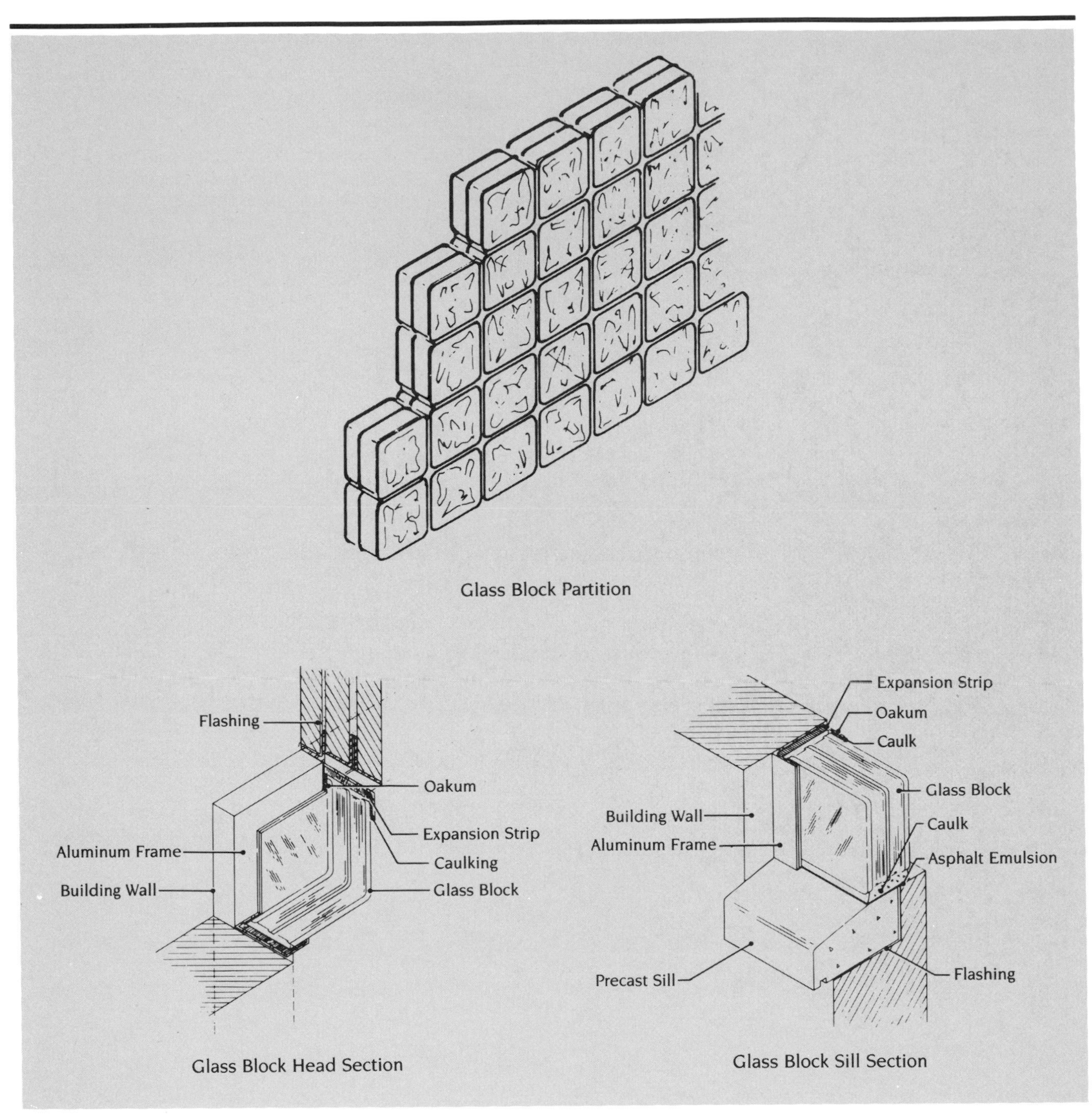
Glass Block Partition
Flashing
Oakum
Expansion Strip
Caulking
Glass Block
Aluminum Frame
Building Wall
Glass Block Head Section
Expansion Strip
Oakum
Caulk
Glass Block
Building Wall
Aluminum Frame
Caulk
Asphalt Emulsion
Flashing
Precast Sill
Glass Block Sill Section

Figure 12.10

Glass Block								
Cost of blocks each, all blocks 4" thick, zone 1 contractor prices.								
Nominal Size (Including Mortar)	Truckload or Carload				Less Than Truckload or Carload			
	Regular	Thinline	Essex	Solar Reflective	Regular	Thinline	Essex	Solar Reflective
6" x 6"	$ 3.20	$2.90	—	—	$ 3.80	$3.85	—	—
8" x 8"	4.70	3.40	$ 4.95	$ 8.40	5.80	5.40	$ 5.40	$ 9.00
Solid 8" x 8"	18.85	—	—	—	—	—	—	—
4" x 12"	—	—	—	—	—	—	—	—
4" x 8"	3.10	2.45	—	—	3.85	3.40	—	—
12" x 12"	11.90	—	—	—	12.90	—	—	—

Size	Per 100 S.F.		Per 1000 Block				
	No. of Block	Mortar 1/4" Joint	Asphalt Emulsion	Caulk	Expansion Joint	Panel Anchors	Wall Mesh
6" x 6"	410 ea.	5.0 C.F.	.17 gal.	1.5 gal.	80 L.F.	20 ea.	500 L.F.
8" x 8"	230	3.6	.33	2.8	140	36	670
12" x 12"	102	2.3	.67	6.0	312	80	1000
Approximate quantity per 100 S.F.			.07 gal.	.6 gal.	32 L.F.	9 ea.	51, 68, 102 L.F.

Figure 12.11

Chapter 13

METALS

The amount of structural steel used in remodeling projects is usually limited. Beams or plates may be employed to reinforce an existing floor system. Beams, channels, or angles may be used to frame a new opening or stairwell. The addition of a mezzanine floor might include more structural steel, requiring columns, beams, open web joists and steel deck. However, the most extensive use of metals in remodeling work is for miscellaneous support and ornamental purposes.

Structural Steel

The installation of structural steel into an existing building can be labor intensive. The steel may have to be moved into the building with dollies or hand carts, hoisted into position with chain falls or come-alongs, raised with a fork lift, or jacked from a scaffold. New connections are either welded (if fire codes permit) or drilled and bolted. Typical bolted steel connections are shown in Figure 13.1. The installation process must be visualized and planned in order to properly estimate the costs. A "complexity factor" may be included for each piece of steel.

The structural steel required to repair, modify, and remodel buildings is often limited to small quantities of columns and beams purchased from local warehouses. Because warehouses sell structural steel shapes such as wide flange, channel, square, and rectangular tubing, and round pipe sections in mill lengths, it is necessary to include the waste of the cut-off material in the estimate, as well as fabrication and cutting costs.

Temporary shoring, jacking, safety nets, work platforms, and needling of the steel into position are additional items to consider. Thought must also be given to the requirements for fire lookout personnel, extinguishers, and fire curtains when cutting and welding torches are used for the dismantling or erection of structural steel.

When drawings are available for a proposed renovation project, they may include: special notes; lintel schedules; size, weight and location of each piece of steel; anchor bolt locations; special fabrication details; and special weldment. These must all be included on the Quantity Sheet.

When performing a structural steel takeoff, the estimator must consider the following: the grade of steel, method of connection, cleaning and painting (type and number of coats), and the number of pieces.

Remodeling and renovation often require "bringing a building up to code" to meet building and fire code requirements. Such work most often includes fireproofing both the existing and new structural elements. Investigation of required fireproofing is essential. Figures

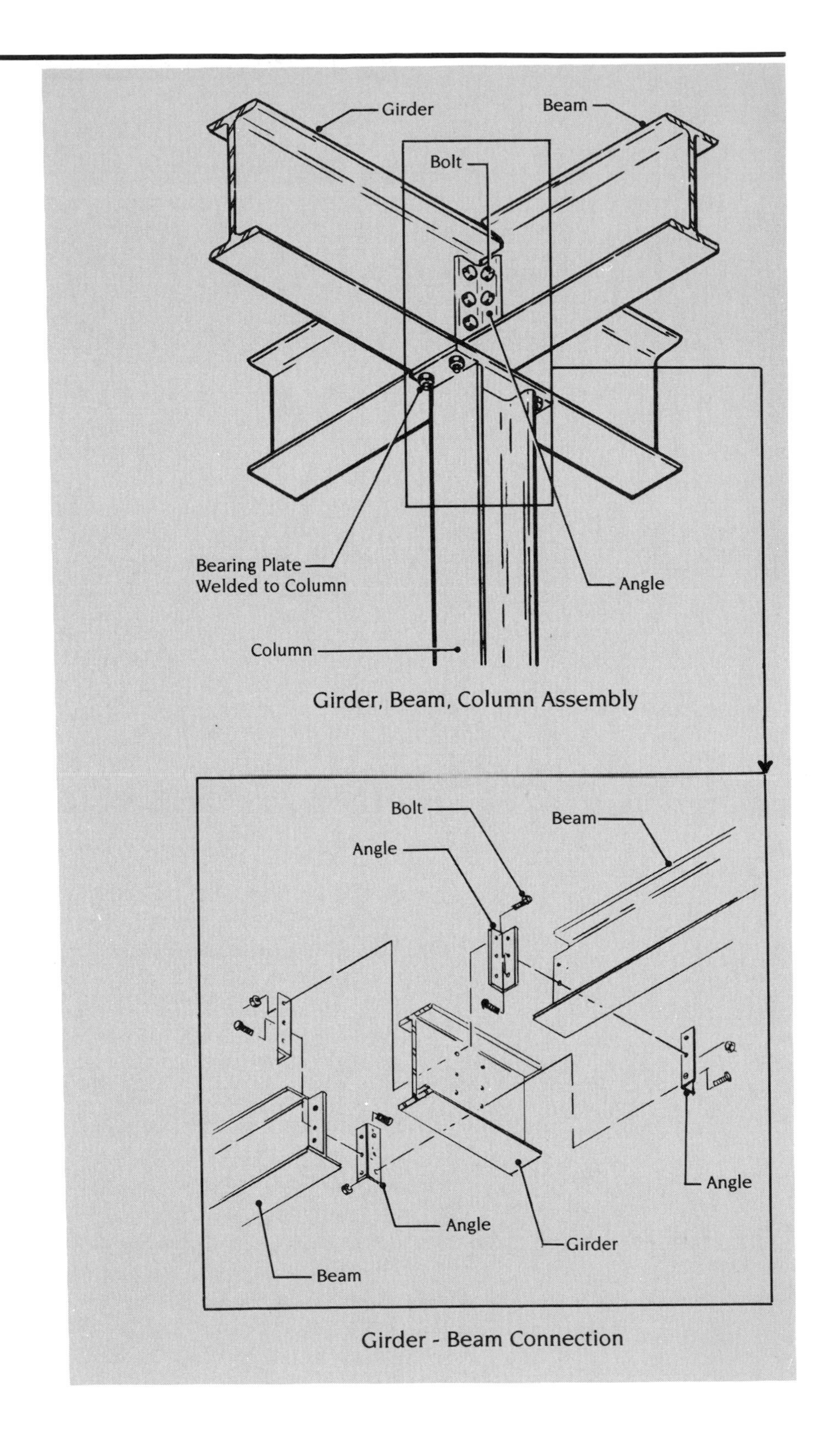
Girder
Beam
Bolt
Bearing Plate
Welded to Column
Angle
Column
Girder, Beam, Column Assembly
Bolt
Beam
Angle
Angle
Angle
Girder
Beam
Girder - Beam Connection

Figure 13.1

13.2a and 13.2b show surface and box areas of structural steel shapes to be used for estimating fireproofing - whether sprayed or drywall.

The estimator must also give consideration to warehouse availability, the type of fabrication required, and fabricator capability and availability. Of equal importance are delivery dates, sequence of deliveries, size of pieces to be delivered, storage area, and access.

Steel Joists and Deck

The estimator often must determine the size and type of required steel joists based on a visual inspection of existing joists. If specifications and drawings are available from the original construction, they will include type of joists, size and length, end bearing conditions, tie joist locations, bridging type and size, number and size of headers required, top and bottom chord extensions, and ceiling extensions.

The estimator must consider the various types and sizes of joists, special attachments, special paint and surface preparation, deliveries, site storage and access, and other variables that must be included in the estimate, but are not called out.

Existing floor and roof systems framed with steel joists are designed to support specified live and dead loads. When a given space is remodeled, the specified loads may be exceeded. Therefore, additional joists may be needed and must be taken off. The joists must be identified by type, size, and length. Special end bearings may also be required. Jacking or temporary shoring may be required to erect the joists.

In renovation, metal decking is usually found in floor or roof systems added to an existing building. Steel-framed mezzanines are an example of a system that normally would include a metal deck. Such a system is illustrated in Figure 13.3, from *Means Interior Cost Data*. The type, depth, gauge, closures, finish, and attachment method of steel deck must be determined and listed.

Again, it is important to consider the quantity, availability, and delivery schedule of the types of deck specified, as well as site storage and access, special requirements, and miscellaneous items necessary to complete the work.

Miscellaneous and Ornamental Metals

Miscellaneous metals can be difficult to identify. The estimator should, therefore, carefully examine the existing conditions and add those items which are neither specified nor shown in sketches and drawings, but are necessary to complete the work. It is also important to determine the type of finish required and whether the metal is to be furnished and erected or furnished only. When required for installation, the field measurements, shop drawings, method of installation, and equipment required to erect must all be noted and given consideration.

Miscellaneous and ornamental metals may be used extensively in remodeling projects for both decorative and functional purposes. The purchase, fabrication, and erection of these items is a specialized trade and a reliable sub-bid should be obtained for this portion of the work.

Surface Areas and Box Areas
W Shapes
Square feet per foot of length

Designation	Case A	Case B	Case C	Case D
W 36x300	9.99	11.40	7.51	8.90
x280	9.95	11.30	7.47	8.85
x260	9.90	11.30	7.42	8.80
x245	9.87	11.20	7.39	8.77
x230	9.84	11.20	7.36	8.73
x210	8.91	9.93	7.13	8.15
x194	8.88	9.89	7.09	8.10
x182	8.85	9.85	7.06	8.07
x170	8.82	9.82	7.03	8.03
x160	8.79	9.79	7.00	8.00
x150	8.76	9.76	6.97	7.97
x135	8.71	9.70	6.92	7.92
W 33x241	9.42	10.70	7.02	8.34
x221	9.38	10.70	6.97	8.29
x201	9.33	10.60	6.93	8.24
x152	8.27	9.23	6.55	7.51
x141	8.23	9.19	6.51	7.47
x130	8.20	9.15	6.47	7.43
x118	8.15	9.11	6.43	7.39
W 30x211	8.71	9.97	6.42	7.67
x191	8.66	9.92	6.37	7.62
x173	8.62	9.87	6.32	7.57
x132	7.49	8.37	5.93	6.81
x124	7.47	8.34	5.90	6.78
x116	7.44	8.31	5.88	6.75
x108	7.41	8.28	5.84	6.72
x 99	7.37	8.25	5.81	6.68
W 27x178	7.95	9.12	5.81	6.98
x161	7.91	9.08	5.77	6.94
x146	7.87	9.03	5.73	6.89
x114	6.88	7.72	5.39	6.23
x102	6.85	7.68	5.35	6.18
x 94	6.82	7.65	5.32	6.15
x 84	6.78	7.61	5.28	6.11
W 24x162	7.22	8.30	5.25	6.33
x146	7.17	8.24	5.20	6.27
x131	7.12	8.19	5.15	6.22
x117	7.08	8.15	5.11	6.18
x104	7.04	8.11	5.07	6.14
x 94	6.16	6.92	4.81	5.56
x 84	6.12	6.87	4.77	5.52
x 76	6.09	6.84	4.74	5.49
x 68	6.06	6.80	4.70	5.45
x 62	5.57	6.16	4.54	5.13
x 55	5.54	6.13	4.51	5.10
W 21x147	6.61	7.66	4.72	5.76
x132	6.57	7.61	4.68	5.71
x122	6.54	7.57	4.65	5.68
x111	6.51	7.54	4.61	5.64
x101	6.48	7.50	4.58	5.61
x 93	5.54	6.24	4.31	5.01
x 83	5.50	6.20	4.27	4.96
x 73	5.47	6.16	4.23	4.92
x 68	5.45	6.14	4.21	4.90
x 62	5.42	6.11	4.19	4.87
x 57	5.01	5.56	4.06	4.60
x 50	4.97	5.51	4.02	4.56
x 44	4.94	5.48	3.99	4.53

Surface Areas and Box Areas
W Shapes
Square feet per foot of length

Designation	Case A	Case B	Case C	Case D
W 18x119	5.81	6.75	4.10	5.04
x106	5.77	6.70	4.06	4.99
x 97	5.74	6.67	4.03	4.96
x 86	5.70	6.62	3.99	4.91
x 76	5.67	6.59	3.95	4.87
x 71	4.85	5.48	3.71	4.35
x 65	4.82	5.46	3.69	4.32
x 60	4.80	5.43	3.67	4.30
x 55	4.78	5.41	3.65	4.27
x 50	4.76	5.38	3.62	4.25
x 46	4.41	4.91	3.52	4.02
x 40	4.38	4.88	3.48	3.99
x 35	4.34	4.84	3.45	3.95
W 16x100	5.28	6.15	3.70	4.57
x 89	5.24	6.10	3.66	4.52
x 77	5.19	6.05	3.61	4.47
x 67	5.16	6.01	3.57	4.43
x 57	4.39	4.98	3.33	3.93
x 50	4.36	4.95	3.30	3.89
x 45	4.33	4.92	3.27	3.86
x 40	4.31	4.89	3.25	3.83
x 36	4.28	4.87	3.23	3.81
x 31	3.92	4.39	3.11	3.57
x 26	3.89	4.35	3.07	3.53
W 14x730	7.61	9.10	5.23	6.72
x665	7.46	8.93	5.08	6.55
x605	7.32	8.77	4.94	6.39
x550	7.19	8.62	4.81	6.24
x500	7.07	8.49	4.68	6.10
x455	6.96	8.36	4.57	5.98
x426	6.89	8.28	4.50	5.89
x398	6.81	8.20	4.43	5.81
x370	6.74	8.12	4.36	5.73
x342	6.67	8.03	4.29	5.65
x311	6.59	7.94	4.21	5.56
x283	6.52	7.86	4.13	5.48
x257	6.45	7.78	4.06	5.40
x233	6.38	7.71	4.00	5.32
x211	6.32	7.64	3.94	5.25
x193	6.27	7.58	3.89	5.20
x176	6.22	7.53	3.84	5.15
x159	6.18	7.47	3.79	5.09
x145	6.14	7.43	3.76	5.05
x132	5.93	7.16	3.67	4.90
x120	5.90	7.12	3.64	4.86
x109	5.86	7.08	3.60	4.82
x 99	5.83	7.05	3.57	4.79
x 90	5.81	7.02	3.55	4.76
x 82	4.75	5.59	3.23	4.07
x 74	4.72	5.56	3.20	4.04
x 68	4.69	5.53	3.18	4.01
x 61	4.67	5.50	3.15	3.98
x 53	4.19	4.86	2.99	3.66
x 48	4.16	4.83	2.97	3.64
x 43	4.14	4.80	2.94	3.61
x 38	3.93	4.50	2.91	3.48
x 34	3.91	4.47	2.89	3.45
x 30	3.89	4.45	2.87	3.43
x 26	3.47	3.89	2.74	3.16
x 22	3.44	3.86	2.71	3.12

Figure 13.2a

Surface Areas and Box Areas
W Shapes
Square feet per foot of length

Designation	Case A	Case B	Case C	Case D
W 12x336	5.77	6.88	3.92	5.03
x305	5.67	6.77	3.82	4.93
x279	5.59	6.68	3.74	4.83
x252	5.50	6.58	3.65	4.74
x230	5.43	6.51	3.58	4.66
x210	5.37	6.43	3.52	4.58
x190	5.30	6.36	3.45	4.51
x170	5.23	6.28	3.39	4.43
x152	5.17	6.21	3.33	4.37
x136	5.12	6.15	3.27	4.30
x120	5.06	6.09	3.21	4.24
x106	5.02	6.03	3.17	4.19
x 96	4.98	5.99	3.13	4.15
x 87	4.95	5.96	3.10	4.11
x 79	4.92	5.93	3.07	4.08
x 72	4.89	5.90	3.05	4.05
x 65	4.87	5.87	3.02	4.02
x 58	4.39	5.22	2.87	3.70
x 53	4.37	5.20	2.84	3.68
x 50	3.90	4.58	2.71	3.38
x 45	3.88	4.55	2.68	3.35
x 40	3.86	4.52	2.66	3.32
x 35	3.63	4.18	2.63	3.18
x 30	3.60	4.14	2.60	3.14
x 26	3.58	4.12	2.58	3.12
x 22	2.97	3.31	2.39	2.72
x 19	2.95	3.28	2.36	2.69
x 16	2.92	3.25	2.33	2.66
x 14	2.90	3.23	2.32	2.65
W 10x112	4.30	5.17	2.76	3.63
x100	4.25	5.11	2.71	3.57
x 88	4.20	5.06	2.66	3.52
x 77	4.15	5.00	2.62	3.47
x 68	4.12	4.96	2.58	3.42
x 60	4.08	4.92	2.54	3.38
x 54	4.06	4.89	2.52	3.35
x 49	4.04	4.87	2.50	3.33
x 45	3.56	4.23	2.35	3.02
x 39	3.53	4.19	2.32	2.98
x 33	3.49	4.16	2.29	2.95
x 30	3.10	3.59	2.23	2.71
x 26	3.08	3.56	2.20	2.68
x 22	3.05	3.53	2.17	2.65
x 19	2.63	2.96	2.04	2.38
x 17	2.60	2.94	2.02	2.35
x 15	2.58	2.92	2.00	2.33
x 12	2.56	2.89	1.98	2.31

Surface Areas and Box Areas
W Shapes
Square feet per foot of length

Designation	Case A	Case B	Case C	Case D
W 8x 67	3.42	4.11	2.19	2.88
x 58	3.37	4.06	2.14	2.83
x 48	3.32	4.00	2.09	2.77
x 40	3.28	3.95	2.05	2.72
x 35	3.25	3.92	2.02	2.69
x 31	3.23	3.89	2.00	2.67
x 28	2.87	3.42	1.89	2.43
x 24	2.85	3.39	1.86	2.40
x 21	2.61	3.05	1.82	2.26
x 18	2.59	3.03	1.79	2.23
x 15	2.27	2.61	1.69	2.02
x 13	2.25	2.58	1.67	2.00
x 10	2.23	2.56	1.64	1.97
W 6x 25	2.49	3.00	1.57	2.08
x 20	2.46	2.96	1.54	2.04
x 15	2.42	2.92	1.50	2.00
x 16	1.98	2.31	1.38	1.72
x 12	1.93	2.26	1.34	1.67
x 9	1.90	2.23	1.31	1.64
W 5x 19	2.04	2.45	1.28	1.70
x 16	2.01	2.43	1.25	1.67
W 4x 13	1.63	1.96	1.03	1.37

Figure 13.2b

Stairs (type of construction, rails, nosing, number of risers and landings), railings (materials, type construction, finishes, protection),

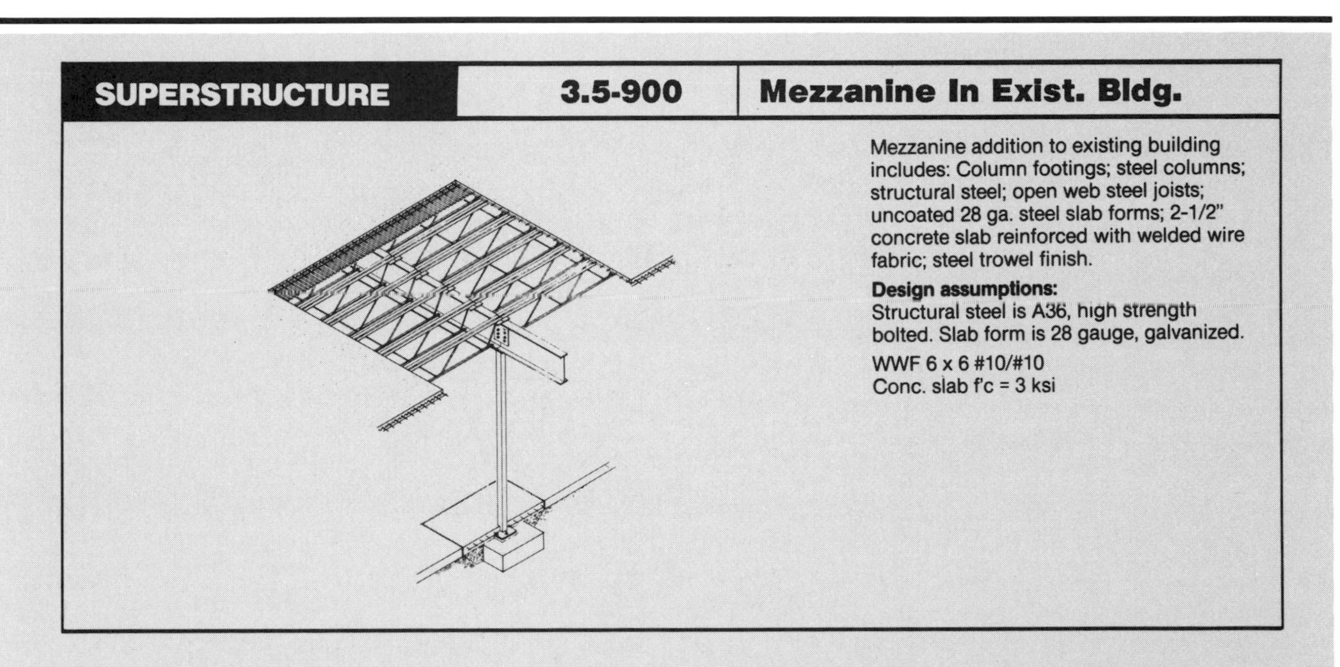

System Components	QUANTITY	UNIT	COST EACH MAT.	COST EACH INST.	COST EACH TOTAL
SYSTEM 03.5-900					
MEZZANINE ADDITION TO EXISTING BUILDING; 100 PSF SUPERIMPOSED LOAD,					
SLAB FORM DECK, MTL. JOISTS, 2.5" CONC. SLAB, 3,000 PSI, 3,000 S.F.					
Concrete footing, 3' sq. (see system 1.1-600)	4.000	C.Y.	1,351.32	5,004.49	6,355.81
Column, 4" x 4" x ¼" x12'	12.000	Ea.	792	648	1,440
Structural steel w 21x50	80.000	L.F.	3,432	528	3,960
W 16x31	80.000	L.F.	2,330.40	429.60	2,760
Open web joists, H series	6.800	Ton	7,854	3,621	11,475
Slab form, steel 28 gauge, galvanized	30.000	C.S.F.	1,080	750	1,830
Concrete 3,000 psi, 2 ½" slab, incl. premium delv. chg.	23.500	C.Y.	2,150.25		2,150.25
Welded wire fabric 6x6 #10/#10 (w1.4/w1.4)	30.000	C.S.F.	264	546	810
Place concrete	23.500	C.Y.		775.50	775.50
Monolithic steel trowel finish	30.000	C.S.F.		1,590	1,590
Curing with sprayed membrane curing compound	30.000	C.S.F.	60.30	136.20	196.50
TOTAL			19,314.27	14,028.79	33,343.06
COST per S.F.			6.44	4.68	11.12

Figure 13.3

gratings (types, sizes, finishes), and attachments must be thought of in the same way as miscellaneous metals.

While making the takeoff, list equipment that can be used to erect fabricated materials. Erection costs cannot be determined without giving consideration to site storage, site access, type and size of equipment required, delivery schedules, and erection sequence. This is one area where the estimator must pay particular attention to minimum labor and equipment charges.

The following is a list of some of the items normally furnished and delivered by a miscellaneous metals supplier.

- Elevator shaft beam separators
- Angle sills with welded-on anchors
- Angle corner-guards with welded-on anchors
- Pipe bollards with welded-on base-plate anchors
- Cast-iron drain grates with frames
- Individual aluminum or steel sleeves for pipe or tube rails
- Cast or extruded abrasive metal nosings for concrete steps
- Templated sleeves welded on a steel flat for continuous pipe or tube-guardrails at balconies or roof
- Transformer vault door frames
- Malleable iron wedge inserts for attached or hung lintels
- Angle frames with welded-on anchors
- Elevator machine-room double-leaf aluminum floor hatches with compensating hinges
- Elevator machine-room ceiling hoist monorail beams
- Roof scuttles
- Stainless steel sleeves for swimming pool ladders and guardrails
- Slab inserts for toilet partitions, operating room lights, and x-ray room ceiling supports
- Loose lintels, 12" longer than the net opening
- Steel stairs, complete as shown on drawings including abrasive extruded nosings, if any, but excluding concrete fill or terrazzo
- Elevator shaft sill angles mounted on inserts, built-in channel door jambs
- All gratings, including any support angles bolted to masonry or concrete or connected to inserts built in by others
- All open-riser ships or engineer's ladders, with diamond plate or grating treads
- Interior ladders to roof scuttles and under hatches
- Exterior ladders with goosenecks from low to high roofs, with or without safety cages as per drawings
- Steel bench supports
- Tube or pipe guardrails in steel, non-ferrous or stainless steel at balconies, roofs, and elsewhere as per drawings, including at interior concrete stairs
- Hung or attached angle lintels for brick supports connected to inserts built in by others
- Catwalks, strutted or suspended, complete with grating walkways, guardrails, and ladder accesses
- Spiral metal angle bases at wooden floor accesses, auditorium stage, book stack accesses, mezzanines, etc.
- Sheet metal angle bases in wooden floor rooms, such as gyms, interior raquetball, squash, wrestling stages, and prosceniums

- Wall handrails at ramps, places other than stairs, hospital hallways, etc.
- Stainless steel swimming pool ladders and pipe guardrails at bleachers
- Toilet partition supports in "Unistrut"
- X-ray machine supports in "Unistrut" or angles
- Operating and autopsy room light "spider-leg" supports
- Monitor supports throughout a hospital
- Hospital linear accelerator supports
- Rolling and fire partitions supports
- Computer room floor supports
- Proscenium grillages
- Acoustic baffle cloud supports
- Motor supports for overhead doors
- Entrance door and other door supports
- Welding of inserts on decking sheets for uses underneath, for ceiling supports
- Projection booth counterweighted port doors for fire protection
- Exterior door saddles, with or without Rixsons
- Interior floor transition door slip-saddles
- Exterior door combination slip-saddles
- Floor, wall, and ceiling non-ferrous expansion joints
- Folding partition supports
- Banquet hall movable partition supports
- Steel or aluminum louvres *not* in contact with any ductwork
- Ornamental metals for glass railings
- Ornamental metals for composite acrylic/wood railings
- Ornamental metals for combination panels and railings
- Ornamental metals for balusters, posts, trillage, and scroll railings
- Non-ferrous expansion joint covers
- Non-ferrous door saddles

Some of the items listed above are illustrated in Figure 13.4. The list of ornamental items included in any interior estimate may be extensive, and the takeoff and pricing can be time consuming. A reliable subcontractor should be contacted and a proper quotation requested.

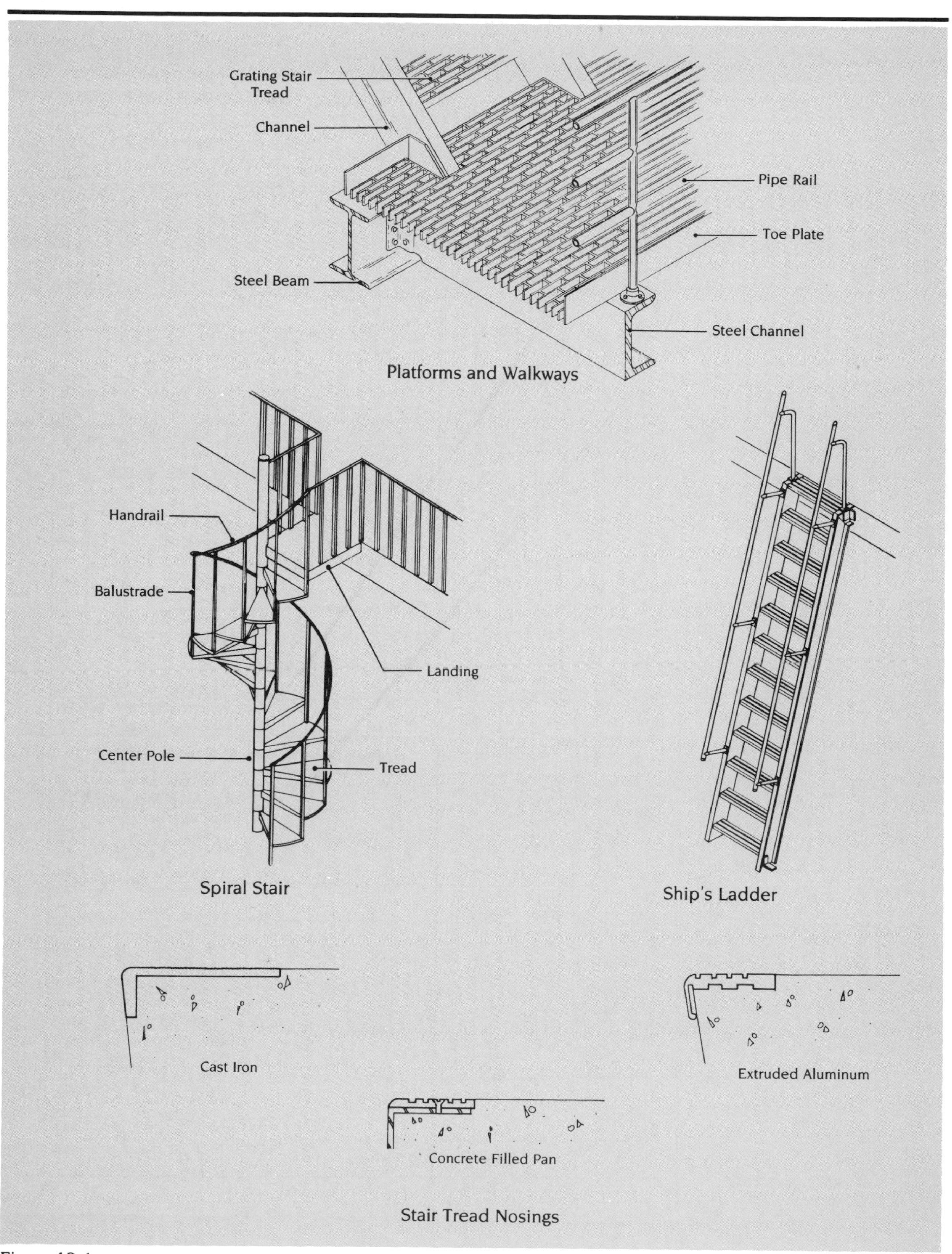
Grating Stair Tread
Channel
Pipe Rail
Toe Plate
Steel Beam
Steel Channel
Platforms and Walkways
Handrail
Balustrade
Landing
Center Pole
Tread
Spiral Stair
Ship's Ladder
Cast Iron
Extruded Aluminum
Concrete Filled Pan
Stair Tread Nosings

Figure 13.4

Chapter 14
WOOD AND PLASTICS

Carpentry work can be broken down into the following categories: rough carpentry, finish carpentry and millwork, and laminated framing and decking. The material prices for carpentry work fluctuate more widely and with greater frequency than is the case with most other building materials. For this reason, when the material list is complete, it is important to obtain *current, local* prices for the lumber. Installation costs greatly depend on productivity. Accurate cost records from past jobs can be most helpful. Since lumber tends to be used extensively in remodeling projects, a careful estimate of this component is essential.

Rough Carpentry

Rough carpentry is most often required in projects where there is extensive renovation, or where the location of partitions, windows, or doors is significantly changed. Lumber is commonly estimated in board feet and purchased in 1,000 board foot quantities. A board foot is the equivalent of 1″ × 12″ × 12″ (nominal) or 3/4″ × 11-1/2″ × 12″ milled (actual). To determine the board feet of a piece of framing, the nominal dimensions can be multiplied, and the result divided by 12. The final result represents the number of board feet per linear foot of that framing size.

Example: 2 × 10 joists
$2 \times 10 = 20$

$$\frac{20}{12} = 1.67 \frac{\text{B.F.}}{\text{L.F.}}$$

The Quantity Sheet for lumber should indicate species, grade, and any type of wood preservative or fire retardant treatment specified or required by code. Floor joists, shown or specified by size and spacing, should be taken off by nominal length and the quantity required. Add for double joists under partitions, headers and cripple joists at openings, overhangs, laps at bearings, and blocking or bridging.

Studs required are noted on the drawings by spacing, usually 16″ on center (O.C.) or 24″ O.C., with the stud size given. The linear feet of like partitions (having the same stud size, height, and spacing) divided by the spacing will give the estimator the approximate number of studs required. Additional studs for openings, corners, double top plates, sole plates, and intersecting partitions must be taken off separately. An allowance for waste should be included (or heights should be recorded as a standard purchased length, e.g., 8′, 10′, 12′, etc.). One "rule of thumb" is to allow one stud for every linear foot of wall, for 16″ O.C. spacing.

Number and size of openings are important takeoff information. Even though there are no studs in these areas, the estimator must take off headers, subsills, king studs, trimmers, cripples, and knee studs. Where bracing and fire blocking are noted, both the type and quantity should be indicated.

Tongue and groove decks of various woods, solid planks, or laminated construction are nominally 2″ to 4″ thick, and are often used with glued laminated beams or heavy timber framing. The square foot method is used to determine quantities and consideration given to non-modular areas for the amount of waste involved. The materials are purchased by board foot measurement. The conversion from square feet to board feet must allow for *net* sizes, as opposed to *board measure*. In this way, loss of coverage due to the tongue and available mill lengths can be taken into account.

Sheathing on Walls can be plywood of different grades and thicknesses, particle board, or solid boards nailed directly to the studs. Plywood may be applied with the grain vertical, horizontal, or rarely, diagonal to the studding. Solid boards are usually nailed diagonally, but can be applied horizontally when lateral forces are not present. For solid board sheathing, add 15% to 20% more material to the takeoff when using tongue and groove (as opposed to square edge) sheathing. Plywood or particle board sheathing can be installed either horizontally or vertically, depending upon wall height and fire code restrictions. When estimating quantities of panel sheathing, the estimator calculates the number of sheets required by measuring the square feet of area to be covered, adding waste, and then dividing by sheet size. Applying these materials diagonally or on non-modular areas creates additional waste. This waste factor must be included in the estimate. For diagonal application of boards, plywood, or wallboard, include an additional 10% to 15% material waste factor.

Subfloors can be CDX-type plywood (with the thickness dependent on the load and span), solid boards laid diagonally or perpendicular to the joists, or tongue and groove planks. The quantity takeoff for subfloors is similar to that for sheathing (previously noted).

Grounds are normally 1″ × 2″ wood strips used for case work or plaster; the quantities are estimated in L.F.

Furring (1″ × 2″ or 3″) wood strips are fastened to wood, masonry, or concrete walls so that wall coverings may be attached thereto. Furring may also be used on the underside of ceiling joists to fasten ceiling finishes. Quantities are estimated by L.F.

Framing Members (studs or joists) are measured in linear feet. The quantity required is based on square feet of surface area (wall, floor, ceiling). The table below can be used to estimate rough quantities of basic framing members. Supplemental framing such as plates, sills, headers and bands, must be added separately.

Spacing of Framing Members	Linear Feet per Square Foot Surface
12" O.C.	1.2 L.F./S.F.
16" O.C.	1.0 L.F./S.F.
24" O.C.	0.8 L.F./S.F.

Additional requirements for rough carpentry, especially those for temporary construction, may not all be directly stated in the plans and specifications. Such items may include blocking, temporary stairs, wood inserts for metal pan stairs, and railings, along with various other requirements for different trades. Temporary construction may also be included in Division 1 – General Requirements.

Typical framing installations are shown in Figures 14.1a, 14.1b and 14.1c.

Laminated Construction

Laminated construction should be listed separately, as it is frequently supplied by a specialty subcontractor. Sometimes the beams are supplied and erected by one subcontractor, and the decking installed by the general contractor or another subcontractor. The takeoff units must be adapted to the system: *square foot* for floors, *linear foot* for members, or *board foot* for lumber. Since the members are factory-fabricated, the plans and specifications must be submitted to a fabricator for takeoff and pricing. Some examples of laminated construction are shown in Figure 14.2.

Finish Carpentry and Millwork

Finish carpentry and millwork – wood rails, paneling, shelves, casements, and cabinetry – are common features, even in buildings that might have no other wood. Upon examination of the plans and specifications, the estimator must determine which items will be built on-site, and which will be fabricated off-site by a millwork subcontractor. Shop drawings are often required for architectural woodwork and are usually included in the subcontract price.

Moldings and Door Trim may be taken off and priced by the "set" or by the linear foot. The common use of pre-hung doors makes it convenient to take off this trim with the doors. Exterior trim, other than door and window trim, should be taken off with the siding, since the details and dimensions are interrelated.

Paneling is taken off by type, finish, and square foot (converted to full sheets). Be sure to list any millwork that would show up on the details. Panel siding and associated trim are taken off by the square foot and linear foot, respectively. Be sure to provide an allowance for waste.

Decorative Beams and Columns that are non-structural should be estimated separately. Decorative trim may be used to wrap exposed structural elements. Particular attention should be paid to the joinery. Long, precise joints are difficult (and expensive) to construct in the field.

Cabinets, Counters, and Shelves are most often priced by the linear foot or by the unit. Job fabricated, prefabricated, and subcontracted work should be estimated separately.

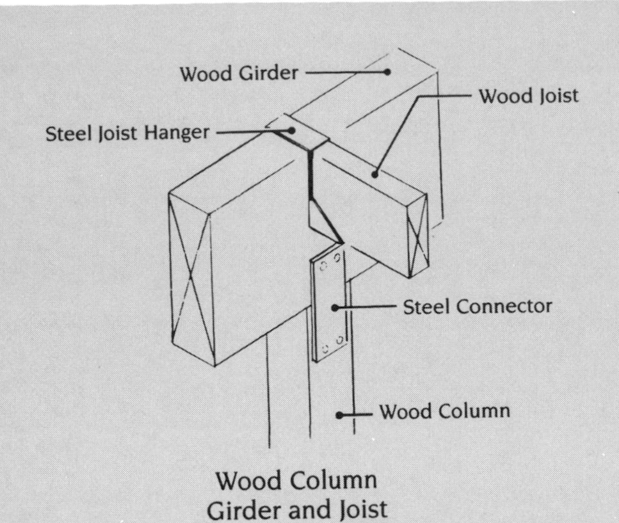

Wood Column
Girder and Joist

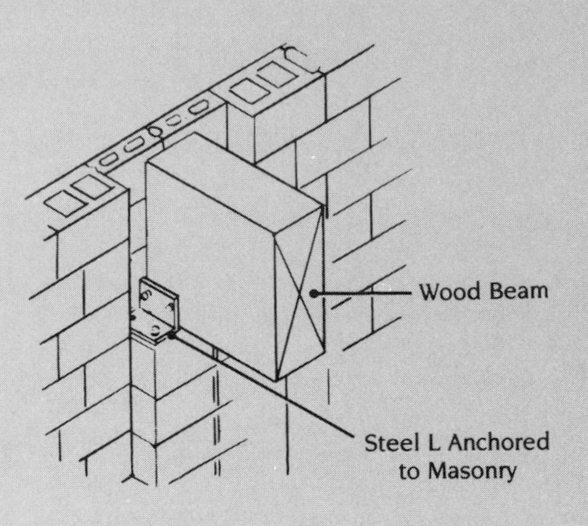

Wood Girder Supported by
Masonry Wall

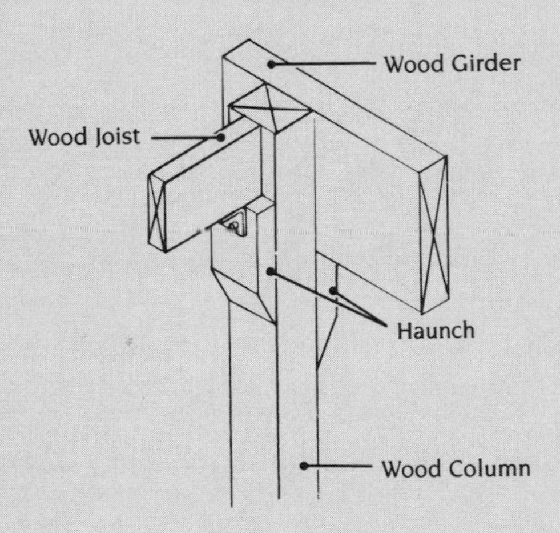

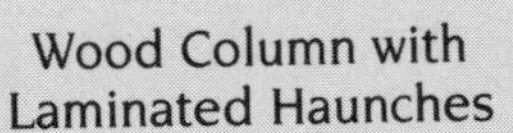

Wood Column with
Laminated Haunches

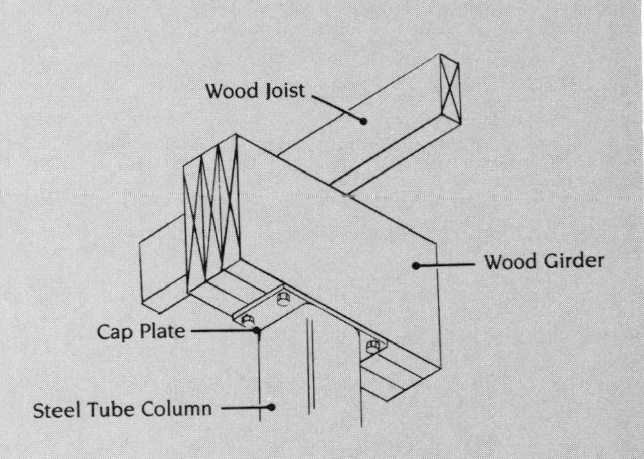

Wood Girder Supported by
Square Tube Column

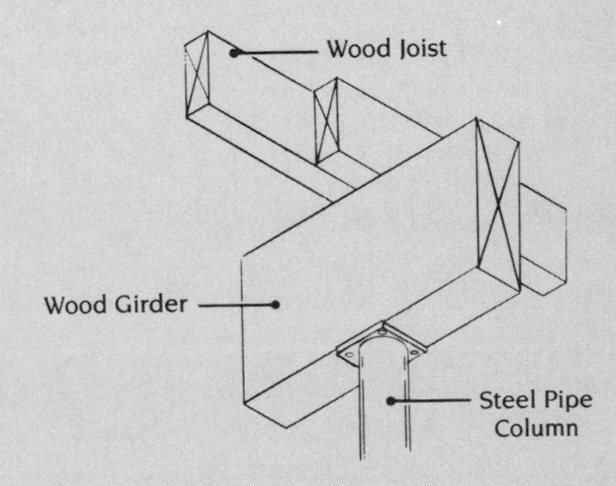

Wood Girder Supported by
Pipe Column

Figure 14.1*a*

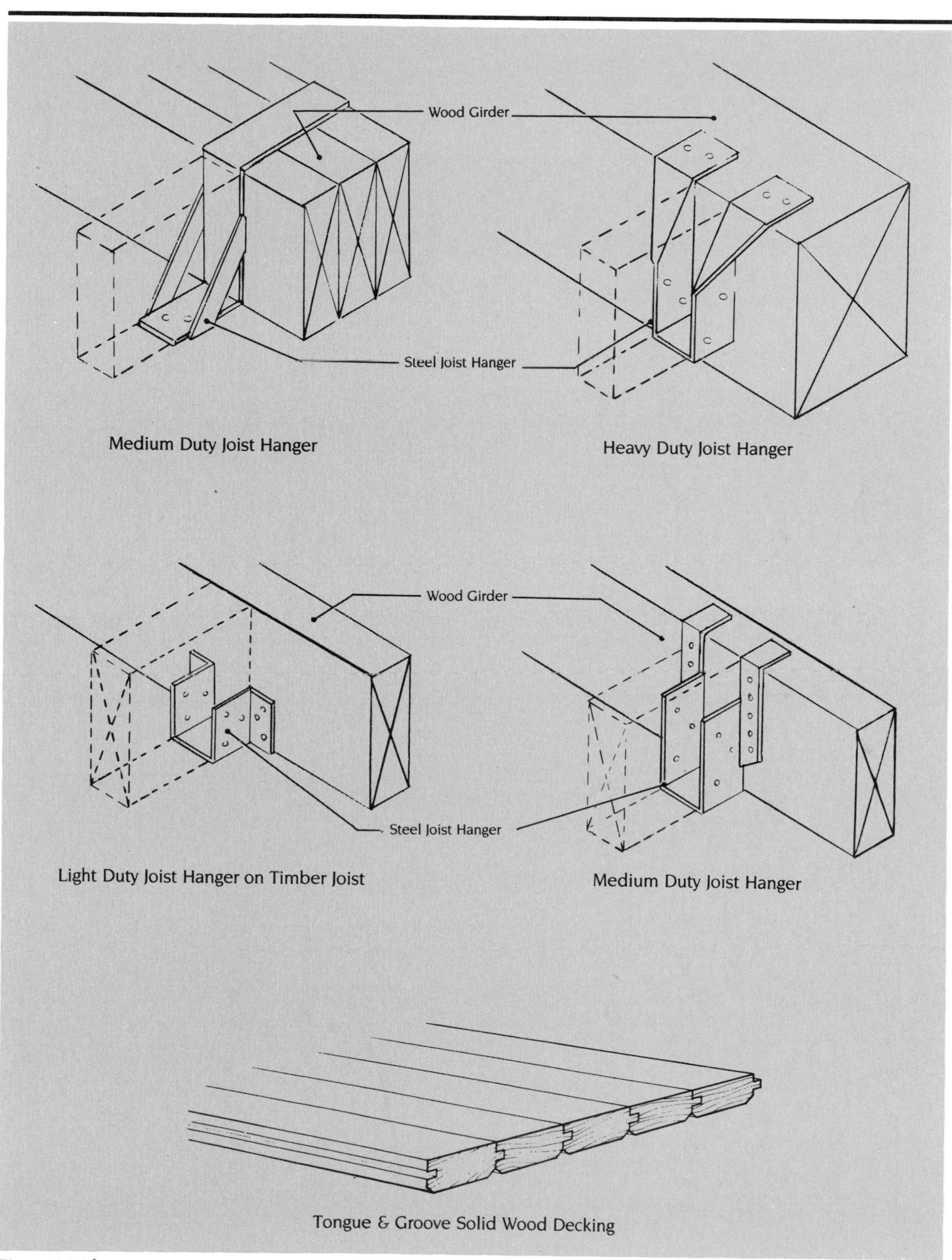
Wood Girder
Steel Joist Hanger
Medium Duty Joist Hanger
Heavy Duty Joist Hanger
Wood Girder
Steel Joist Hanger
Light Duty Joist Hanger on Timber Joist
Medium Duty Joist Hanger
Tongue & Groove Solid Wood Decking

Figure 14.1*b*

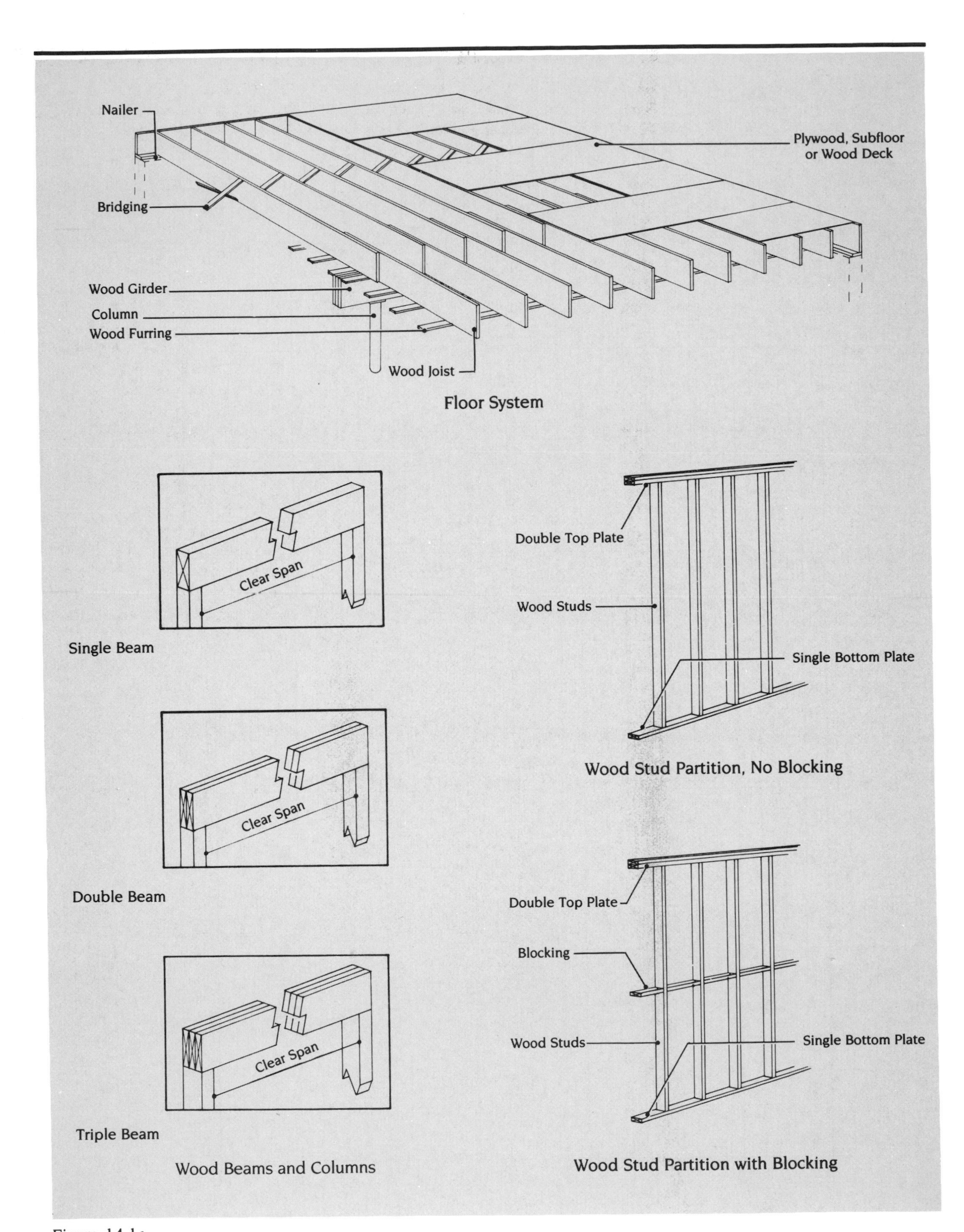
Nailer
Plywood, Subfloor or Wood Deck
Bridging
Wood Girder
Column
Wood Furring
Wood Joist
Floor System
Clear Span
Single Beam
Clear Span
Double Beam
Clear Span
Triple Beam
Wood Beams and Columns
Double Top Plate
Wood Studs
Single Bottom Plate
Wood Stud Partition, No Blocking
Double Top Plate
Blocking
Wood Studs
Single Bottom Plate
Wood Stud Partition with Blocking

Figure 14.1c

Stairs should be estimated by individual component unless accurate, complete system costs have been developed from previous projects. Typical components and units for estimating are shown in Figure 14.3.

General Rule

A general rule for budgeting millwork is that total costs can be two to three times the cost of the materials. Millwork is often ordered and purchased directly by the owner. When installation is the responsibility of the contractor, costs for handling, storage, and protection, as well as those for installation, should be included. Typical finish carpentry and millwork items are illustrated in Figures 14.4a and 14.4b.

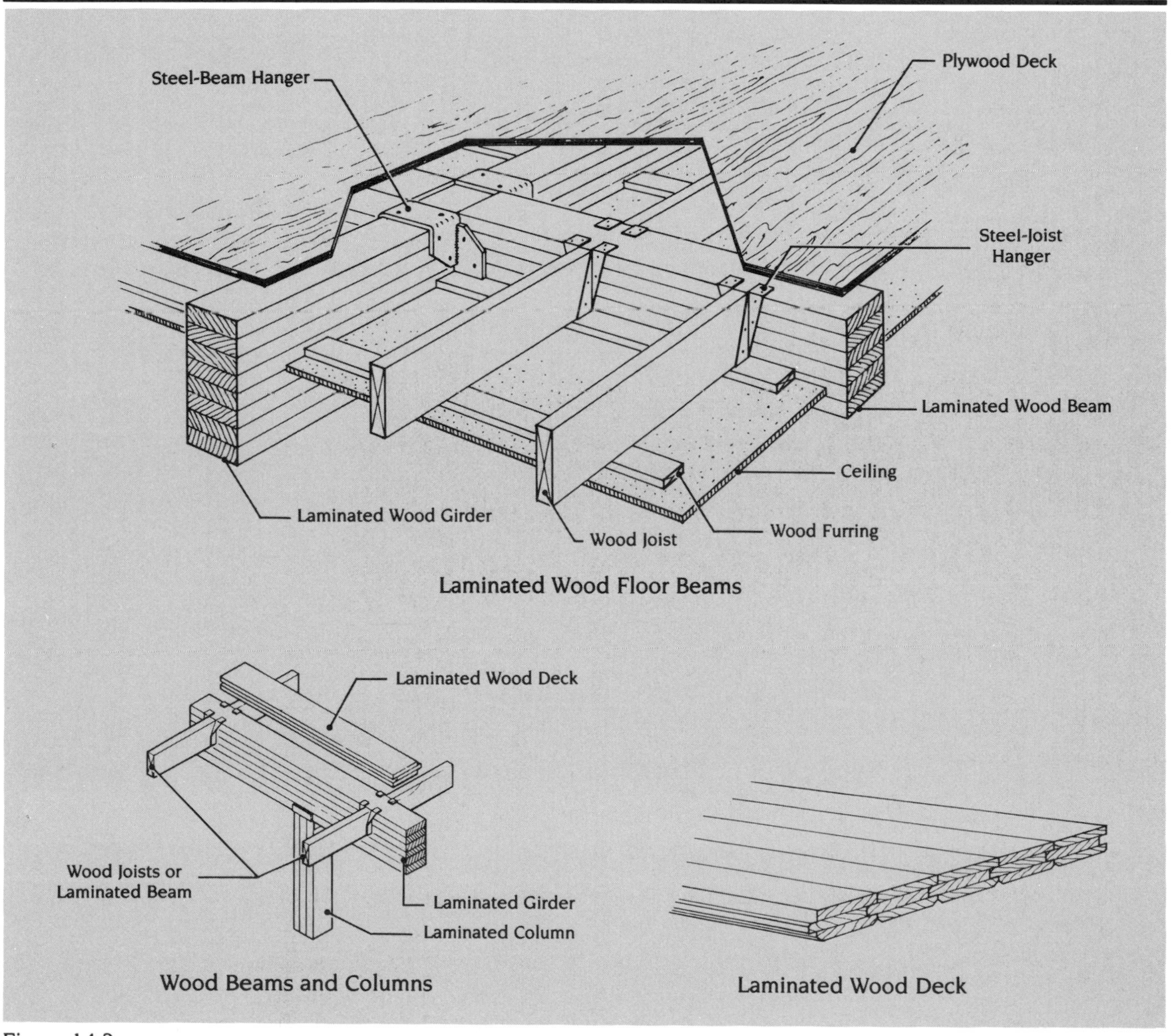

Figure 14.2

064 | Architectural Woodwork

064 300 | Stairwork & Handrails

				DAILY	MAN-		BARE COSTS				TOTAL	
			CREW	OUTPUT	HOURS	UNIT	MAT.	LABOR	EQUIP.	TOTAL	INCL O&P	
306	4000	Residential, wood, oak treads, prefabricated	2 Carp	1.50	10.670	Flight	630	230		860	1,050	306
	4200	Built in place	"	.44	36.360	"	750	780		1,530	2,075	
	4400	Spiral, oak, 4'-6" diameter, unfinished, prefabricated,										
	4500	incl. railing, 9' high	2 Carp	1.50	10.670	Flight	2,775	230		3,005	3,425	
	9000	Minimum labor/equipment charge	"	3	5.330	Job		115		115	185	
308	0010	STAIR PARTS Balusters, turned, 30" high, pine, minimum	1 Carp	28	.286	Ea.	3.50	6.10		9.60	13.65	308
	0100	Maximum		26	.308		5	6.60		11.60	16.05	
	0300	30" high birch balusters, minimum (84)		28	.286		5	6.10		11.10	15.30	
	0400	Maximum		26	.308		7	6.60		13.60	18.25	
	0600	42" high, pine balusters, minimum		27	.296		4.40	6.35		10.75	15	
	0700	Maximum		25	.320		5.75	6.85		12.60	17.30	
	0900	42" high birch balusters, minimum		27	.296		6	6.35		12.35	16.80	
	1000	Maximum		25	.320		10	6.85		16.85	22	
	1050	Baluster, stock pine, 1-1/16" x 1-1/16"		240	.033	L.F.	.50	.71		1.21	1.70	
	1100	1-5/8" x 1-5/8"		220	.036	"	.98	.78		1.76	2.33	
	1200	Newels, 3-1/4" wide, starting, minimum		7	1.140	Ea.	30	24		54	72	
	1300	Maximum		6	1.330		200	29		229	265	
	1500	Landing, minimum		5	1.600		40	34		74	99	
	1600	Maximum		4	2		210	43		253	300	
	1800	Railings, oak, built-up, minimum		60	.133	L.F.	4.25	2.85		7.10	9.25	
	1900	Maximum		55	.145		11.50	3.11		14.61	17.65	
	2100	Add for sub rail		110	.073		1.75	1.56		3.31	4.42	
	2110											
	2300	Risers, Beech, 3/4" x 7-1/2" high	1 Carp	64	.125	L.F.	4.05	2.68		6.73	8.75	
	2400	Fir, 3/4" x 7-1/2" high		64	.125		1.10	2.68		3.78	5.50	
	2600	Oak, 3/4" x 7-1/2" high		64	.125		3.70	2.68		6.38	8.35	
	2800	Pine, 3/4" x 7-1/2" high		66	.121		1.10	2.59		3.69	5.35	
	2850	Skirt board, pine, 1" x 10"		55	.145		1.45	3.11		4.56	6.60	
	2900	1" x 12"		52	.154		1.75	3.29		5.04	7.20	
	3000	Treads, oak, 1-1/16" x 9-1/2" wide, 3' long		18	.444	Ea.	16.50	9.50		26	33	
	3100	4' long		17	.471		23	10.05		33.05	41	
	3300	1-1/16" x 11-1/2" wide, 3' long		18	.444		23	9.50		32.50	41	
	3400	6' long		14	.571		39	12.25		51.25	63	
	3600	Beech treads, add					40%					
	3800	For mitered return nosings, add				L.F.	2.25			2.25	2.48	
	9000	Minimum labor/equipment charge	1 Carp	3	2.670	Job		57		57	92	
310	0010	RAILING Custom design, architectural grade, hardwood, minimum	1 Carp	38	.211	L.F.	10	4.51		14.51	18.25	310
	0100	Maximum		30	.267		30	5.70		35.70	42	
	0300	Stock interior railing with spindles 6" O.C., 4' long		40	.200		24	4.28		28.28	33	
	0400	8' long		48	.167		22	3.57		25.57	30	
	9000	Minimum labor/equipment charge		3	2.670	Job		57		57	92	

064 400 | Misc Ornamental Items

			CREW	DAILY OUTPUT	MAN-HOURS	UNIT	MAT.	LABOR	EQUIP.	TOTAL	TOTAL INCL O&P	
402	0011	COLUMNS										402
	0050	Aluminum, round colonial, 6" diameter	2 Carp	80	.200	V.L.F.	6.20	4.28		10.48	13.70	
	0100	8" diameter		62.25	.257		8	5.50		13.50	17.65	
	0200	10" diameter		55	.291		11.40	6.25		17.65	23	
	0250	Fir, stock units, hollow round, 6" diameter		80	.200		14.80	4.28		19.08	23	
	0300	8" diameter		80	.200		16.10	4.28		20.38	25	
	0350	10" diameter		70	.229		18	4.89		22.89	28	
	0400	Solid turned, to 8' high, 3-1/2" diameter		80	.200		4	4.28		8.28	11.25	
	0500	4-1/2" diameter		75	.213		6.15	4.57		10.72	14.10	
	0600	5-1/2" diameter		70	.229		8.15	4.89		13.04	16.80	
	0800	Square columns, built-up , 5" x 5"		65	.246		5.50	5.25		10.75	14.50	
	0900	Solid, 3-1/2" x 3-1/2"		130	.123		4.40	2.63		7.03	9.05	

For expanded coverage of these items see *Means Interior Cost Data 1989*

85

Figure 14.3

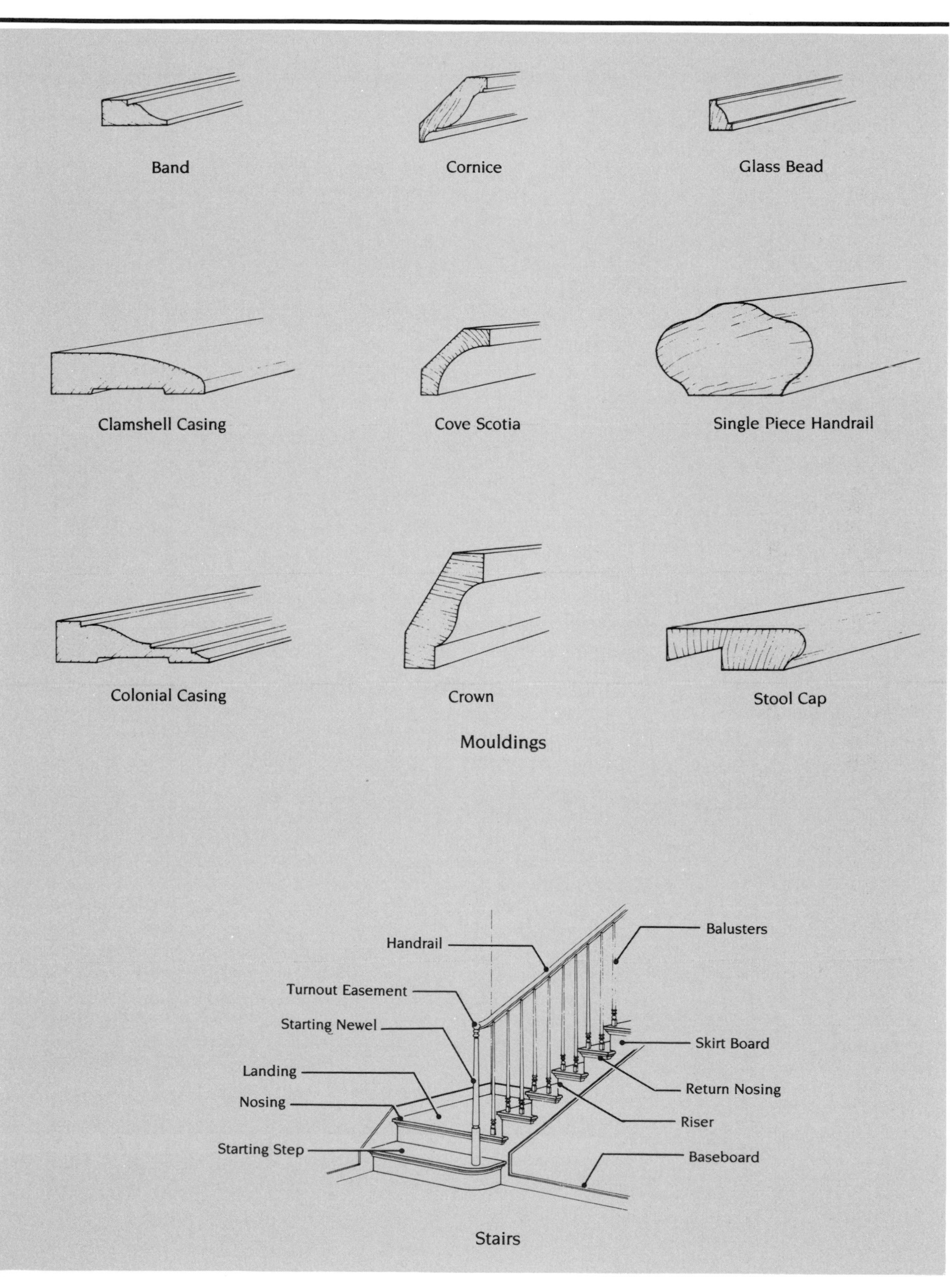
Band
Cornice
Glass Bead
Clamshell Casing
Cove Scotia
Single Piece Handrail
Colonial Casing
Crown
Stool Cap
Mouldings
Handrail
Balusters
Turnout Easement
Starting Newel
Skirt Board
Landing
Return Nosing
Nosing
Riser
Starting Step
Baseboard
Stairs

Figure 14.4a

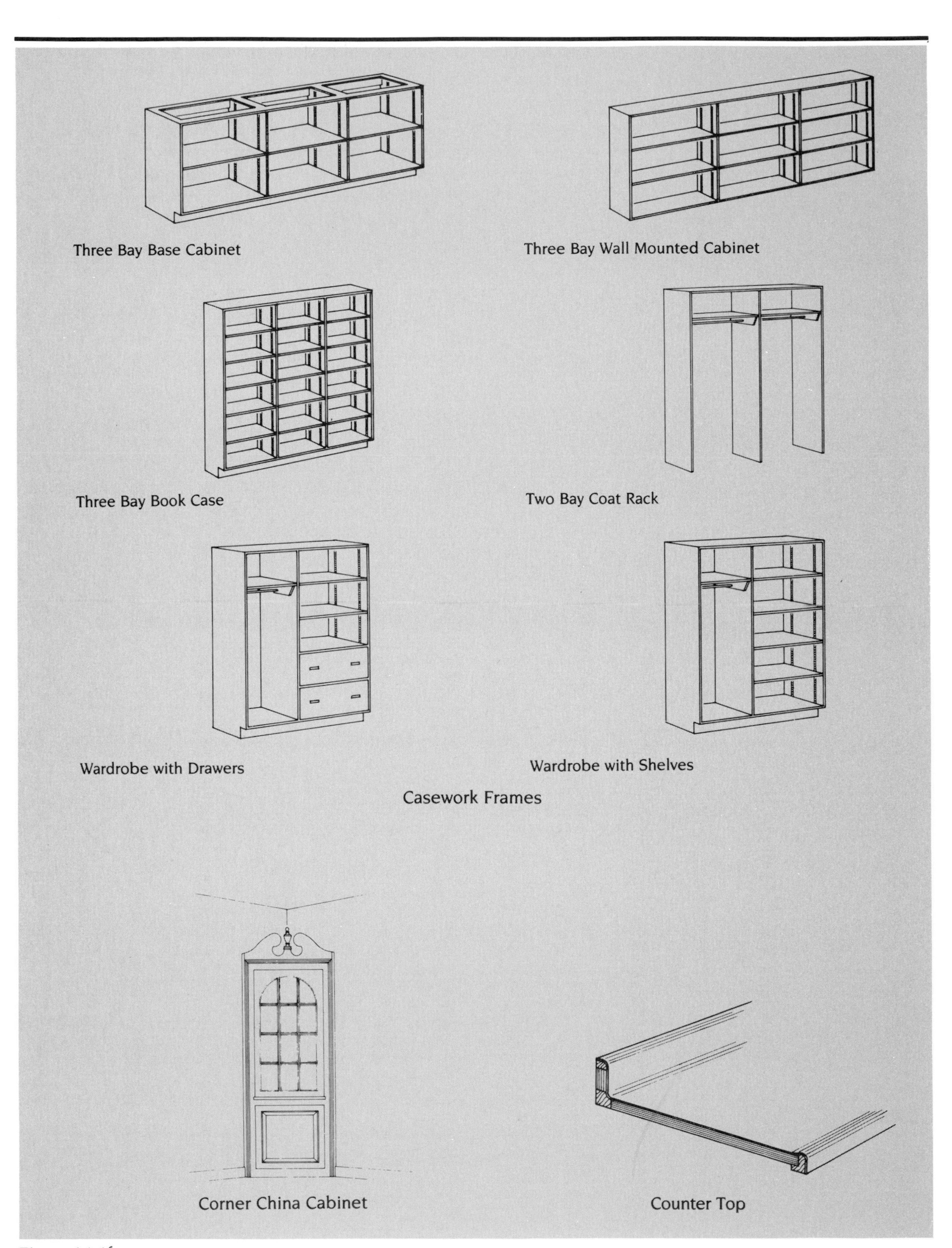
Three Bay Base Cabinet
Three Bay Wall Mounted Cabinet
Three Bay Book Case
Two Bay Coat Rack
Wardrobe with Drawers
Wardrobe with Shelves
Casework Frames
Corner China Cabinet
Counter Top

Figure 14.4*b*

Chapter 15

MOISTURE AND THERMAL PROTECTION

This division includes materials for sealing the outside of a building—for protection against moisture and air infiltration, as well as insulation and associated accessories. When reviewing the plans and specifications, and when inspecting the site, the estimator should visualize the construction process, and thus determine all probable areas where these materials will be required. The technique used for quantity takeoff depends on the specific materials and installation methods.

Waterproofing

- Dampproofing
- Vapor Barriers
- Caulking and Sealants
- Sheet and Membrane
- Integral Cement Coatings

A distinction should be made between dampproofing and waterproofing. Dampproofing is used to inhibit the migration of moisture or water vapor. In most cases, dampproofing will not stop the flow of water (even at minimal pressures). Waterproofing, on the other hand, consists of a continuous, impermeable membrane and is used to prevent or stop the flow of water.

Dampproofing

Dampproofing usually consists of one or two bituminous coatings applied to foundation walls from the bottom of the footings to approximately the finished grade line. The areas involved are calculated based on the total height of the dampproofing and the length of the wall. After separate areas are figured and added together to provide a total square foot area, a unit cost per square foot can be selected for the type of material, the number of coats, and the method of application specified for the building.

Waterproofing Below Grade

Waterproofing at or below grade with elastomeric sheets or membranes is estimated on the same basis as dampproofing, with two basic exceptions. First, the installed unit costs for the elastomeric sheets do not include bonding adhesive or splicing tape, which must be figured as an additional cost. Second, the membrane waterproofing under slabs must be estimated separately from the higher cost installation on walls. In all cases, unit costs are per square foot of covered surface.

For walls below grade, protection board is often specified to prevent damage to the waterproofing membrane when the excavation is backfilled. Rigid foam insulation installed outside of the barrier may also serve a protective function. Metallic coating material may be applied to floors or walls, usually on the interior or dry side, after the masonry surface has been prepared (usually by chipping) for bonding to the new material. The unit cost per square foot for these materials depends on the thickness of the material, the orientation of the area to be covered, and the preparation required. In many cases, these materials must be applied in locations where access is difficult and under the control of others. The estimator should make an allowance for delays caused by this problem.

Caulking and Sealants

Caulking and sealants are most often required on the exterior of the building and for certain conditions on the interior. In most cases, caulking and sealing are done to prevent water and/or air from entering a building. Caulking and sealing are usually specified at joints, expansion joints, control joints, door and window frames, and in places where dissimilar materials meet over the surface of the building exterior. To estimate the installed cost of this type of material, two things must be determined. First, the estimator must note (from the specifications) the kind of material to be used for each caulking or sealing job. Second, the size of the joints to be caulked or sealed must be measured, with attention given to any requirements for backer rods. With this information, the estimator can select the applicable cost per linear foot and multiply it by the total length in feet. The result is an estimated cost for each kind of caulking or sealing on the job. Caulking and sealing may often be overlooked as incidental items. They may, in fact, represent a significant cost, depending on the type of construction and the quantity.

Insulation

- Batt or Roll
- Blown-in
- Board (Rigid and Semi-rigid)
- Cavity Masonry
- Perimeter Foundation
- Poured in Place
- Reflective
- Roof
- Sprayed

Insulation is primarily used to reduce heat transfer through the exterior enclosure of the building. The choice of insulation type and form varies according to its location in the structure and the size of the space it occupies. Major categories of insulation include mineral granules, fibers and foams, vegetable fibers and solids, and plastic foams. These materials may be required around foundations, on or inside walls, and under roofing. Many different details of the drawings must be examined in order to determine types, methods, and quantities of insulation. The cost of insulation depends on the type of material, its form (loose, granular, batt or boards), its thickness in inches, the method of installation, and the total area in square feet.

Specifying by "R" Value

It is becoming popular to specify insulation by "R" value only. The estimator may have a choice of materials, given a required "R" value and a certain cavity space (which may dictate insulation thickness). For example, the specifications may require an "R" value of 11, and only 2" of wall cavity is available for the thickness of the insulation. From Figures 15.1 and 15.2, it is seen that only rigid urethane (line 072-116-2560) meets the design criteria. Note that if more cavity space were available, 3-1/2" non-rigid fiberglass (line 072-118-0420) would be a much less expensive alternative. The estimator may have to do some comparison shopping to find the least expensive material for the specified "R" value and thickness. Installation costs may vary from one material to another. Also, wood blocking, furring, and/or nailers are often required to match the insulation thickness in some instances.

Associated Costs

Working with the above data, the estimator can accurately select the installed cost per square foot and estimate the total cost. The estimate for insulation should also include associated costs, such as cutting and patching for difficult installation, or requirements for air vents and other accessories.

Insulating for Sound

Insulation is not only used for controlling heat transfer. It is also specified for use in internal walls and ceilings for controlling *sound* transfer. Although the noise reduction coefficient of batt insulation is not as great as specialized sound attenuation blankets, the costs are considerably less.

Shingles

Most residences and many smaller types of commercial buildings have sloping roofs covered with some form of shingle or watershed material. The materials used in shingles vary from the more common granular-covered asphalt and fiberglass units to wood, metal, clay, concrete, or slate.

The first step in estimating the cost of a shingle roof is to determine the material specified, shingle size and weight, and installation method. With this information, the estimator can select the accurate installed cost of the roofing material.

Determining Roof Area

In a sloping roof deck, the ridge and eave lengths, as well as the ridge to eaves dimension, must be known or measured before the actual roof area can be calculated. When the plan dimensions of the roof are known and the sloping dimensions are not known, the actual roof area can still be estimated, providing the slope of the roof is known. Figure 15.3 is a table of multipliers that can be used for this purpose. The roof slope is given in both the inches of rise per foot of horizontal run, and in the degree of slope, which allows direct conversion of the horizontal plan dimension into the dimension on the slope.

072 | Insulation

		072 100 | Building Insulation	CREW	DAILY OUTPUT	MAN-HOURS	UNIT	BARE COSTS MAT.	LABOR	EQUIP.	TOTAL	TOTAL INCL O&P	
115	1100	Finish coat, 5 mils, elastomeric aliphatic, 200 S.F./gal.				Gal.	35			35	39	115
	9000	Minimum labor/equipment charge	A-1	2	4	Job		67	26	93	135	
116	0010	**WALL INSULATION, RIGID**										116
	0040	Fiberglass, 1.5#/C.F., unfaced, 1" thick, R4.1	1 Carp	1,000	.008	S.F.	.22	.17		.39	.52	
	0060	1-½" thick, R6.2		1,000	.008		.39	.17		.56	.70	
	0080	2" thick, R8.3		1,000	.008		.50	.17		.67	.82	
	0120	3" thick, R12.4		800	.010		.78	.21		.99	1.20	
	0370	3#/C.F., unfaced, 1" thick, R4.3		1,000	.008		.56	.17		.73	.89	
	0390	1-½" thick, R6.5		1,000	.008		.84	.17		1.01	1.20	
	0400	2" thick, R8.7		890	.009		1.17	.19		1.36	1.60	
	0420	2-½" thick, R10.9		800	.010		1.40	.21		1.61	1.88	
	0440	3" thick, R13		800	.010		1.68	.21		1.89	2.19	
	0520	Foil faced, 1" thick, R4.3		1,000	.008		1.06	.17		1.23	1.44	
	0540	1-½" thick, R6.5		1,000	.008		1.34	.17		1.51	1.75	
	0560	2" thick, R8.7		890	.009		1.62	.19		1.81	2.09	
	0580	2-½" thick, R10.9		800	.010		1.90	.21		2.11	2.43	
	0600	3" thick, R13		800	.010		3.05	.21		3.26	3.70	
	0670	6#/C.F., unfaced, 1" thick, R4.3		1,000	.008		1	.17		1.17	1.37	
	0690	1-½" thick, R6.5		890	.009		1.50	.19		1.69	1.96	
	0700	2" thick, R8.7		800	.010		2	.21		2.21	2.54	
	0721	2-½" thick, R10.9		800	.010		2.57	.21		2.78	3.17	
	0741	3" thick, R13		730	.011		3.08	.23		3.31	3.76	
	0821	Foil faced, 1" thick, R4.3		1,000	.008		1.45	.17		1.62	1.87	
	0840	1-½" thick, R6.5		890	.009		1.97	.19		2.16	2.48	
	0850	2" thick, R8.7		800	.010		2.45	.21		2.66	3.04	
	0880	2-½" thick, R10.9		800	.010		2.90	.21		3.11	3.53	
	0900	3" thick, R13	↓	730	.011	↓	3.20	.23		3.43	3.90	
	1000											
	1500	Foamglass, 1-½" thick, R2.64	1 Carp	800	.010	S.F.	1.52	.21		1.73	2.02	
	1550	2" thick, R5.26	"	730	.011	"	2.15	.23		2.38	2.74	
	1600	Isocyanurate, 4' x 8' sheet, foil faced, both sides										
	1610	½" thick, R3.9	1 Carp	800	.010	S.F.	.23	.21		.44	.60	
	1620	⅝" thick, R4.5		800	.010		.26	.21		.47	.63	
	1630	¾" thick, R5.4		800	.010		.30	.21		.51	.67	
	1640	1" thick, R7.2		800	.010		.36	.21		.57	.74	
	1650	1-½" thick, R10.8		730	.011		.53	.23		.76	.96	
	1660	2" thick, R14.4		730	.011		.69	.23		.92	1.14	
	1670	3" thick, R21.6		730	.011		1.06	.23		1.29	1.54	
	1680	4" thick, R28.8		730	.011		1.41	.23		1.64	1.93	
	1700	Perlite, 1" thick, R2.77		800	.010		.39	.21		.60	.77	
	1750	2" thick, R5.55		730	.011		.73	.23		.96	1.18	
	1900	Polystyrene, extruded blue, 2.2#/C.F., ¾" thick, R4		800	.010		.41	.21		.62	.79	
	1940	1-½" thick, R8.1		730	.011		.67	.23		.90	1.11	
	1960	2" thick, R10.8		730	.011		.88	.23		1.11	1.34	
	2100	Molded bead board, white, 1" thick, R3.85		800	.010		.17	.21		.38	.53	
	2120	1-½" thick, R5.6		730	.011		.27	.23		.50	.67	
	2140	2" thick, R7.7		730	.011		.34	.23		.57	.75	
	2350	Sheathing, insulating foil faced fiberboard, ⅜" thick		670	.012		.22	.26		.48	.65	
	2510	Urethane, no paper backing, ½" thick, R2.9		800	.010		.26	.21		.47	.63	
	2520	1" thick, R5.8		800	.010		.47	.21		.68	.86	
	2540	1-½" thick, R8.7		730	.011		.66	.23		.89	1.10	
	2560	2" thick, R11.7		730	.011		.88	.23		1.11	1.34	
	2710	Fire resistant, ½" thick, R2.9		800	.010		.31	.21		.52	.68	
	2720	1" thick, R5.8		800	.010		.63	.21		.84	1.04	
	2740	1-½" thick, R8.7		730	.011		.81	.23		1.04	1.27	
	2760	2" thick, R11.7		730	.011	↓	1.08	.23		1.31	1.56	
	9000	Minimum labor/equipment charge	↓	4	2	Job		43		43	69	

89

Figure 15.1

072 | Insulation

		072 100 Building Insulation	CREW	DAILY OUTPUT	MAN-HOURS	UNIT	BARE COSTS MAT.	BARE COSTS LABOR	BARE COSTS EQUIP.	BARE COSTS TOTAL	TOTAL INCL O&P	
118	0010	WALL OR CEILING INSUL., NON-RIGID										118
	0040	Fiberglass, kraft faced, batts or blankets										
	0060	3-½" thick, R11, 11" wide (90)	1 Carp	1,150	.007	S.F.	.22	.15		.37	.48	
	0080	15" wide		1,600	.005		.22	.11		.33	.41	
	0100	23" wide		1,600	.005		.25	.11		.36	.45	
	0140	6" thick, R19, 11" wide		1,000	.008		.36	.17		.53	.67	
	0160	15" wide		1,350	.006		.36	.13		.49	.60	
	0180	23" wide		1,600	.005		.36	.11		.47	.57	
	0200	9" thick, R30, 15" wide		1,150	.007		.50	.15		.65	.79	
	0220	23" wide		1,350	.006		.50	.13		.63	.75	
	0240	12" thick, R38, 15" wide		1,000	.008		.72	.17		.89	1.07	
	0260	23" wide		1,350	.006		.72	.13		.85	1	
	0400	Fiberglass, foil faced, batts or blankets										
	0420	3-½" thick, R11, 15" wide	1 Carp	1,600	.005	S.F.	.27	.11		.38	.47	
	0440	23" wide		1,600	.005		.27	.11		.38	.47	
	0460	6" thick, R19, 15" wide		1,350	.006		.39	.13		.52	.63	
	0480	23" wide		1,600	.005		.39	.11		.50	.60	
	0500	9" thick, R30, 15" wide		1,150	.007		.56	.15		.71	.86	
	0550	23" wide		1,350	.006		.56	.13		.69	.82	
	0800	Fiberglass, unfaced, batts or blankets										
	0820	3-½" thick, R11, 15" wide	1 Carp	1,350	.006	S.F.	.20	.13		.33	.42	
	0830	23" wide		1,600	.005		.20	.11		.31	.39	
	0860	6" thick, R19, 15" wide		1,150	.007		.33	.15		.48	.60	
	0880	23" wide		1,350	.006		.33	.13		.46	.57	
	0900	9" thick, R30, 15" wide		1,000	.008		.54	.17		.71	.87	
	0920	23" wide		1,150	.007		.54	.15		.69	.83	
	0940	12" thick, R38, 15" wide		1,000	.008		.67	.17		.84	1.01	
	0960	23" wide		1,150	.007		.67	.15		.82	.98	
	1300	Mineral fiber batts, kraft faced										
	1320	3-½" thick, R13	1 Carp	1,600	.005	S.F.	.28	.11		.39	.48	
	1340	6" thick, R19		1,600	.005		.44	.11		.55	.66	
	1380	10" thick, R30		1,350	.006		.75	.13		.88	1.03	
	1900	For foil backing, add					.06			.06	.07	
	9000	Minimum labor/equipment charge	1 Carp	4	2	Job		43		43	69	
		072 200 Roof & Deck Insulation										
203	0010	ROOF DECK INSULATION										203
	0030	Fiberboard, mineral, 1" thick, R2.78	1 Rofc	800	.010	S.F.	.27	.20		.47	.64	
	0080	1-½" thick, R4		800	.010		.44	.20		.64	.82	
	0100	2" thick, R5.26		800	.010		.61	.20		.81	1.01	
	0300	Fiberglass, in 3' x 4' or 4' x 8' sheets										
	0400	15/16" thick, R3.3	1 Rofc	1,000	.008	S.F.	.39	.16		.55	.70	
	0460	1-1/16" thick, R3.8		1,000	.008		.50	.16		.66	.82	
	0600	1-5/16" thick, R5.3		1,000	.008		.62	.16		.78	.95	
	0650	1-⅝" thick, R5.7		1,000	.008		.73	.16		.89	1.08	
	0700	1-⅞" thick, R7.7		1,000	.008		.78	.16		.94	1.13	
	0800	2-¼" thick, R8		800	.010		.78	.20		.98	1.20	
	0900	Fiberglass and urethane composite, 3' x 4' sheets										
	1000	1-11/16" thick, R11.1	1 Rofc	1,000	.008	S.F.	.62	.16		.78	.95	
	1200	2" thick, R14.3		800	.010		.73	.20		.93	1.14	
	1300	2-⅝" thick, R18.2		800	.010		.96	.20		1.16	1.40	
	1500	Foamglass, 2' x 4' sheets, rectangular										
	1510	1-½" thick R3.95	1 Rofc	800	.010	S.F.	1.63	.20		1.83	2.13	
	1520	2" thick R5.26		800	.010		2.06	.20		2.26	2.61	
	1530	3" thick R7.89		700	.011		2.44	.23		2.67	3.07	
	1540	4" thick R10.53		700	.011		4.33	.23		4.56	5.15	
	1600	Tapered 1/16", ⅛" or ¼" per foot				B.F.	1.10			1.10	1.21	
	1650	Perlite, 2' x 4' sheets										

90

Figure 15.2

Converting Area to Units

After the roof area has been estimated in square feet, it must be divided by 100 to convert it into roofing squares (the conventional "unit" for roofing – one square equals 100 square feet). To determine the quantity of shingles required for hips or ridges, add one square for each 100 linear feet of hips and/or ridges.

Allowance for Waste

When the total squares of roofing have been calculated, the estimator should make an allowance for waste based on the design of the roof. A minimum allowance of 3% to 5% is needed if the roof has two straight sides with two gable ends and no breaks. At the other extreme, any roof with several valleys, hips, and land ridges may need a waste allowance of 15% or more to cover the excess cutting required.

Factors for Converting Inclined to Horizontal

Roof Slope	Approx. Angle	Factor	Roof Slope	Approx. Angle	Factor
Flat	0	1.000	12 in 12	45.0	1.414
1 in 12	4.8	1.003	13 in 12	47.3	1.474
2 in 12	9.5	1.014	14 in 12	49.4	1.537
3 in 12	14.0	1.031	15 in 12	51.3	1.601
4 in 12	18.4	1.054	16 in 12	53.1	1.667
5 in 12	22.6	1.083	17 in 12	54.8	1.734
6 in 12	26.6	1.118	18 in 12	56.3	1.803
7 in 12	30.3	1.158	19 in 12	57.7	1.873
8 in 12	33.7	1.202	20 in 12	59.0	1.943
9 in 12	36.9	1.250	21 in 12	60.3	2.015
10 in 12	39.8	1.302	22 in 12	61.4	2.088
11 in 12	42.5	1.357	23 in 12	62.4	2.162

Example:

[20′ (1.302) 90′] 2 = 4,687.2 S.F. = 46.9 Sq.

OR

[40′ (1.302) 90′] = 4,687.2 S.F. = 46.9 Sq.

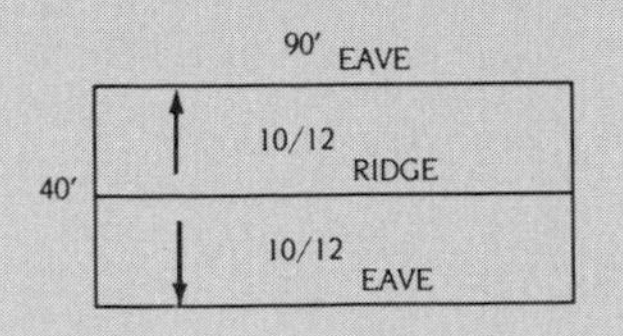

Figure 15.3

Related Work

Accessories that are part of a shingle roof include drip edges and flashings at chimneys, dormers, skylights, vents, valleys, and walls. These are items necessary to complete the roof and should be included in the estimate for the shingles.

Other Roofing Materials and Siding

In addition to shingles, many types of roofing and siding are used on commercial and industrial buildings. These are made of several kinds of material and are available in many forms for both roofing and siding. These include panels, sheets, membranes, and boards.

The materials used in roofing and siding panels include: aluminum, mineral fiber-cement, epoxy, fibrous glass, steel, vinyl, many types of synthetic sheets and membranes, coal tar, asphalt, tar felt and asphalt felt. Most of the latter materials are used in job-fabricated, built-up roofs and as backing for other materials, such as shingles. The basic data required for estimating either roofing or siding includes the specification of the material, the supporting structure, the method of installation, and the area to be covered. When selecting the current unit price for these materials, the estimator must remember that basic installed unit costs are *per square foot* for siding and *per square* for roofing. The major exceptions to this general rule are prefabricated roofing panels and single-ply roofing, which are priced per square foot.

Single-ply Roofs

Since the early 1970's the use of single-ply roofing (SPR) membranes in the construction industry has been on the rise. Market surveys have recently shown that of all the single-ply systems being installed, about one in three is on new construction. Use of SPR is also increasing in renovation and remodeling. Materially, these roofs are more expensive than other, more conventional roofs. However, labor costs are much lower because of faster installation. Re-roofing represents the largest market for single-ply roofing today. Single-ply roof systems are normally installed in one of the following ways.

Loose-laid and Ballasted

Generally this is the easiest type of single-ply roof to install. Some special consideration must be given, however, when flashing is attached to the roof. The membrane is typically fused together at the seams, stretched out flat and ballasted with stone (1-1/2" @ 10-12 PSF) to prevent wind blow-off. This extra load must be considered during design stages. It is particularly important if re-roofing over an existing built-up roof that already weighs 10-15 PSF. A slip-sheet or vapor barrier is sometimes required to separate the new roof from the old.

Partially-adhered

This method of installation uses a series of bar or point attachments which adhere the membrane to a substrate. The membrane manufacturer typically specifies the method to be used based on the material and substrate. Partially-adhered systems do not use ballast material. A slip-sheet may be required.

Fully-adhered

This is generally the most time-consuming of the single-plies to install, because these roofs employ a contact cement, cold adhesive, or hot bitumen to adhere the membrane uniformly to the substrate. Only manufacturer-approved insulation board or substrate should be used to receive the membrane. No ballast is required.

The most common single-ply materials available can be classified in three basic categories:

- Thermo-Setting: EDPM, Neoprene, and PIB
- Thermo-Plastic: Hypalon, PVC, and CPE
- Composites: Modified Bitumen

Each has its own requirements and performance characteristics. Most are available for all three installation methods.

Single-ply roof systems are available from many sources. However, most, if not all, manufacturers sell their materials only to franchised installers. As a result, there may be only one source for a price in any given area. Read the specifications carefully. Estimate the system required, exactly as specified; substitutes are usually not allowed.

Sheet Metal

- Copper and Stainless Steel
- Gutters and Downspouts
- Edge Cleats and Gravel Stops
- Flashings
- Trim
- Miscellaneous

Sheet metal work included in this division is limited to that used on roofs or sidewalls of buildings, usually on the exterior exposed to the weather. Many of the items covered are wholly or partially prefabricated, with labor added for installation. Several are materials that require labor added for on-site fabrication; this cost must be estimated separately.

Pricing shop-made items such as downspouts, drip edges, expansion joints, gravel stops, gutters, reglets, and termite shields requires that the estimator determine the material, size, and shape of the fabricated section, and the linear feet of the item.

The cost of items like copper roofing and metal flashing is estimated in a similar manner, except that unit costs are per square foot. Some roofing systems, particularly single-ply, require flashing materials that are unique to that roofing system.

Roofing materials like monel, stainless steel, and zinc copper alloy are also estimated by the same method, except that the unit costs are per square (100 square feet). Prefabricated items like strainers and louvers are priced on a cost-per-unit basis. Adhesives are priced by the gallon. The installed cost of roofing adhesives depends on the cost per gallon and the coverage per gallon. With trowel grade adhesive, the coverage varies from a light coating at 25 S.F. per gallon to a heavy coating at 10 S.F. per gallon. With most flashing work, the asphalt adhesive covers an average of 15 S.F. per gallon for each layer or course. Many specifications state coverage of special materials like adhesives. This information should be used as the basis for the estimate.

Roof Accessories

- Hatches
- Skylights
- Vents
- Snow Guards

Roof accessories must be considered as part of the complete weatherproofing system. Standard size accessories, such as ceiling, roof and smoke vents or hatches, and snow guards, are priced per installed unit. Accessories that must be fabricated to meet project specifications may be priced per square foot, per linear foot or per unit. Skylight costs, for example, are listed by the square foot, with unit costs decreasing in steps as the nominal size of individual units increases. Some types of skylights and hatches are shown in Figure 15.4.

Skyroofs are priced on the same basis, but due to the many variations in the shape and construction of these units, costs are per square foot of surface area. These costs will vary with the size and type of unit, and in many cases, maximum and minimum costs give the estimator a range of prices for different design variations. Because there are many types and styles, the estimator must determine the exact specifications for the skyroof being priced. The accuracy of the total cost figure will depend entirely on the selection of the proper unit cost and calculation of the skyroof area. Skyroofs are becoming widely used in the industry and the work is growing more and more specialized. Often a particular manufacturer is specified. Specialty installing subcontractors are often factory-authorized and required to perform the installation to maintain warranties and waterproof integrity.

Accessories such as roof drains, plumbing vents, and duct penetrations are usually installed by other appropriate subcontractors. However, costs for flashing and sealing these items are often included by the roofing subcontractor.

When estimating Division 7, associated costs must be included for items that may not be directly stated in the specifications. Placement of materials, for example, may require the use of a crane or conveyors. Pitch pockets, sleepers, pads, and walkways may be required for rooftop equipment. Blocking and cant strips, and items associated with different trades must also be coordinated. Once again, the estimator must visualize the construction process.

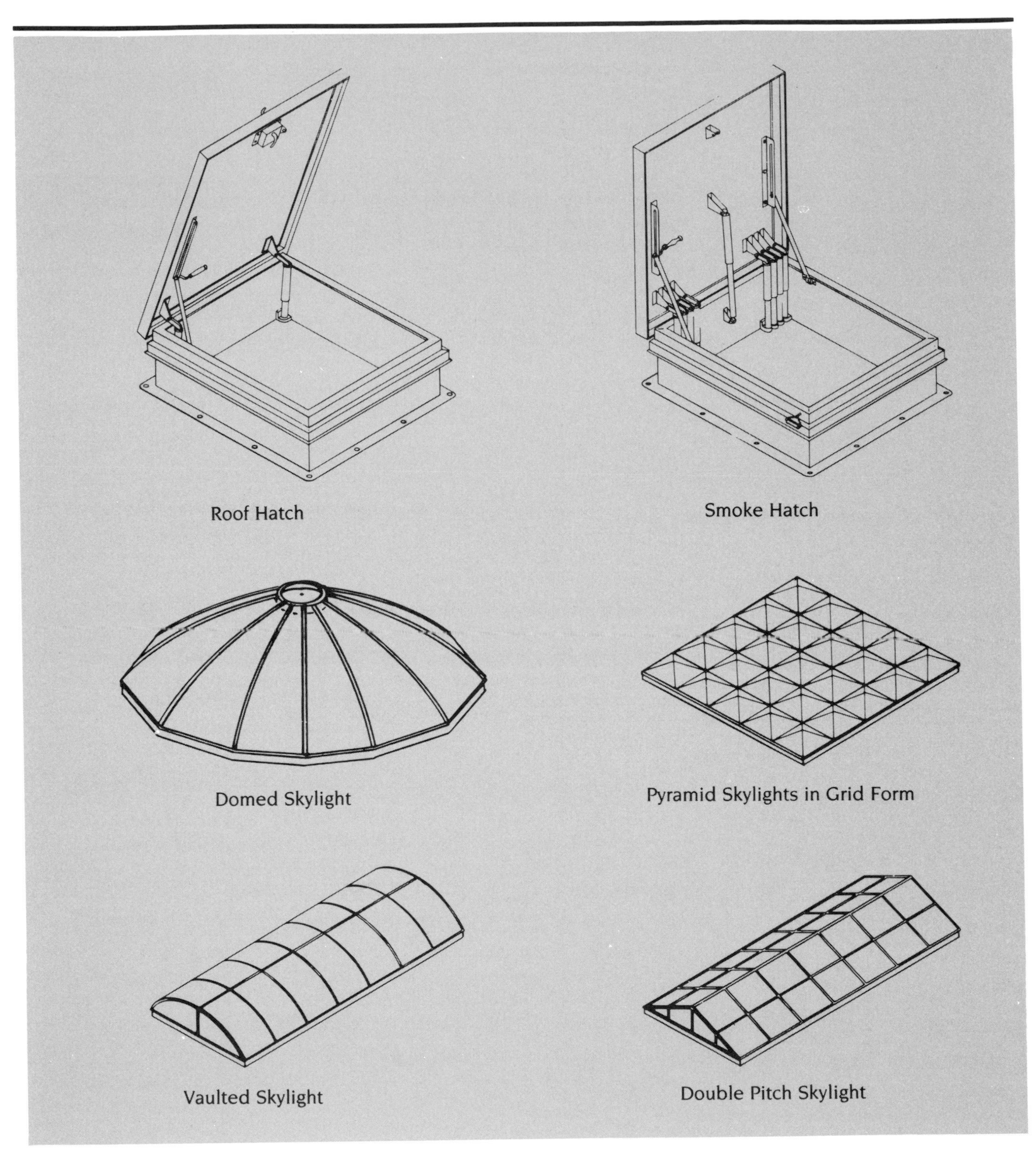
Roof Hatch
Smoke Hatch
Domed Skylight
Pyramid Skylights in Grid Form
Vaulted Skylight
Double Pitch Skylight

Figure 15.4

Chapter 16

DOORS, WINDOWS, AND GLASS

The cost of doors, windows, and glass can represent a sizeable portion of a total remodeling project. Windows and doors require hardware, an additional material cost. Any one door assembly (door, frame, hardware) can be a combination of the variable features listed below:

Door	Frame	Hardware
Size	Size	Lockset
Thickness	Throat	Passage set
Wood-type	Wood-type	Panic bar
Metal gauge	Metal gauge	Closer
Laminate	Casing	Hinges
Hollow-core type	Stops	Stops
Solid-core material	Fire rating	Bolts
Fire rating	Knock-down	Finish
Finish	Welded	Plates

When available, most architectural plans and specifications usually include door, window, and hardware schedules, listing the items needed for the required combinations. The estimator should use these schedules and details in conjunction with the plans to avoid duplication or omission of units when determining the quantities. The schedules should identify the location, size, and type of each unit. Schedules should also include information regarding the frame, fire rating, hardware, and special notes. If no such schedules are included, the estimator should prepare them in order to provide an accurate quantity takeoff. Figure 16.1 is an example of a schedule that may be prepared by the estimator. Most suppliers prepare separate schedules; each must be approved by the architect or owner.

A proper door schedule on the architectural drawings identifies each opening in detail. The estimator should define each opening using the schedule and any other pertinent data. Installation information should be carefully reviewed in the specifications.

For the quantity takeoff, the numbers of all similar doors and frames should be combined. Each should be checked off on the plans to ensure that none have been left out. An easy and obvious

Means Forms

DOOR
AND FRAME SCHEDULE

PAGE ____ OF ____

PROJECT ____ ARCHITECT ____ DATE ____

LOCATION ____ OWNER ____ BY ____

	DOOR NO.	DOOR							FRAME					FIRE RATING		HARDWARE		REMARKS
		SIZE			MAT.	TYPE	GLASS	LOUVER	MAT.	TYPE	DETAILS					SET NO.	KEYSIDE ROOM NO.	
		W	H	T							JAMB	HEAD	SILL	LAB	CON			

Figure 16.1

double-check is to count the total number of openings, making certain that two doors and only one frame have been included where double doors are used. Important details to check for both door and frame are the following:

- Material
- Gauge
- Size
- Core Material
- Fire Rating Label
- Finish
- Style

Wood Doors

Wood doors are manufactured in either flush or paneled designs, and are separated into three grades: *architectural/commercial*, *residential*, and *decorator*. A wide variety of frames are available for remodeling installations in metal, pine, hardwood, and for various partition thicknesses. Some doors are available pre-hung in frames for quick installation.

Architectural or Commercial Wood Doors

This is the type of door most often specified in commercial renovation. The stiles are made of hardwood, and the core is dense and of hot-bonded construction. They feature thick face veneers that are exterior-glued and matched in their grain patterns. Because of durability, this grade of door often carries a lifetime warranty.

Residential Wood Doors

These doors are chosen for low-frequency use where economy is a primary consideration. The stiles are manufactured from soft wood; the core from low density materials. The face veneers are thin, interior-glued, and broken in their grain patterns.

Decorator Wood Doors

These doors are manufactured from solid wood and are usually hand carved. Because of the choice woods used and the special craftsmanship required in their production, their cost is several times that of similar size architectural wood doors.

Flush Doors

The cores of these doors are produced in varying densities: hollow, particle-board, or veneer core. Lauan mahogany, birch, oak, or other hardwood veneers are used for facings. Synthetic veneers, created from a medium-density overlay or high-pressure plastic laminate, may serve as an alternate choice to natural wood veneers. Flush wood doors may be fire-rated. Door swings and wood door details are shown in Figure 16.2.

Fire Doors

The estimator must pay particular attention to fire door specifications when performing the quantity takeoff. It is important to determine the exact type of door required. Figure 16.3 is a table describing various types of fire doors. Please note that a "B" label door can be one of four types. If the plans or door schedule do not specify exactly which temperature rise is required, the estimator should consult the architect or local building inspector. Many building and fire codes also require that frames and hardware at fire

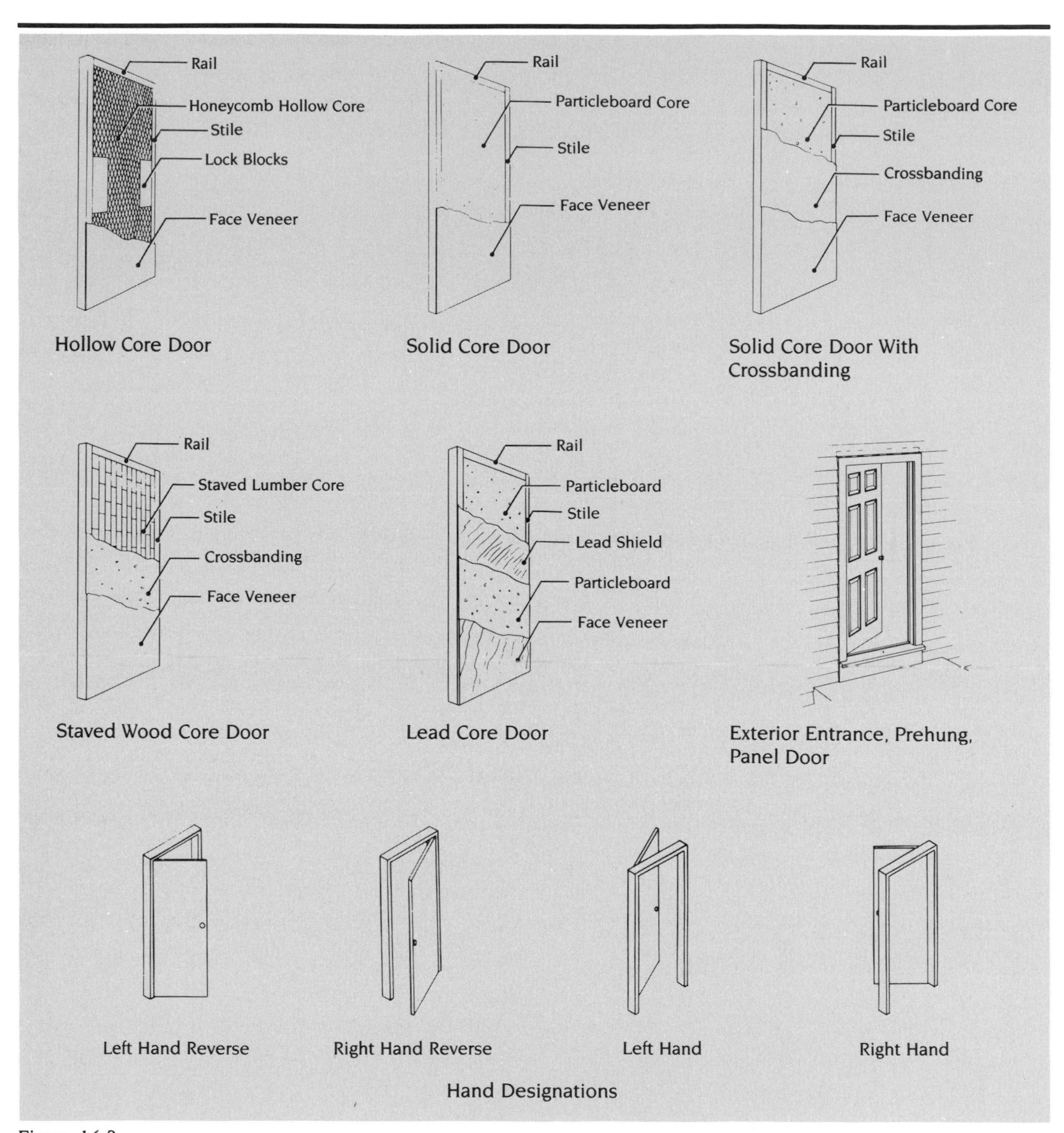

Rail
Honeycomb Hollow Core
Stile
Lock Blocks
Face Veneer
Hollow Core Door
Rail
Particleboard Core
Stile
Face Veneer
Solid Core Door
Rail
Particleboard Core
Stile
Crossbanding
Face Veneer
Solid Core Door With Crossbanding
Rail
Staved Lumber Core
Stile
Crossbanding
Face Veneer
Staved Wood Core Door
Rail
Particleboard
Stile
Lead Shield
Particleboard
Face Veneer
Lead Core Door
Exterior Entrance, Prehung, Panel Door
Left Hand Reverse
Right Hand Reverse
Left Hand
Right Hand
Hand Designations

Figure 16.2

doors also be fire-rated and labeled as such. When determining quantities, the estimator must also include any glass (usually wired) or special inserts to be installed in fire doors (or in any doors).

Metal Doors and Frames

Hollow Metal Doors

Hollow metal doors are available in stock or custom fabrication, flush or embossed, with glazing or louvres, labeled or unlabeled, and in various steel gauges and core fills. Stock doors may be supplied for low-, moderate-, or high-frequency use from some manufacturers. The doors are available in widths of 2′ to 4′ and heights varying from 6′-6″ to 10′ or more. They may be used as single doors, in pairs with

Fire Door

Classification	Time Rating (as Shown on Label)	Temperature Rise (as Shown on Label)	Maximum Glass Area
3 Hour fire doors (A) are for use in openings in walls separating buildings or dividing a single building into the areas.	3 Hr. (A) 3 Hr. (A) 3 Hr. (A) 3 Hr. (A)	30 Min. 250°F Max 30 Min. 450°F Max 30 Min. 650°F Max *	None
1½ Hour fire doors (B) and (D) are for use in openings in 2 Hour enclosures of vertical communication through buildings (stairs, elevators, etc.) or in exterior walls which are subject to severe fire exposure from outside of the building. 1 Hour fire doors (B) are for use in openings in 1 Hour enclosures of vertical communication through buildings (stairs, elevators, etc.)	1½ Hr. (B) 1½ Hr. (B) 1½ Hr. (B) 1½ Hr. (B) 1 Hr. 1½ Hr. (D) 1½ Hr. (D) 1½ Hr. (D) 1½ Hr. (D)	30 Min. 250°F Max 30 Min. 450°F Max 30 Min. 650°F Max * 30 Min. 250°F Max 30 Min. 250°F Max 30 Min. 450°F Max 30 Min. 650°F Max *	100 square inches per door None
3/4 Hour fire doors (C) and (E) are for use in openings in corridor and room partitions or in exterior walls which are subject to moderate fire exposure from outside of the building.	3/4 Hr. (C)	**	1296 Square
	3/4 Hr. (E)	**	720 square inches per light
1/2 Hour fire doors and 1/3 Hour fire doors are for use where smoke controls is a primary consideration and are for the protection of openings in partitions between a habitable room and a corridor when the wall has a fire-resistance rating of not more than one hour.	1/2 Hr. 1/3 Hr.	** **	No limit

*The labels do not record any temperature rise limits. This means that the temperature rise on the unexposed face of the door at the end of 30 minutes of test is in excess of 650°F.

**Temperature rise is not recorded.

Figure 16.3

both leaves active, and in pairs with one active leaf, including an astragal. Bi-fold hollow metal doors are available for specified applications.

Hollow metal doors are reinforced at the stress points and pre-mortised for the hardware required for the door application. Hollow metal, labelled fire doors can be supplied stock or custom-manufactured with A, B, C, D, E labels, with 3/4 to 3-hour ratings, depending on the glass area, height and width restrictions, and maximum expected temperature rise (shown in Figure 16.3). The door types shown in Figure 16.4 are examples of typical metal fire doors. Code requirements for fire doors and ratings vary from state to state and often from city to city.

Hollow Metal Frames

Hollow metal frames may be supplied in 14, 16, or 18-gauge galvanized or plain steel in knock-down standard frames or welded customized frames that can be fabricated to satisfy most design conditions. Frames with borrowed lights, transoms, or cased openings are available in stick components from some manufacturers.

Frames may be wraparound (enclosing the wall) or may butt up against the opening. A wraparound frame may terminate into the enclosed wall when it is covered by a finish such as plaster, or the frame may return along the enclosed wall when it is exposed, as in drywall construction. Frames are sometimes supplied in two pieces to suit varied wall thicknesses, or in one piece to satisfy standard wall thicknesses. Frames are normally reinforced at stress points and are prepared for hinges and strikes. Anchors to attach the frame to the wall are supplied to suit wall construction requirements. Custom frames normally require a hardware schedule and templates to produce required shop drawings and to accomplish fabrication. Typical hollow metal frames for various applications are shown in Figure 16.5.

Special Doors

Metal Access Panels and Doors

These doors are available in steel or stainless steel for fire-rated or non-fire-rated applications. Panels are fabricated for flush installations in drywall (both skim-coated or taped), for masonry and tile applications, for plastered walls and ceilings, and for acoustical ceilings. These doors are available in stock sizes and types to suit most applications.

Blast Doors

These doors are available in standard designs, and may also be custom-designed to withstand specified pressures and to resist penetration.

Cold-storage Doors

This type of door is available in standard designs in wood, steel, fiberglass, plastic, and stainless steel – for all types of cold storage requirements. These doors are manufactured to provide insulation for cool zones, coolers, and freezers for manual, air, electric, or hydraulic operations. These doors may operate in any of the following ways: sliding, vertical lift, bi-parting overhead, and single- and double-swing.

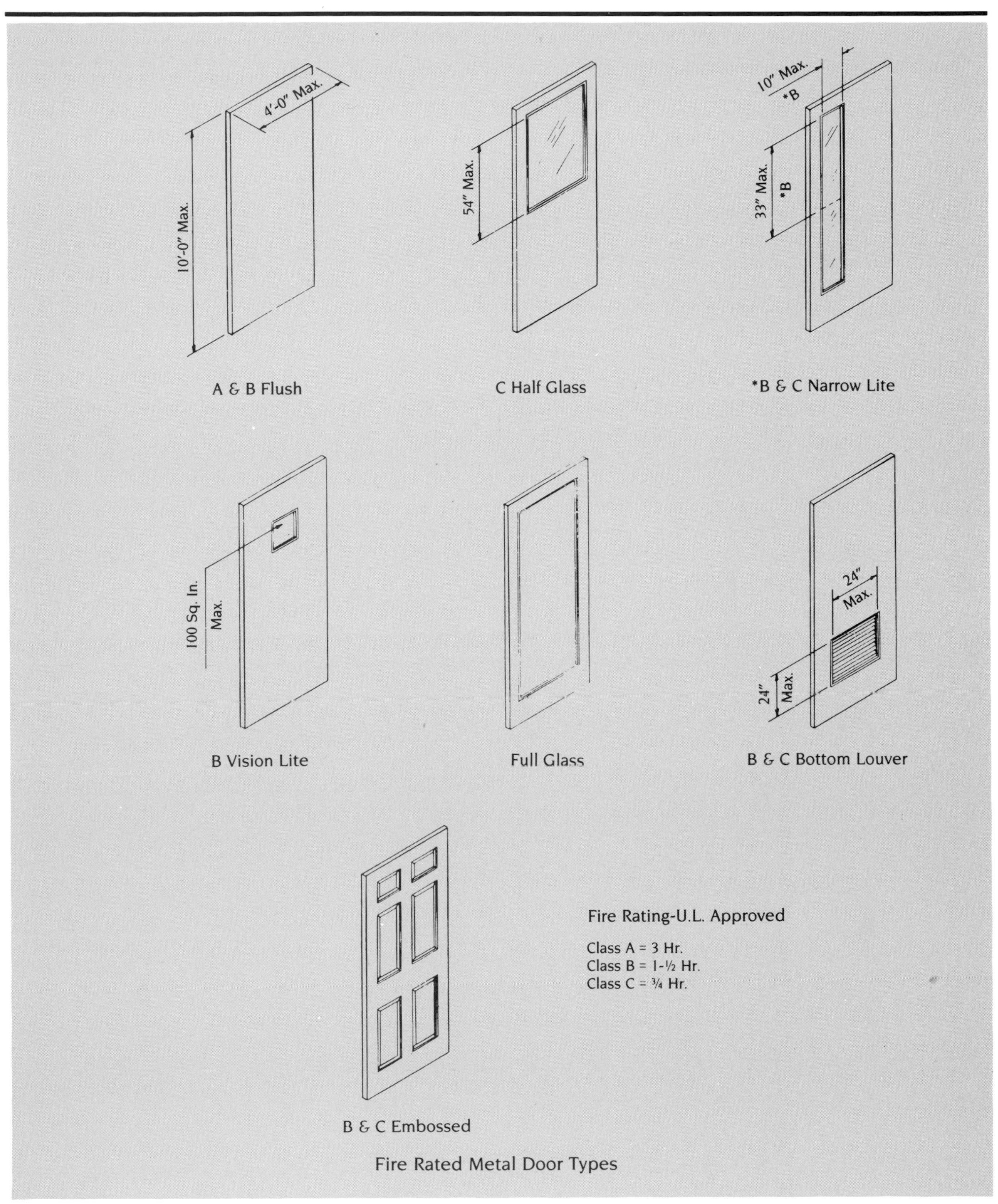
4'-0" Max.
10'-0" Max.
A & B Flush
54" Max.
C Half Glass
10" Max.
*B
33" Max.
*B
*B & C Narrow Lite
100 Sq. In.
Max.
B Vision Lite
Full Glass
24"
Max.
24"
Max.
B & C Bottom Louver
B & C Embossed
Fire Rating-U.L. Approved
Class A = 3 Hr.
Class B = 1-½ Hr.
Class C = ¾ Hr.
Fire Rated Metal Door Types

Figure 16.4

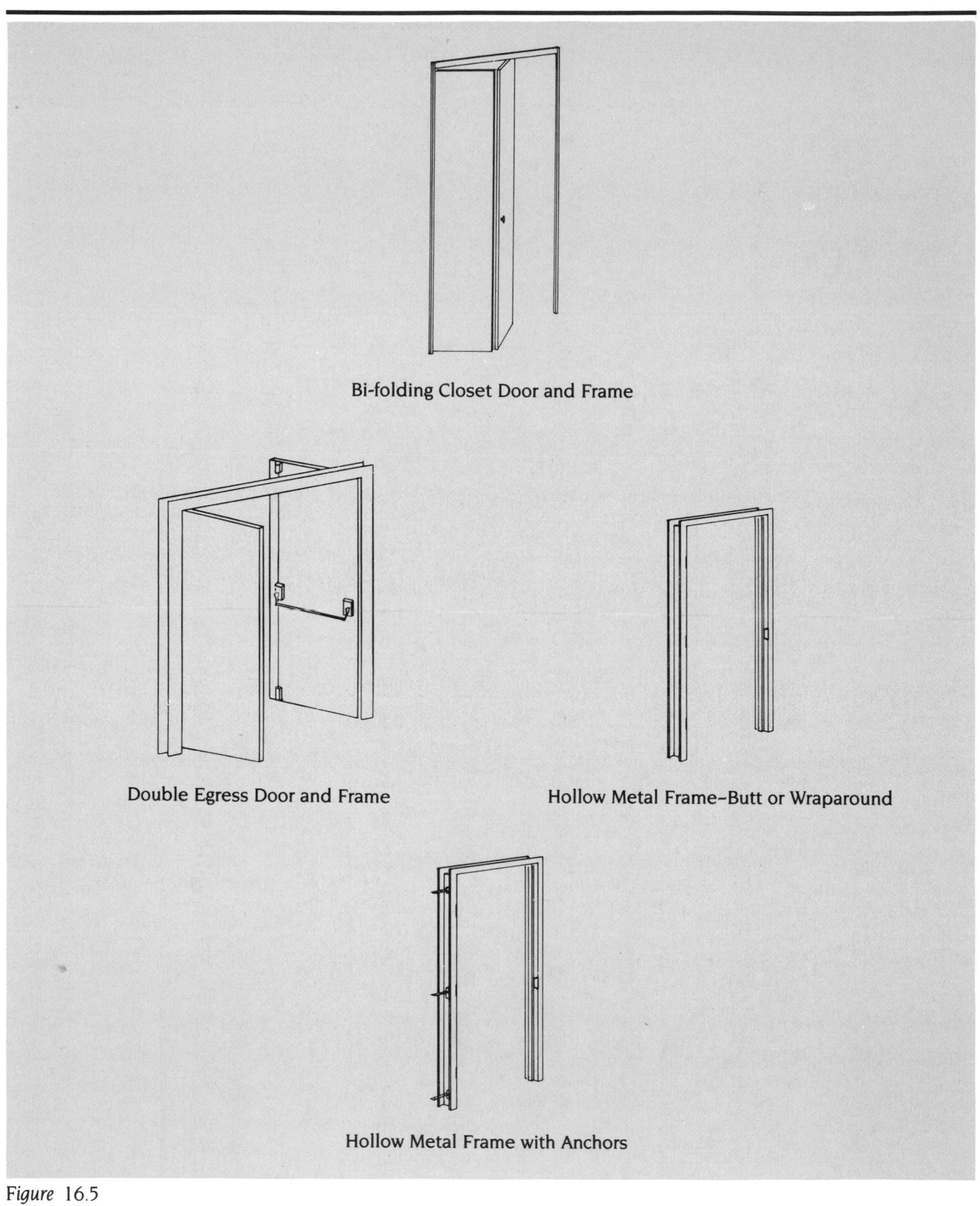
Bi-folding Closet Door and Frame
Double Egress Door and Frame
Hollow Metal Frame–Butt or Wraparound
Hollow Metal Frame with Anchors

Figure 16.5

Pass windows (vertical and horizontal sliding), rotating shelf windows, ticket windows, cashier doors, and coin and cash trays are available in aluminum, steel, or stainless steel. Pass windows, roll-up shutters, and projection booth shutters are available, labelled or non-labelled, with fusible links to suit most applications and requirements. A roll-up shutter and a roll-up gate are shown in Figure 16.6.

Glass and Glazing

Glass and glazing in remodeling construction most often includes interior glazed partitions, entrances, and storefronts. Interior glazed partitions are commonly constructed of tubular aluminum framing. Glazing subcontractors may estimate such partitions by measuring and pricing the length of each component of the frame separately. Examples are shown in Figures 16.7 and 16.8. The glass can be plate, tempered, safety, tinted, insulated, or combinations thereof, depending on project and code requirements. Glass is estimated by the square foot or by united length (length plus width). Glass doors

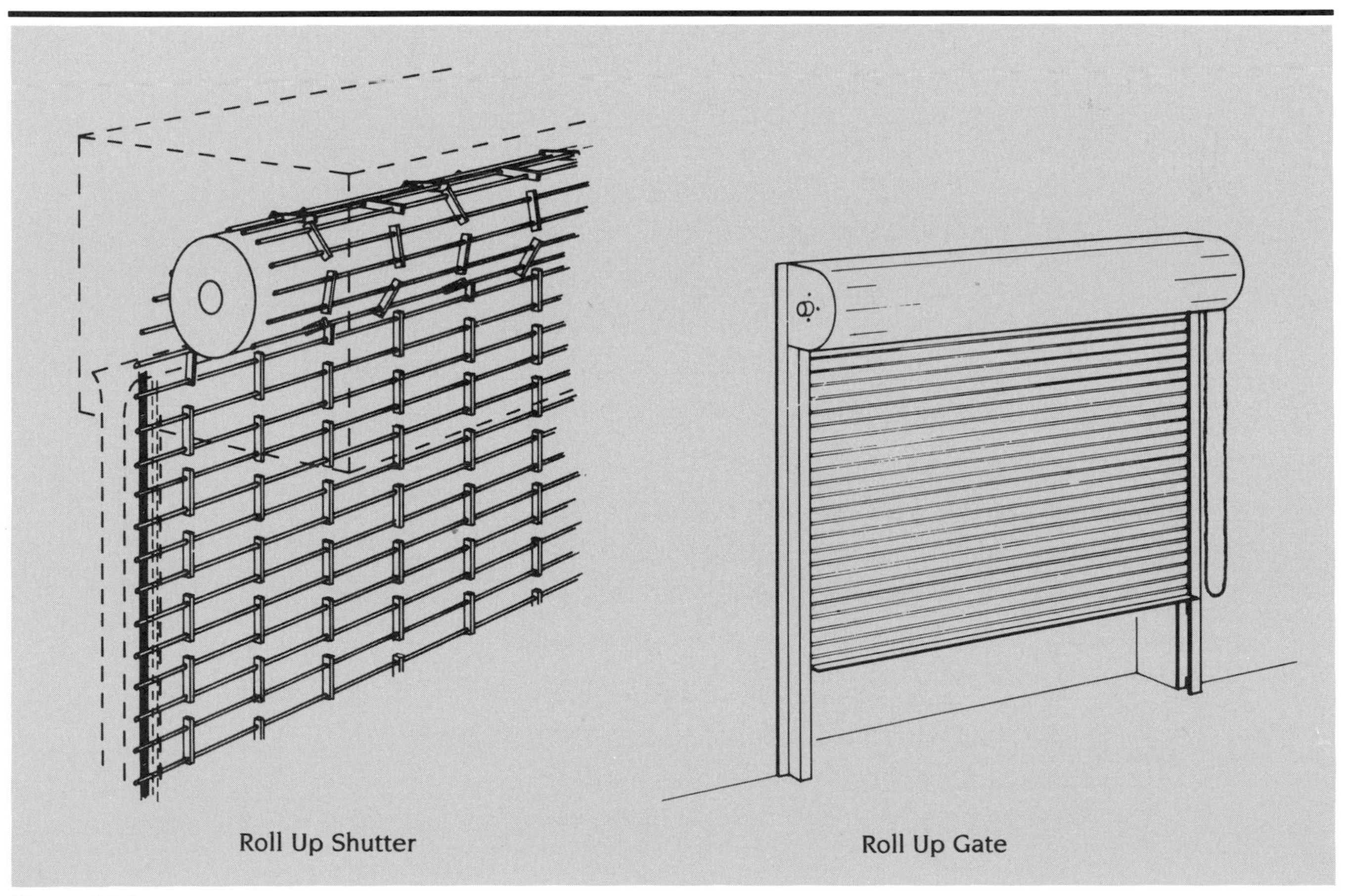

Figure 16.6

and hardware may also be estimated separately. Another method of estimating is to use complete system prices, whether by square foot (Figure 16.9) or by the opening.

Entrances and storefronts are almost all special designs and combinations of unit items to fit a unique situation. The estimator should submit the plans and specifications to a specialty installer for takeoff and pricing. Typical entrances are shown in Figure 16.10. The general procedure for the installer's takeoff is:

For stationary units:

- Determine the height and width of each like unit.
- Determine the linear feet of intermediate, horizontal, and vertical members, rounded to the next higher foot.
- Determine the number of joints.

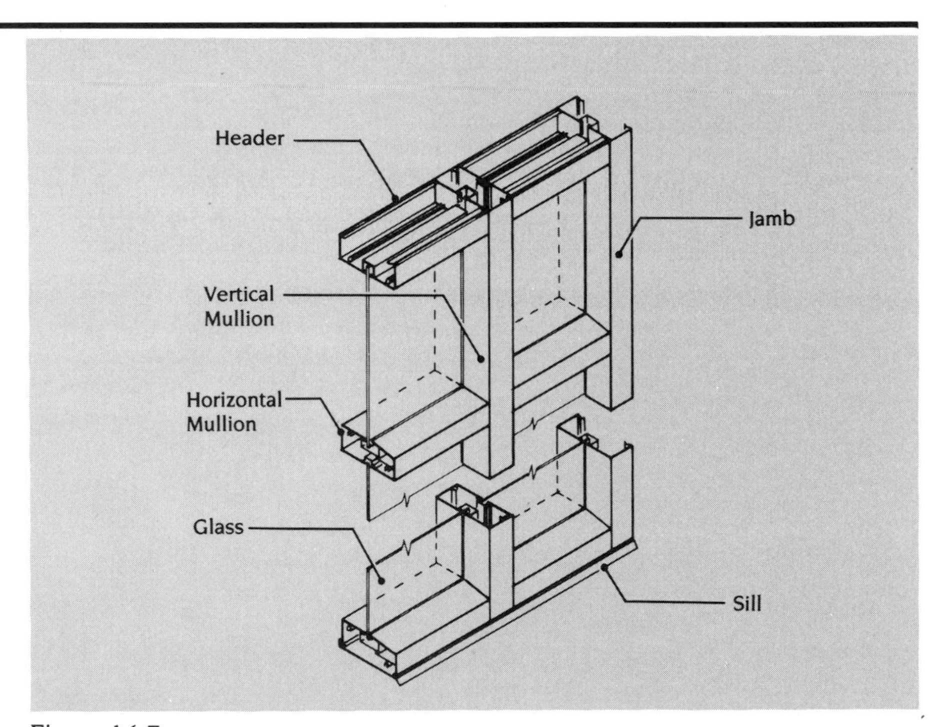

Figure 16.7

088 | Glazing

		088 100 | Glass	CREW	DAILY OUTPUT	MAN-HOURS	UNIT	BARE COSTS MAT.	BARE COSTS LABOR	BARE COSTS EQUIP.	BARE COSTS TOTAL	TOTAL INCL O&P	
132	3370	Reflective or tinted, add				S.F.	2.70			2.70	2.97	132
	9000	Minimum labor/equipment charge	1 Glaz	2	4	Job		88		88	140	
160	0010	**REFLECTIVE GLASS** ¼″ float with fused metallic oxide, tinted	2 Glaz	115	.139	S.F.	6.95	3.07		10.02	12.45	160
	0500	¼″ float glass with reflective applied coating		115	.139		4.50	3.07		7.57	9.75	
	2000	Solar film on glass, not including glass, minimum		180	.089		1.05	1.96		3.01	4.23	
	2050	Maximum		225	.071		2.05	1.57		3.62	4.72	
176	0010	**WINDOW GLASS** Clear float, stops, putty bed, ⅛″ thick		480	.033		2.60	.74		3.34	4.01	176
	0500	3/16″ thick, clear		480	.033		2.75	.74		3.49	4.18	
	0600	Tinted		480	.033		3.10	.74		3.84	4.56	
	0700	Tempered		480	.033		5.25	.74		5.99	6.95	
	2000	Replace broken window lite, ⅛″ glass (9 S.F. maximum)	1 Glaz	48	.167		2.50	3.68		6.18	8.50	
	2100	¼″ plate (16 S.F. maximum)	"	48	.167		2.65	3.68		6.33	8.70	
	9000	Minimum labor/equipment charge	2 Glaz	5	3.200	Job		71		71	110	

		088 400 | Plastic Glazing	CREW	DAILY OUTPUT	MAN-HOURS	UNIT	BARE COSTS MAT.	BARE COSTS LABOR	BARE COSTS EQUIP.	BARE COSTS TOTAL	TOTAL INCL O&P	
408	0010	**POLYCARBONATE** Clear, masked, cut sheets, ⅛″ thick	2 Glaz	170	.094	S.F.	2.80	2.08		4.88	6.35	408
	0500	3/16″ thick		165	.097		3.75	2.14		5.89	7.50	
	1000	¼″ thick		155	.103		4.70	2.28		6.98	8.75	
	1500	⅜″ thick		150	.107		7.80	2.35		10.15	12.25	
	9000	Minimum labor/equipment charge	1 Glaz	2	4	Job		88		88	140	

089 | Glazed Curtain Wall

		089 200 | Glazed Curtain Wall	CREW	DAILY OUTPUT	MAN-HOURS	UNIT	BARE COSTS MAT.	BARE COSTS LABOR	BARE COSTS EQUIP.	BARE COSTS TOTAL	TOTAL INCL O&P	
204	0010	**TUBE FRAMING** For window walls and store fronts, aluminum, stock										204
	0020											
	0050	Plain tube frame, mill finish, 1-¾″ x 1-¾″	2 Glaz	90	.178	L.F.	3.50	3.92		7.42	10	
	0150	1-¾″ x 4″		90	.178		6.35	3.92		10.27	13.15	
	0200	1-¾″ x 4-½″		90	.178		6.75	3.92		10.67	13.60	
	0250	2″ x 6″		90	.178		9.75	3.92		13.67	16.90	
	0350	4″ x 4″		90	.178		8.95	3.92		12.87	16	
	0400	4-½″ x 4-½″		90	.178		11.40	3.92		15.32	18.70	
	0450	Glass bead		240	.067		1.50	1.47		2.97	3.96	
	1000	Flush tube frame, mill finish, ¼″ glass, 1-¾″ x 4″, open header		80	.200		4.75	4.41		9.16	12.15	
	1050	Open sill		82	.195		4.80	4.30		9.10	12.05	
	1100	Closed back header		83	.193		5.95	4.25		10.20	13.20	
	1150	Closed back sill		85	.188		6.50	4.15		10.65	13.65	
	1200	Vertical mullion, one piece		75	.213		6.50	4.70		11.20	14.55	
	1250	Two piece		73	.219		7.15	4.83		11.98	15.45	
	1300	90° or 180° vertical corner post		75	.213		11.35	4.70		16.05	19.85	
	1400	1-¾″ x 4-½″, open header		80	.200		5.25	4.41		9.66	12.70	
	1450	Open sill		82	.195		5.30	4.30		9.60	12.60	
	1500	Closed back header		83	.193		6.25	4.25		10.50	13.55	
	1550	Closed back sill		85	.188		7.75	4.15		11.90	15.05	
	1600	Vertical mullion, one piece		75	.213		6.60	4.70		11.30	14.65	
	1650	Two piece		73	.219		7.75	4.83		12.58	16.10	
	1700	90° or 180° vertical corner post		75	.213		12.30	4.70		17	21	
	2000	Flush tube frame, mill fin. for ins. glass, 2″ x 4-½″, open header		75	.213		6.25	4.70		10.95	14.25	
	2050	Open sill		77	.208		6.60	4.58		11.18	14.45	
	2100	Closed back header		78	.205		7.75	4.52		12.27	15.65	

For expanded coverage of these items see *Means Interior Cost Data 1989* 137

Figure 16.8

083 | Special Doors

		083 750 Swing Doors	CREW	DAILY OUTPUT	MAN-HOURS	UNIT	BARE COSTS MAT.	LABOR	EQUIP.	TOTAL	TOTAL INCL O&P	
752	2050	7′ wide	2 Carp	3.80	4.210	Pr.	1,030	90		1,120	1,275	752
	9000	Minimum labor/equipment charge	"	2	8	Job		170		170	275	
754	0010	GLASS, SWING Tempered, ½″ thick, incl. hardware, 3′ x 7′ opening	2 Glaz	2	8	Opng.	1,810	175		1,985	2,275	754
	0100	6′ x 7′ opening	↓	1.40	11.430	"	3,495	250		3,745	4,250	
	9000	Minimum labor/equipment charge	↓	2	8	Job		175		175	275	

		083 800 Sound Retardant Doors	CREW	DAILY OUTPUT	MAN-HOURS	UNIT	MAT.	LABOR	EQUIP.	TOTAL	TOTAL INCL O&P	
804	0010	ACOUSTICAL Incl. framed seals, 3′ x 7′, wood, 27 STC rating	F-2	1.50	10.670	Ea.	425	230	10	665	845	804
	0100	Steel, 40 STC rating		1.50	10.670		2,190	230	10	2,430	2,775	
	0200	45 STC rating		1.50	10.670		2,350	230	10	2,590	2,950	
	0300	48 STC rating		1.50	10.670		2,465	230	10	2,705	3,100	
	0400	52 STC rating	↓	1.50	10.670	↓	2,640	230	10	2,880	3,275	
	9000	Minimum labor/equipment charge	F-1	4	2	Job		43	1.88	44.88	71	

		083 900 Screen And Storm Doors	CREW	DAILY OUTPUT	MAN-HOURS	UNIT	MAT.	LABOR	EQUIP.	TOTAL	TOTAL INCL O&P	
904	0010	STORM DOORS & FRAMES Aluminum, residential,										904
	0020	combination storm and screen										
	0400	Clear anodic coating, 6′-8″ x 2′-6″ wide	F-2	15	1.070	Ea.	145	23	1	169	195	
	0420	2′-8″ wide		14	1.140		150	24	1.07	175.07	205	
	0440	3′-0″ wide	↓	14	1.140		160	24	1.07	185.07	215	
	0500	For 7′-0″ door, add					5%					
	1000	Mill finish, 6′-8″ x 2′-6″ wide	F-2	15	1.070		135	23	1	159	185	
	1020	2′-8″ wide		14	1.140		135	24	1.07	160.07	190	
	1040	3′-0″ wide	↓	14	1.140		140	24	1.07	165.07	195	
	1100	For 7′-0″ door, add					5%					
	1500	White painted, 6′-8″ x 2′-6″ wide	F-2	15	1.070		148	23	1	172	200	
	1520	2′-8″ wide		14	1.140		152	24	1.07	177.07	210	
	1540	3′-0″ wide	↓	14	1.140		154	24	1.07	179.07	210	
	1600	For 7′-0″ door, add				↓	5%					
	2000	Wood door & screen, see division 082-078										
	2020											
	9000	Minimum labor/equipment charge	F-1	4	2	Job		43	1.88	44.88	71	

084 | Entrances and Storefronts

		084 100 Aluminum	CREW	DAILY OUTPUT	MAN-HOURS	UNIT	BARE COSTS MAT.	LABOR	EQUIP.	TOTAL	TOTAL INCL O&P	
103	0010	BALANCED DOORS Hdwre & frame, alum. & glass, 3′ x 7′, econ.	2 Sswk	.90	17.780	Ea.	2,375	415		2,790	3,325	103
	0150	Premium		.70	22.860	"	4,150	530		4,680	5,475	
	9000	Minimum labor/equipment charge	↓	1	16	Job		370		370	645	
105	0010	STOREFRONT SYSTEMS Aluminum frame, clear ⅜″ plate glass,										105
	0020	incl. 3′ x 7′ door with hardware (400 sq. ft. max. wall)										
	0500	Wall height to 12′ high, commercial grade	2 Glaz	150	.107	S.F.	11.55	2.35		13.90	16.40	
	0600	Institutional grade		130	.123		14.95	2.71		17.66	21	
	0700	Monumental grade		115	.139		21.50	3.07		24.57	28	
	1000	6′ x 7′ door with hardware, commercial grade		135	.119		10.65	2.61		13.26	15.80	
	1100	Institutional grade		115	.139		15.95	3.07		19.02	22	
	1200	Monumental grade	↓	100	.160		22.25	3.53		25.78	30	
	1500	For bronze anodized finish, add					17%					
	1600	For black anodized finish, add				↓	25%					

122

For expanded coverage of these items see *Means Interior Cost Data 1989*

Figure 16.9

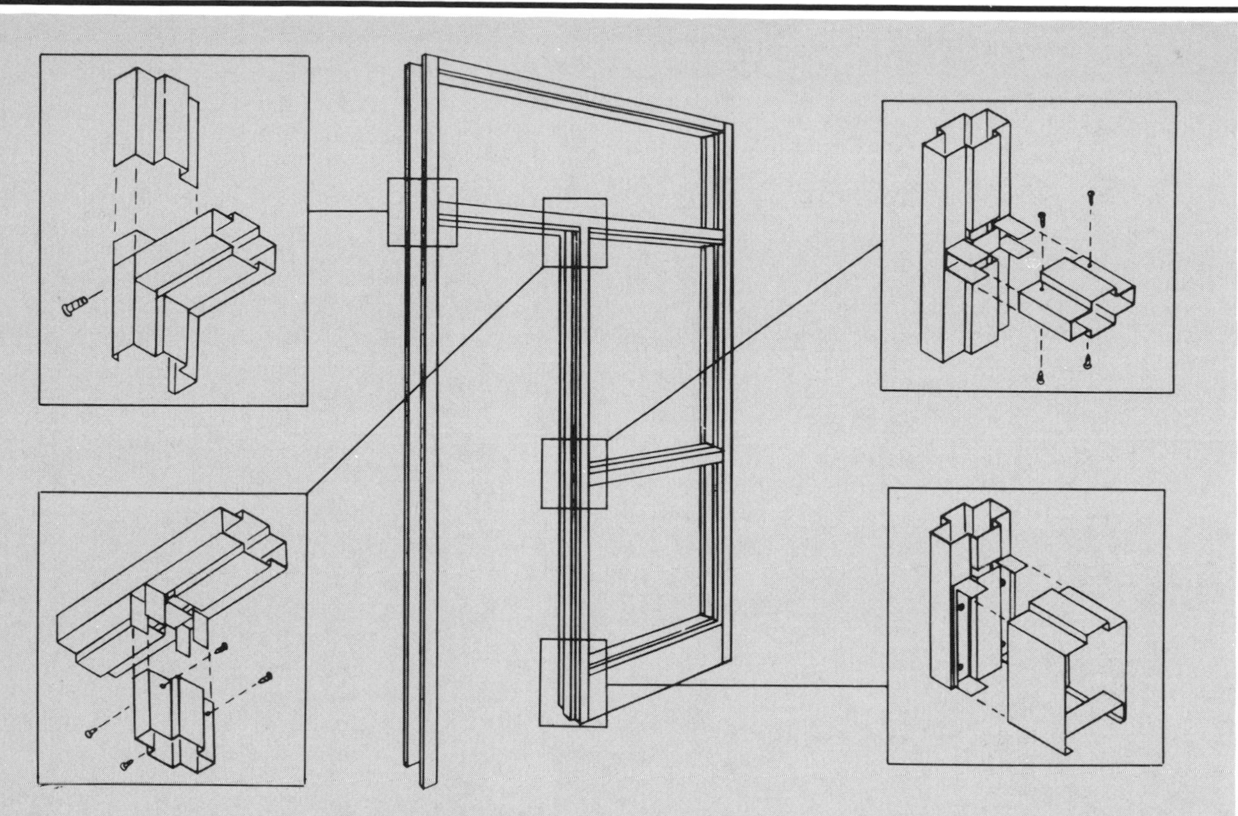

Pre-Engineered, "Stick" System Entrance

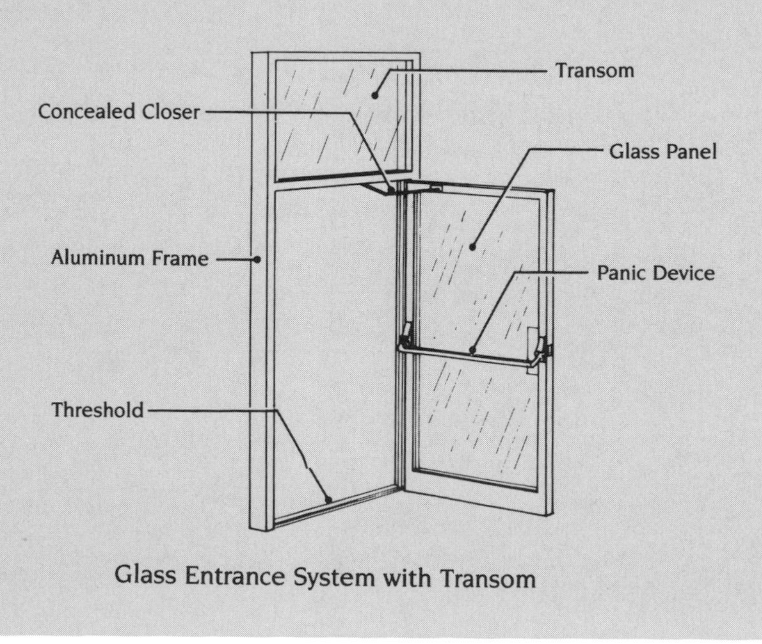

Glass Entrance System with Transom

Figure 16.10

For entrance units:

- Determine the number of joints
- Determine special frame hardware per unit.
- Determine special door hardware per unit.
- Determine thresholds and closers.

Curtain Walls

A curtain wall is a non-structural facade consisting of panels in a wide variety of materials and constructions. A curtain wall is held in place in a metal frame by caulking, gaskets, and sealants. The curtain wall can be prefabricated to the following degree depending on the type.

- Stick, in which all components are field assembled
- Panel and mullion, in which the panels are prefabricated into frames and field-connected to mullions
- Total panel systems, in which the mullions are pre-assembled into the panels

Figure 16.11 shows different types of wall systems.

Finish Hardware

Finish hardware is the construction industry term for the devices used to operate doors, windows, drawers, shutters, closets, and cabinets. This category includes such items as hinges, latches, locks, panic devices, security and detection systems, astragals, and weatherstripping. Some examples are shown in Figure 16.12a and 16.12b. In a typical building, the finish hardware accounts for between 2% and 3% of the total job cost. Consequently, the difference between economy and quality hardware can mean a 1% difference in the total building cost.

A hardware specialist will often prepare a schedule and specify the hardware that is to be used for each opening. There are two general classifications of hardware: *Builder's*, and *Commercial*. *Builder's* is generally used for residential construction.

Another hardware specification is based on frequency of use: *heavy*, *light*, or *medium*. The size, weight, and material of a door and its frame, for example, dictate the size and number of hinges. Building codes and security requirements determine the selection of the proper fire barrier and electronic hardware in different remodeling situations.

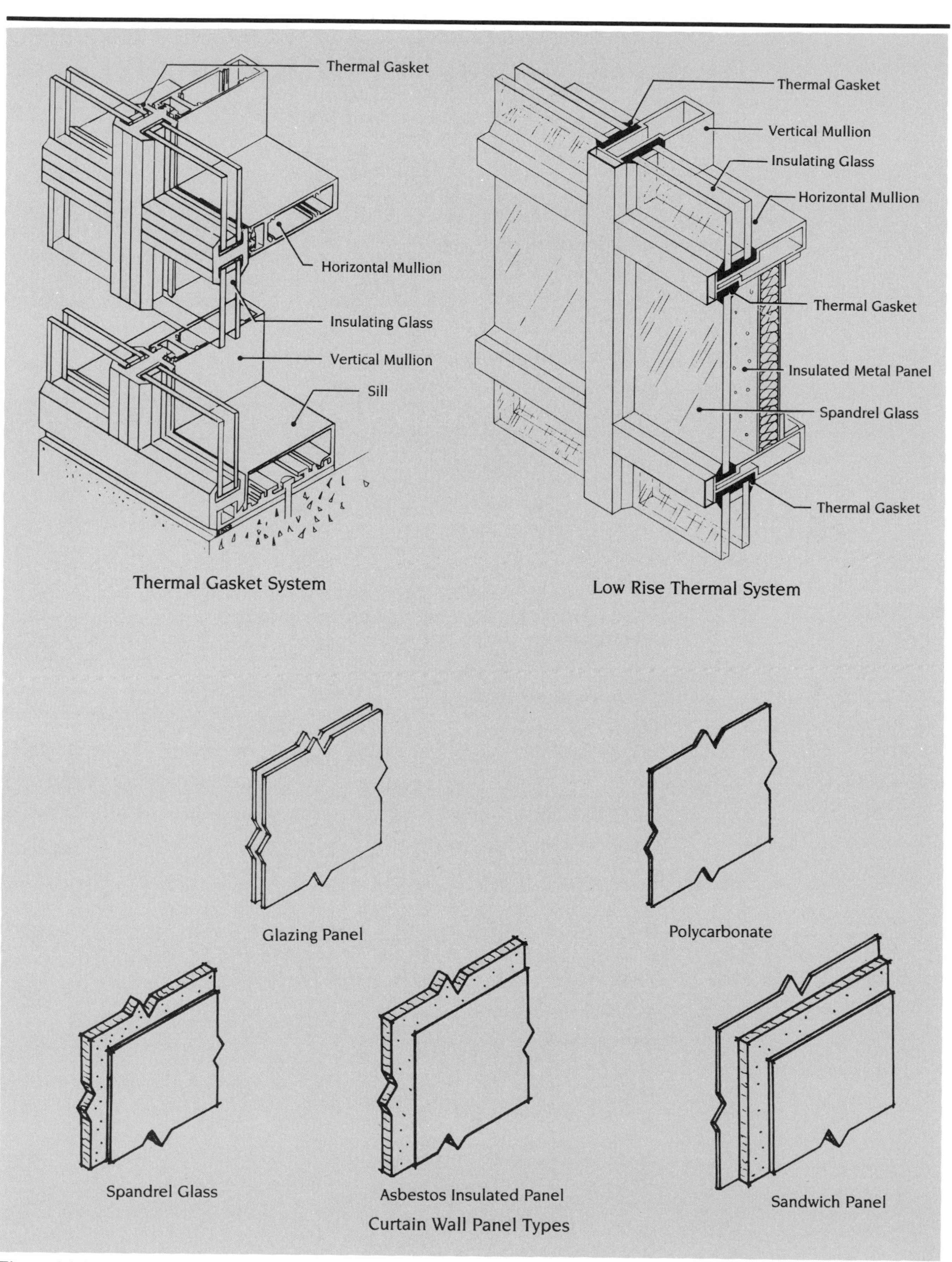
Thermal Gasket
Horizontal Mullion
Insulating Glass
Vertical Mullion
Sill
Thermal Gasket System
Thermal Gasket
Vertical Mullion
Insulating Glass
Horizontal Mullion
Thermal Gasket
Insulated Metal Panel
Spandrel Glass
Thermal Gasket
Low Rise Thermal System
Glazing Panel
Polycarbonate
Spandrel Glass
Asbestos Insulated Panel
Sandwich Panel
Curtain Wall Panel Types

Figure 16.11

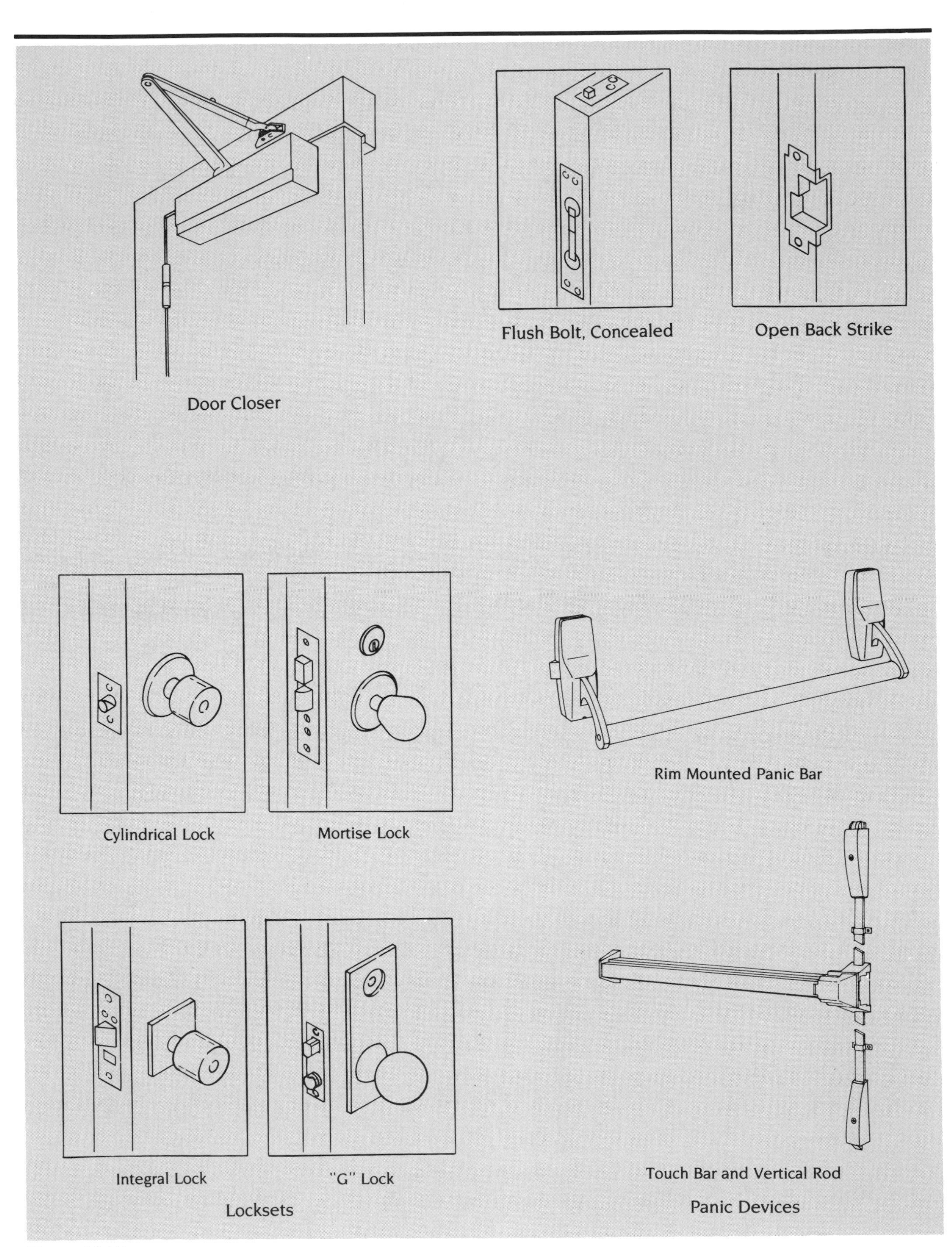

Figure 16.12*a*

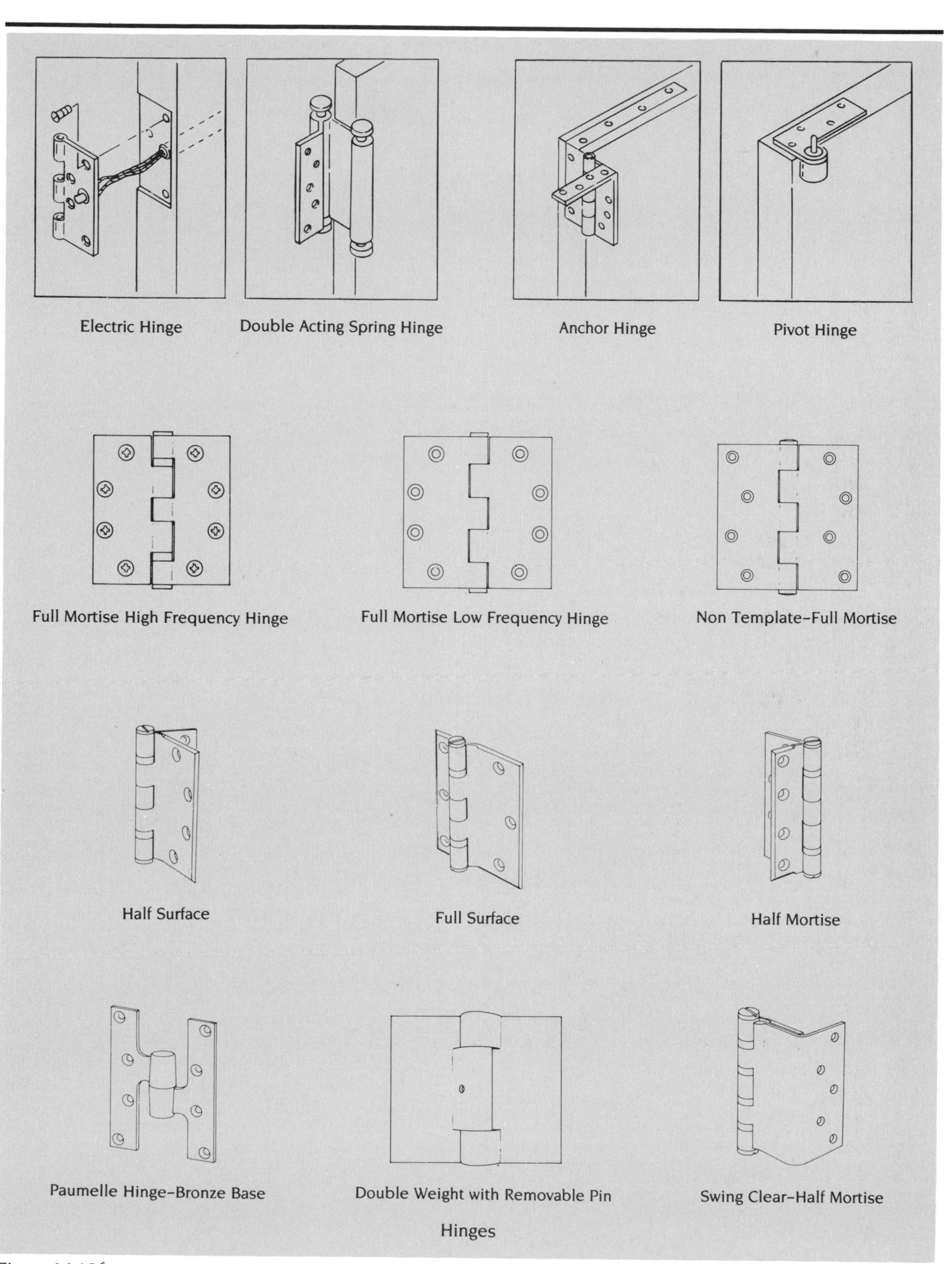

Figure 16.12*b*

Chapter 17

FINISHES

This chapter contains descriptions of the basic types of finishes, with appropriate estimating methods. Materials used to finish walls, ceilings, doors, windows, and trimwork can include any of a number of available products. Most building codes (and specifications) require strict adherence to standards for protection against maximum fire, flame spread, and smoke generation characteristics. In today's fireproof and fire-resistant types of construction, some finish materials may be the only combustibles used in a building project. These combustible materials may have to be treated for fire retardancy, at an additional cost. The estimator must be sure that all materials meet the specified requirements and that any necessary fireproofing be included in the estimate for interior finishes.

Lath and Plaster

The different types of plaster work require different pricing strategies. Large open areas of continuous walls or ceilings involve considerably less labor per unit of area than small areas, repair work, or intricate work such as archways, curved walls, cornices, and window returns. Gypsum and metal lath are most often used as sub-bases. However, plaster may be applied directly on masonry, concrete, and, in some restoration work, wood. In the latter cases, a bonding agent may be specified. Illustrations of plaster partitions and metal lath are shown in Figure 17.1.

Number and Type of Coatings

The number of coats of plaster may also vary. Traditionally, a scratch coat is applied to the substrate. A brown coat is then applied two days later, and the finish, smooth coat seven days after the brown coat. Currently, the systems most often used are two-coat and one-coat (imperial plaster on "blueboard"). Textured surfaces, with and without patterns, may be required. The many variables in plaster work make it difficult to develop "system" prices. Each project, and even areas within each project, must be examined individually.

Sequence of Work

The quantity takeoff should proceed in the normal construction sequence: furring (or studs), lath, plaster, and accessories. Studs, furring, and/or ceiling suspension systems, whether wood or steel, should be taken off separately. Responsibility for the installation of these items should be clearly noted. Depending on local work practices, lathers may or may not install studs or furring. These materials are usually estimated by the piece or linear foot, and sometimes by the square foot.

Determining Quantities

Lath is traditionally estimated by the square yard for both gypsum and metal lath and, more recently, by the square foot. Usually, a 5% allowance for waste is included. Casing bead, corner bead, and other accessories are measured by the linear foot. An extra foot of surface area should be allowed for each linear foot of corner or stop.

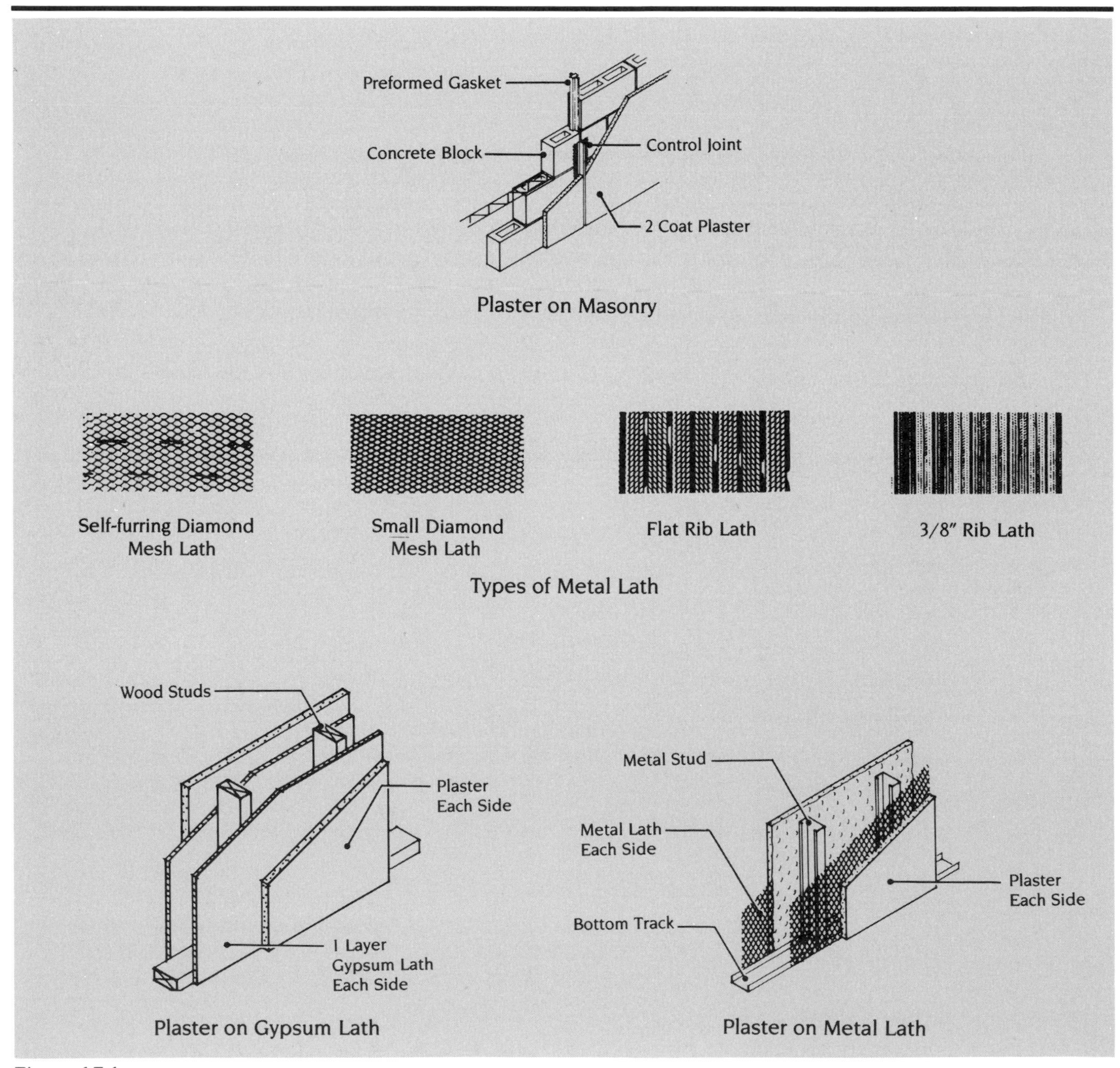

Figure 17.1

Although wood plaster grounds are usually installed by carpenters, they should be measured when taking off the plaster requirements.

Plastering is also traditionally measured by the square yard. Deductions for openings vary by preference–from a zero deduction to 50% of all openings over two feet in width. Some estimators deduct a percentage of the total yardage for openings. One extra square foot of wall area should be allowed for each linear foot of inside or outside corner located below the ceiling level. The areas of small radius work should be doubled.

Plaster quantities are determined by measuring surface area (walls, ceilings, etc.). The estimator must consider the complexity and the intricacy of the work, as well as the quality. There are two basic quality categories:

- **Ordinary quality** is used for commercial purposes. Waves 1/8″ to 3/16″ in 10 feet are acceptable. Angles and corners must be fairly true.
- **First quality** requires that variations be less than 1/16″ in 10 feet. Labor costs for first quality work are approximately 20% more than those for ordinary plastering.

Drywall

Framing

With the advent of light gauge metal framing, tin snips are becoming as important a tool to the carpenter as the circular saw. Metal studs and framing are usually installed by the drywall subcontractor. The estimator should make sure that studs (and other framing–whether metal or wood) are not included twice by different subcontractors. In some drywall systems, such as shaftwall, the framing is integral and installed simultaneously with the drywall panels. Typical drywall partition systems are shown in Figure 17.2.

Metal studs are manufactured in various widths (1-5/8″, 2-1/2″, 3-5/8″, 4″, and 6″) and in various gauges, or metal thicknesses. They may be used for both load-bearing and non-load-bearing partitions, depending on design criteria and code requirements. Metal framing is particularly useful as a replacement for structural wood, since the latter material is often prohibited due to its combustible quality. Metal studs, track, and accessories are purchased by the linear foot, and usually stocked in 8′ to 16′ lengths, by 2′ increments. For large orders, metal studs can be purchased in any length up to 20′.

For estimating, light gauge metal framing is taken off by the linear foot or by the square foot of wall area of each type. Different wall types–(with different stud widths, stud spacing, or drywall requirements)–should each be taken off separately, especially if estimating by the square foot.

Fasteners

Metal studs can be installed very quickly. Depending on the specification, they may have to be fastened to the track with self-tapping screws, tack welds or clips, or may not have to be prefastened. Each requirement will affect the labor costs. Fasteners such as self-tapping screws, clips and powder-actuated studs are very expensive, though labor-saving. These costs must be included.

Drywall

Drywall may be purchased in various thicknesses – 1/4″ to 1″ – and in various sizes - 2′ × 8′ to 4′ × 20′. Different types include *standard, fire-resistant, water-resistant, blueboard, coreboard,* and *pre-finished.* There are many variables and possible combinations of sizes and types. While the installation cost of 5/8″ standard drywall may be the same as that of 5/8″ fire-resistant drywall, these two types (and all others) should be taken off separately. The takeoff will be used for purchasing; material costs will vary.

Because drywall is used in such large quantities, current, local prices should always be checked. A variation of a few cents per square foot can amount to many thousands of dollars over a whole project.

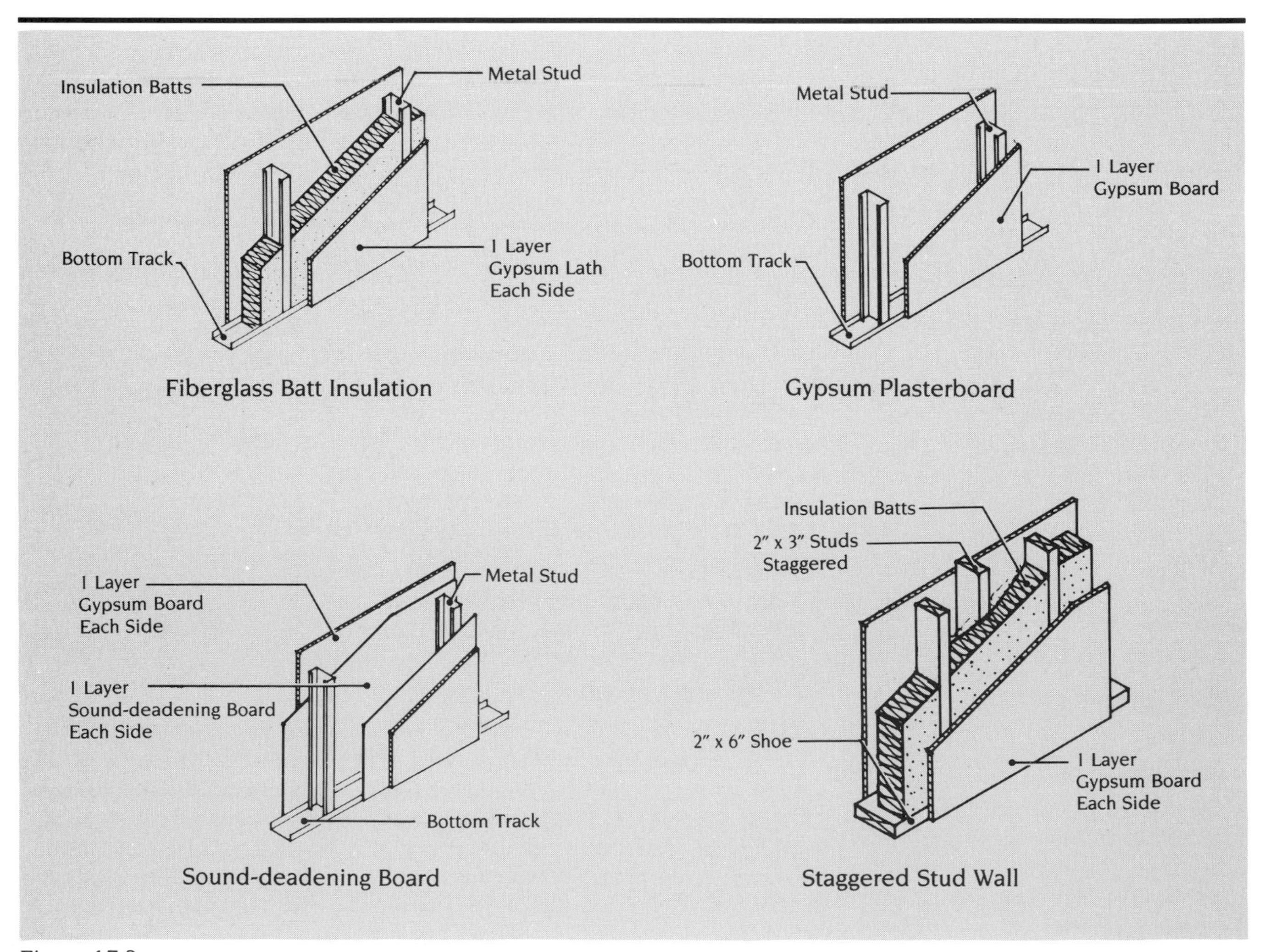

Figure 17.2

Firewalls

Fire-resistant drywall provides an excellent design advantage in creating relatively lightweight, easy-to-install firewalls (as opposed to masonry walls). As with any type of drywall partition, the variations are numerous. The estimator must be very careful to take off the appropriate firewalls exactly as specified.

Even more important, the contractor must **build** the firewalls exactly as specified. There is a tremendous potential for liability in the event the system fails. For example, a metal stud partition with two layers of 1/2" fire-resistant drywall on each side may constitute a *two-hour* partition (when all other requirements such as staggered joints, taping, sealing openings, etc., are met). If a *one-hour* partition is called for, the estimator cannot assume that one layer of 1/2" fire-resistant drywall on each side of a metal stud partition will suffice. Alone, it does not.

When left to choose the appropriate assembly (given the rating required), the estimator must be sure that the system has been tested and approved for use—by Underwriters Laboratory as well as local building and fire codes and responsible authorities. In all cases, the drywall (and studs) for firewalls must extend completely from the deck below to the underside of the deck above, covering the area above and around any and all obstructions. All penetrations must be protected.

In the past, structural members—such as beams or columns—to be fireproofed had to be "wrapped" with a specified number of layers of fire-resistant drywall. This is a labor-intensive and expensive task. With the advent of spray-on fireproofing, structural members can be much more easily protected. This type of work is usually performed by a specialty subcontractor. Takeoff and pricing are done by square foot of surface area. (See Figure 13.2 for surface and boxed areas of structural steel members.)

Soundproofing

When walls are specified for minimal sound transfer, the same continuous, unbroken construction used for firewalls is required. Soundproofing specifications may include additional accessories and related work. Resilient channels attached to studs, mineral fiber batts, and staggered studs may all be used. In order to develop high noise reduction coefficients, double stud walls may be required with sheet lead between double or triple layers of drywall. (Sheet lead may also be required for X-ray installations.) Caulking is required at all joints and seams. All openings must be specially framed with double, "broken" door and window jambs.

Shaftwall

Cavity shaftwall, developed for a distinct design advantage, is another drywall assembly which should be estimated separately. While firewalls require equal protection from both sides, shaftwall (used at vertical openings such as elevators and utility chases) can be installed completely from one side. Special track, studs (C-H or double E type) and drywall (usually 1" thick and 2' wide coreboard) are used. These items should be priced separately from other drywall partition components. Figure 17.3 illustrates several types of cavity shaftwall.

Material Handling

Because of the size and weight of drywall, costs for material handling and loading should be included with installation and material costs. Using larger sheets (manufactured up to 4′ × 20′) may require less taping and finishing, but these sheets may each weigh well in excess

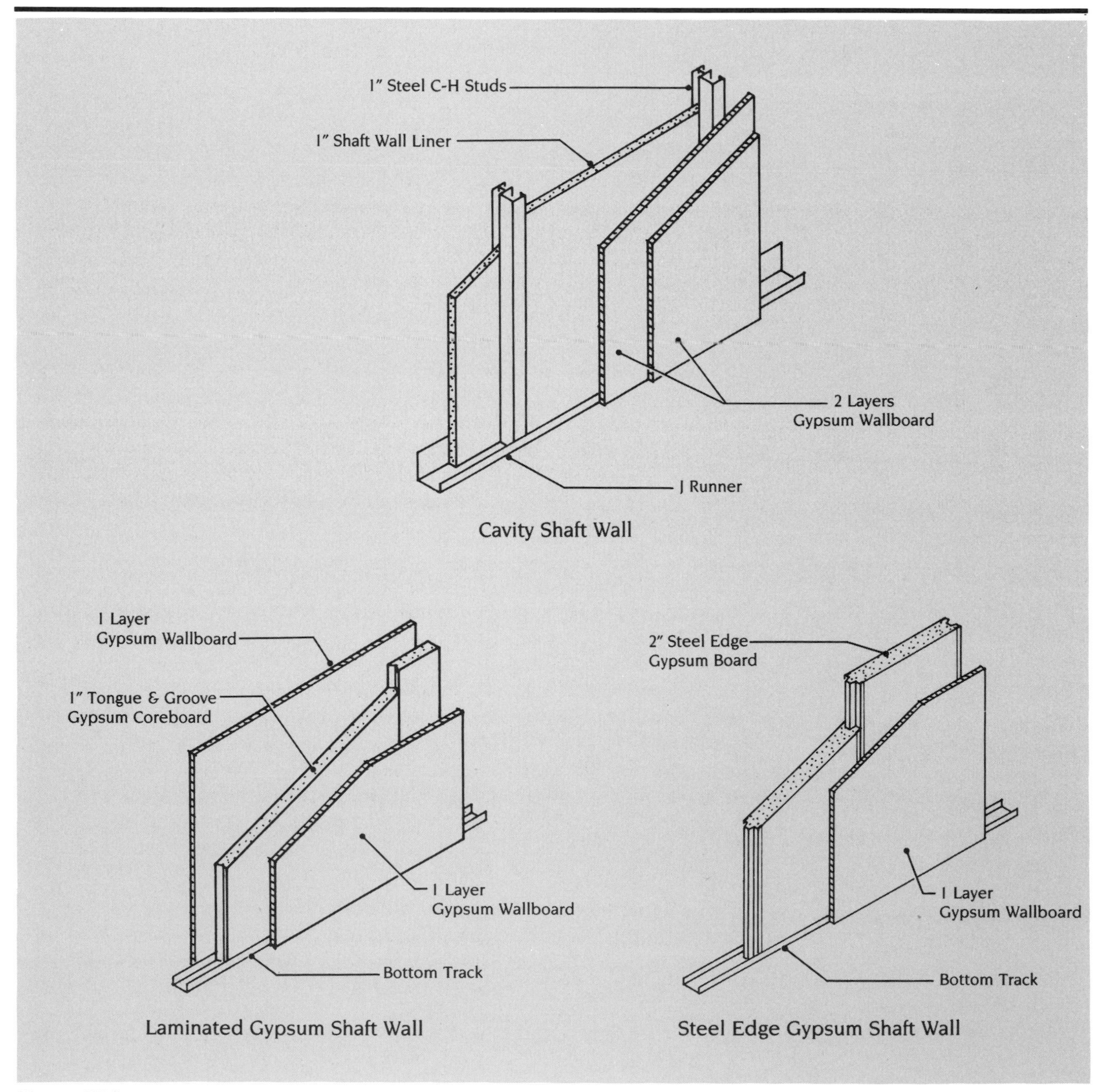

Figure 17.3

of 100 pounds and are awkward to handle. The weight of drywall is also a factor to be taken into account (for distribution) when loading a job in existing buildings, so that allowable floor loads are not exceeded. *All* material handling involves costs that must be included.

Pricing

As with plaster work, open spans of drywall should be priced differently from small, intricate areas that require much cutting and patching. Similarly, areas with many corners or curves require higher finishing costs than open walls or ceilings. Corners, both inside and outside, should be estimated by the linear foot, in addition to the square feet of surface area.

Although difficult because of variations, the estimator may be able to develop historical systems or assemblies prices for metal studs, drywall, and taping and finishing, similar to those shown in Figure 17.4. When using systems, whether complete or partial, the estimator must be sure that the system, as specified, is exactly the same as that one for which the costs were developed. For example, a cost has been developed for 5/8" fire resistant drywall, taped and finished. The project specifications require a firewall with two layers of the 5/8" drywall on each side of the studs. The "system" cost for each layer can be used, when only one of the two layers is to be taped and finished. The more detailed the breakdown and delineation, the less chance for error.

Tile

Ceramic tile and tile manufactured from other special use materials provide an almost endless source of interior wall, floor, and countertop coverings. Some of the more common applications of ceramic and other tile include: toilet rooms, tubs, steam rooms, swimming pools, and other related installations where easily cleaned, water repellent, and durable surfaces are required.

Available Styles and Materials

Ceramic and other tiles are manufactured from several basic materials in a wide range of shapes, sizes, and finishes. Ceramic tiles are manufactured from clay, porcelain, or cement. Metal and plastic are the most commonly used materials for hard surface tiles that are not labelled as *ceramic*. Floor tiles may be ceramic or they may be brick pavers, quarry tiles, or terra cotta tiles. Tile shapes and sizes vary from 1" squares to variously-sized rectangles, mosaics, patterned combinations, hexagons, octagons, valencia, wedges, and circles. They are available in multi-colored designs and mural sets or with individual pieces embossed with designs or pictures. Tiles with special designs and logos may also be custom-manufactured. Standard and custom ceramic tiles are produced with glazed and unglazed surfaces with bright, matte, non-slip, and textured finishes.

Installation

Proper installation methods for ceramic tiles are determined by the location and the type of backing surface, which may vary from stud or masonry walls to wood or concrete floors. The tiling material may be placed in individual pieces, or installed in factory-prepared back-mounted and ungrouted sections or sheets which cover two square feet or more as a unit. Tiling material is also available in

INTERIOR CONSTR. | 06.1-594 | Partitions, Metal Stud, NLB

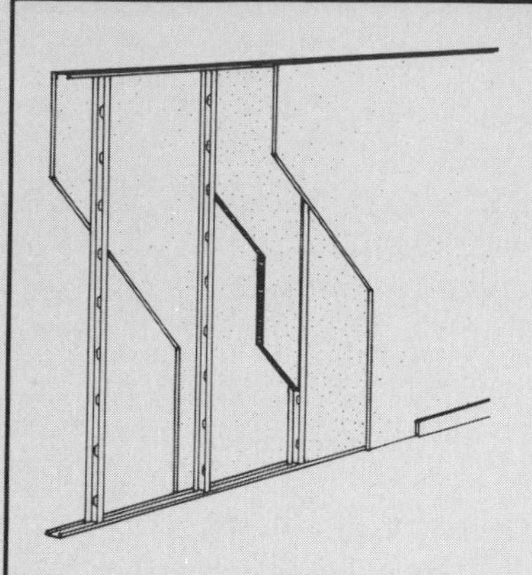

This page illustrates and describes a non-load bearing metal stud partition system including metal studs with runners, gypsum plasterboard, taped and finished, insulation, baseboard and painting. Lines 06.1-594-04 thru 10 give the unit price and total price per square foot for this system. Prices for alternate non-load bearing metal stud partition systems are on Line Items 06.1-594-13 thru 23. Both material quantities and labor costs have been adjusted for the system listed.

Factors: To adjust for job conditions other than normal working situations use Lines 06.1-594-29 thru 40.

Example: You are to install the system and cut and patch to match existing construction. Go to Line 06.1-594-30 and apply these percentages to the appropriate MAT. and INST. costs.

LINE NO.	DESCRIPTION	QUANTITY	COST PER S.F.		
			MAT.	INST.	TOTAL
01	**Non-load bearing metal studs, including top & bottom runners, ⅝″ drywall,**				
02	**Taped, finished and painted 2 faces, insulation, painted baseboard.**				
03					
04	Metal studs, 25 ga., 3-⅝″ wide, 24″ O.C.	1 S.F.	.30	.55	.85
05	Gypsum drywall, ⅝″ thick	2 S.F.	.53	.55	1.08
06	Taping & finishing	2 S.F.	.11	.55	.66
07	Insulation, 3-½″ fiberglass batts	1 S.F.	.30	.17	.47
08	Baseboard, painted	.2 L.F.	.17	.33	.50
09	Painting, roller work, 2 coats	2 S.F.	.20	.54	.74
10	TOTAL	S.F.	1.61	2.69	4.30
11					
12	For alternate metal stud systems:				
13	Non-load bearing, 25 ga., 24″ O.C., 2-½″ wide	S.F.	1.56	2.68	4.24
14	6″ wide		1.74	2.71	4.45
15	16″ O.C., 2-½″ wide		1.62	2.82	4.44
16	3-⅝″ wide		1.67	2.84	4.51
17	6″ wide		1.84	2.86	4.70
18	20 ga., 24″ O.C., 2-½″ wide		1.73	2.68	4.41
19	3-⅝″ wide		1.89	2.70	4.59
20	6″ wide		1.98	2.71	4.69
21	16″ O.C., 2-½″ wide		1.83	2.82	4.65
22	3-⅝″ wide		2.04	2.84	4.88
23	6″ wide	↓	2.15	2.85	5
24					
25					
26					
27					
28					
29	Cut & patch to match existing construction, add, minimum		2%	3%	
30	Maximum		5%	9%	
31	Dust protection, add, minimum		1%	2%	
32	Maximum		4%	11%	
33	Material handling & storage limitation, add, minimum		1%	1%	
34	Maximum		6%	7%	
35	Protection of existing work, add, minimum		2%	2%	
36	Maximum		5%	7%	
37	Shift work requirements, add, minimum			5%	
38	Maximum			30%	
39	Temporary shoring and bracing, add, minimum		2%	5%	
40	Maximum		5%	12%	
41					
42					

278

Figure 17.4

factory-prepared sections and patterns in which the tiles have been preset and pre-grouted with silicone rubber.

The two recognized installation methods are the thick set, or mud set method and the thin set method, in which the tile is directly adhered to the base, sub-base, or wall material. In thick set *floor* installations, Portland cement and sand are placed and screeded to a thickness of 3/4" to 1-1/4". For thick set *wall* installations, Portland cement, sand, and lime are placed on the backing surface and troweled to a thickness of 3/4" to 1'. With both floor and wall placement, the mortar may be reinforced with metal lath or mesh, and can be backed with impervious membranes. The tiles may be placed on and adhered to the mortar bed while it remains plastic, or they may be placed after the bed has cured, and adhered with a thin bond coat of Portland cement with sand additives. The thin set method of tile installation requires specially prepared mortars and adhesives on properly prepared surfaces. Latex Portland cement, epoxy mortar, epoxy emulsion mortar, epoxy adhesive, and organic adhesives are some of these specialized thin set preparations.

The grouting of the placed and adhered tile material is the final critical step in the installation process, as it ensures the sealing of the joints between the tiles and affects the appearance of the installation. The choice of grouting material to be used depends on the type of tile material and the conditions of its exposure. Some of the available grouting mixtures include Portland cement grouts with additives, mastic grout, furan resin grout, epoxy grout, and silicone rubber grout. The recommendations of the manufacturer should be carefully followed in all aspects of the grouting operation. A typical ceramic tile installation is shown in Figure 17.5.

Pricing

Tile is usually taken off and priced per square foot. Linear features such as bullnose trim and cove base are estimated per linear foot. Specialties (accessories) are taken off as *each*.

Ceilings

In addition to exposed structural systems that are painted or stained, there are three basic types of ceilings:

- Directly applied acoustical tile
- Gypsum board, or plaster
- Suspended systems, hung below the supporting superstructure to allow space for ductwork, piping, and lighting systems

Suspension Systems

Suspension systems usually consist of aluminum or steel main runners of *light*, *intermediate*, or *heavy duty* classification, spaced two, three, four, or five feet on center with snap-in cross tees usually available in one- to five-foot lengths. Main runners are usually hung from the supporting structure with tie wire or metal straps spaced as required by the system. The runners and cross tees may be exposed, natural, or painted: or they may be concealed (enclosed by the ceiling tiles). Drywall suspension systems consist of main and cross tee members with flanges to allow the gypsum board to be screwed to the supports.

Acoustical Tiles

Acoustical ceiling tiles are available in mineral fiber with many patterns and textures. The most common sizes are 1′ × 1′, 2′ × 2′, and 2′ × 4′. The face may be perforated, fissured, textured, or plastic-covered. In addition to mineral fiber, ceiling tiles are also available in perforated steel, stainless steel, or aluminum to allow for easy cleaning in high humidity or kitchen areas. To retain the acoustical characteristics, these panels are filled with sound-absorbing pads. Specialty metal systems are also available as linear ceilings in varying widths in many finishes and configurations. Integrated ceiling systems combine a suspension grid, air handling, lighting, and ceiling tiles into one modular system.

Gypsum Board or Plaster

Drywall ceilings may be applied directly to furring strips attached to the supporting structure, or a compatible suspension system. They may be painted, sprayed with acoustical material, or covered with acoustical tiles. Plaster ceilings can be applied directly on self-furring or gypsum lath. Suspended plaster ceilings attach to cold-rolled channels hung from the supporting structure and covered with metal or gypsum lath. Figures 17.6a and b show typical ceiling applications.

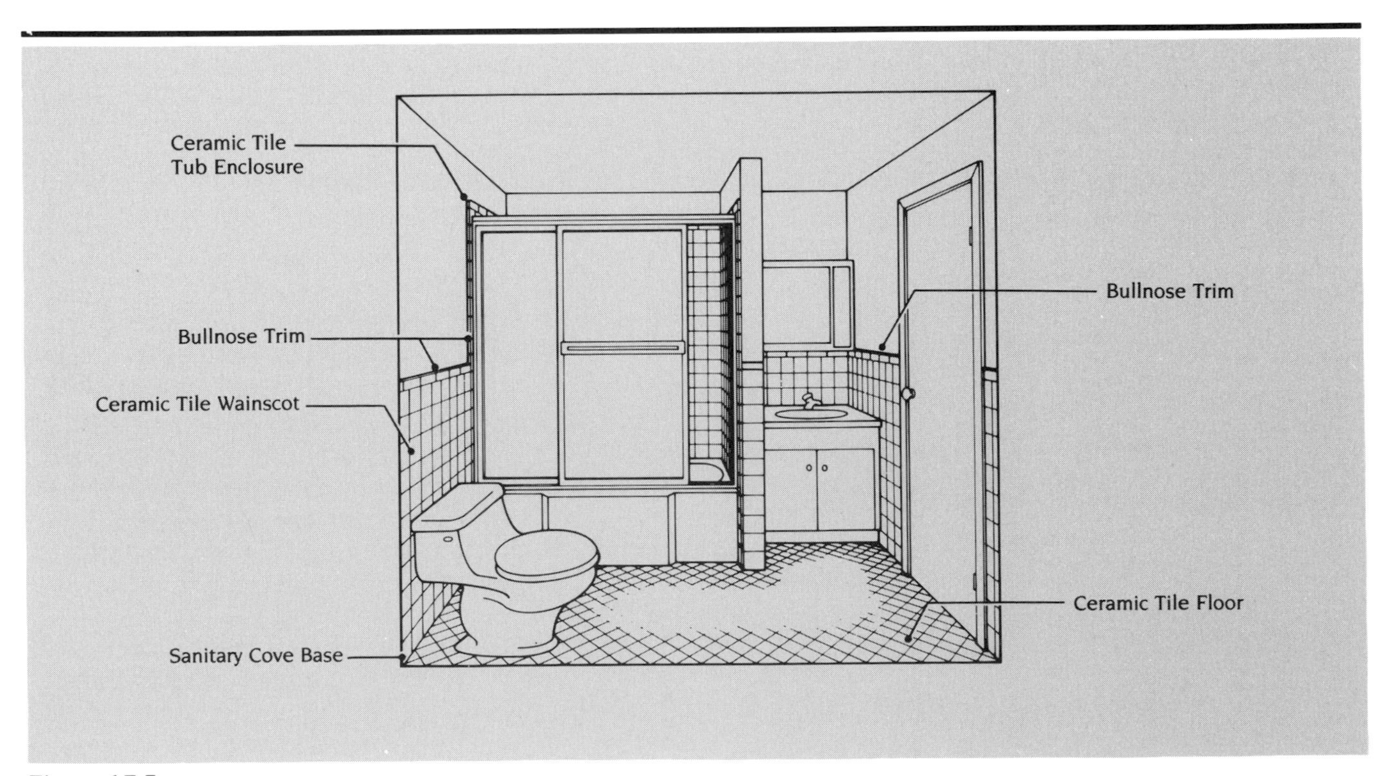

Figure 17.5

Fire Resistance

Fire resistance (rated in *hours*) for the various ceiling systems is evaluated as part of the floor or structural floor and ceiling assembly. To achieve a rating in hours, fire dampers, light fixtures, grilles, and diffusers are evaluated along with the structure, suspension system, and ceiling.

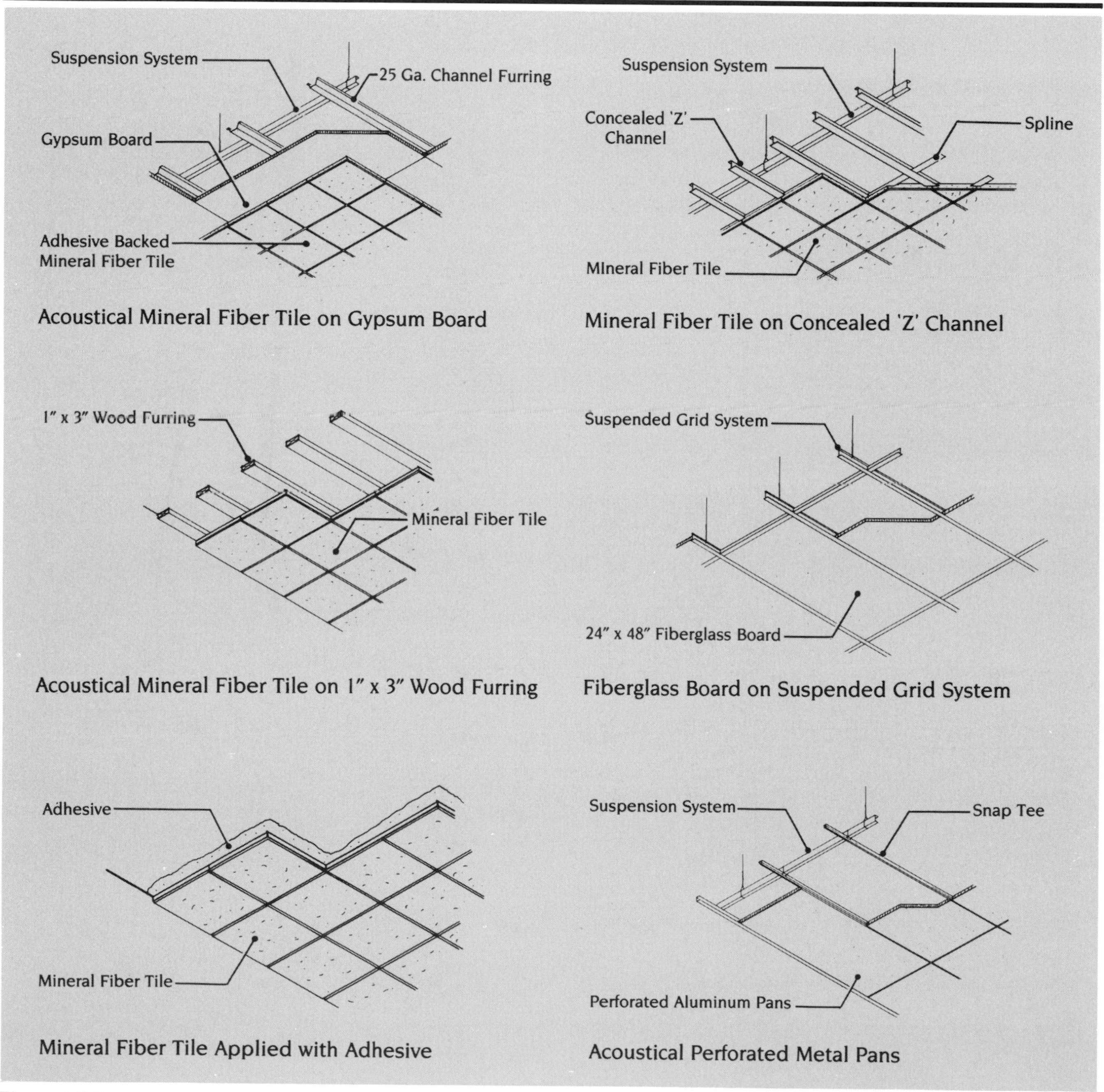

Figure 17.6a

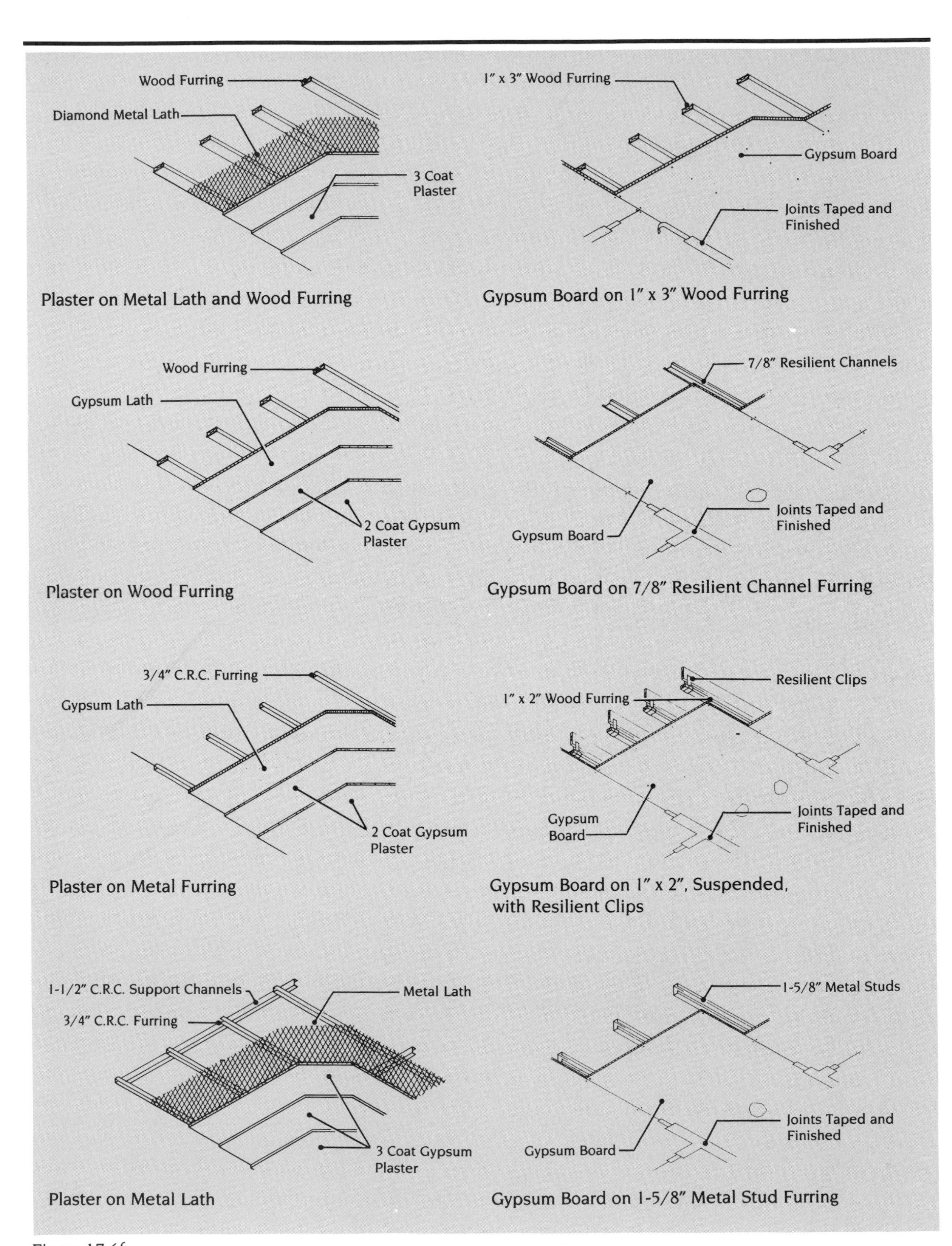

Wood Furring
Diamond Metal Lath
3 Coat Plaster
Plaster on Metal Lath and Wood Furring
1" x 3" Wood Furring
Gypsum Board
Joints Taped and Finished
Gypsum Board on 1" x 3" Wood Furring
Wood Furring
Gypsum Lath
2 Coat Gypsum Plaster
Plaster on Wood Furring
7/8" Resilient Channels
Gypsum Board
Joints Taped and Finished
Gypsum Board on 7/8" Resilient Channel Furring
3/4" C.R.C. Furring
Gypsum Lath
2 Coat Gypsum Plaster
Plaster on Metal Furring
Resilient Clips
1" x 2" Wood Furring
Gypsum Board
Joints Taped and Finished
Gypsum Board on 1" x 2", Suspended, with Resilient Clips
1-1/2" C.R.C. Support Channels
3/4" C.R.C. Furring
Metal Lath
3 Coat Gypsum Plaster
Plaster on Metal Lath
1-5/8" Metal Studs
Gypsum Board
Joints Taped and Finished
Gypsum Board on 1-5/8" Metal Stud Furring

Figure 17.6*b*

Sound barriers for acoustical deadening usually consist of materials with mass that deflects sound. They may be masonry, metal with sandwich insulation, leaded vinyl, fiberglass batts, or sheet lead.

Determining Quantities

Suspended ceilings of acoustical tiles, drywall, or plaster are usually priced per S.F. Some system costs include all or part of the suspension grid. Others include only the panels, in which case suspension costs must be estimated separately. Patterned and angled ceilings require more material and labor to install.

There is little waste when installing grid systems (usually less than 5%), because pieces can be butted and joined. Waste for tile, however, can range from as low as 5% for large open areas, to as high as 30%-40% for small areas and rooms. Waste for tile may depend on grid layout as well as room dimensions. Figure 17.7 demonstrates that for the same size room, the layout of a typical 2′ × 4′ grid has a significant effect on generated waste of ceiling tile. Since most textures and patterns on ceiling tile are aligned in one direction, pieces cannot be turned 90 degrees (to the specified alignment) to reduce waste.

Certain tile types, such as tegular (recessed), require extra labor for cutting and fabrication at edge moldings. Soffits, fascias and "boxouts" should be estimated separately due to extra labor, material waste, and special attachment techniques. Costs should also be added for unusually high numbers of tiles to be specially cut for items such as sprinkler heads, diffusers, or telepoles.

Handling Costs

While the weight of ceiling tile is not the primary consideration that it is with drywall, some material handling and storage costs will still be incurred and must be included. Acoustical tile is very bulky and cumbersome, as well as fragile, and must be protected from damage before installation.

Carpeting

Carpeting is manufactured in various combinations of size, material, and texture. For commercial carpeting, rolled goods are most commonly available in 54″ and 12′ widths, and the standard size of carpet tiles is 18″ × 18″ or 24″ × 24″. Carpet may be tufted, woven, or fusion-bonded. The surface texture may be level loop or multi-level loop, cut level pile, velvet cut pile or a combination of cut and loop pile. Materials may be wool, nylon (different brand name types), or acrylic. Carpets may be of one fabric or a blend.

Determining Quantities

If a custom color or pattern is required and the quantity exceeds 100 S.Y., there is generally no additional cost for manufacturing, provided the manufacturer has the appropriate equipment to construct the carpet. If the quantity of custom carpet required is less than 100 S.Y., and is a custom pattern or color, there may be an additional charge of up to 20%.

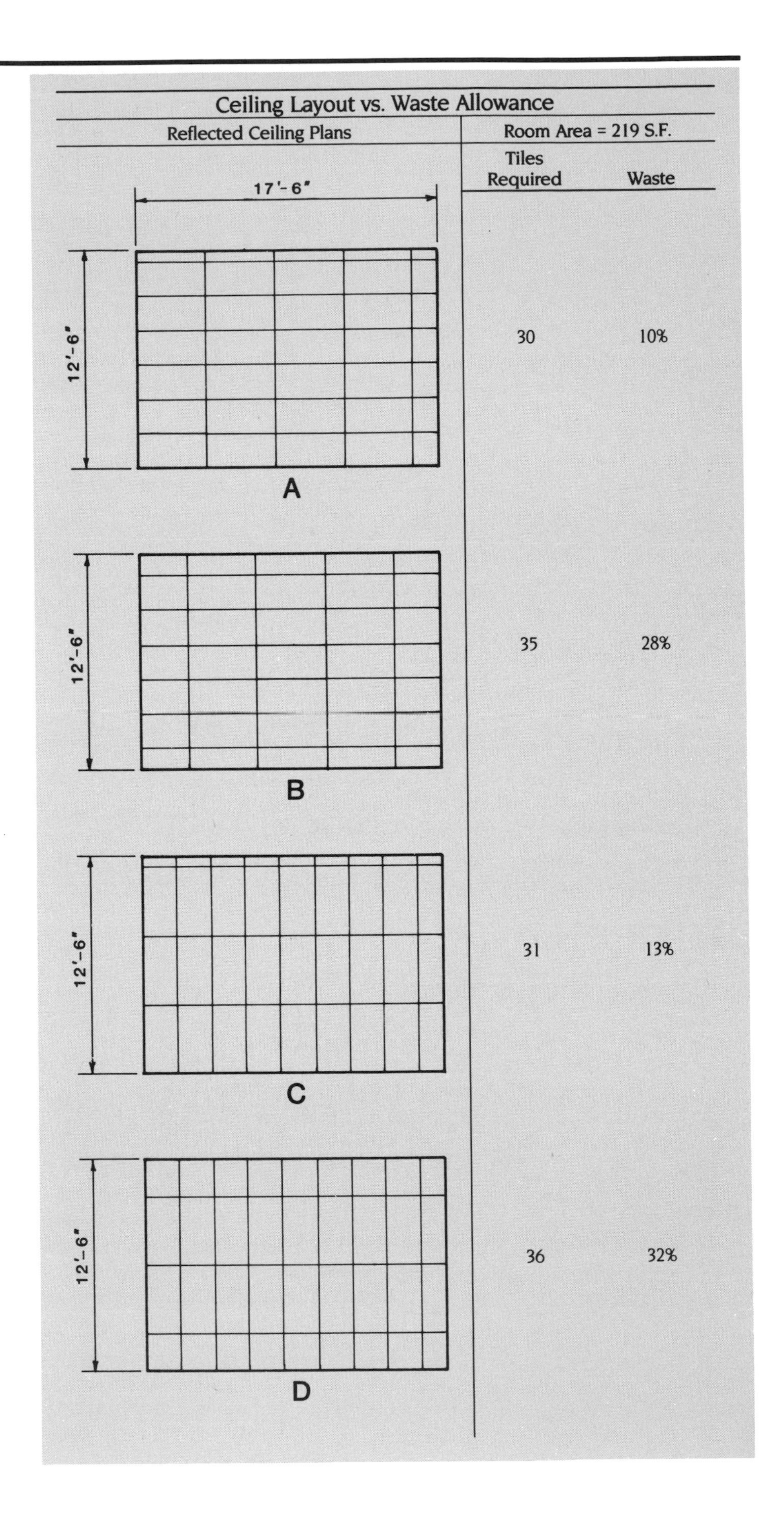
Ceiling Layout vs. Waste Allowance
Reflected Ceiling Plans
Room Area = 219 S.F.
Tiles Required
Waste
17'-6"
12'-6"
A
30
10%
12'-6"
B
35
28%
12'-6"
C
31
13%
12'-6"
D
36
32%

Figure 17.7

Carpet material may be specified or measured by its face weight in ounces per square yard, its pile height, its density, or stitches per inch. The following formula is commonly used for determining the carpet's density.

$$\text{Density (oz./S.Y.)} = \frac{\text{Face weight, oz./S.Y.} \times 36}{\text{Pile height, in.}}$$

Tufted carpets are usually available in 12 or 15 foot widths. Woven carpets are manufactured in variable widths. A variety of backing materials, including polypropelene, jute, cotton, rayon, and polyester, provide support for the surface material. The methods used for integrating the surface materials and attaching them to the backing include tufting, weaving, fusion bonding, knitting, and needle punching. A second backing material of foam, urethane, or sponge rubber may be applied to the standard carpet backing to serve as an attached pad.

When rolled or broadloom carpet with a large pattern repeat is specified, up to 15% additional carpet should be ordered to allow for pattern matching. The average waste for pattern matching is 10%-12%. To determine the quantity of carpet actually required, the estimator should prepare a seam diagram. This involves using an actual scale floor plan and showing where all the carpet seams will occur when the carpet is rolled on the floor.

Waste will inevitably occur where there are cut-outs or protruding corners. However, the diagram will provide an accurate determination of the amount of carpet that must be ordered, and acceptable locations for seams. To estimate the total square yards of carpet required, the following equation can be used, based on the total lineal feet of carpet measured on the seam diagram:

$$\frac{\text{Total L.F. of rolled carpet} \times 12}{9} = \text{Total S.Y. of Carpet Required}$$

Installation

The method of installation is usually determined by such variables as the size of the area to be covered, the amount and type of traffic it will carry, and the style of carpeting. Depending on these conditions, carpet may be installed over separate felt, sponge rubber, or urethane foam pads, or it may be directly adhered to the sub-floor. Wide expanses and carpet locations that carry heavy or wheeled traffic usually require direct glue-down installation to prevent wrinkling, bulging, and movement of the carpet. If padding is required in a glue-down installation, it must consist of an attached pad of synthetic latex, polyurethane foam, or similar material. Carpet may also be installed over a separate pad and then stretched and edge fastened. This method of installation is not recommended for large areas or heavy traffic situations.

Carpet Tiles

With the development of flat undercarpet cable systems for power, telephone, and data line wiring, the use of modular carpet tiles has increased greatly. Modular carpet tiles, when placed directly over the undercarpet cable run, provide easy access to the wiring system.

Because taps can be made at any point along the route of the undercarpet system, facilities relocation can be accomplished conveniently, quickly, and economically.

Sizes for carpet tiles are exact. Since they fit together with factory edges, which are clean, the seam does not show. Because the manufactured sizes are small and modular, there is less waste with carpet tiles than with broadloom or rolled carpet. However, seven to ten percent extra should be added to ensure proper color matching for future replacement.

To minimize waste due to dye lot variation, all carpeting should be ordered at the same time. The carpet should also be installed in the numerical sequence identified on the rolls by the carpet manufacturer.

Composition Flooring

Composition floors are seamless coverings manufactured from epoxy, polyester, acrylic, and polyurethane resins. Various aggregates or color chips may be combined with the resins to produce decorative effects, or special coatings may be applied to enhance the functional effectiveness of a floor's surface. Composition floors are normally installed in special-use situations where other flooring materials do not meet sanitary, safety, or durability standards.

In most cases, the materials that are combined to form composition flooring are determined by and designed for the specific application. Some of the available composition flooring systems include: terrazzo (for decorative floors), wood toppings and sealers, concrete toppings and sealers, waterproof and chemical-resistant floors, gymnasium sports surfaces, interior and exterior sports surfaces, and conductive floors. The materials employed in these systems are applied by trowel, roller, squeegee, notched trowel, spray gun, brush, or similar tools. After the flooring materials have been applied and allowed to cure, they may be sanded and/or sealed to produce the desired texture and surface finish. Composition floors are normally estimated and installed by trained and approved specialty contractors.

Resilient Floors

Resilient floors are designed for situations where durability and low maintenance are primary considerations. The various resilient flooring materials and their installation formats include: asphalt tiles, cork tiles, polyethylene in rolls, polyvinyl chloride in sheets, rubber in tiles and sheets, vinyl composition tiles, and vinyl in sheets and tiles. All of these materials may be manufactured with or without resilient backing and, except for the polyethylene rolls, they are available in a wide range of colors, designs, textures, compositions, and styles. Rubber and vinyl accessories, which are designed to complement any type of flooring material, include: bases, beveled edges, thresholds, corner guards, stair treads, risers, nosings, and stringer covers. Examples are illustrated in Figure 17.8.

Manufacturers' recommendations should be carefully followed for the installation details of any resilient floor. Generally, any concrete floor surface should be dry, clean, and free from depressions and adhered droppings. Curing and separating compounds should also be thoroughly removed from the surface before the resilient floor is

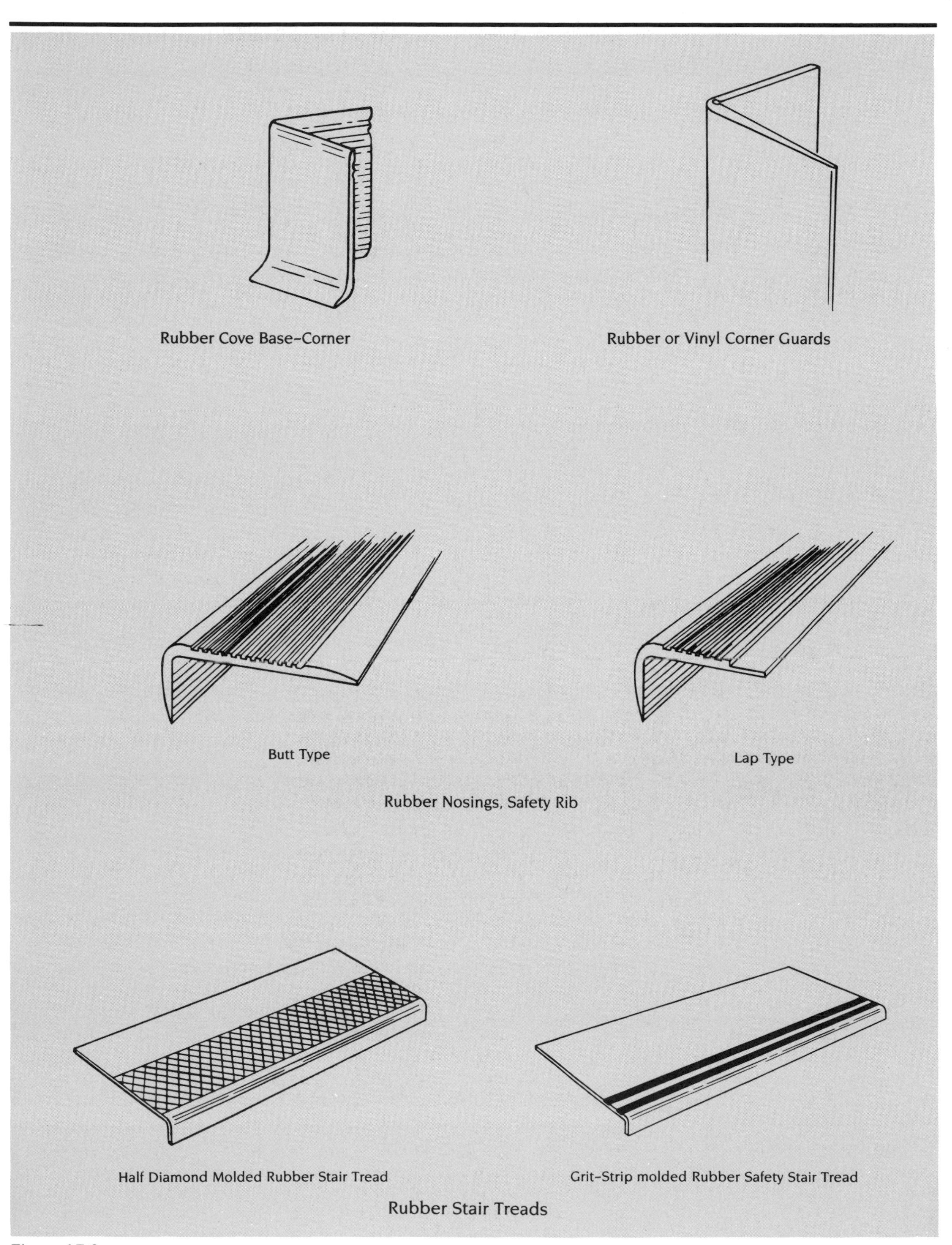
Rubber Cove Base-Corner
Rubber or Vinyl Corner Guards
Butt Type
Lap Type
Rubber Nosings, Safety Rib
Half Diamond Molded Rubber Stair Tread
Grit-Strip molded Rubber Safety Stair Tread
Rubber Stair Treads

Figure 17.8

placed. Special consideration is required for installations on slabs or wood surfaces that are located below grade level or above low crawl spaces.

The floor thickness specified is important to note, as there are different thickness grades for commercial and residential use. Resilient floors are usually estimated and priced per square foot; base and stair stringers are priced per linear foot.

Due to the thinness and flexibility of resilient goods, defects in the subfloor easily "telegraph" through the material. Consequently, the subfloor material and the quality of surface preparation are very important. Subcontractors will often make contracts conditional on a smooth, level subfloor. Surface preparation, which can involve chipping and patching, grinding or washing, is often an "extra." Some surface preparation will invariably be required and an allowance for this work should be included in the estimate. This is especially true in renovation where, in extreme cases, the floor may have to be leveled with a complete application of special lightweight leveling concrete.

Terrazzo

Terrazzo flooring materials provide many options to produce colorful, durable, and easily cleaned floors. Conventional ground and polished terrazzo floors employ granite, marble, glass, and onyx chips in a choice of specialized matrices. Well-graded gravel and other stone materials may be used to create different textures when added to the mix. Precast terrazzo tiles and bases, which are normally installed finished and polished on a cement sand base, are also available in many color combinations and aggregates.

Conventional installation of terrazzo flooring involves mixing the aggregate with one of three commonly used bonding matrices and placing the mix in sections defined by divider strips. The matrices employed to bond the aggregate include: cementitious matrices, which consist of natural or colored Portland cement; cement with an acrylic additive; and resinous matrices, which consist of epoxy and polyester. The divider strips, which are manufactured from zinc, brass, or colored plastic, provide for expansion and are, therefore, positioned over breaks in the substrate and at critical locations where movement is expected in large sections of flooring. The divider strips are also used to terminate pours, to act as leveling guides, and to permit changes in the aggregate mix, section-to-section, to create designs or patterns.

The terrazzo flooring materials may be installed on a cement sand underbed, or may be applied in the thin set method directly to the concrete slab. Sand-cushioned terrazzo employs three layers of material: a 1/4" thick sand cushion placed on the slab and covered by an isolation membrane; a mesh-reinforced underbed of approximately 1-3/4" thickness; and a 1/2" thick terrazzo topping (see Figure 17.9). Bonded terrazzo consists of two layers of material: a 1-3/4" thick underbed placed on the slab and a 1/2" thick terrazzo topping. Monolithic terrazzo flooring is comprised of a single 1/2" thick layer of terrazzo topping applied directly to the slab after control joints have been saw-cut into the slab and the divider strips grouted into place. With the thin set method, the terrazzo mix is applied in a single 1/4″ thick layer directly on the concrete slab. The surface of stair treads and risers may be poured in place with a 1/2″ thick layer of terrazzo topping on a 3/4″ underbed, or they may be installed as precast terrazzo pieces on a 3/4″ thick underbed.

Wood Floors

Wood flooring material is manufactured in three common formats: solid or laminated planks; solid or laminated parquet; and end-grain blocks, which are available in individual pieces or in pre-assembled strips. Commonly employed hardwood flooring materials include ash, beech, cherry, mahogany, oak, pecan, teak, and walnut, as well as exotic hardwood species, such as ebony, karpa wood, rosewood, and zebra wood. Cedar, fir, pine, and spruce are among the popular softwood flooring materials. End-grain block flooring is manufactured from alder, fir, hemlock, mesquite, oak, and yellow pine.

Strip Flooring

This material is supplied in several different milling formats and combinations, including tongue-and-groove and matched, square-edged, and jointed with square edges and splines. The installation of tongue-and-groove flooring usually requires blind nailing or fastening by means of metal attachment clips, while square-edged flooring is usually face-fastened. Strip floors may be installed over wood-framed subfloors of planks or plywood sheets which have been fastened to the floor joists. A layer of building paper should be laid between the subfloor and the finish strip floor to provide a comfort cushion and to reduce noise. If the strip floor is placed over a concrete surface, a subfloor of exterior plywood should be fastened directly to the slab and protected from moisture by a polyethylene vapor barrier between the concrete and the subfloor. The plywood subfloor may also be fastened to lapped sleepers which

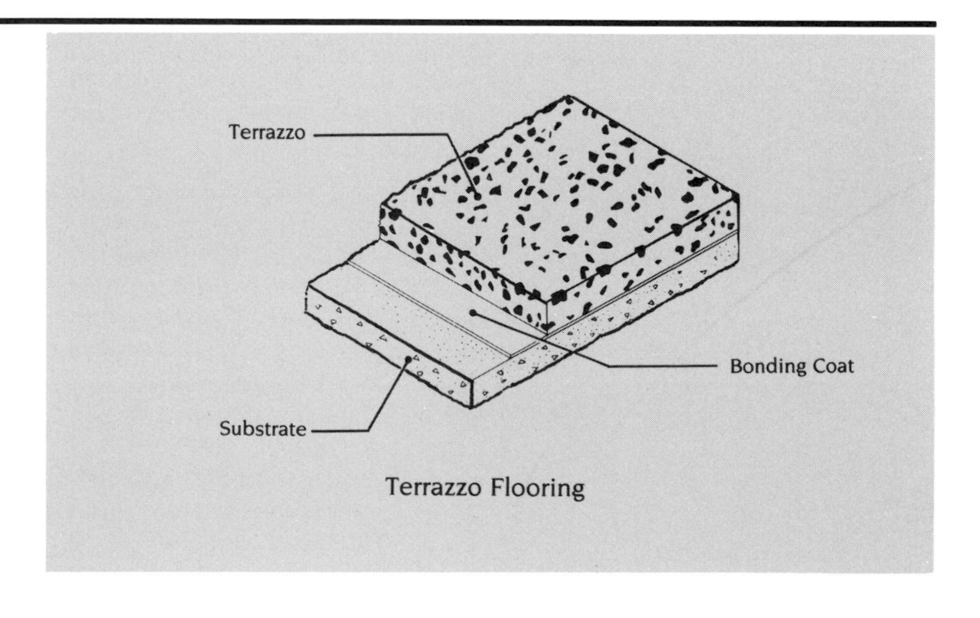

Terrazzo Flooring

Figure 17.9

have been imbedded in asphalt floor mastic on the slab and draped with the polyethylene vapor barrier before the subfloor is placed.

Parquet Floors

These floors are prefabricated in panels of various sizes which are milled with square edges, tongue-and-groove edges, or splines. The panels may also contain optional adhered backings which protect the flooring material from moisture, add comfort to the walking area, provide insulation, and deaden sound. Adhesives are normally employed to attach the parquet flooring panels to firm, level subfloors of concrete, wood, plywood, particle board, resilient tile, or terrazzo. Some parquet floor manufacturers also supply feature strips and factory-fashioned moldings to cover the required expansion space at the edges of the floor.

Wood Block Floors

Wood block floors for industrial application are manufactured in individual blocks and pre-assembled strips. The surface dimensions of individual blocks are 3" by 6", 4" by 6", and 4" by 8", with nominal thicknesses ranging from 1-1/2" to 4". The blocks are normally installed in a layer of pitch applied to a concrete floor. The finish coating of pitch or similar material is then squeezed into the joints between the blocks to provide additional fastening strength. A sealer is applied to provide surface finish. End-grain strip block flooring is placed in mastic adhesive which has been troweled into a dampproofing membrane that covers the concrete subfloor. After the strip block flooring material has been laid, it is sanded, filled, and finished with penetrating oil.

Wood Gymnasium or Sports Floors

This type of floor usually requires specialized installation methods because of its unique function and large size. These floors may be placed over sleepers that are installed on cushions or pads, or they may be laid directly over a resilient base material or plywood sub-base on cushioned pads. If the flooring material is placed on a sleeper support system, metal clips may be applied for fastening. If it is placed on a plywood subfloor, direct nailing is normally used for fastening. Because large floor areas, such as gymnasiums, require a wide expansion space at the edges, the placement and type of closure strip installed to cover the space deserve special consideration. Wood floors are usually priced per square foot of coverage required.

Painting and Finishing

Painting and finishing are required to protect interior and exterior surfaces against wear and corrosion and to provide a coordinated finish appearance on the protected material. Normally, paints can be classified by their binders or vehicles. **Alkyds** are oil-modified resin used to manufacture fast-drying enamels. **Chlorinated rubber** produces coatings that are resistant to alkalies, acids, chemicals, and water. **Catalyzed epoxies** are two-part coatings that produce a hard film resistant to abrasion, traffic, chemicals, and cleaning. **Epoxy esters** are epoxies (modified with oil) that dry by oxidation, but are not as hard as catalyzed epoxies. **Latex binders** (commonly polyvinylacetate, acrylics, or vinyl acetate-acrylics) are binders mixed with a water base. **Silicone alkyd binders** are oil modified to produce coatings with heat resistance and high-gloss retention. **Urethanes** are isocynate polymers that are modified with drying oils or alkyds. **Vinyl coating solutions** are plasticized copolymers of vinyl chloride and vinyl acetate dissolved in strong solvents. **Zinc coatings** are primers high in zinc dust content dispersed in various vehicles to provide coatings to protect steel from oxidation.

Surface Preparation

In order to achieve successful painting, the surface to be painted must first be cleaned. Preparation methods for painting include solvent cleaning, hand-tool cleaning, (including wire brushing, scraping, chipping, and sanding), power-tool cleaning, white metal blasting (including removal of all rust and scale), commercial blasting (to remove oil, grease, dirt, rust), and scaling or brushing off blast (to remove oil, grease, dirt, and loose rust). These surface preparation processes can be costly and must be included in the estimate where required.

Primers, Sealers, and Topcoats

Fillers, primers, and sealers are manufactured for use with paint or clear coatings for metal, wood plaster, drywall, and masonry. Sanding fillers are available for clear coatings and block fillers for masonry construction.

Topcoats or finish coats should be specified and applied to meet service requirements. Finishes can be supplied for *normal* use and atmospheric conditions, *hard service* areas, and *critical* areas. Finishes can be furnished in gloss, semi-gloss, low lustre, eggshell, or flat. Intumescent coatings (expandable paints) are available to meet fire-retardant requirements.

Meeting Specified Requirements

As with most finishes, architects and designers will be particularly insistent about adherence to specifications and specified colors for painting. This is not an area in which to cut corners in estimating or performance of the work. Usually, the specifications clearly define acceptable materials, manufacturers, preparation, and application methods for each different type of surface to be painted. Many samples of various colors and/or finishes may be required for approval by the architect or owner before final decisions are made.

Determining Quantities

The materials and methods for painting should be included in the specifications. Areas to be painted are usually defined on a Room Finish Schedule and taken off from the plans and elevations in square feet of surface area. Odd-shaped and special items can be converted to an equivalent wall area. The table in Figure 17.10 includes suggested conversion factors for various types of surfaces. While the factors in Figure 17.10 are used to determine quantities of equivalent wall surface areas, the appropriate, specified application method must be used for pricing.

Pricing

The choice of application method will have a significant impact on the final cost. Spraying is very fast, but the costs of masking the areas to be protected may offset the savings. Oversized rollers may be used to increase production. Brushwork, on the other hand, is

Balustrades:		1 Side x 4
Blinds:	Plain	Actual area x 2
	Slotted	Actual area x 4
Cabinets:	Including interior	Front area x 5
Downspouts and Gutters:		Actual area x 2
Drop Siding:		Actual area x 1.1
Cornices:	1 Story	Actual area x 2
	2 Story	Actual area x 3
	1 Story Ornamental	Actual area x 4
	2 Story Ornamental	Actual area x 6
Doors:	Flush	Actual area x 1.5
	Two panel	Actual area x 1.75
	Four Panel	Actual area x 2.0
	Six Panel	Actual area x 2.25
Door Trim:		LF x 0.5
Fences:	Chain Link	1 side x 3 for both sides
	Picket	1 side x 4 for both sides
Gratings:		1 side x 0.66
Grilles:	Plain	1 side x 2.0
	Lattice	Actual area x 2.0
Mouldings:	Under 12" Wide	1 SF/LF
Open Trusses:		Length x Depth x 2.5
Pipes:	Up to 4"	1 SF per LF
	4" to 8"	2 SF per LF
	8" to 12"	3 SF per LF
	12" to 16"	4 SF per LF
	Hangers Extra	
Radiators:		Face area x 7
Sanding and Puttying:	Quality Work	Actual area x 2
	Average Work	Actual area x 0.5%
	Industrial	Actual area x 0.25%
Shingle Siding:		Actual area x 1.5
Stairs:		No. of risers x 8 widths
Tie Rods:		2 SF per LF
Wainscoting, Paneled:		Actual area x 2
Walls and Ceilings:		Length x Width no deductions for less than 100 SF
Window Sash:		1 LF of part = 1 SF

Figure 17.10

labor intensive. The specifications often include (or restrict) certain application methods. Typical coverage and man-hour rates are shown in Figure 17.11 for different application methods. These figures include masking and protection of adjacent surfaces.

Depending on local work rules, some unions require that painters be paid higher rates for spraying, and even for roller work. Higher rates also tend to apply for structural steel painting, for high work and for the application of fire-retardant paints. The estimator should determine which restrictions may apply.

Wall Coverings

Wall coverings are manufactured, printed, or woven in burlaps, jutes, weaves, grasses, paper, leather, vinyl, silks, stipples, corks, foils, sheets, cork tiles, flexible wood veneers, and flexible mirrors. Wallpaper, vinyl wall coverings, and woven coverings are usually available in different weights, backings, and quality. Surface preparation and adhesive selection are important considerations in placing wall coverings.

Determining Quantities

Wall coverings are usually estimated by the number of rolls. Single rolls contain approximately 36 S.F. This figure is used to determine the number of rolls required. Wall coverings are, however, usually sold in double or triple roll bolts.

Painting

Item	Coat	One Gallon Covers			In 8 Hrs. Man Covers			Man Hours per 100 S.F.		
		Brush	Roller	Spray	Brush	Roller	Spray	Brush	Roller	Spray
Paint wood siding	prime others	275 S.F. 300	250 S.F. 275	325 S.F. 325	1150 S.F. 1600	1400 S.F. 2200	4000 S.F. 4000	.695 .500	.571 .364	.200 .200
Paint exterior trim	prime 1st 2nd	450 525 575	— — —	— — —	650 700 750	— — —	— — —	1.230 1.143 1.067	— — —	— — —
Paint shingle siding	prime others	300 400	285 375	335 425	1050 1200	1700 2000	2800 3200	.763 .667	.470 .400	.286 .250
Stain shingle siding	1st 2nd	200 300	190 275	220 325	1200 1300	1400 1700	3200 4000	.667 .615	.571 .471	.250 .200
Paint brick masonry	prime 1st 2nd	200 300 375	150 250 340	175 320 400	850 1200 1300	1700 2200 2400	4000 4400 4400	.941 .364 .615	.471 .364 .333	.200 .182 .182
Paint interior plaster or drywall	prime others	450 500	425 475	550 550	1600 1400	2500 3000	4000 4000	.500 .571	.320 .267	.200 .200
Paint interior doors and windows	prime 1st 2nd	450 475 500	— — —	— — —	1300 1150 1000	— — —	— — —	.333 .696 .800	— — —	— — —

Figure 17.11

The area to be covered is measured by multiplying the length times height of wall above the baseboards in order to get the square footage of each wall. This figure is divided by 30 to obtain the number of single rolls, allowing 6 S.F. of waste per roll. One roll should be deducted for every two door openings. Two pounds of dry paste make about three gallons of ready-to-use adhesive, which will cover about 36 single rolls of light-to-medium weight paper, or 14 rolls of heavyweight paper. Application labor costs vary with the quality, pattern, and type of joint required.

With vinyls and grass cloths requiring no pattern match, a waste allowance of 10% is normal (approximately 3.5 S.F. per roll). Wall coverings which require a pattern match may have about 25%-30% waste, or 9-11 S.F. per roll. Waste can run as high as 50%-60% on wall coverings with a large, bold, or intricate pattern repeat.

Commercial wallcoverings are available in widths from 21" to 54", and in lengths from 5-1/3 yards (single roll) to 100-yard bolts. To determine quantities, independent of width, the linear (perimeter) footage of walls to be covered should be measured. The linear footage should then be divided by the width of the goods, to determine the number of "strips" or drops. The number of strips per bolt or package can be determined by dividing the length per bolt by the ceiling height.

$$\frac{\text{Linear Footage of Walls}}{\text{Width of Goods}} = \text{No. of Strips Required}$$

$$\frac{\text{Length of Bolt (roll)}}{\text{Ceiling Height}} = \text{No. of Strips (whole no.) per Bolt (roll)}$$

Finally, divide the quantity of strips required by the number of strips per bolt (roll) in order to determine the required amount of material, using the same waste allowances as above.

$$\frac{\text{No. of Strips Required}}{\text{No. of Strips per Bolt(roll)}} = \text{No. of Bolts (rolls)}$$

Surface preparation costs for wall covering must also be included. If the wall covering is to be installed over new surfaces, the walls must be treated with a wall sizing, shellac, or primer coat for proper adhesion. For existing surfaces, scraping, patching, and sanding may be necessary. Requirements will be included in the specifications.

Chapter 18

SPECIALTIES

Specialties include prefinished, manufactured items that are usually installed at the end of a project when other finish work is complete. Following is a partial list of items that may be included in Division 10–Specialties, of the specifications.

- Bathroom Accessories
- Bulletin Boards
- Chutes
- Control Boards
- Directory Boards
- Display Cases
- Key Cabinets
- Lockers
- Mailboxes
- Medicine Cabinets
- Partitions: folding accordion
  folding leaf
  hospital
  moveable office
  operable
  portable
  shower
  toilet
  woven wire
- Part Bins
- Projection Screens
- Security Gates
- Shelving
- Signs
- Telephone Enclosures
- Turnstiles

A thorough review of the drawings and specifications is necessary to be sure that all items are accounted for. The estimator should list each type of item and the recommended manufacturers. Often, no substitutes are allowed. Each type of item is then counted. Takeoff units will vary with different items.

Quotations and bids should be solicited from local suppliers and specialty subcontractors. The estimator must include all appropriate shipping and handling costs. When a specialty item is particularly large, job-site equipment may be needed for placement or installation.

The estimator should pay particular attention to the construction requirements which are necessary to Division 10 work, but are included in other divisions. Almost all items in Division 10 require some form of base, or backing, or preparation work for proper installation. These requirements may or may not be included in the construction documents, but are usually listed in the manufacturers' recommendations for installation. It is often stated in the General Conditions of the specifications that "the contractor shall install all products according to manufacturers' recommendations" – a catch-all phrase that places responsibility on the contractor (and estimator).

Preparation costs prior to the installation of specialty items may, in some cases, exceed the costs of the items themselves. The estimator must visualize the installation in order to anticipate all of the work and costs.

Partitions

Folding partitions, operable walls, and relocatable partitions are manufactured in a variety of sizes, shapes, and finishes. Operating partitions include folding accordion, folding leaf, (both shown in Figure 18.1), or individual panel systems. These units may be operated by hand or power. Relocatable partitions include the portable type, which are designed for frequent relocation, and the demountable type, for infrequent relocation.

Operating partitions are supported by aluminum, steel, or wood framing members. The panels are usually filled with sound-insulation material, as most partitions are rated by their sound reduction qualities. Panel skin materials include aluminum, composition board, fabric, or wood. The panels may be painted or covered with carpeting, fabric, plastic laminate, vinyl, or wood paneling. Large operating partitions are generally installed by factory specialists after the supporting members and framing have been supplied and erected by the building contractor.

Toilet Partitions, Dressing Compartments, and Screens

Toilet partitions, dressing compartments, and privacy screens are manufactured in a variety of materials, finishes, and colors. They are available in many stock sizes for both regular and handicapped-equipped installations. These partitions may be custom-fabricated to fit special size or use requirements, and may be supported from the floor, braced overhead, or hung from the ceiling or wall. Available finish materials include marble, painted metal, plastic laminate, porcelain enamel, and stainless steel. Various types of toilet partitions are shown in Figure 18.2.

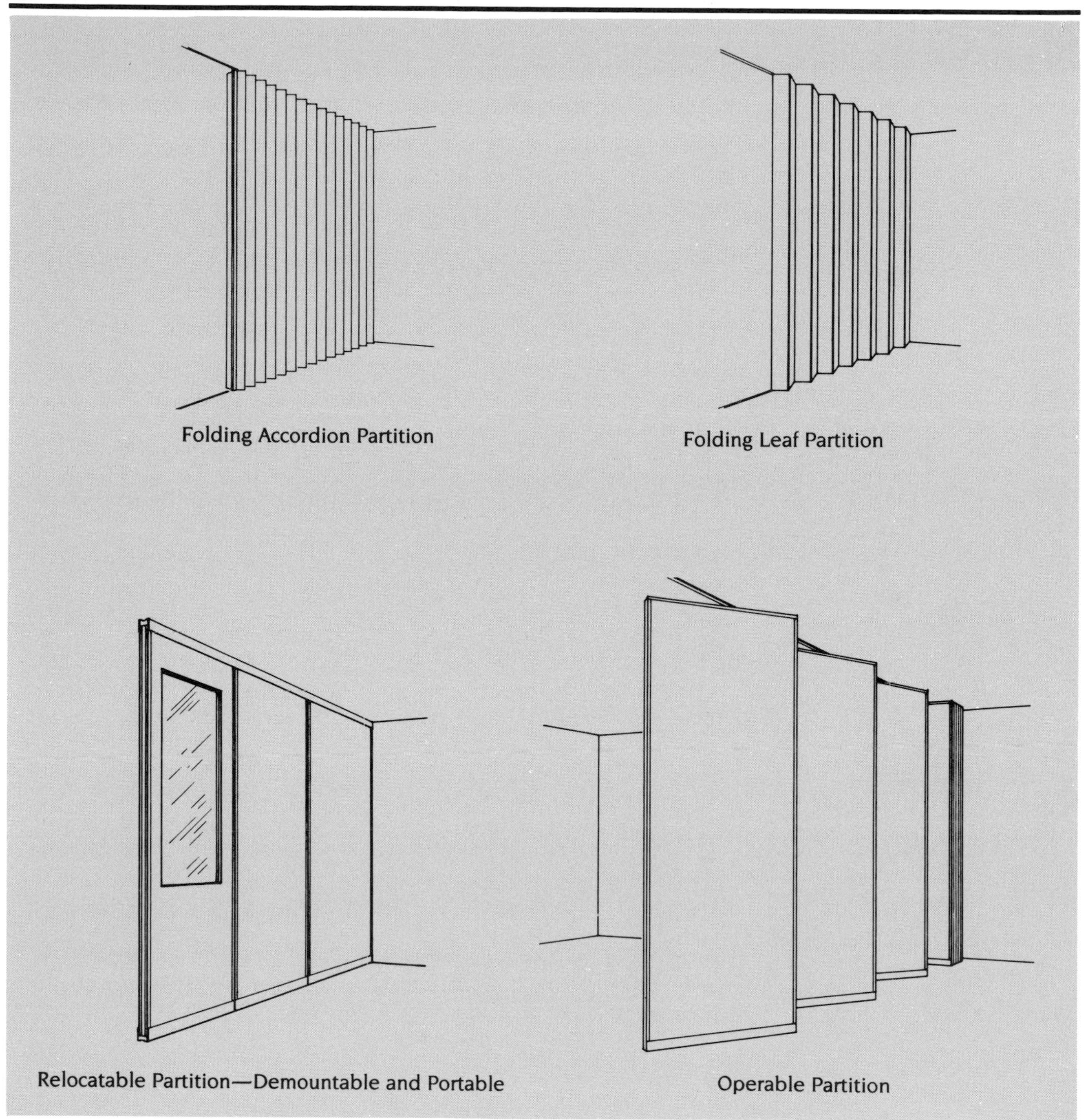
Folding Accordion Partition
Folding Leaf Partition
Relocatable Partition—Demountable and Portable
Operable Partition

Figure 18.1

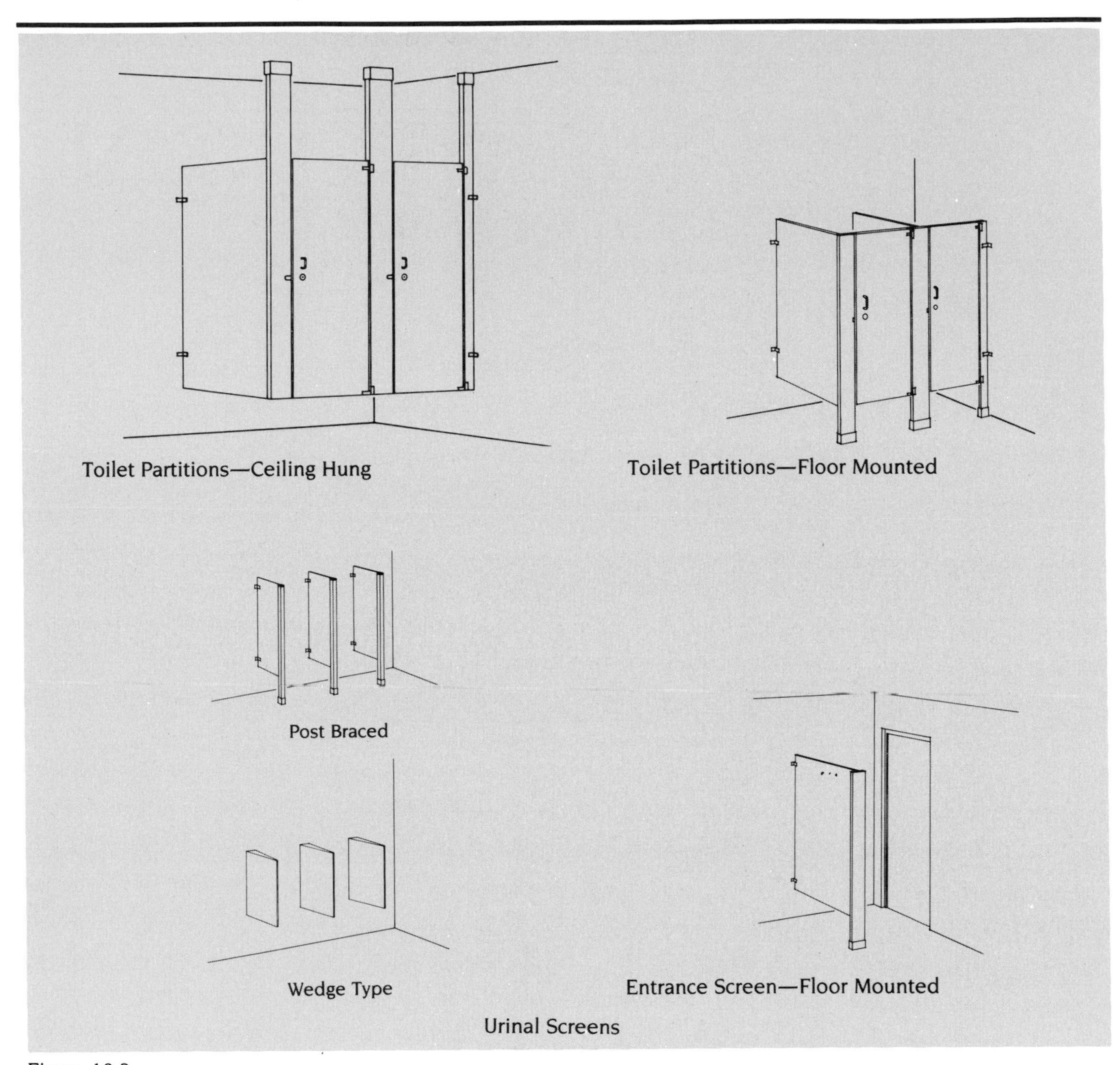
Toilet Partitions—Ceiling Hung
Toilet Partitions—Floor Mounted
Post Braced
Wedge Type
Entrance Screen—Floor Mounted
Urinal Screens

Figure 18.2

Chapter 19

ARCHITECTURAL EQUIPMENT

Architectural equipment includes the permanent fixtures that cause the space to function as designed – book stacks for libraries, vaults for banks. The construction documents may specify that the owner will purchase architectural equipment directly, and that the contractor will install it. In such cases, the estimator must include the installation cost of the equipment. If architectural equipment is furnished by the owner, it is common practice to add about 10% of the materials cost into the estimate. This allowance covers handling costs associated with the materials. Often the contractor is responsible for receipt, storage, and protection of these owner-purchased items until they are installed and accepted.

The following items might be included in this category:

- Appliances
- Bank Equipment
- Baking Equipment
- Church Equipment
- Checkout Counter
- Darkroom Equipment
- Dental Equipment
- Detention Equipment
- Health Club Equipment
- Kitchen Equipment
- Laboratory Equipment
- Laundry Equipment
- Medical Equipment
- Movie Equipment
- Refrigerator Food Cases
- Safe
- Sauna
- School Equipment
- Stage Equipment
- Steam Bath
- Vocational Shop Equipment
- Waste Handling Equipment
- Wine Vault

Like specialties (described in Chapter 18), architectural equipment must also be evaluated to determine what is required from other divisions for its successful installation. Some possibilities are:

- Concrete Work
- Miscellaneous Metal Supports
- Rough Carpentry and Backing
- Mechanical Coordination
- Electrical Requirements

Division 11 – Architectural Equipment includes equipment that can be packaged and delivered complete or partially assembled by the factory. Also, some items can or must be purchased and installed by an authorized factory representative. The estimator must investigate these variables in order to include adequate costs.

Chapter 20
FURNISHINGS

Numerous furnishings may be specified in the plans for remodeling projects. Some may be mass produced, while others may be custom-made, one-of-a-kind items. Furnishings may be supplied by the owner and installed by the vendor, supplied by the owner and installed by the contractor, or carried under a specified allowance.

Before beginning a furniture estimate, the estimator should carefully review both the furnishings drawings and the complete set of construction documents and specifications. Items such as casework, wall art, or banquette seating in a restaurant may not be specified in the furnishings drawings, but will be found in the construction documents. A systematic method should be used to ensure that all furnishings items are accounted for. The full set of drawings could be scanned and any item that may affect the furnishings costs marked with a colored pencil. Use of a standardized form can also aid in accounting for all items in the project.

A system for labeling each type or style component is also helpful For example, all tables of a particular type or style could be labeled T-1, another style T-2, and yet another T-3. The estimator could then take off each style table specified, and record them by category and quantity. T-1:43 would indicate 43 tables of the specific type and style denoted by T-1. This coding system is helpful both for quick reference and for accounting purposes.

Furniture can also be estimated using the *systems estimating method*. This method involves the takeoff, not by individual items, but by work stations or groupings.

Furniture

Office Furniture

Office materials are generally made of wood, metal, or plastic laminate. Differences in the construction of drawer glides, detailing, and the configuration of drawers and compartments distinguish the various products available. The estimator should verify the fact that two apparently identical desks (from two manufacturers) are actually the same. The construction can differ greatly and may affect the cost appreciably.

Chairs vary in style and quality. Office seating is usually ergonomic in design, but each style may offer different features. The specifications should be carefully reviewed, as each style chair within a given price range may have unique options. For example, lounge chairs vary in price based on construction quality and covering fabric.

Tables for office use may range from specific name designer tables to generic drum and sled base tables. The design and style called for will determine competitiveness of bidding. Designer furniture, for example, items that are patented by a specific designer, may be available through one specific dealership. Generic items, on the other hand, such as sled base and drum tables, can be obtained through many furniture dealerships and bidding for these items may be competitive.

Conference tables are usually fabricated by specialty manufacturers. They may be made of wood, plastic laminate, or stone (marble or granite). Prices should be obtained from conference table manufacturers. There may be added costs for shipping, crating, and delivery, due to special size or special handling requirements. Bases for conference tables may vary greatly in type and style. Before adding the cost of shipping, delivery, and installation, the base specifications should be noted, since size can affect these costs significantly.

The quality, style, and price of office furnishings available may vary greatly. Therefore, the estimator should determine exactly what is being specified by technical description if not by specific manufacturer. Any "grey areas" should be clarified by the specifier, as bidding can be competitive.

Hotel Furniture

Hotel furnishings include bed frames, headboards, mattresses, benches, desks, chairs, dressers, guest tables, mirrors, sleep sofas, and cocktail and end tables. Specifications should be reviewed to determine the requirements for each of these items. Large quantities of hotel furniture may be ordered from a hotel furniture manufacturer, with quantity discounts if no special tooling is required. However, availability of the specified items should be verified, so that bids are not based on discontinued items. Substitutions could otherwise incur additional costs.

Restaurant Furniture

The manufacture and sale of restaurant furnishings can often be a competitive business. Since the cost of chairs and tables of all types, styles, sizes and finishes can vary greatly, the amount of detail specified should first be determined. The specified quality of upholstered seating and finishes may affect costs significantly, especially if the fabric or finish is available from only a single manufacturer. Fabric may be COM (Customer's Own Material), ordered from a fabric manufacturer and delivered to a furniture manufacturer, who then covers a specific style of chair with the material.

The cost of tables varies considerably, depending on the quality specified. Restaurant table tops and bases are generally sold separately, and, therefore, should be priced separately. The type of top and base required should be determined from the specifications. Top materials may be wood, glass, metal, or stone, with various edge and top finishes. Bases may be made of wood, metal, or stone. Tops and bases are usually shipped separately, the top having predrilled holes for installation of the base. An additional cost for attaching the table to the base should also be considered when figuring installation cost.

Dormitory Furniture

Dormitory furniture includes all bookcases, beds, chairs, bureaus, desks, dressing units, mattresses, mirrors, night stands, and wardrobes. These units are usually made of wood, plastic laminate, and/or metal, or combinations thereof.

When estimating the cost of dormitory furniture, material and installation requirements should be calculated separately. Additional material, such as hardware, or additional labor may be necessary to install these units.

Library Furniture

Library furniture includes book stacks, attendant desks, book display items, book trucks, card catalogues, study carrels (both single and double face), chairs, charge desks, dictionary stands, exhibit cases, globe stands, newspaper and magazine racks, and tables for card catalogue reference and study use. Bids for these types of furnishings should be obtained from a specialty contractor, since special detailing may be required in many of the pieces of casework to meet the library's functional requirements.

School Furniture

School furniture includes chairs and desks made of molded plastic or wood and metal. Estimates should be obtained from local furniture dealers specializing in school furniture.

Furnishings Items

The following section describes items often encountered in furniture estimates. These items are often ordered through a commercial furniture dealership, an interior designer, or an architectural office with purchasing services. Since costs may vary greatly, quotations for specific items and quantities should be obtained. These quotes are often only good for a limited period of time, and any time limit should be noted.

Stack and Folding Chairs

Chairs may be metal, wood, or plastic. Before estimating the number of chairs required, the estimator can find out from the manufacturer how many chairs can be stored/stacked on each dolly, or carrier. This information can then be used to determine the number of dollies and carriers required. If upholstered chairs are specified, the fabric should be determined. This cost can be significant, especially for stacking chairs used for hotel ballroom projects, since a large quantity of these chairs are usually required. Floor glides for chairs differ depending on whether they are intended for use with carpet or hard floors. The cost of the glides could affect the cost of the chairs. Another factor is the requirement for tablet arms, and whether these devices, if required, are to be operable or fixed. Chair finishes also vary – from lacquered wood to painted metal. Therefore, the estimator should check the specifications for all finishes prior to determining cost.

Booths

Booths, also known as *fixed seating*, are available in a variety of sizes and shapes. The cost varies according to the type of seating specified. For example, banquette upholstered seating for restaurant use differs in price from other types of fixed seating. The latter consist only of one-piece plastic chairs and plastic-laminated table-top units such as those in fast food restaurants. When banquette seating is called for, the estimator should check the specifications and drawings to verify the exact requirements. In many cases banquette seating should be estimated by a specialty contractor, especially where curved units are required, since these are not manufactured, standard items.

Multiple Seating

Multiple seating is used in airports, hotel lobbies, and reception areas. Costs vary depending on the construction specifications. Numerous types, styles, and materials are available. Therefore, the specifications and drawings should be carefully reviewed to determine what is required.

Lecture Hall Seating

Lecture hall seating requires pedestal-mounted (floor-mounted) seats. The cost for this kind of seating is affected by the following factors: the amount and quality of upholstery required, whether a veneer surface is called for, whether the tablet arm is to be retractable or fixed, and the type of material the shell is to be made of (wood, metal, or plastic). Design and style are other considerations that may affect the cost of lecture hall seating.

Upholstery

Upholstery may be used in interior projects on chairs or as wall covering. Upholstery materials include various blends of nylons, polyester, silk, or wool. To estimate the cost of upholstery, the specifications should be carefully reviewed to determine the type of weave, weight, color, pattern requirements, and fabric width. There may be dye lot variations in various rolls. Therefore, the quantity of material needed must be carefully estimated to ensure that an adequate amount of material from the same dye lot is available for the job.

Fabric treatments may include flame-retardant, acrylic backing, paper backing, and stain protection. When estimating the cost for fabric treatments, costs for testing of fabric/treatment compatibility should also be included.

Folding Tables

Folding tables are used in conference centers, hotels, schools, and wherever large assemblies of people are expected for dining or conference purposes. Shapes may be rectangular or circular. Sizes vary from conventional depths of 18″, 30″, and 36″, and lengths of 48″, 60″, 72″ and 96″. Rounds are usually 60″ and 72″ in diameter. Carts for storage and moving these items should be included in the estimate.

Files & File Systems

There are grades of filing systems for different uses, ranging from *light residential* to *heavy administrative* use. Files may be two to four drawer, vertical or lateral. The specifications should be carefully checked since the specifier has determined the appropriate type of file for each function. Floor loading capacities must also be considered.

Where large, mobile filing systems are used with floor-mounted tracks in filing banks, or for file systems on rollers or rotating banks, a specialty supplier and/or contractor should be consulted for accurate pricing. The price of these systems may vary. The specialty contractor is experienced in installing these file systems (and, therefore, familiar with the costs). A structural engineer should be consulted whenever excessive loads are to be placed on a floor not specifically designed for such a purpose.

Cabinets and Countertops

Cabinets for kitchen, hospital, and residential use are generally prefabricated in fixed sizes, ready for installation. Plastic laminated countertops are available in a variety of widths and may be stock or custom made. Laminated tops are usually cut and installed by a specialty contractor skilled in assembling laminated products, since special post forming machines are required to make plastic laminated curved surfaces. Costs should be adjusted, however, when small quantities are ordered.

Blinds

Blinds may be either vertical or horizontal, in a variety of finishes including vinyl, metal, fabric, and wood. Slat sizes and the type of operation may vary. The estimator should verify all options in the specifications. The finish for both sides should be checked, since the two sides may differ and this may affect the cost. Unique installations, such as door-hung blinds that require "hold down clips," or blinds in sloped skylights that require special side rails to keep the blinds in place, should be verified with a local specialty contractor to ensure accurate estimates.

Panels and Dividers

Panels and dividers for office use vary in type, style, and size. They may be powered or non-powered, with a fabric edge or hard edge of metal, plastic, or wood. The acoustical rating of panels varies with each manufacturer. Available colors and types of connectors also vary.

Coat Racks & Wardrobes

Coat racks and wardrobes are usually wood or metal, with components that vary from project to project. The specifications should be checked to determine the height, width, number of hangers required, and the amount of hat storage required above. (Hat storage is indicated by the number and width of shelves at the top of the unit). Finishes should be verified. If custom colors are required, a specialty manufacturer should be consulted, as there may be a surcharge for custom colors.

Floor Mats

Floor mats are usually made to order. Size, material, and installation should be considered when estimating. Material for floor mats may be recessed units made of rubber, aluminum, or synthetics. Floor mat sizes should be confirmed and field-measured prior to ordering. Tile units, for example, come in boxes with a certain amount per box. The specific manufacturer should be consulted to determine the number of boxes required. Installation procedures may vary, from mechanical fasteners to glue, and should be confirmed with the manufacturer.

Ash/Trash Receivers

Trash receivers come in a variety of shapes and sizes with various options. Ash/Trash receivers may be simple sand urns, or units with mechanical self-cleaning tops. The estimator should check the furnishings specifications for the type of operation required and other options, as well as finishes, as these have an impact on the cost of the unit.

Packing, Crating, Insurance, Shipping, and Delivery

There is often an added cost for packing, crating, insurance, shipping, and delivery of furnishings. This added cost should be anticipated for each furnishings item where applicable. The specifications dictate how the furniture is to be packed and crated. Otherwise, large pieces of furniture might be blanket-wrapped at no additional cost and put on a truck without the extra protection of a crate or proper packing. Large pieces of furniture, such as hotel and office items like chairs and desks, should be packed and crated. The extra charge varies depending on the size of the piece of furniture. Since crating and additional insurance charges are extra, they apply to items that are of "high value." The specifications should be checked for this requirement.

Shipping costs for furnishings can generally be estimated at 10% to 15% of the furniture cost. Furnishings may be delivered with an FOB (Freight on Board) designation, and are drop-shipped at the receiving dock.

Delivery time can vary from project to project. Delivery time for small, stock orders will be 30 to 120 days. For medium-sized office and hotel projects, or for custom orders, delivery can be from 60 to 120 days. Therefore, for each furniture item, delivery time should be anticipated. Many manufacturers dealing in office furnishings offer a quick ship/delivery program, but this is usually only for certain basic stock items offered in limited quantities, since they do not usually store large inventories of office furniture.

Delivery charges are generally extra. For example, if furnishings are for the fourth floor of an office building, the contract may be only for delivery to the loading dock of the address specified. Provisions for storage and handling should be made ahead of time for large quantities of furnishings, and the costs appropriately included.

Chapter 21
SPECIAL CONSTRUCTION

A partial list of items in the Special Construction section, Division 13, of the specifications and drawings appears below. Subcontractor quotations should be carefully evaluated to make sure that all required items are included. Some of the materials may have to be supplied and/or installation performed by other suppliers or subcontractors.

- Acoustical Enclosures
- Air Curtains
- Air-Supported Structures
- Greenhouses
- Integrated Ceilings
- Pedestal Access Floors
- Refrigerators - Walk-in
- Shielding
- Sports Courts
- Swimming Pools
- Storage Tanks
- Vault Front

It is a good idea to review this portion of the project with the subcontractor(s) to determine both the exact scope of the work and those items that are not covered by the quotation. If the subcontractor requires services such as excavation, unloading, or other temporary work, then these otherwise excluded items must be included elsewhere in the estimate.

Often, manufacturers of these products require that only trained personnel install the product. Otherwise, material and performance warranties may be voided.

The specialty subcontractor can provide more detailed information concerning the system. The more detailed the estimator's knowledge of a system, the easier it may be to subdivide the costs into material, labor, and equipment to fully identify the direct cost of the specialty item for future purposes.

Chapter 22
CONVEYING SYSTEMS

Conveying systems used in remodeling may include, but are not limited to, the following:

- Correspondence Lifts
- Dumbwaiters
- Elevators
- Escalators
- Material Handling Conveyers
- Motorized Car Distribution Systems
- Moving Ramps and Walks
- Parcel Lifts
- Pneumatic Tube Systems
- Vertical Conveyers

Current quotations from competent contractors should be obtained for all of the items listed above, if specified and shown on the plans. Budget costs are available and should, when used properly, allow sufficient money to cover material and installation costs for new buildings. Installation in existing buildings must be priced for each individual project. It should be noted that new hydraulic elevators with telescoping shafts have been developed to combat some of the problems encountered when installing elevators in existing buildings. A checklist for the various systems is shown in Figure 22.1.

The following illustrations show typical conveying systems used in renovation projects. Figures 22.2a and 22.2b are schematic drawings of typical conveying systems. Figures 22.3a and 22.3b illustrate an elevator selective cost sheet from *Means Building Construction Cost Data*, 1989. This chart can be used to develop budget costs for various types of elevators, based on specific project requirements. Note the number of variables that can significantly affect the cost of elevators.

Conveying Systems

Dumbwaiters:

- capacity
- floors
- speed
- size
- stops
- finish

Elevators:

- hydraulic or electric
- capacity
- floors
- stops
- finish
- door type
- special requirements
- geared or gearless
- size
- number required
- speed
- machinery location
- signals

Material handling systems:

- automated
- non-automated

Moving stairs and walks:

- capacity
- floors
- story height
- finish
- incline angle
- size
- number required
- speed
- machinery location
- special requirements

Pneumatic tube systems:

- automatic
- size
- length
- manual
- stations
- special requirements

Vertical conveyer:

- automatic
- non-automatic

Figure 22.1

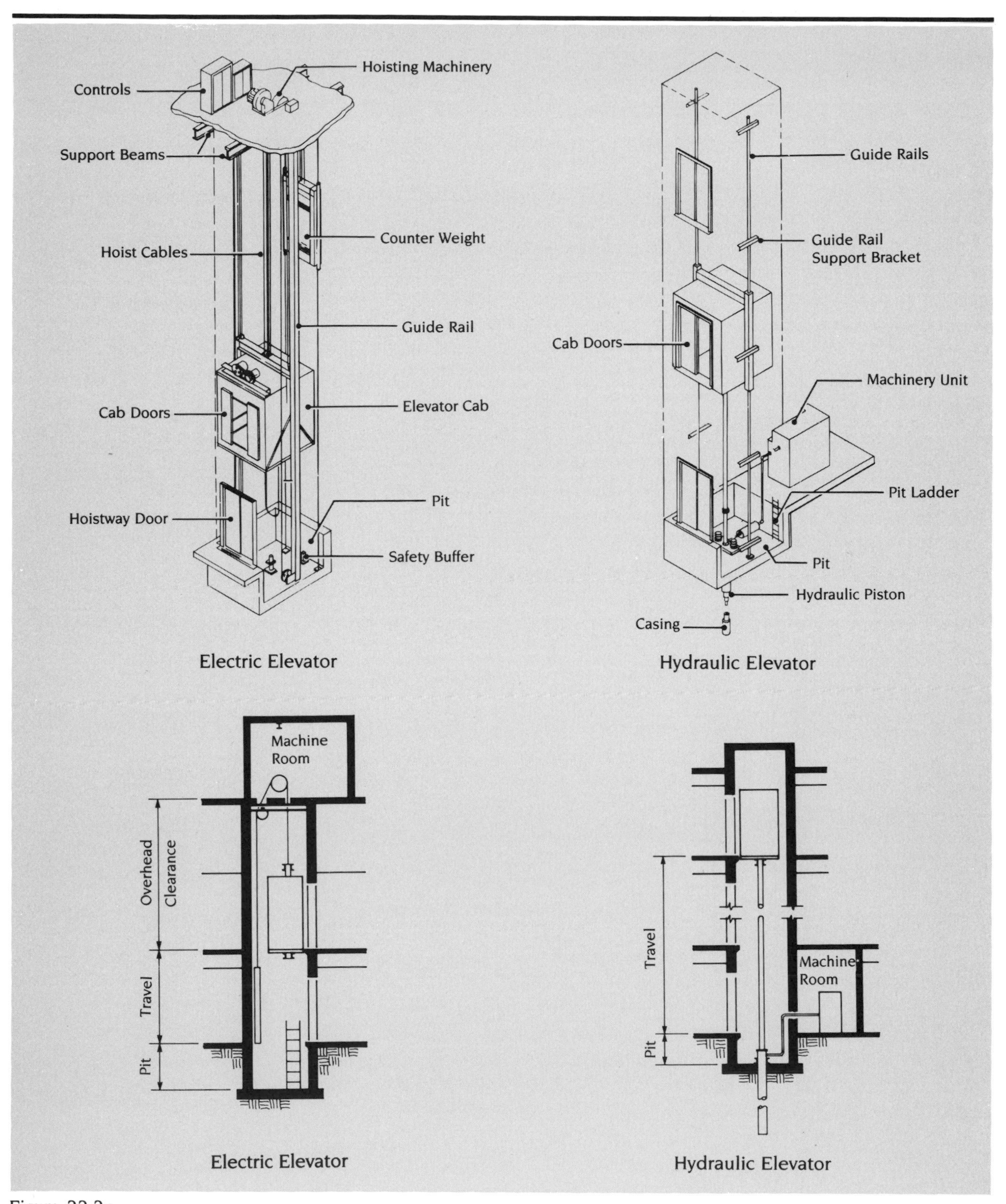
Controls
Hoisting Machinery
Support Beams
Hoist Cables
Counter Weight
Guide Rail
Cab Doors
Elevator Cab
Hoistway Door
Pit
Safety Buffer
Electric Elevator
Guide Rails
Guide Rail
Support Bracket
Cab Doors
Machinery Unit
Pit Ladder
Pit
Hydraulic Piston
Casing
Hydraulic Elevator
Machine
Room
Overhead
Clearance
Travel
Pit
Electric Elevator
Travel
Pit
Machine
Room
Hydraulic Elevator

Figure 22.2a

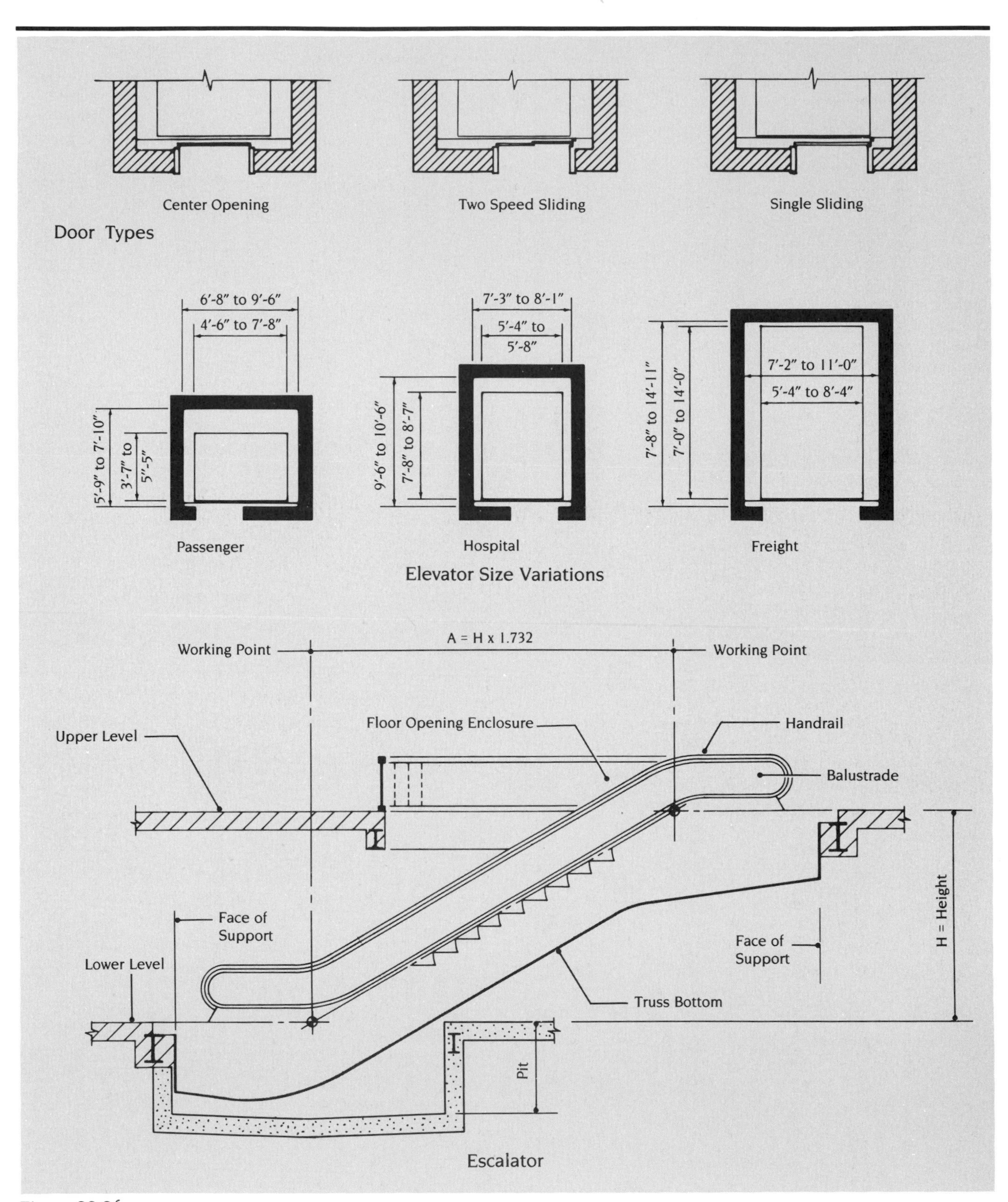

Center Opening
Two Speed Sliding
Single Sliding
Door Types
6'-8" to 9'-6"
4'-6" to 7'-8"
5'-9" to 7'-10"
3'-7" to 5'-5"
Passenger
7'-3" to 8'-1"
5'-4" to 5'-8"
9'-6" to 10'-6"
7'-8" to 8'-7"
Hospital
7'-8" to 14'-11"
7'-0" to 14'-0"
7'-2" to 11'-0"
5'-4" to 8'-4"
Freight
Elevator Size Variations
A = H x 1.732
Working Point
Working Point
Floor Opening Enclosure
Handrail
Upper Level
Balustrade
H = Height
Face of Support
Face of Support
Lower Level
Truss Bottom
Pit
Escalator

Figure 22.2*b*

CIRCLE REFERENCE NUMBERS

(123) Elevator Selective Costs (Div. 142-010)

	Passenger		Freight		Hospital	
A. Base Unit	Hydraulic	Electric	Hydraulic	Electric	Hydraulic	Electric
Capacity	1500 Lb.	2000 Lb.	2000 Lb.	4000 Lb.	3500 Lb.	3500 Lb.
Speed	50 F.P.M.	100 F.P.M.	25 F.P.M.	50 F.P.M.	50 F.P.M.	100 F.P.M.
#Stops/Travel Ft.	2/20	4/40	2/20	4/40	2/20	4/40
Push Button Oper.	Yes	Yes	Yes	Yes	Yes	Yes
Telephone Box & Wire	"	"	"	"	"	"
Emergency Lighting	"	"	No	No	"	"
Cab	Painted Steel	Painted Steel	Painted Steel	Painted Steel	S.S. Wainscot, Baked	Enamel Above
Cove Lighting	Yes	Yes	No	No	Yes	Yes
Floor	V.A.T.	V.A.T.	Wood w/Safety Treads	Wood w/Safety Treads	V.A.T.	V.A.T.
Doors, & Speedside Slide	Yes	Yes	No	No	Yes	Yes
Gates, Manual	No	No	Yes	Yes	No	No
Signals, Lighted Buttons	Car Only	Car Only	In Use Light	In Use Light	Car and Hall	Car and Hall
O.H. Geared Machine	N.A.	Yes	N.A.	Yes	N.A.	Yes
Variable Voltage Contr.	"	"	"	"	"	"
Emergency Alarm	"	"	"	"	"	"
Class "A" Loading	"	N.A.	Yes	"	"	N.A.
Base Cost	$40,100	$55,000	$33,000	$51,000	$45,000	$66,500
B. Capacity Adjustment						
2,000 Lb.	$ 2,500					
2,500	4,800	$ 2,600	$ 2,500			
3,000	5,600	3,000	2,900			
3,500	7,000	4,200	4,500			
4,000	7,400	5,100	5,900		$ 4,500	$ 3,000
4,500	8,600	6,000	6,100		5,150	4,000
5,000	9,600	7,300	7,500	$ 2,700	7,500	5,200
6,000			9,000	4,250		
7,000			11,000	4,250		
8,000			13,200	13,000		
10,000			15,500	15,500		
12,000			19,250	19,250		
16,000			22,500	22,500		
20,000			26,500	27,600		
C. Travel Over Base	$ 535 V.L.F.	$ 185 V.L.F.	$ 640 V.L.F.	$ 185 V.L.F.	$ 640 V.L.F.	$ 185 V.L.F.
D. Additional Stops	$ 5,025 Ea.	$ 5,025 Ea.	$ 4,500 Ea.	$ 4,500 Ea.	$ 5,400 Ea.	$ 5,200 Ea.
E. Speed Adjustment						
50 F.P.M.			$ 575			
75	$ 575		900	$ 1,975	$ 650	
100	1,100		1,400	3,075	1,110	
125	1,400		2,575	4,200	1,600	
150	3,675		3,500	5,350	2,550	
175			4,000	6,400	3,200	
200		$ 4,950		7,000		$ 5,800
Geared 250		6,950		8,200		7,950
4 Flrs. 300		8,550		10,200		9,550
Min. 350		9,800		11,700		10,600
400		10,700		12,800		11,700
500		25,700		16,400		26,700
Gearless 600		29,900		18,000		29,900
10 Flrs. 700		31,000		21,000		30,900
Min. 800		38,500		23,500		38,500
1,000		Spec. Applic.		Spec. Applic.		Spec. Applic.
1,200		"		"		"
F. Other Than Class "A" Loading						
"B"			$ 1,450	$ 1,450		
"C-1"			2,400	2,400		
"C-2"			1,550	1,550		
"C-3"			3,000	3,000		

Figure 22.3*a*

CIRCLE REFERENCE NUMBERS

(123) Elevator Selective Costs (cont.)

	Passenger	Freight	Hospital
G. Options			
1. Controls			
Automatic, 2 car group	$ 3,750		$ 3,750
3 car group	6,300		6,300
4 car group	7,500		7,500
5 car group	9,200		9,200
6 car group	13,000		13,000
Emergency, fireman service	2,500		2,500
Intercom service	1,700		1,700
Selective collective, single car	2,600		2,600
Duplex car	4,100		4,100
2. Doors			
Center opening, 1 speed	$ 875		$ 875
2 speed	1,200		1,200
Rear opening-opposite front	—		5,500
Side opening, 2 speed	1,200		1,200
Freight, bi-parting	—	$3,000	—
Power operated door and gate	—	6,900	—
3. Emergency power switching, automatic	$ 2,225		$ 2,225
Manual	1,175		1,175
4. Finishes based on 3500# cab			
Ceilings, acrylic panel	$ 225		—
Aluminum egg crate	310		$ 230
Doors, stainless steel	475		230
Floors, carpet, class "A"	100		—
Epoxy	215		230
Quarry tile	175		230
Slate	200		—
Steel plate	—	$ 580	—
Textured rubber	90		65
Walls, plastic laminate	350		280
Stainless steel	765		600
Return at door	430		430
Steel plate, 1/4" x 4' high, 14 ga. above	—	1,350	—
Entrance, doors, baked enamel	325		325
Stainless steel	450		450
Frames, baked enamel	300		300
Stainless steel	600		600
5. Maintenance contract - 12 months	$ 3,200	$2,000	$ 3,500
6. Signal devices			
Hall lantern, each	$ 375	$ 375	$ 375
Position indicator, car or lobby	270	270	270
Add for over three each	65	65	65
7. Specialties			
High speed, heavy duty door opener	$ 600		$ 600
Variable voltage, O.H. gearless machine	27,000 - 60,000		27,000 - 60,000
Basement installed geared machine	6,800	$6,800	6,800

Figure 22.3b

Chapter 23

MECHANICAL

It is doubtful that the remodeling estimator can spend the time or is capable of preparing a realistic unit price estimate for the mechanical trades. A reliable sub-bid should be sought for the mechanical portion of the interior work as soon as specific requirements are known. As a check, or for preliminary budgets, costs can be quickly calculated using the systems approach to mechanical estimating. A systems estimate is basically accomplished by counting the fixtures in the plumbing portion, establishing the class of fire protection required, and determining the heat and air conditioning load, source, and method of distribution. Appropriate systems costs are applied based on determined quantities and/or the size and occupancy use of the project.

For renovation projects, costs for the mechanical portion of the estimate must be estimated by using good judgment and past experience. The "cutting and patching" may be extensive, and existing conditions may severely restrict normal work procedures; both factors can significantly affect cost.

Mechanical work is usually broken down into three basic systems:

- Plumbing
- Fire Protection
- Heating, Ventilating and Air Conditioning

Plumbing

In order to determine a good budget estimate for plumbing, the estimator must count the different fixture types and also determine other specific requirements. Systems costs can be applied per bathroom as shown in Figure 23.1, from *Means Repair & Remodeling Cost Data*, 1989. Note that listed in the "systems components" section are quantities for rough-in, partitions, and accessories, and the fixture itself.

Prior to or during design development, the estimator may be required to determine or anticipate plumbing fixture requirements for the proposed project. Figure 23.2, (from *Means Plumbing Cost Data*, 1989) adapted from the BOCA Plumbing Code, can be used to determine fixture requirements at this stage. For final design purposes, local building officials must be consulted to ensure compliance with applicable codes and requirements.

For many remodeling projects, plumbing stacks already exist in the space. Toilet rooms and other fixtures are usually located close to this piping. However, when the plumbing fixtures are located away from the stacks or if the stacks are to be part of the work being estimated, costs must be determined and included. For small quantities, costs for pipe (linear feet) and fittings (each) can be

MECHANICAL	08.1-710	Plumbing - Public Restroom

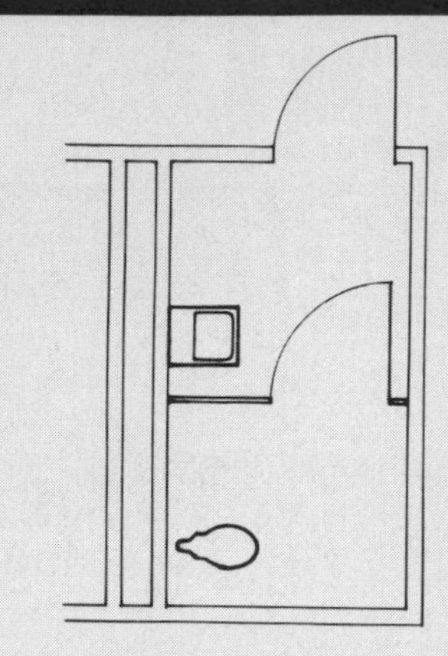

This page illustrates and describes a women's public restroom system including a water closet, lavatory, accessories, and service piping. Lines 08.1-710-04 thru 10 give the unit price and total price on a cost each basis for this system. Prices for alternate women's public restroom systems are on Line Items 08.1-710-14, thru 17. Both material quantities and labor costs have been adjusted for the system listed.

Factors: To adjust for job conditions other than normal working situations use Lines 08.1-710-29 thru 40.

Examples: You are to install the system and protect surrounding area from dust. Go to Line 08.1-710-31 and apply these percentages to the MAT. and INST. costs.

LINE NO.	DESCRIPTION	QUANTITY	COST EACH MAT.	INST.	TOTAL
01	**Public women's restroom incl. water closet, lavatory, accessories and**				
02	**Necessary service piping to install this system in one wall.**				
03					
04	Water closet, wall mounted, one piece	1 Ea.	271.70	93.30	365
05	Rough-in waste and vent for water closet	1 Set	147.84	262.16	410
06	Lavatory, 20" x 18" P.E. cast iron with accessories	1 Ea.	143	67	210
07	Rough-in waste and vent for lavatory	1 Set	122.46	322.54	445
08	Partition, painted metal between walls, floor mounted, access.	1 Ea.	372.90	118.10	491
09	Accessories	1 Set	564.30	70.70	635
10	TOTAL	System	1622.20	933.80	2556
11					
12					
13	For alternate size restrooms:				
14	Two water closets, two lavatories	System	2737.30	1779.70	4517
15					
16	For each additional water closet over 2, add	System	687.94	442.06	1130
17	For each additional lavatory over 2, add	"	438.16	417.84	856
18					
19					
20					
21					
22					
23					
24	Note: PLUMBING APPROXIMATIONS				
25	WATER CONTROL: water meter, backflow preventer,				
26	Shock absorbers, vacuum breakers, mixer....10 to 15% of fixtures				
27	PIPE AND FITTINGS: .... 30 to 60% of fixtures				
28					
29	Cut & patch to match existing construction, add, minimum		2%	3%	
30	Maximum		5%	9%	
31	Dust protection, add, minimum		1%	2%	
32	Maximum		4%	11%	
33	Equipment usage curtailment, add, minimum		1%	1%	
34	Maximum		3%	10%	
35	Material handling & storage limitation, add, minimum		1%	1%	
36	Maximum		6%	7%	
37	Protection of existing work, add, minimum		2%	2%	
38	Maximum		5%	7%	
39	Shift work requirements, add, minimum			5%	
40	Maximum			30%	
41					
42					

302

Figure 23.1

PLUMBING	C8.1-400	Individual Fixture Systems

Table 8.1-401 Minimum Plumbing Fixture Requirements

TYPE OF BUILDING/USE	WATER CLOSETS		URINALS		LAVATORIES		BATHTUBS OR SHOWERS		DRINKING FOUNTAIN	OTHER
	Persons	Fixtures	Persons	Fixtures	Persons	Fixtures	Persons	Fixtures	Fixtures	Fixtures
Assembly Halls Auditoriums Theater Public assembly	1-100 101-200 201-400	1 2 3	1-200 201-400 401-600	1 2 3	1-200 201-400 401-750	1 2 3			1 for each 1000 persons	1 service sink
	Over 400 add 1 fixt. for ea. 500 men: 1 fixt. for ea. 300 women		Over 600 add 1 fixture for each 300 men		Over 750 add 1 fixture for each 500 persons					
Assembly Public Worship	300 men 150 women	1 1	300 men	1	men women	1 1			1	
Dormitories	Men: 1 for each 10 persons Women: 1 for each 8 persons		1 for each 25 men, over 150 add 1 fixture for each 50 men		1 for ea. 12 persons 1 separate dental lav. for each 50 persons recom.		1 for ea. 8 persons For women add 1 additional for each 30. Over 150 persons add 1 for each 20.		1 for each 75 persons	Laundry trays 1 for each 50 serv. sink 1 for ea. 100
Dwellings Apartments and homes	1 fixture for each unit				1 fixture for each unit		1 fixture for each unit			
Hospitals Indiv. Room Ward Waiting room	8 persons	1 1 1			10 persons	1 1 1	20 persons	1 1	1 for 100 patients	1 service sink per floor
Industrial Mfg. plants Warehouses	1-10 11-25 26-50 51-75 76-100	1 2 3 4 5	0-30 31-80 81-160 161-240	1 2 3 4	1-100 over 100	1 for ea. 10 1 for ea. 15	1 Shower for each 15 persons subject to excessive heat or occupational hazard		1 for each 75 persons	
	1 fixture for each additional 30 persons									
Public Buildings Businesses Offices	1-15 16-35 36-55 56-80 81-110 111-150	1 2 3 4 5 6	Urinals may be provided in place of water closets but may not replace more than 1/3 required number of men's water closets		1-15 16-35 36-60 61-90 91-125	1 2 3 4 5			1 for each 75 persons	1 service sink per floor
	1 fixture for ea. additional 40 persons				1 fixture for ea. additional 45 persons					
Schools Elementary	1 for ea. 30 boys 1 for ea. 25 girls		1 for ea. 25 boys		1 for ea. 35 boys 1 for ea. 35 girls		For gym or pool shower room 1/5 of a class		1 for each 40 pupils	
Schools Secondary	1 for ea. 40 boys 1 for ea. 30 girls		1 for ea. 25 boys		1 for ea. 40 boys 1 for ea. 40 girls		For gym or pool shower room 1/5 of a class		1 for each 50 pupils	

Figure 23.2

estimated and can be included separately. For budgeting large projects, percentage multipliers, and rules of thumb, such as those shown in Figures 23.3 and 23.4, can be useful.

Fire Protection

Square foot costs for fire protection systems should be developed from past projects for budget purposes. These costs may be based on the relative hazards of occupancy - *light*, *ordinary*, and *extra*—and on the type of system. Some requirements of the different level hazards are shown in Figure 23.5.

Consideration must also be given to special or unusual requirements. For example, many architects specify that sprinkler heads must be located in the center of ceiling tiles. Each head may require extra elbows and nipples for precise location. Recessed heads are more expensive. Special dry pendent heads are required in areas subject to freezing. When installing a sprinkler system in an existing structure, a completely new water service may be required in addition to the existing domestic water service. These are just a few examples of requirements that may necessitate an adjustment of square foot costs.

In addition to the hazard, the **type** of sprinkler system is the most significant factor affecting cost. The size of the system (square footage) and the number of floors served by one system also affect the cost. This cost variation is illustrated along with the components of a typical wet pipe sprinkler system in Figure 23.6.

Plumbing Approximations for Quick Estimating

Water Control Water Meter; Backflow Preventer; Shock Absorbers; Vacuum Breakers; Mixer.	10 to 15% of Fixtures
Pipe And Fittings:	30 to 60% of Fixtures

Note: Lower percentage for compact buildings or larger buildings with plumbing in one area.
Larger percentage for large buildings with plumbing spread out.
In extreme cases pipe may be more than 100% of fixtures.
Percentages **do not** include special purpose or process piping.

Plumbing Labor:

1 & 2 Story Residential	Rough-in Labor = 80% of Materials
Apartment Buildings	Rough-in Labor = 90 to 100% of Materials

Labor for handling and placing fixtures is approximately 25 to 30% of fixtures.

Quality/Complexity Multiplier (For all installations)

Economy installation, add	0 to 5%
Good quality, medium complexity, add	5 to 15%
Above average quality and complexity, add	15 to 25%

Figure 23.3

Wet Pipe Systems

Wet Pipe Systems employ automatic sprinklers attached to a piping system. The pipes contain water and are connected to a water supply so that water discharges immediately from sprinklers activated by a fire.

Dry Pipe Systems

These systems employ automatic sprinklers attached to piping that contains air under pressure. As the air is released by the opening of sprinklers, the water pressure opens a valve known as a dry pipe valve. The water then flows into the piping system and out the opened sprinklers.

Pre-Action Systems

Like the dry and wet pipe systems, automatic sprinklers are again used for the pre-action systems. The sprinklers are attached to piping containing air that may or may not be under pressure. There is also a supplemental heat-responsive system of generally more sensitive characteristics than the automatic sprinklers themselves, installed in the same areas as the sprinklers. Actuation of the heat-responsive system, as from a fire, opens a valve which permits water to fill the sprinkler piping system and to be discharged from any sprinklers that may open.

Deluge Systems

These systems employ open sprinklers attached to piping that is connected to a water supply through a valve. The valve is opened by the operation of a heat-responsive system installed in the same areas as the sprinklers. When this valve opens, water flows into the piping system and discharges from all sprinklers at once.

Labor Hours to Install Plumbing Fixtures

Item	Rough-In	Set	Total Hours	Item	Rough-In	Set	Total Hours
Bathtub	5	5	10	Shower head only	2	1	3
Bathtub and shower, cast iron	6	6	12	Shower drain	3	1	4
Fire hose reel and cabinet	4	2	6	Shower stall, slate		15	15
Floor drain to 4 inch diameter	3	1	4	Slop sink	5	3	8
Grease trap, single, cast iron	5	3	8	Test 6 fixtures			14
Kitchen gas range		4	4	Urinal, wall	6	2	8
Kitchen sink, single	4	4	8	Urinal, pedestal or floor	6	4	10
Kitchen sink, double	6	6	12	Water closet and tank	4	3	7
Laundry tubs	4	2	6	Water closet and tank, wall hung	5	3	8
Lavatory wall hung	5	3	8	Water heater, 45 gals. gas, automatic	5	2	7
Lavatory pedestal	5	3	8	Water heater, 65 gals. gas, automatic	5	2	7
Shower and stall	6	4	10	Water heater, electric, plumbing only	4	2	6

Figure 23.4

Combined Dry Pipe and Pre-Action Sprinkler Systems

In these systems, automatic sprinklers are attached to piping containing air under pressure. There is also a supplemental heat-responsive system that is more sensitive than the automatic sprinklers themselves, installed in the same areas as the sprinklers. Activation of the heat-responsive system also opens approved air exhaust valves at the end of the feed main which facilitates the filling of the system with water. This process usually precedes the opening of sprinklers. The heat-responsive system also serves as an automatic fire alarm system.

Limited Water Supply Systems

These systems employ automatic sprinklers. They conform to standards, but are supplied by a pressure tank of limited capacity.

Sprinkler System Classification

Rules for installation of sprinkler systems vary depending on the classification of occupancy falling into one of three categories as follows:

Light Hazard Occupancy	Ordinary Hazard Occupancy	Extra Hazard Occupancy
The protection area allotted per sprinkler should not exceed 200 S.F. with the maximum distance between lines and sprinklers on lines being 15′. The sprinklers do not need to be staggered. Branch lines should not exceed eight sprinklers on either side of a cross main. Each large area requiring more than 100 sprinklers and without a sub-dividing partition should be supplied by feed mains or risers sized for ordinary hazard occupancy.	The protection area allotted per sprinkler shall not exceed 130 S.F. of noncombustible ceiling and 120 S.F. of combustible ceiling. The maximum allowable distance between sprinkler lines and sprinklers on line is 15′. Sprinklers shall be staggered if the distance between heads exceed 12′. Branch lines should not exceed eight sprinklers on either side of a cross main.	The protection area allotted per sprinkler shall not exceed 90 S.F. of noncombustible ceiling and 80 S.F. of combustible ceiling. The maximum allowable distance between lines and between sprinklers on lines is 12′. Sprinklers on alternate lines shall be staggered if the distance between sprinklers on lines exceeds 8′. Branch lines should not exceed six sprinklers on either side of a cross main.
Included in this group are:	**Included in this group are:**	**Included in this group are:**
Auditoriums Churches Clubs Educational Hospitals Institutional Libraries (except large stack rooms) Museums Nursing Homes Offices Residential Restaurants Schools Theaters	Automotive garages Bakeries Beverage manufacturing Bleacheries Boiler houses Canneries Cement plants Clothing factories Cold storage warehouses Dairy products manufacturing Distilleries Dry cleaning Electric generating stations Feed mills Grain elevators Ice manufacturing Laundries Machine shops Mercantiles Paper mills Printing and publishing Shoe factories Warehouses Wood product assembly	Aircraft hangars Chemical works Explosives manufacturing Linseed manufacturing Linseed oil mills Oil refineries Paint shops Shade cloth manufacturing Solvent extracting Varnish works Volatile flammable liquid manufacturing & use

Figure 23.5

MECHANICAL | 08.2-910 | Fire Sprinkler Systems, Wet

This page illustrates and describes a wet type fire sprinkler system. Lines 08.2-910-04 thru 15 give the unit cost of a system for two 2,000 square foot buildings. Lines 08.2-910-19 thru 26 give the square foot costs for alternate systems. Both material quantities and labor costs have been adjusted for the system listed.

Factors: To adjust for conditions other than normal working conditions, use Lines 08.2-910-29 thru 40.

LINE NO.	DESCRIPTION	QUANTITY	COST EACH		
			MAT.	INST.	TOTAL
01	**Wet pipe fire sprinkler system, ordinary hazard, open area to 2000 S.F.**				
02	**On one floor.**				
03					
04	4" OS & Y valve	1 Ea.	165	180	345
05	Wet pipe alarm valve	1 Ea.	610.50	279.50	890
06	Water motor alarm	1 Ea.	123.20	76.80	200
07	3" check valve	1 Ea.	110	185	295
08	Pipe riser, 4" diameter	10 L.F.	60.39	189.61	250
09	Water gauges and trim	1 Set	913	562	1475
10	Electric fire horn	1 Ea.	29.70	44.30	74
11	Sprinkler head supply piping	168 L.F.	368.66	1204.84	1573.50
12	Pipe fittings	1 L.S.	146.61	1139.39	1286
13	Sprinkler heads	16 Ea.	54.91	313.09	368
14	Fire department connection	1 Ea.	251.90	113.10	365
15	TOTAL	System	2833.87	4287.63	7121.50
16					
17			COST PER S.F.		
18			MAT.	INST.	TOTAL
19	Ordinary hazard, one floor, area to 2000 S.F./floor	S.F.	1.42	2.14	3.56
20	For each additional floor, add per floor		.32	1.42	1.74
21	Area to 3200 S.F./floor		1.01	1.93	2.94
22	For each additional floor, add per floor		.32	1.48	1.80
23	Area to 5000 S.F./ floor		.84	1.84	2.68
24	For each additional floor, add per floor		.40	1.55	1.95
25	Area to 8000 S.F./ floor		.60	1.51	2.11
26	For each additional floor, add per floor	↓	.33	1.33	1.66
27					
28					
29	Cut & patch to match existing construction, add, minimum		2%	3%	
30	Maximum		5%	9%	
31	Dust protection, add, minimum		1%	2%	
32	Maximum		4%	11%	
33	Equipment usage curtailment, add, minimum		1%	1%	
34	Maximum		3%	10%	
35	Material handling & storage limitation, add, minimum		1%	1%	
36	Maximum		6%	7%	
37	Protection of existing work, add, minimum		2%	2%	
38	Maximum		5%	7%	
39	Shift work requirements, add, minimum			5%	
40	Maximum			30%	
41					
42					

310

Figure 23.6

Chemical Systems

Chemical systems use halon, carbon dioxide, dry chemicals, or high expansion foam as selected for special requirements. The chemical agent may extinguish flames by excluding oxygen, interrupting chemical action of the oxygen uniting with fuel, or sealing and cooling the combustion center.

Firecycle Systems

Firecycle systems are fixed fire protection sprinkler systems utilizing water as the extinguishing agent. These are time-delayed, recycling, preaction-type systems which automatically shut the water off when heat is reduced below the detector-operating temperature, and turn the water back on when that temperature is exceeded. The system senses a fire condition through a closed circuit electrical detector system which automatically controls water flow to the fire. Batteries supply up to a 90-hour emergency power supply for system operation. The piping system is dry (until water is required) and is monitored with pressurized air. Should any leak in the system piping occur, an alarm will sound, but water will not enter the system until heat is sensed by a firecycle detector.

Heating, Ventilating, and Air Conditioning

As with fire protection, square foot (or systems) costs can be developed for HVAC (heating, ventilating, and air conditioning) by keeping records from past projects. It is recommended that the estimator obtain quotations or detailed estimates from experienced engineers or subcontractors whenever possible. HVAC is a specialized trade; specific knowledge is required for a proper estimate. However, budgets can be developed based on square foot, cubic foot, or systems costs. Such costs will vary based on the type of system and the use of the occupied space, as well as the size of the space. Different types of systems are illustrated in Figures 23.7a and 23.7b.

For preliminary budgets, prior to or during design development, the estimator can determine rough heating and cooling requirements in order to determine costs. The tables in Figure 23.8 can be used to determine heat loss (in BTU's per hour). This figure, for all practical purposes, is the capacity required of the heat source (e.g., boiler, furnace). Figure 23.9 can be used to determine cooling requirements in tons for 45 types of building uses. (One ton of cooling equals 12,000 BTU's per hour.) When heating and cooling systems are combined, as with rooftop systems, the cooling capacity is used to determine the size of the system. When the system capacity has been calculated, budget costs can be determined.

For most remodeling projects, the main heating and cooling systems will already be in place. However, distribution (ductwork, hot water baseboard, fan coil units, etc.) usually depends on the final design details and is often executed as part of the renovation. Computer rooms usually require complete, new, and often independent cooling systems. (Computer rooms rarely require supplemental heating.)

Computer Rooms

Computer rooms impose special requirements on air conditioning systems. A prime requirement is *reliability*, due to the potential monetary loss that could be incurred by a system failure. A second basic requirement is the tolerance of *control* with which temperature and humidity are regulated, and dust eliminated. Because the air conditioning system reliability is so vital, the additional cost of reserve capacity and redundant components is often justified.

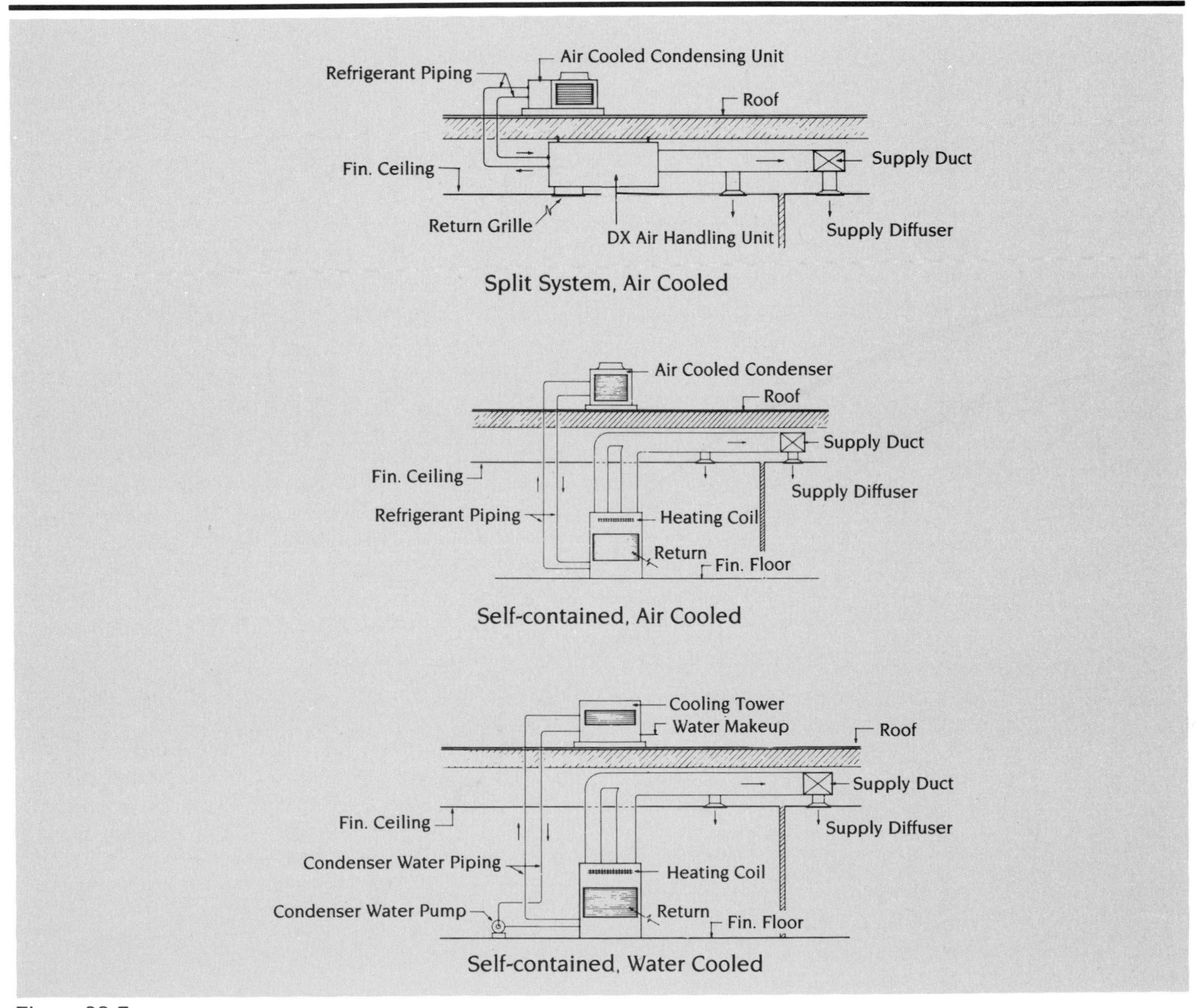

Figure 23.7*a*

Computer areas may be environmentally controlled by one of three methods as follows:

- **Self-contained Units**. These are units built to high standards of performance and reliability, and usually contain alarms and controls to indicate component operation failure, the need to change filters, etc. It should be remembered that these units occupy interior space that is relatively expensive to build, and that all alterations and service of the equipment will also have to be accomplished within the computer area.

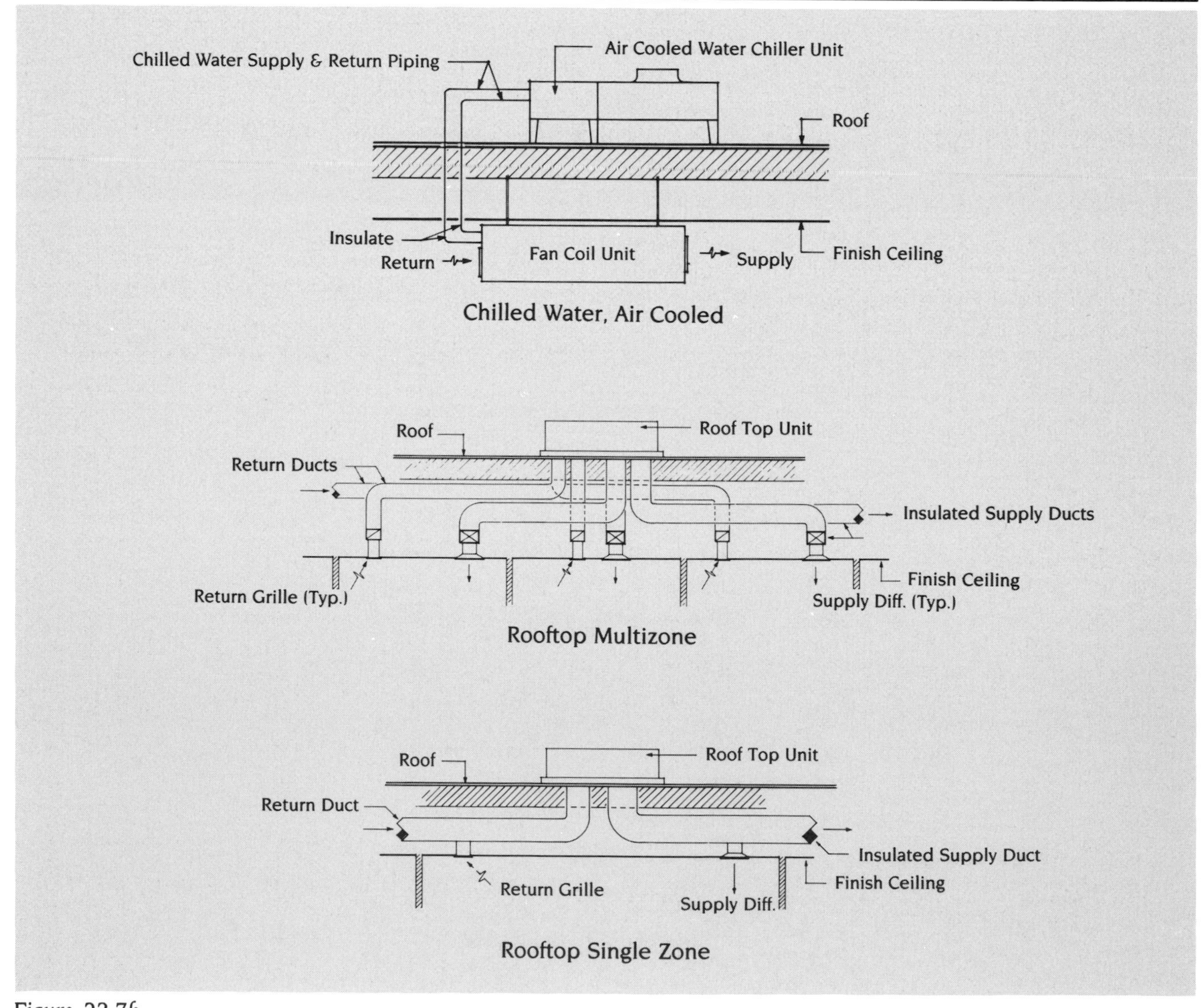

Figure 23.7b

Factors for Determining Heat Loss for Various Types of Buildings

General: While the most accurate estimates of heating requirements would naturally be based on detailed information about the building being considered, it is possible to arrive at a reasonable approximation using the following procedure:

1. Calculate the cubic volume of the room or building.
2. Select the appropriate factor from Table A. Note that the factors apply only to inside temperatures listed in the first column and to 0°F outside temperature.
3. If the building has bad north and west exposures, multiply the heat loss factor by 1.1
4. If the outside design temperature is other than 0°F, multiply the factor from Table A by the factor from Table B.
5. Multiply the cubic volume by the factor selected from Table A. This will give the estimated BTU per hour heat loss which must be made up to maintain inside temperature.

Table A
Factories & Industrial Plants General Office Area 70°F

Conditions	Qualifications	Loss Factor*
One Story	Skylight in Roof	6.2
	No Skylight in Roof	5.7
Multiple Story	Two Story	4.6
	Three Story	4.3
	Four Story	4.1
	Five Story	3.9
	Six Story	3.6
All Walls Exposed	Flat Roof	6.9
	Heated Space Above	5.2
One Long Warm Common Wall	Flat Roof	6.3
	Heated Space Above	4.7
Warm Common Walls on Both Long Sides	Flat Roof	5.8
	Heated Space Above	4.1
Warehouses 60°F		
All Walls Exposed	Skylights in Roof	5.5
	No Skylights in Roof	5.1
	Heated Space Above	4.0
One Long Warm Common Wall	Skylight in Roof	5.0
	No Skylight in Roof	4.9
	Heated Space Above	3.4
Warm Common Walls on Both Long Sides	Skylight in Roof	4.7
	No Skylight in Roof	4.4
	Heated Space Above	3.0

*Note: This table tends to be conservative particularly for new buildings designed for minimum energy consumption.

Table B - Outside Design Temperature Correction Factor (for Degrees Fahrenheit)

Outside Design Temp.	50	40	30	20	10	0	-10	-20	-30
Correction Factor	0.29	0.43	0.57	0.72	0.86	1.00	1.14	1.28	1.43

Figure 23.8

- **Decentralized Air Handling Units**. In operation, these units are similar to the self-contained units, except that cooling capability comes from remotely located refrigeration equipment as refrigerant or chilled water. Since no compressors or refrigerating equipment are needed in the air units, they are smaller and require less service than self-contained units.
- **Central System Supply**. In this system, the cooling is obtained from a central source which, since it is not located within the computer room, may have excess capacity and permit greater flexibility without interfering with the computer components. System performance criteria must still be met. This type of system may provide less control than independent units.

Air Conditioning Requirements

BTU's per Hour per S.F. of Floor Area and S.F. per Ton of Air Conditioning

Type Building	BTU per S.F.	S.F. per Ton	Type Building	BTU per S.F.	S.F. per Ton	Type Building	BTU per S.F.	S.F. per Ton
Apartments, Individual	26	450	Dormitory, Rooms	40	300	Libraries	50	240
Corridors	22	550	Corridors	30	400	Low Rise Office, Exterior	38	320
Auditoriums & Theaters	666	18*	Dress Shops	43	280	Interior	33	360
Banks	50	240	Drug Stores	80	150	Medical Centers	28	425
Barber Shops	48	250	Factories	40	300	Motels	28	425
Bars & Taverns	133	90	High Rise Office-Ext. Rms.	46	263	Office (small suite)	43	280
Beauty Parlors	66	180	Interior Rooms	37	325	Post Office, Individual Office	42	285
Bowling Alleys	68	175	Hospitals, Core	43	280	Central Area	46	260
Churches	600	20*	Perimeter	46	260	Residences	20	600
Cocktail Lounges	68	175	Hotel, Guest Rooms	44	275	Restaurants	60	200
Computer Rooms	141	85	Public Spaces	55	220	Schools & Colleges	46	260
Dental Offices	52	230	Corridors	30	400	Shoe Stores	55	220
Dept. Stores, Basement	34	350	Industrial Plants, Offices	38	320	Shop'g. Ctrs., Super Markets	34	350
Main Floor	40	300	General Offices	34	350	Retail Stores	48	250
Upper Floor	30	400	Plant Areas	40	300	Specialty Shops	60	200

*Persons per ton

12,000 BTU = 1 ton of air conditioning

Figure 23.9

Chapter 24
ELECTRICAL

For most remodeling projects, electrical power is supplied—at least to a panel at or near the space to be constructed. In some cases, a new electric service and feeder distribution system may be required. A new service may include charges to be paid to the local utility, expensive switchgear, and, in some cases, transformers (exterior or within an interior vault). A feeder distribution system involves runs of conduit and large wire to electrical panels located throughout a building (to each floor or tenant space). This type of work should be estimated by an experienced engineer or electrical contractor. The project estimator should nevertheless be familiar with the various types of distribution systems and methods.

Distribution Systems

For many projects an engineered specialty distribution network may be in place prior to the start of the remodeling work. Such networks are often modular and may be installed during initial construction of the building. In industrial or manufacturing space, these networks may be cable tray or bus duct systems (shown in Figure 24.1). These systems are most advantageous where flexibility for power usage and ease of installation are most important. In commercial office buildings, in-place distribution networks may include underfloor raceway systems, trench duct, and cellular concrete floor raceway systems. These are illustrated in Figure 24.2.

If there is no "built-in" distribution system in place within the space to be constructed, certain options exist. Conventional wiring consists primarily of the following types of raceways and/or conductors:

- Rigid galvanized steel conduit
- Aluminum conduit
- EMT (thin wall conduit)
- BX (armored cable)
- Romex (non-metallic sheathed cable)

The choice depends on national and local electrical, building, and fire codes, and on occupant requirements. Figure 24.3 demonstrates how each of the different types of wiring can affect the project costs.

Undercarpet Systems

In recent years, a new type of distribution system has been developed which offers the advantages of the underfloor raceway and trench duct systems, but allows the flexibility of conventional methods—custom installation when the remodeling work is performed, and ease of change during operation for specific occupant requirements. This new type is the undercarpet power, data, and communication system.

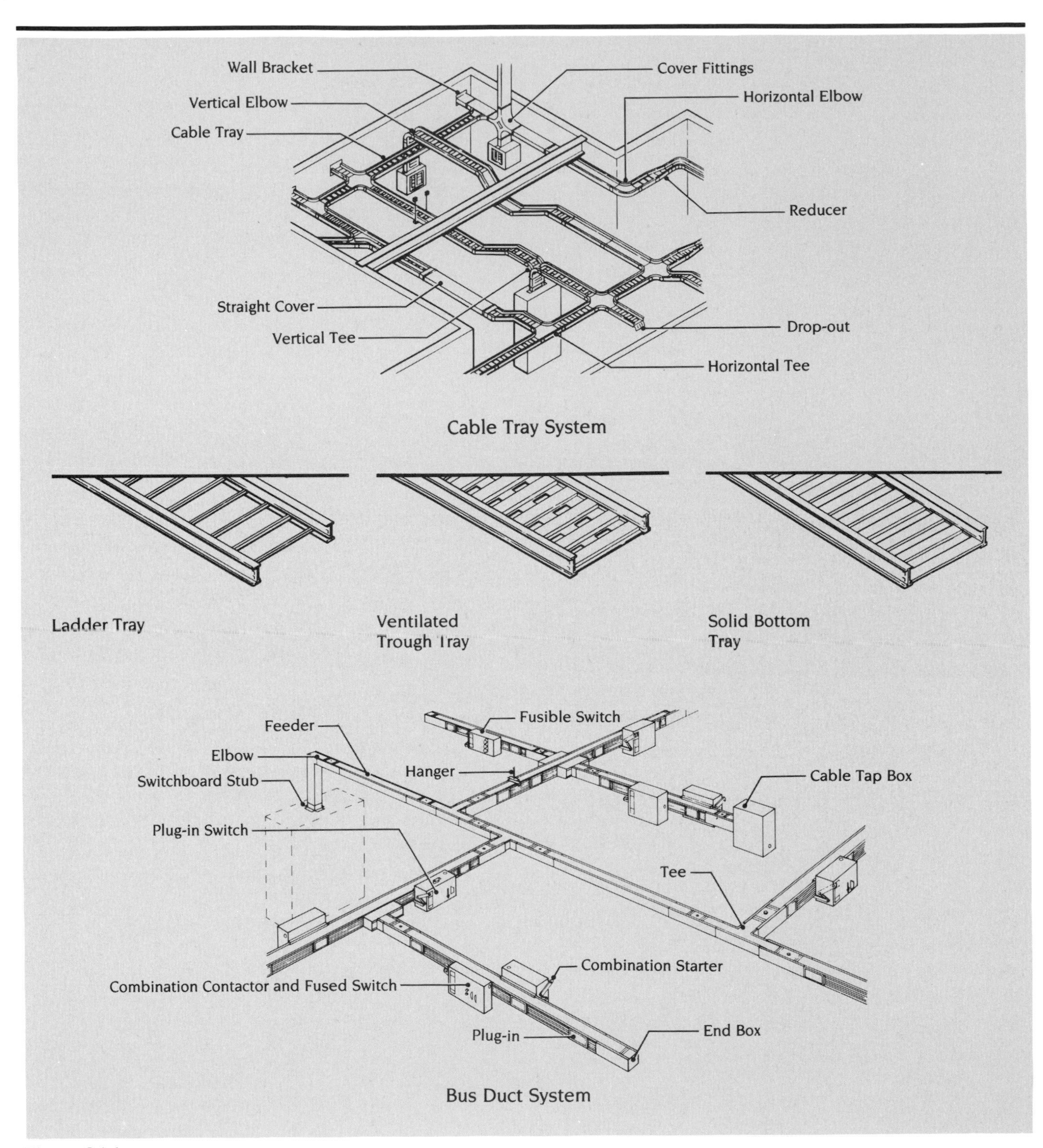
Wall Bracket
Cover Fittings
Vertical Elbow
Horizontal Elbow
Cable Tray
Reducer
Straight Cover
Vertical Tee
Drop-out
Horizontal Tee
Cable Tray System
Ladder Tray
Ventilated Trough Tray
Solid Bottom Tray
Feeder
Fusible Switch
Elbow
Hanger
Switchboard Stub
Cable Tap Box
Plug-in Switch
Tee
Combination Starter
Combination Contactor and Fused Switch
Plug-in
End Box
Bus Duct System

Figure 24.1

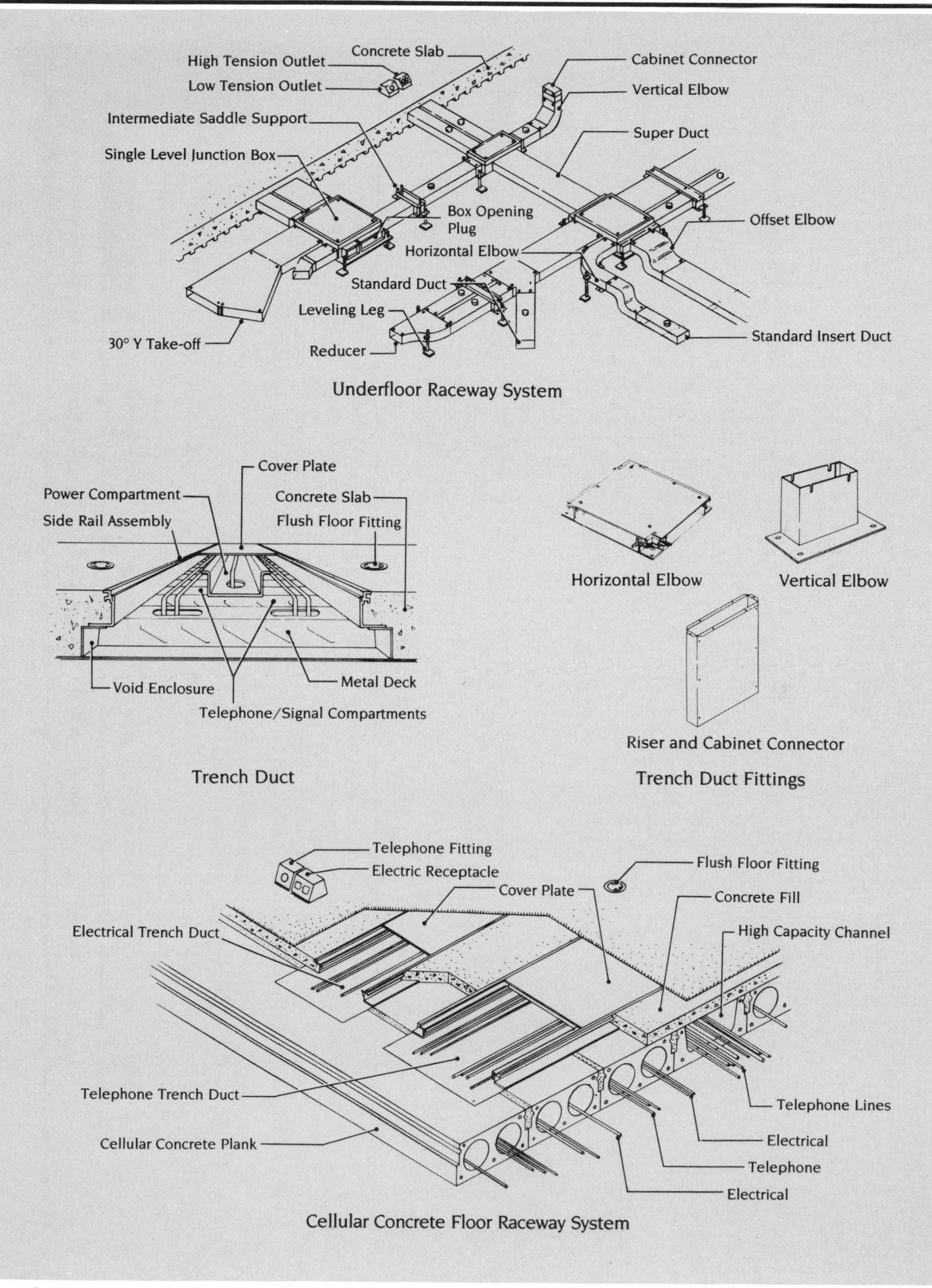
Concrete Slab
High Tension Outlet
Low Tension Outlet
Cabinet Connector
Vertical Elbow
Intermediate Saddle Support
Super Duct
Single Level Junction Box
Box Opening Plug
Offset Elbow
Horizontal Elbow
Standard Duct
Leveling Leg
30° Y Take-off
Reducer
Standard Insert Duct
Underfloor Raceway System
Cover Plate
Power Compartment
Concrete Slab
Side Rail Assembly
Flush Floor Fitting
Void Enclosure
Metal Deck
Telephone/Signal Compartments
Trench Duct
Horizontal Elbow
Vertical Elbow
Riser and Cabinet Connector
Trench Duct Fittings
Telephone Fitting
Electric Receptacle
Flush Floor Fitting
Cover Plate
Concrete Fill
Electrical Trench Duct
High Capacity Channel
Telephone Trench Duct
Telephone Lines
Cellular Concrete Plank
Electrical
Telephone
Electrical
Cellular Concrete Floor Raceway System

Figure 24.2

168 100 | Special Systems

170

	Description	CREW	DAILY OUTPUT	MAN-HOURS	UNIT	BARE COSTS MAT.	BARE COSTS LABOR	BARE COSTS EQUIP.	BARE COSTS TOTAL	TOTAL INCL O&P
4140	For GFI see line 4300 below									
4150	Decorator style, 15 amp recpt., type NM cable	1 Elec	14.55	.550	Ea.	6.45	13.35		19.80	27
4170	Type MC cable		12.31	.650		10.70	15.75		26.45	36
4180	EMT & wire		5.33	1.500		13.40	36		49.40	70
4200	With #12/2 type NM cable		12.31	.650		7.60	15.75		23.35	32
4220	Type MC cable		10.67	.750		11.60	18.20		29.80	40
4230	EMT & wire		4.71	1.700		14.35	41		55.35	78
4250	20 amp recpt. #12/2 type NM cable		12.31	.650		10.90	15.75		26.65	36
4270	Type MC cable		10.67	.750		13.85	18.20		32.05	43
4280	EMT & wire		4.71	1.700		18	41		59	82
4300	GFI, 15 amp recpt., type NM cable		12.31	.650		26	15.75		41.75	53
4320	Type MC cable		10.67	.750		30	18.20		48.20	61
4330	EMT & wire		4.71	1.700		33	41		74	99
4350	GFI with #12/2 type NM cable		10.67	.750		27	18.20		45.20	57
4370	Type MC cable		9.20	.870		31	21		52	66
4380	EMT & wire		4.21	1.900		34	46		80	105
4400	20 amp recpt., #12/2 type NM cable		10.67	.750		29	18.20		47.20	60
4420	Type MC cable		9.20	.870		33	21		54	68
4430	EMT & wire		4.21	1.900		35	46		81	110
4500	Weather-proof cover for above receptacles, add		32	.250		3.28	6.05		9.33	12.80
4550	Air conditioner outlet, 20 amp-240 volt recpt.									
4560	30' of #12/2, 2 pole circuit breaker									
4570	Type NM cable	1 Elec	10	.800	Ea.	24	19.40		43.40	56
4580	Type MC cable		9	.889		29	22		51	65
4590	EMT & wire		4	2		40	49		89	120
4600	Decorator style, type NM cable		10	.800		26	19.40		45.40	58
4620	Type MC cable		9	.889		31	22		53	67
4630	EMT & wire		4	2		41	49		90	120
4650	Dryer outlet, 30 amp-240 volt recpt., 20' of #10/3									
4660	2 pole circuit breaker									
4670	Type NM cable	1 Elec	6.41	1.250	Ea.	40	30		70	90
4680	Type MC cable		5.71	1.400		41	34		75	97
4690	EMT & wire		3.48	2.300		38	56		94	125
4700	Range outlet, 50 amp-240 volt recpt., 30' of #8/3									
4710	Type NM cable	1 Elec	4.21	1.900	Ea.	67	46		113	145
4720	Type MC cable		4	2		65	49		114	145
4730	EMT & wire		2.96	2.700		54	66		120	160
4750	Central vacuum outlet		6.40	1.250		30	30		60	79
4770	Type MC cable		5.71	1.400		35	34		69	90
4780	EMT & wire		3.48	2.300		45	56		101	135
4800	30 amp-110 volt locking recpt., #10/2 circ. bkr.									
4810	Type NM cable	1 Elec	6.20	1.290	Ea.	43	31		74	95
4820	Type MC cable		5.40	1.480		45	36		81	105
4830	EMT & wire		3.20	2.500		42	61		103	140
4900	Low voltage outlets									
4910	Telephone recpt., 20' of 4/C phone wire	1 Elec	26	.308	Ea.	4.80	7.45		12.25	16.60
4920	TV recpt., 20' of RG59U coax wire, F type connector	"	16	.500	"	7.30	12.15		19.45	26
4950	Door bell chime, transformer, 2 buttons, 60' of bellwire									
4970	Economy model	1 Elec	11.50	.696	Ea.	42	16.85		58.85	72
4980	Custom model		11.50	.696		85	16.85		101.85	120
4990	Luxury model, 3 buttons		9.50	.842		185	20		205	235
6000	Lighting outlets									
6050	Wire only (for fixture) type NM cable	1 Elec	32	.250	Ea.	3.20	6.05		9.25	12.75
6070	Type MC cable		24	.333		6.30	8.10		14.40	19.20
6080	EMT & wire		10	.800		8.50	19.40		27.90	39
6100	Box (4") and wire (for fixture), type NM cable		25	.320		5.50	7.75		13.25	17.85
6120	Type MC cable		20	.400		9.80	9.70		19.50	26
6130	EMT & wire		11	.727		11.90	17.65		29.55	40

170

For expanded coverage of these items see *Means Electrical Cost Data 1989*

231

Figure 24.3

Undercarpet systems are an alternative to conventional round cable for wiring commercial and industrial offices. They provide a method of distributing power almost anywhere on the floors without having to channel through underfloor ducts, walls, or ceilings.

The flat, low-profile design of this type of system allows for its installation directly on top of wood, concrete, composition, or ceramic floors. It is then covered with carpet tiles (18" to 30" square) which allow for change at any time.

The basic elements of undercarpet systems include three groupings of components:

- Specialized flat, low profile cable
- Transition fittings, which house the round-to-flat conductor connections at the supply end of the cable
- Floor fittings, which house the flat-to-round connections and provide various access configurations at the other end of the cable

Undercarpet power, telephone, and data systems are illustrated in Figure 24.4.

Undercarpet systems involve specialized materials and installation methods. Specific manufacturers should be consulted for material costs. Installation costs (for other than rough budget purposes) should be estimated by experienced subcontractors.

Lighting and Power

The following types of lighting may be included in a remodeling estimate:

- Fluorescent
- Incandescent
- High Intensity Discharge
- Emergency Lights and Power

Typical fluorescent and incandescent fixtures are illustrated in Figure 24.5. Typical high intensity discharge fixtures are shown in Figure 24.6. A graphic relationship of the light output versus power usage for the different types is shown in Figure 24.7.

Fluorescent Lighting

A fluorescent lamp consists of a hot cathode in a phosphor-coated tube which contains inert gas and mercury vapor. When energized, the cathode causes a mercury arc to produce ultraviolet light and fluorescence on the phosphor coating of the tube. The color of the light varies according to the type of phosphor used in the coating. Fluorescent lamps are high in efficiency, and with limited switching on and off, they have a life in excess of 20,000 hours. A ballast is required in the lamp circuit to limit the current. Ballasts are required in various watt-saving types and can be matched with special energy-saving lamps. Special ballasts are required for dimming.

Fluorescent tubes are produced in many different wattages, sizes, and types. One manufacturer lists lamps of 4 watts to 215 watts with lengths of 6" to 96". Three basic types of fluorescent lamps are currently manufactured: *preheat*, *instant start*, and *rapid start*. The preheat lamp, which is the oldest type, requires a starter. The instant start lamp, or *slimline*, was developed after the preheat type. The rapid start lamp, which is most commonly used today, operates at 425 mA. High

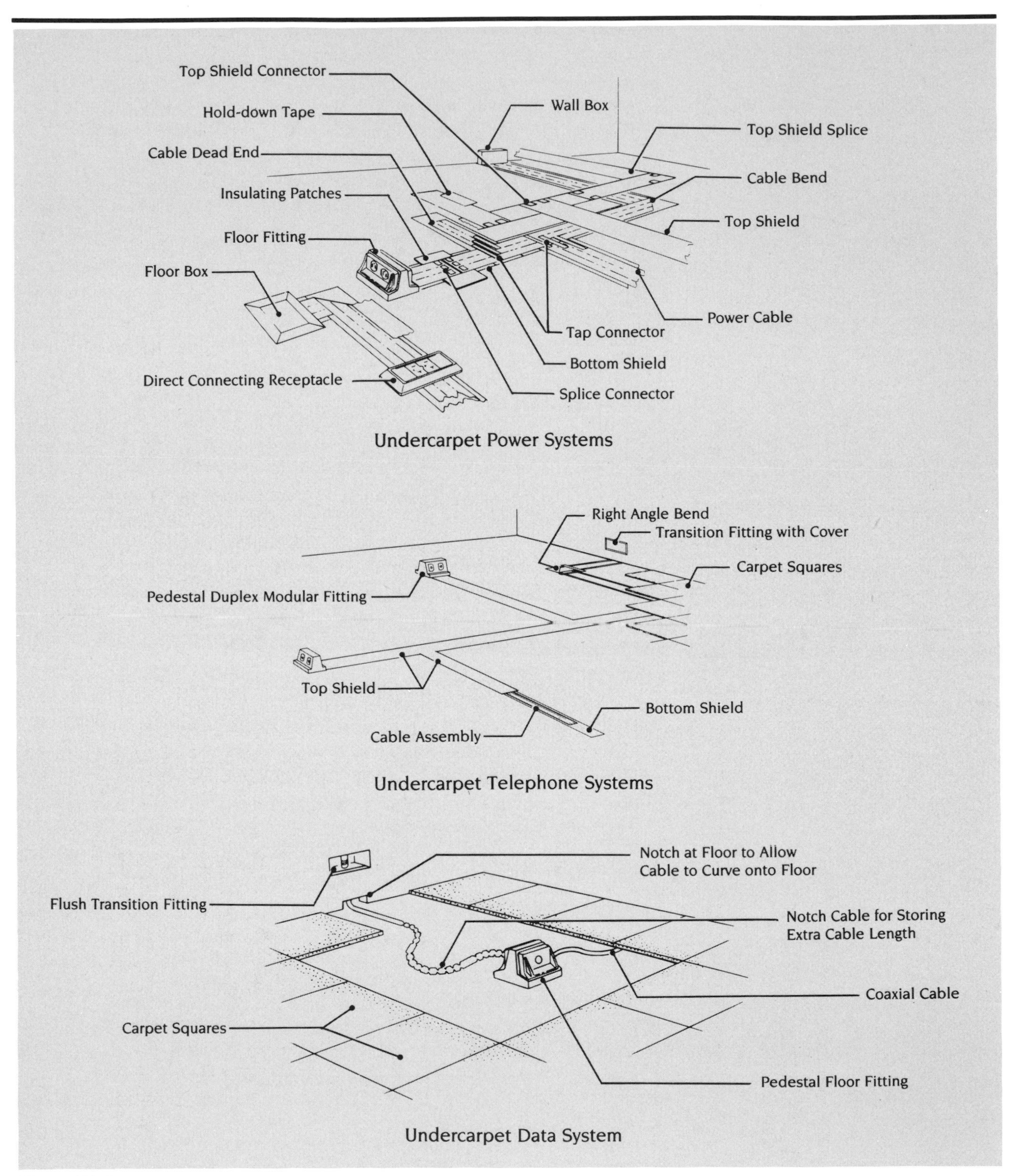

Top Shield Connector
Hold-down Tape
Cable Dead End
Insulating Patches
Floor Fitting
Floor Box
Direct Connecting Receptacle
Wall Box
Top Shield Splice
Cable Bend
Top Shield
Power Cable
Tap Connector
Bottom Shield
Splice Connector
Undercarpet Power Systems
Right Angle Bend
Transition Fitting with Cover
Carpet Squares
Pedestal Duplex Modular Fitting
Top Shield
Bottom Shield
Cable Assembly
Undercarpet Telephone Systems
Notch at Floor to Allow
Cable to Curve onto Floor
Flush Transition Fitting
Notch Cable for Storing
Extra Cable Length
Coaxial Cable
Carpet Squares
Pedestal Floor Fitting
Undercarpet Data System

Figure 24.4

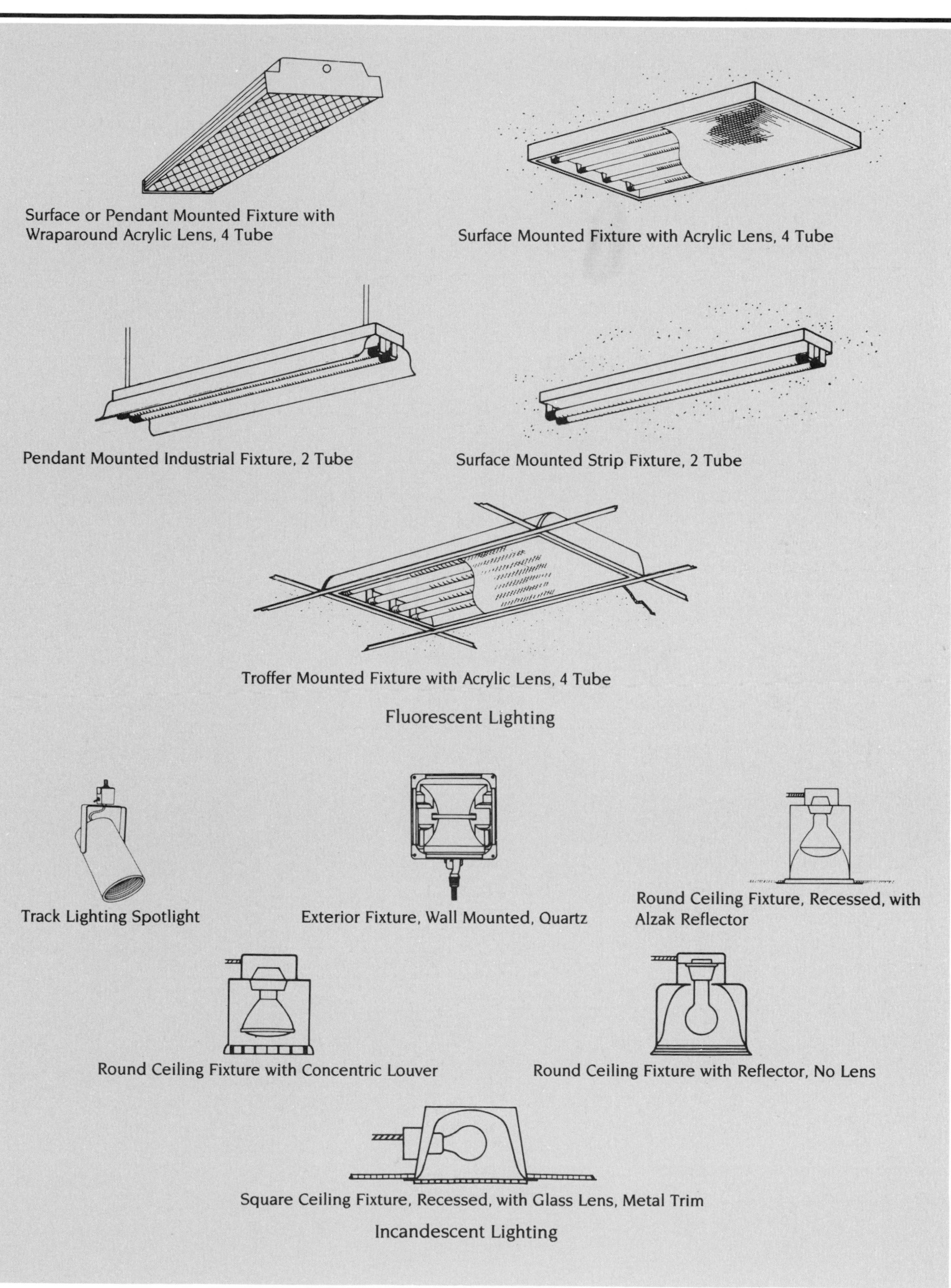

Surface or Pendant Mounted Fixture with Wraparound Acrylic Lens, 4 Tube
Surface Mounted Fixture with Acrylic Lens, 4 Tube
Pendant Mounted Industrial Fixture, 2 Tube
Surface Mounted Strip Fixture, 2 Tube
Troffer Mounted Fixture with Acrylic Lens, 4 Tube
Fluorescent Lighting
Track Lighting Spotlight
Exterior Fixture, Wall Mounted, Quartz
Round Ceiling Fixture, Recessed, with Alzak Reflector
Round Ceiling Fixture with Concentric Louver
Round Ceiling Fixture with Reflector, No Lens
Square Ceiling Fixture, Recessed, with Glass Lens, Metal Trim
Incandescent Lighting

Figure 24.5

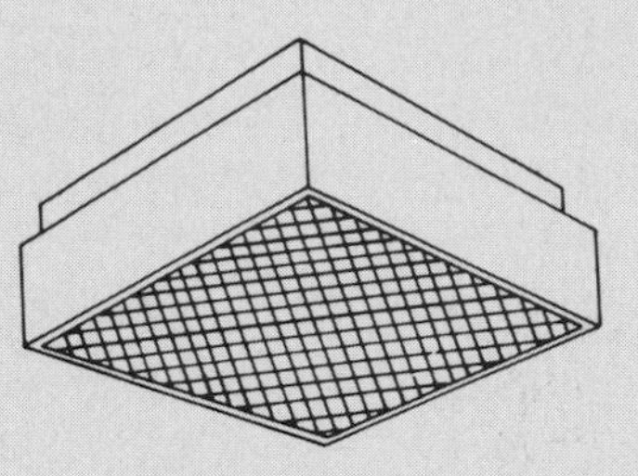

Mercury Vapor Ceiling Fixture, Recessed, Integral Ballast

High Pressure Sodium Fixture, Round, Surface

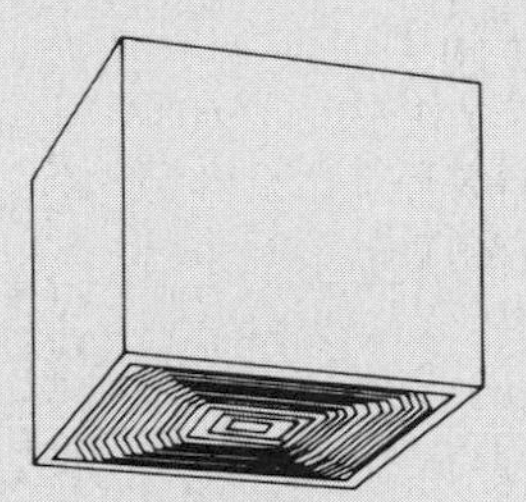

Mercury Vapor Fixture, Surface Mounted

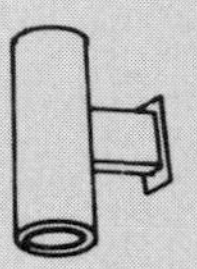

High Pressure Sodium, Round, Wall Mounted

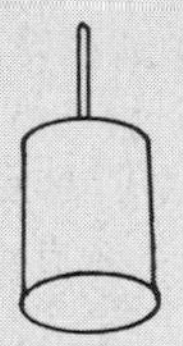

Mercury Vapor Fixture, Round, Pendent

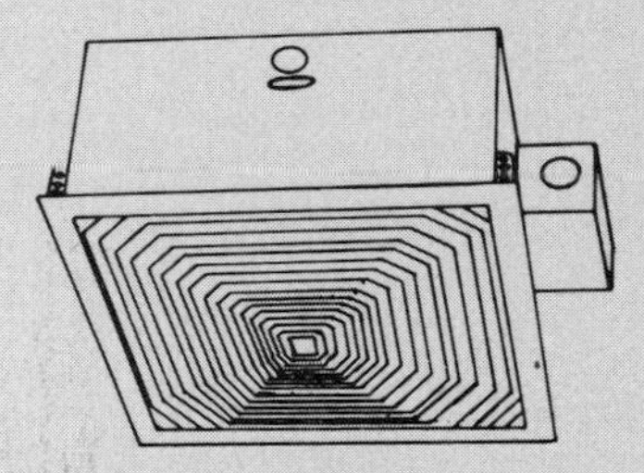

Vaporproof, High Pressure Sodium Fixture, Recessed

Mercury Vapor Fixture, Square, Pendent Mounted

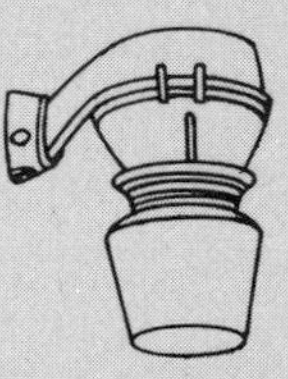

Vaporproof, High Pressure Sodium Fixture, Wall Mounted

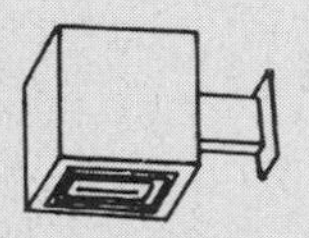

Mercury Vapor Fixture, Square, Wall Mounted

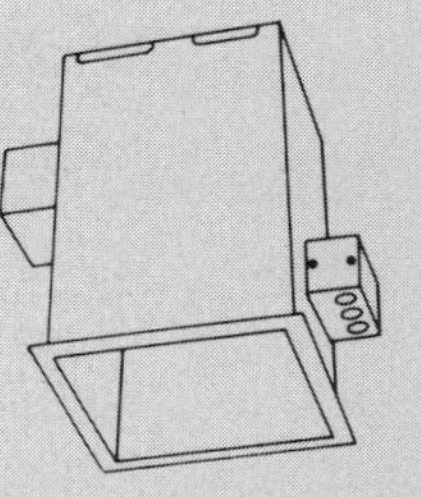

Metal Halide Fixture, Square, Recessed

H.I.D. Lighting

Figure 24.6

output lamps operate at 800 mA; very high output, at 1500 mA. Because the ballasts used in high output and very high output lamps tend to be noisy, these types of fluorescent lamps are not recommended for use in quiet areas.

Incandescent Lighting

An incandescent lamp is a glass bulb which contains a tungsten filament with a mixture of argon and nitrogen gas. The base of the bulb is usually capped with a screw base made of brass or aluminum. Incandescent lamps are versatile sources of light, as they are manufactured in many different sizes, shapes, wattages, and base configurations. Some of these variations include bulbs with clear, frosted, and hard glass (for weatherproof applications); aluminized reflectors, wide, narrow, and spot beam pre-focused; and three-way wattage switching. For general applications, incandescent lamps are rated from 2 watts to 1500 watts, but some street lighting lamps may be rated as high as 15,000 watts.

Along with the advantage of variety of lamp sizes and special features, the relatively small size of incandescents allows them to be fit easily into the design of the fixtures that hold them. They are low

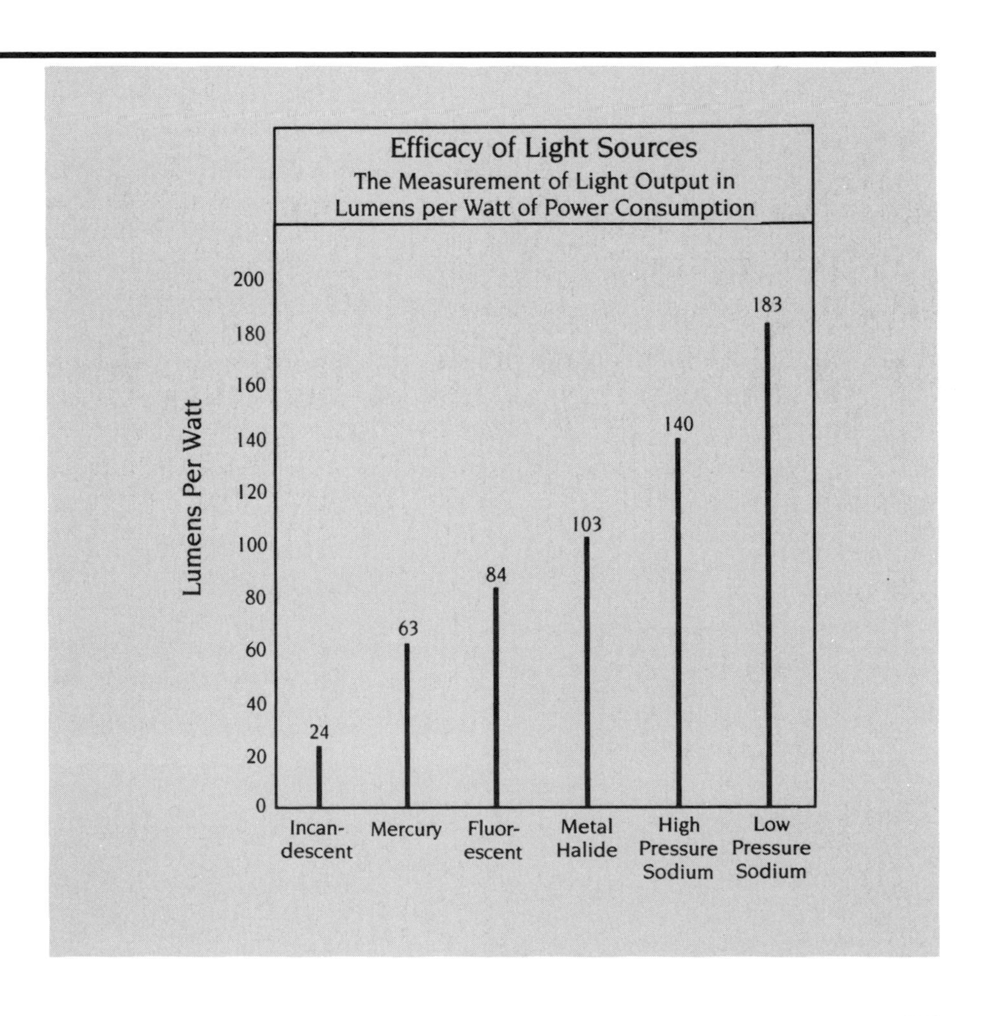

Figure 24.7

in cost, soft in color, and easy to use with dimmers. The disadvantages of incandescent lamps include relatively short life, usually less than 1000 hours, and higher energy consumption than fluorescent and mercury vapor lamps (shown in Figure 24.7). If the incandescent lamp is used in a system with a higher voltage than that recommended by the manufacturer, the life of the lamp decreases significantly. An excess of just 10 volts above the recommended voltage can reduce the lamp's life considerably.

The quartz lamp, or *tungsten halogen lamp*, is a special type of incandescent lamp that consists of a quartz tube with various configurations. Some quartz lamps are simple double-ended tubes, while others include a screw base or are mounted inside an R- or PAR-shaped bulb. A quartz lamp maintains maximum light output throughout its life, which varies according to its type and size from 2000 to 4000 hours. Generally, quartz lamps are more energy efficient than regular incandescents, but their purchase price is higher. Quartz lamps are available in sizes from 36 watts to 1500 watts.

Fixtures for regular incandescent lamps vary widely, depending on their function, location, and desired appearance. Simple lampholders, decorative multi-lamp chandeliers, down lights, spotlights, accent wall lights, and track lights are just a few of the many types of fixtures. Some of the fixtures used for quartz lamps include exterior flood, track, accent, and emergency lights, to name a few.

High Intensity Discharge Lighting

High intensity discharge lighting (HID) types include *mercury vapor*, *metal halide*, and *high and low pressure sodium lamps*. HID lamps are usually installed to light large indoor and outdoor areas, such as factories, gymnasiums, sports complexes, parking lots, building perimeters, streets, and highways.

A mercury vapor lamp is a glass bulb containing high pressure mercury vapor. It works on the same principle as do fluorescent lamps, except that the pressure of the mercury vapor within the bulb is much higher than that of a standard fluorescent tube. A metal halide lamp is basically a mercury vapor lamp with modifications in its arc tube arrangement. Generally, the color, as well as the efficiency, is better in this type of lamp when compared to conventional mercury vapor models.

High pressure sodium lamps differ from mercury lamps primarily in the type of vapor contained in the arc tube. These lamps use a mixture of sodium, mercury, and xenon to produce a slightly yellow color. Like mercury lamps, sodium lamps are efficient in energy consumption and high-rated lamp life.

The low pressure sodium lamp is more efficient than the high pressure sodium type, but its intense yellow color makes it unsuitable for indoor use. This type of lamp is used primarily for lighting large outdoor areas, such as roadways, parking lots, and places that require security lighting.

Estimating Lighting

Most projects include many different types and sizes of lighting fixtures. Lighting fixtures on the drawing can be organized if each type is assigned a letter symbol. A lighting schedule is prepared to list each type with a description and specifications. A sample lighting schedule is shown in Figure 24.8.

There are three basic ways to estimate the costs for lighting and other branch circuit power requirements (receptacles, switches, special circuits). The most accurate is a detailed unit price estimate. This involves counting each fixture or wiring device, measuring the lengths of each type of raceway and conductor, and obtaining material prices and labor productivity rates in order to determine costs. The detail and time required limits this type of estimate to subcontractors, primarily for bidding purposes.

A second method is to use square foot costs that may be developed from past projects. While this method is fast, it does not allow for variation in fixture or device density unless records are kept for different types of occupancy (e.g., office, school, etc.).

In recent years, most state and local governments have adopted energy codes which, in effect, limit the maximum power usage based on the type of occupancy. These limits are usually defined as the

Type	Manufacturer & Catalog #	Fixture	Type	Lamps Qty	Lamps Volts	Watts	Mounting	Remarks
A	Meansco #7054	2'x4' Troffer	F-40 CW	4	277	40	Recessed	Acrylic Lens
B	Meansco #7055	1'x4' Troffer	F-40 CW	2	277	40	Recessed	Acrylic Lens
C	Meansco #7709	6"x4'	F-40 CW	1	277	40	Surface	Acrylic Wrap
D	Meansco #7710	6"x8' Strip	F96T12 CW	1	277	40	Surface	
E	Meansco #7900A	6"x4'	F-40	1	277	40	Surface	Mirror Light
F	Kingston #100A	6"x4'	F-40	1	277	40	Surface	Acrylic Wrap
G	Kingston #110C	' Strip	F-40 CW	1	277	40	Surface	
H	Kingston #3752	Wallpack	HPS	1	277	150	Bracket	W/Photo Cell
I	Kingston #201-202	Floodlight	HPS	1	277	400	Surface	2' Below Fascia
K	Kingston #203		HPS	1	277	100	Wall Bracket	
L	Meansco #8100	Exit Light	1-13W 20W	1	120	13	Surface	
			T6-1/2	2	6½	20		
M	Meansco #9000	Battery Unit	Sealed Beam	2	12	18	Wall Mount	12 Volt Unit

Figure 24.8

maximum wattage per square foot of area. Besides type of building use, different types of spaces within a building may have different requirements: office areas, bathrooms, hallways, etc. Figure 24.9 shows typical allowances based on type of occupancy and use. Local codes and officials should be consulted.

The estimator can use these power usage requirements to the best advantage when determining costs. Device and lighting costs can be determined based on wattage per square foot. This third method of estimating can account for fixture and device density variations based on different types of occupancy and use.

General Lighting Loads by Occupancies	
Type of Occupancy	Unit Load per S.F. (Watts)
Armories and Auditoriums	1
Banks	5
Barber Shops and Beauty Parlors	3
Churches	1
Clubs	2
Court Rooms	2
*Dwelling Units	3
Garages — Commercial (storage)	1/2
Hospitals	2
*Hotels and Motels, including apartment houses without provisions for cooking by tenents	2
Industrial Commercial (loft) Buildings	2
Lodge Rooms	1½
Office Buildings	5
Restaurants	2
Schools	3
Stores	3
Warehouses (storage)	1/4
*In any of the above occupancies except one-family dwellings and individual dwelling units of multi-family dwellings:	
Assembly Halls and Auditoriums	1
Halls, Corridors, Closets	1/2
Storage Spaces	1/4

Lighting Limit (Connected Load) for Listed Occupancies: New Building Proposed Energy Conservation Guideline	
Type of Use	Maximum Walls per S.F.
Interior:	
Category A: Classrooms, office areas, automotive mechanical areas, museums, conference rooms, drafting rooms, clerical areas, laboratories, merchandising areas, kitchens, examining rooms, book stacks, athletic facilities.	3.00
Category B: Auditoriums, waiting areas, spectator areas, restrooms, dining areas, transportation terminals, working corridors in prisons and hospitals, book storage areas, active inventory storage, hospital bedrooms, hotel and motel bedrooms, enclosed shopping mall concourse areas, stairways.	1.00
Category C: Corridors, lobbies, elevators, inactive storage areas.	0.50
Category D: Indoor parking.	0.25
Exterior:	
Category E: Building perimeter: wall-wash, facade, canopy.	5.00 (per linear foot)
Category F: Outdoor parking.	0.10

Figure 24.9

Chapter 25

USING MEANS REPAIR & REMODELING COST DATA

Users of *Means Repair & Remodeling Cost Data* are chiefly interested in obtaining quick, reasonable, average prices for remodeling, renovation and repair work. This is the primary purpose of the annual book–to eliminate guesswork when pricing unknowns. Many persons use the cost data, whether for bids or verification of quotations or budgets, without being fully aware of how the prices are obtained and derived. Without this knowledge, this resource is not being used to the fullest advantage.

In addition to the basic cost data, the book also contains a wealth of information to aid the estimator, the contractor, the designer, and the owner, to better plan and manage remodeling projects. Productivity data is provided in order to assist with scheduling. National labor rates are analyzed. Tables and charts for location and time adjustments are included and help the estimator tailor the prices to a specific location. The costs in *Means Repair & Remodeling Cost Data* consist of over 13,000 unit price line items, as well as prices for thousands of construction assemblies. The unit price information, organized according to the Construction Specification Institute's MASTERFORMAT divisions, also provides an invaluable checklist to the construction professional to ensure that all required items are included in a project.

Format and Data

The major portion of *Means Repair & Remodeling Cost Data* is the **Unit Price section.** This is the primary source of unit cost data and is organized according to the CSI MASTERFORMAT. This index was developed by representatives of all parties concerned with the building construction industry and has been accepted by the American Institute of Architects (AIA), the Associated General Contractors of America, Inc. (AGC), the Construction Specifications Institute, Inc. (CSI) and Construction Specifications Canada (CSC).

CSI MASTERFORMAT Divisions:

Division 1 – General Requirements
Division 2 – Site Work
Division 3 – Concrete
Division 4 – Masonry
Division 5 – Metals
Division 6 – Wood & Plastics
Division 7 – Thermal & Moisture Control
Division 8 – Doors and Windows
Division 9 – Finishes
Division 10 – Specialties
Division 11 – Equipment
Division 12 – Furnishings
Division 13 – Special Construction
Division 14 – Conveying Systems
Division 15 – Mechanical
Division 16 – Electrical

The Assemblies section contains over 2,000 costs for related assemblies, or systems. Components of the assemblies are fully detailed and accompanied by illustrations. Included also are factors used to adjust costs for restrictive existing conditions. The assemblies cost data is organized according to the twelve Uniformat divisions.

Division 1 – Foundations
Division 2 – Substructures
Division 3 – Superstructure
Division 4 – Exterior Closure
Division 5 – Roofing
Division 6 – Interior Construction
Division 7 – Conveying
Division 8 – Mechanical
Division 9 – Electrical
Division 10 – General Conditions
Division 11 – Special
Division 12 – Site Work

The Reference section contains tables and reference charts, alternate pricing methods, technical data, and design information. It also provides estimating procedures and explanations of cost development which support and supplement the unit price and assemblies cost data.

The Appendix includes City Cost Indexes, representing the compilation of construction data for 162 major U.S. and Canadian cities. Cost adjustment factors are given for each city, by trade, relative to the national average. A comprehensive index and list of abbreviations are also included.

Source of Costs

The prices presented in *Means Repair & Remodeling Cost Data* are national averages. Material and equipment costs are developed through annual contact with manufacturers, dealers, distributors, and contractors throughout the United States. Means' staff of engineers is constantly updating prices and keeping abreast of changes and fluctuations within the industry. Labor rates are the national average of each trade as determined from union agreements from thirty major U.S. cities. Throughout the calendar year, as new wage agreements are negotiated, labor costs should be factored accordingly.

Following is a list of factors and assumptions on which the costs presented in *Means Repair & Remodeling Cost Data* have been based:

- **Quality**: The costs are based on methods, materials, and workmanship in accordance with U.S. Government standards and represent good, sound construction practice.
- **Overtime**: The costs as presented include *no* allowance for overtime. If overtime or premium time is anticipated, labor costs must be factored accordingly.
- **Productivity**: The daily output and man-hour figures are based on an eight-hour workday, during daylight hours. The chart in Figure 25.1 shows that as the number of hours worked per day (over eight) increases, and as the days per week (over five) increase, production efficiency decreases.

Days per Week	Hours per Day	Production Efficiency					Payroll Cost Factors	
		1 Week	2 Weeks	3 Weeks	4 Weeks	Average 4 Weeks	@ 1½ Times	@ 2 Times
5	8	100%	100%	100%	100%	100%	100%	100%
	9	100	100	95	90	96.25	105.6	111.1
	10	100	95	90	85	91.25	110.0	120.0
	11	95	90	75	65	81.25	113.6	127.3
	12	90	85	70	60	76.25	116.7	133.3
6	8	100	100	95	90	96.25	108.3	116.7
	9	100	95	90	85	92.50	113.0	125.9
	10	95	90	85	80	87.50	116.7	133.3
	11	95	85	70	65	78.75	119.7	139.4
	12	90	80	65	60	73.75	122.2	144.4
7	8	100	95	85	75	88.75	114.3	128.6
	9	95	90	80	70	83.75	118.3	136.5
	10	90	85	75	65	78.75	121.4	142.9
	11	85	80	65	60	72.50	124.0	148.1
	12	85	75	60	55	68.75	126.2	152.4

Figure 25.1

- **Size of Project**: Costs in *Means Repair & Remodeling Cost Data* are based on projects which cost approximately $5,000 to $500,000. Large residential projects are also included.
- **Local Factors**: Weather conditions, season of the year, local union restrictions, and unusual building code requirements can all have a significant impact on construction costs. The availability of a skilled labor force, sufficient materials and even adequate energy and utilities will also affect costs. These factors vary in impact and do not necessarily depend on location. They must be reviewed for each project in every area.

In presenting prices in *Means Repair & Remodeling Cost Data*, certain rounding rules are employed to make the numbers easy to use without significantly affecting accuracy. The rules are used consistently and are as follows:

Prices From	To	Rounded to Nearest
0.01	5.00	0.01
5.01	20.00	0.05
20.01	100.00	1.00
100.01	1,000.00	5.00
1,000.01	10,000.00	25.00
10,000.01	50,000.00	100.00
50,000.01	up	500.00

Unit Price Section

The Unit Price section of *Means Repair & Remodeling Cost Data* contains a great deal of information in addition to the unit cost for each construction component. Figure 25.2 is a typical page, showing costs for drywall. Note that prices are included for several types of drywall, each based on the type of installation and finishing requirements. In addition, approximate crews, workers, and productivity data are indicated. The information and cost data is broken down and itemized in this way to provide for the most detailed pricing possible.

Within each individual line item, there is a description of the **construction component**, information regarding **typical crews** designated to perform the work, and **productivity** shown as daily output and as man-hours. Costs are presented as "bare," or unburdened, as well as with mark-ups for overhead and profit. Figure 25.3 is a graphic representation of how to use the Unit Price section as presented in *Means Repair & Remodeling Cost Data*.

Line Numbers

Every construction item in the Means unit price cost data has a unique line number. This line number acts as an "address" so that each item can be quickly located and/or referenced. The numbering system is based on the CSI MASTERFORMAT division and Mediumscope classification. In Figure 25.2, note the bold number in reverse type, "092". This number represents the major subdivision, in this case "Drywall," of the major CSI Division 9—Finishes. All 16 divisions are organized in this manner. Within each subdivision, the data is broken down into Mediumscope designations and major classifications. The major classifications are listed alphabetically within the Mediumscope headings and are designated by bold type for both numbers and descriptions. Each item, or line, is further

092 | Lath, Plaster and Gypsum Board

092 600 | Gypsum Board Systems

			CREW	DAILY OUTPUT	MAN-HOURS	UNIT	BARE COSTS MAT.	BARE COSTS LABOR	BARE COSTS EQUIP.	BARE COSTS TOTAL	TOTAL INCL O&P
602	1400	On beams, columns, or soffits, standard, no finish included	2 Carp	675	.024	S.F.	.29	.51		.80	1.13
	1450	With thin coat plaster finish		475	.034		.38	.72		1.10	1.58
	1600	Fire resistant, no finish included		675	.024		.33	.51		.84	1.18
	1700	With thin coat plaster finish		475	.034		.42	.72		1.14	1.62
	3000	½" thick, on walls or ceilings, standard, no finish included		1,900	.008		.20	.18		.38	.51
	3100	With thin coat plaster finish		875	.018		.29	.39		.68	.95
	3300	Fire resistant, no finish included		1,900	.008		.24	.18		.42	.55
	3400	With thin coat plaster finish		875	.018		.33	.39		.72	.99
	3450	On beams, columns, or soffits, standard, no finish included		675	.024		.30	.51		.81	1.14
	3500	With thin coat plaster finish		475	.034		.39	.72		1.11	1.59
	3700	Fire resistant, no finish included		675	.024		.34	.51		.85	1.19
	3800	With thin coat plaster finish		475	.034		.43	.72		1.15	1.63
	5000	⅝" thick, on walls or ceilings, fire resistant, no finish included		1,900	.008		.24	.18		.42	.55
	5100	With thin coat plaster finish		875	.018		.33	.39		.72	.99
	5500	On beams, columns, or soffits, no finish included		675	.024		.34	.51		.85	1.19
	5600	With thin coat plaster finish		475	.034		.43	.72		1.15	1.63
	6000	For high ceilings, over 8' high, add		3,060	.005		.10	.11		.21	.29
	6500	For over 3 stories high, add per story	↓	6,100	.003	↓	.05	.06		.11	.15
	9000	Minimum labor/equipment charge	1 Carp	2	4	Job		86		86	135
608	0010	DRYWALL Gypsum plasterboard, nailed or screwed to studs,									
	0100	unless otherwise noted									
	0150	⅜" thick, on walls, standard, no finish included	2 Carp	2,000	.008	S.F.	.19	.17		.36	.48
	0200	On ceilings, standard, no finish included		1,800	.009		.19	.19		.38	.51
	0250	On beams, columns, or soffits, no finish included	↓	675	.024	↓	.29	.51		.80	1.13
	0270										
	0300	½" thick, on walls, standard, no finish included	2 Carp	2,000	.008	S.F.	.20	.17		.37	.49
	0350	Taped and finished		965	.017		.25	.35		.60	.84
	0400	Fire resistant, no finish included		2,000	.008		.24	.17		.41	.54
	0450	Taped and finished		965	.017		.29	.35		.64	.89
	0500	Water resistant, no finish included		2,000	.008		.30	.17		.47	.60
	0550	Taped and finished		965	.017		.35	.35		.70	.95
	0600	Prefinished, vinyl, clipped to studs	↓	900	.018	↓	.60	.38		.98	1.27
	0650										
	1000	On ceilings, standard, no finish included	2 Carp	1,800	.009	S.F.	.20	.19		.39	.53
	1050	Taped and finished		765	.021		.25	.45		.70	.99
	1100	Fire resistant, no finish included		1,800	.009		.24	.19		.43	.57
	1150	Taped and finished		765	.021		.29	.45		.74	1.04
	1200	Water resistant, no finish included		1,800	.009		.30	.19		.49	.64
	1250	Taped and finished		765	.021		.35	.45		.80	1.10
	1500	On beams, columns, or soffits, standard, no finish included		675	.024		.30	.51		.81	1.14
	1550	Taped and finished		475	.034		.35	.72		1.07	1.54
	1600	Fire resistant, no finish included		675	.024		.34	.51		.85	1.19
	1650	Taped and finished		475	.034		.39	.72		1.11	1.59
	1700	Water resistant, no finish included		675	.024		.40	.51		.91	1.25
	1750	Taped and finished		475	.034		.45	.72		1.17	1.65
	2000	⅝" thick, on walls, standard, no finish included		2,000	.008		.24	.17		.41	.54
	2050	Taped and finished		965	.017		.29	.35		.64	.89
	2100	Fire resistant, no finish included		2,000	.008		.26	.17		.43	.56
	2150	Taped and finished		965	.017		.31	.35		.66	.91
	2200	Water resistant, no finish included		2,000	.008		.34	.17		.51	.65
	2250	Taped and finished		965	.017		.39	.35		.74	1
	2300	Prefinished, vinyl, clipped to studs	↓	900	.018	↓	.67	.38		1.05	1.35
	2350										
	3000	On ceilings, standard, no finish included	2 Carp	1,800	.009	S.F.	.24	.19		.43	.57
	3050	Taped and finished		765	.021		.29	.45		.74	1.04
	3100	Fire resistant, no finish included		1,800	.009		.26	.19		.45	.59
	3150	Taped and finished	↓	765	.021	↓	.31	.45		.76	1.06

For expanded coverage of these items see *Means Interior Cost Data 1989*

141

Figure 25.2

HOW TO USE UNIT PRICE PAGES

Important
Prices in this section are listed in two ways: as bare costs and as costs including overhead and profit of the installing contractor. In most cases, if the work is to be subcontracted, it is best for a general contractor to add an additional 10% to the figures found in the column titled **"TOTAL INCL. O&P"**.

(86) Circle Reference Number
These reference numbers refer to charts, tables, estimating data, cost derivations and other information which may be useful to the user of this book. This information is located in the Reference Section of this book.

Crew F-2

Crew No.	Bare Costs		Incl. Subs O & P		Cost Per Man-hour	
Crew F-2	Hr.	Daily	Hr.	Daily	Bare Costs	Incl. O&P
2 Carpenters	$21.40	$342.40	$34.35	$549.60	$21.40	$34.35
Power Tools		15.00		16.50	.93	1.03
16 M.H., Daily Totals		$357.40		$566.10	$22.33	$35.38

Unit
The unit of measure listed here reflects the material being used in the line item. For example: headers over openings are defined in linear feet (L.F.).

Bare Costs are developed as follows for line no. **061-128-2050**
Mat. is **Bare Material Cost ($.57)**
Labor for Crew F2 = Man-hour Cost (**$21.40**) × Man-hour Units (**.047**) = **$1.01**
Equip. for Crew F2 = Equip. Hour Cost (**$.93**) × Man-hour Units (**.047**) = **$.04**
Total = **Mat. Cost ($.57)** + **Labor Cost ($1.01)** + **Equip. Cost ($.04)** = **$1.62** each.
(**Note:** When a Crew is indicated Equipment and Labor costs are derived from the Crew Tables. See example above.)

Productivity
The daily output represents typical total daily amount of work that the designated crew will produce. Man-hours are a unit of measure for the labor involved in performing a task. To derive the total man-hours for a task, multiply the quantity of the item involved times the man-hour figure shown.

Total Costs Including O&P are developed as follows:
Mat. is **Bare Material Cost** + 10% = **$.57** + **$.06** = **$.63**
Labor for Crew F2 = Man-hour Cost (**$34.35**) × Man hour Units (**.047**) = **$1.62**
Equip. for Crew F2 = Equip. Hour Cost (**$1.03**) × Man-hour Units (**.047**) = **$.05**
Total = **Mat. Cost ($.63)** + **Labor Cost ($1.62)** + **Equip. Cost ($.05)** = **$2.30**
(**Note:** Where a crew is indicated, Equipment and Labor costs are derived from the Crew Tables. See example at top of this page. **"Total"** line costs are rounded.)

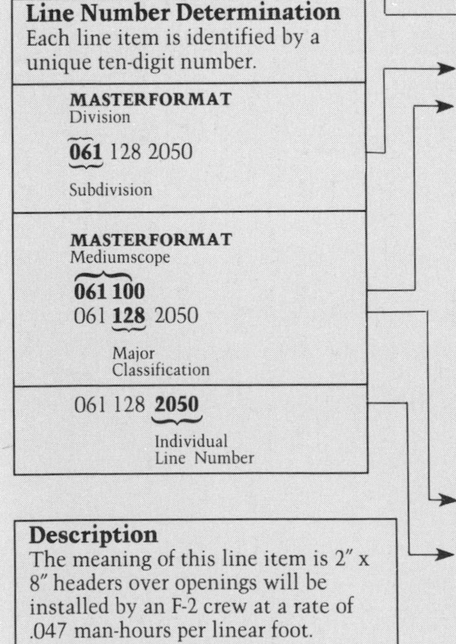

061 | Rough Carpentry

	061 100 Wood Framing		Crew	Daily Output	Man-Hours	Unit	Bare Costs Mat.	Bare Costs Labor	Bare Costs Equip.	Bare Costs Total	Total Incl O&P	
122	4460	4" x 8"	F-2	300	.053	L.F.	1.88	1.14	.05	3.07	3.96	122
	4480	4" x 10"	" "	260	.062	"	2.35	1.32	.06	3.73	4.77	
	9000	Minimum labor/equipment charge	1 Carp	4	2	Job		43		43	69	
124	0010	FRAMING, SLEEPERS										124
	0020											
	0100	On concrete, treated, 1" x 2"	F-2	2,350	.007	L.F.	.09	.15	.01	.25	.34	
	0150	1" x 3"		2,000	.008		.14	.17	.01	.32	.43	
	0200	2" x 4"		1,500	.011		.34	.23	.01	.58	.75	
	0250	2" x 6"		1,300	.012		.51	.26	.01	.78	1	
	9000	Minimum labor/equipment charge	1 Carp	4	2	Job		43		43	69	
126	0010	FRAMING, SOFFITS & CANOPIES										126
	0020											
	1000	Canopy or soffit framing , 1" x 4"	F-2	900	.018	L.F.	.13	.38	.02	.53	.77	
	1040	1" x 8"		750	.021		.28	.46	.02	.76	1.06	
	1100	2" x 4"		620	.026		.26	.55	.02	.83	1.20	
	1140	2" x 8"		500	.032		.57	.68	.03	1.28	1.76	
	1200	3" x 4"		500	.032		.57	.68	.03	1.28	1.76	
	1240	3" x 10"		300	.053		1.43	1.14	.05	2.62	3.46	
	9000	Minimum labor/equipment charge	1 Carp	4	2	Job		43		43	69	
128	0010	FRAMING, WALLS										128
	0020											
	2000	Headers over openings, 2" x 6" (86)	F-2	360	.044	L.F.	.39	.95	.04	1.38	2	
	2050	2" x 8"		340	.047		.57	1.01	.04	1.62	2.30	
	2110	2" x 10"		320	.050		.78	1.07	.05	1.90	2.63	
	2150	2" x 12"		300	.053		.91	1.14	.05	2.10	2.89	
	2200	4" x 12"		190	.084		2.29	1.80	.08	4.17	5.50	
	2250	6" x 12"		140	.114		3.43	2.45	.11	5.99	7.80	
	5000	Plates, untreated, 2" x 3"		850	.019		.20	.40	.02	.62	.89	
	5020	2" x 4"		800	.020		.26	.43	.02	.71	1	
	5040	2" x 6"		750	.021		.39	.46	.02	.87	1.18	
	5120	Studs, 8' high wall, 2" x 3"		1,200	.013		.20	.29	.01	.50	.69	

Figure 25.3

defined by an individual number. As shown in Figure 25.3, the full line number for each item consists of: a major CSI subdivision number–a major classification number–an item line number. Each full line number describes a unique construction element. For example, in Figure 25.2, the line number for 1/2" drywall, on walls, taped and finished, is 092-608-0350.

Line Description

Each line has a text description of the item for which costs are listed. The description may be self-contained and all-inclusive. Or, if indented, the complete description for a line depends on the information provided above. All indented items are delineations (by size, color, material, etc.) or breakdowns of previously described items. An index is provided in the back of *Means Repair & Remodeling Cost Data* to aid in locating particular items.

Crew

For each construction element (each line item), a minimum typical crew is designated as appropriate to perform the work. The crew may include one or more trades, foremen, craftsmen and helpers, and any equipment required for proper installation of the described item. If an individual trade installs the item using only hand tools, the smallest efficient number of tradesmen will be indicated (1 Carp, 2 Elec, etc.). Abbreviations for trades are shown in Figure 25.4. If more than one trade is required to install the item and/or if powered equipment is needed, a crew number will be designated (B-5, D-3, etc.). A complete listing of crews is presented in the Foreword pages of *Means Repair & Remodeling Cost Data*. On these pages, each crew is broken down into the following components:

- Number and type of workers designated.
- Number, size, and type of any equipment required.
- Hourly labor costs listed two ways: *bare*–base rate including fringe benefits; and including installing contractor's overhead and profit–*billing rate*. (See Figure 25.4 from the inside back cover of *Means Repair & Remodeling Cost Data* for labor rate information).
- Daily equipment costs, based on the weekly equipment rental cost divided by 5 days per week, plus the hourly operating cost, times 8 hours. This cost is listed two ways: as a bare cost and with a 10 percent mark-up to cover handling and management costs.
- Labor and equipment are broken down further into *cost per man-hour* for labor, and cost per man-hour for the equipment.
- The total daily man-hours for the crew.
- The total bare cost for the crew, including equipment.
- The total daily cost of the crew, including the installing contractor's overhead and profit.

The total daily cost of the required crew is used to calculate the unit installation cost for each item (for both bare costs and total costs including overhead and profit).

The crew designation does not mean that this is the only crew that can perform the work. Crew size and content have been developed and chosen based on practical experience and feedback from contractors. These designations represent a labor and equipment make-up commonly found in the industry. The most appropriate crew

Installing Contractor's Overhead & Profit

Below are the **average** installing contractor's percentage mark-ups applied to base labor rates to arrive at typical billing rates.

Column A: Labor rates are based on union wages averaged for 30 major U.S. cities. Base rates including fringe benefits are listed hourly and daily. These figures are the sum of the wage rate and employer-paid fringe benefits such as vacation pay, employer-paid health and welfare costs, pension costs, plus appropriate training and industry advancement funds costs.

Column B: Workers' Compensation rates are the national average of state rates established for each trade.

Column C: Column C lists average fixed overhead figures for all trades. Included are Federal and State Unemployment costs set at 6.2%; Social Security Taxes (FICA) set at 7.65%; Builder's Risk Insurance costs set at 0.34%; and Public Liability costs set at 1.55%. All the percentages except those for Social Security Taxes vary from state to state as well as from company to company.

Column D and E: Percentages in Columns D and E are based on the presumption that the installing contractor has annual billing of $500,000 and up. Overhead percentages may increase with smaller annual billing. The overhead percentages for any given contractor may vary greatly and depend on a number of factors, such as the contractor's annual volume, engineering and logistical support costs, and staff requirements. The figures for overhead and profit will also vary depending on the type of job, the job location, and the prevailing economic conditions. All factors should be examined very carefully for each job.

Column F: Column F lists the total of columns B, C, D, and E.

Column G: Column G is Column A (hourly base labor rate) multiplied by the percentage in Column F (O&P percentage).

Column H: Column H is the total of Column A (hourly base labor rate) plus Column G (Total O&P).

Column I: Column I is Column H multiplied by eight hours.

Abbr.	Trade	A		B	C	D	E	F	G	H	I
		Base Rate Incl. Fringes		Workers' Comp. Ins.	Average Fixed Overhead	Overhead	Profit	Total Overhead & Profit		Rate with O & P	
		Hourly	Daily					%	Amount	Hourly	Daily
Skwk	Skilled Workers Average (35 trades)	$21.90	$175.20	12.5%	15.7%	16.0%	15.0%	59.2%	$12.95	$34.85	$278.80
	Helpers Average (5 trades)	16.60	132.80	13.8				60.5	10.05	26.65	213.20
	Foremen Average, Inside (50¢ over trade)	22.40	179.20	12.5				59.2	13.25	35.65	285.20
	Foremen Average, Outside ($2.00 over trade)	23.90	191.20	12.5				59.2	14.15	38.05	304.40
Clab	Common Building Laborers	16.85	134.80	13.9				60.6	10.20	27.05	216.40
Asbe	Asbestos Workers	23.85	190.80	11.6				58.3	13.90	37.75	302.00
Boil	Boilermakers	23.90	191.20	7.9				54.6	13.05	36.95	295.60
Bric	Bricklayers	22.10	176.80	11.3				58.0	12.80	34.90	279.20
Brhe	Bricklayer Helpers	17.10	136.80	11.3				58.0	9.90	27.00	216.00
Carp	Carpenters	21.40	171.20	13.9				60.6	12.95	34.35	274.80
Cefi	Cement Finishers	20.80	166.40	7.9				54.6	11.35	32.15	257.20
Elec	Electricians	24.25	194.00	5.2				51.9	12.60	36.85	294.80
Elev	Elevator Constructors	24.15	193.20	6.8				53.5	12.90	37.05	296.40
Eqhv	Equipment Operators, Crane or Shovel	22.40	179.20	8.8				55.5	12.45	34.85	278.80
Eqmd	Equipment Operators, Medium Equipment	21.65	173.20	8.8				55.5	12.00	33.65	269.20
Eqlt	Equipment Operators, Light Equipment	20.60	164.80	8.8				55.5	11.45	32.05	256.40
Eqol	Equipment Operators, Oilers	18.45	147.60	8.8				55.5	10.25	28.70	229.60
Eqmm	Equipment Operators, Master Mechanics	23.10	184.80	8.8				55.5	12.80	35.90	287.20
Glaz	Glaziers	22.05	176.40	10.4				57.1	12.60	34.65	277.20
Lath	Lathers	21.60	172.80	8.4				55.1	11.90	33.50	268.00
Marb	Marble Setters	21.95	175.60	11.3				58.0	12.75	34.70	277.60
Mill	Millwrights	22.15	177.20	8.4				55.1	12.20	34.35	274.80
Mstz	Mosaic and Terrazzo Workers	21.45	171.60	7.0				53.7	11.50	32.95	263.60
Pord	Painters, Ordinary	20.10	160.80	10.4				57.1	11.50	31.60	252.80
Psst	Painters, Structural Steel	20.85	166.80	34.9				81.6	17.00	37.85	302.80
Pape	Paper Hangers	20.25	162.00	10.4				57.1	11.55	31.80	254.40
Pile	Pile Drivers	21.50	172.00	21.9				68.6	14.75	36.25	290.00
Plas	Plasterers	21.10	168.80	11.0				57.7	12.15	33.25	266.00
Plah	Plasterer Helpers	17.35	138.80	11.0				57.7	10.00	27.35	218.80
Plum	Plumbers	24.45	195.60	6.2				52.9	12.95	37.40	299.20
Rodm	Rodmen (Reinforcing)	23.15	185.20	22.0				68.7	15.90	39.05	312.40
Rofc	Roofers, Composition	19.75	158.00	25.7				72.4	14.30	34.05	272.40
Rots	Roofers, Tile & Slate	19.85	158.80	25.7				72.4	14.35	34.20	273.60
Rohe	Roofer Helpers (Composition)	14.65	117.20	25.7				72.4	10.60	25.25	202.00
Shee	Sheet Metal Workers	24.05	192.40	8.8				55.5	13.35	37.40	299.20
Spri	Sprinkler Installers	25.25	202.00	6.7				53.4	13.50	38.75	310.00
Stpi	Steamfitters or Pipefitters	24.55	196.40	6.2				52.9	13.00	37.55	300.40
Ston	Stone Masons	22.00	176.00	11.3				58.0	12.75	34.75	278.00
Sswk	Structural Steel Workers	23.25	186.00	26.8				73.5	17.10	40.35	322.80
Tilf	Tile Layers (Floor)	21.50	172.00	7.0				53.7	11.55	33.05	264.40
Tilh	Tile Layer Helpers	17.05	136.40	7.0				53.7	9.15	26.20	209.60
Trlt	Truck Drivers, Light	17.45	139.60	11.8				58.5	10.20	27.65	221.20
Trhv	Truck Drivers, Heavy	17.60	140.80	11.8				58.5	10.30	27.90	223.20
Sswl	Welders, Structural Steel	23.25	186.00	26.8				73.5	17.10	40.35	322.80
Wrck	*Wrecking	16.80	134.40	28.1				74.8	12.55	29.35	234.80

*Not included in Averages.

Figure 25.4

for a given task is best determined based on particular project requirements. Unit costs may vary if crew sizes or content are significantly changed.

Unit

The unit column (see Figures 25.2 and 25.3) identifies the component for which the costs have been calculated. It is this "unit" on which unit price estimating is based. The units as used represent standard estimating and quantity takeoff procedures. However, the estimator should always check to be sure that the units taken off are the same as those priced. A list of standard abbreviations is included at the back of *Means Repair & Remodeling Cost Data*.

Bare Costs

The four columns listed under *Bare Costs - Material, Labor, Equipment*, and *Total* - represent the actual cost of construction items to the contractor. In other words, bare costs are those which *do not* include the overhead and profit of the installing contractor, whether for a subcontractor or a general contracting company using its own crews.

Material costs are based on the national average contractor purchase price delivered to the job site. Delivered costs are assumed to be within a 20-mile radius of metropolitan areas. No sales tax is included in the material prices because of variation from state to state.

The prices are based on quantities that would normally be purchased for remodeling projects costing $5,000 to $500,000. Prices for small quantities must be adjusted accordingly. If more current costs for materials are available for the appropriate location, it is recommended that unit costs be adjusted to reflect any cost difference.

Labor costs are calculated by multiplying the bare labor cost per man-hour times the number of man-hours, from the *Man-Hours* column. The *bare labor rate* is determined by adding the base rate plus fringe benefits. The base rate is the actual hourly wage of a worker, as used in figuring the payroll. It is this figure from which employee deductions are taken (federal withholding, FICA, state withholding). Fringe benefits include all employer-paid benefits, above and beyond the payroll amount (employer-paid health, vacation pay, pension, profit-sharing). The *Bare Labor Cost* is, therefore, the actual amount that the contractor must pay directly for construction workers. Figure 25.4 shows labor rates for the 35 standard construction trades plus skilled worker, helper, and foreman averages. These rates are the averages of union wage agreements effective January 1 of the year of publication, from 30 major cities in the United States. The *Bare Labor Cost* for each trade, as used in *Means Repair & Remodeling Cost Data* is shown in column "A" as the base rate including fringes. Refer to the *Crew* column to determine the rate used to calculate the *Bare Labor Cost* for a particular line item.

Equipment costs are calculated by multiplying the *Bare Equipment Cost* per man-hour, from the appropriate *Crew* listing, times the man-hours in the *Man-Hours* column.

Total Bare Costs

This column simply represents the arithmetic sum of the bare material, labor and equipment costs. This total is the average cost to the contractor for the particular item of construction, furnished and installed, or "in place." No overhead and/or profit is included.

Total Including Overhead and Profit

The prices in the *Total Including Overhead and Profit* column represent the total cost of an item including the installing contractor's overhead and profit. The installing contractor could be either the general contractor or a subcontractor. If these costs are used for an item to be installed by a subcontractor, the general contractor should include an additional percentage (usually 10% to 20%) to cover the expenses of supervision and management.

The costs in this column are the arithmetic sum of the following three calculations:

- Bare Material Cost plus 10%.
- Labor Cost, including fixed overhead, overhead, and profit, per man-hour.
- Equipment Costs plus 10% per man-hour, times the number of man-hours. The Labor and Equipment Costs, including overhead and profit, are found in the appropriate crew listings. The overhead and profit percentage factor for labor is obtained from column F in Figure 25.4.

The following items are included in the increase for fixed overhead, overhead, and profit:

- **Worker's Compensation Insurance** rates vary from state to state and are tied into the construction trade safety records in that particular state. Rates also vary by trade according to the hazard involved. The proper authorities will most likely keep the contractor well informed of the rates and obligations.
- **State and Federal Unemployment Insurance** rates are adjusted by a merit rating system according to the number of former employees applying for benefits. Contractors who find it possible to offer a maximum of steady employment can enjoy a reduction in the unemployment tax rate.
- **Employer-Paid Social Security (FICA)** is adjusted annually by the federal government. It is a percentage of an employee's salary up to a maximum annual contribution.
- **Builder's Risk and Public Liability** insurance rates vary according to the trades involved and the state in which the work is done.

Overhead is an average percentage to be added for office or operating overhead. This is the cost of doing business. The percentages are presented as national averages by trade as shown in Figure 25.4. Note that the operating overhead costs are applied to *labor only* in *Means Repair & Remodeling Cost Data*.

Profit is the fee (usually a percentage) added by the contractor to offer both a return on investment and an allowance to cover the risk involved in the type of construction being bid. The profit percentage may vary from 4% on large, straightforward projects to as much as 25% on smaller, high-risk jobs. Profit percentages are directly affected by economic conditions, the expected number of bidders, and the estimated risk involved in the project. For estimating purposes, *Means Repair & Remodeling Cost Data* assumes 10% on labor costs to be a reasonable average profit factor.

Assemblies Section

Means' assemblies data is divided into twelve *Uniformat* divisions, which reorganize the components of construction into logical groupings. The Assemblies, or Systems approach was devised to provide quick and easy methods for estimating, even when only preliminary design data is available.

The groupings, or systems, are presented in such a way that the estimator can easily vary components within the systems, as well as substitute one system for another. This flexibility is extremely useful when adapting to budget, design, or other considerations. Figure 25.5 shows how the data is presented in the Assemblies section.

Each assembly is illustrated and accompanied by a detailed description. The book lists the components and sizes of each system, usually in the order of construction. Alternates for the most commonly variable components are also listed. Each individual component is found in the Unit Price section. If an alternate component (not listed in the assembly) is required, it can easily be substituted.

Quantity

A unit of measure is established for each system. For example, partition systems are measured by the square foot of wall area, and doors are measured by "each." Within each system, the components are measured by industry standard, using the same units as in the Unit Price section.

Material

The cost of each component in the *Material* column is the *Bare Material Cost*, plus 10% handling, for the unit and quantity as defined in the *Quantity* column.

Installation

Installation costs as listed in the *Systems* pages contain both labor and equipment costs. The labor rate includes the *Bare Labor Cost* plus the installing contractor's fixed overhead and profit (shown in Figure 25.4). The equipment rate is the *Bare Equipment Cost*, plus 10%.

Reference Section

Throughout the Unit Price section are circled reference numbers. These numbers serve as footnotes, referring the reader to illustrations, charts, and estimating reference tables in the Reference Section of the book. Figures 25.6 & 25.7 show two examples of the kinds of information provided in circle reference numbers. The reference tables explain the development of unit costs for many items. They also provide design criteria for many types of construction to aid the designer/estimator in making appropriate choices.

City Cost Indexes

The unit prices in *Means Repair & Remodeling Cost Data* are national averages. When they are to be applied to a particular location, these prices must be adjusted to local conditions. Means has developed the City Cost Indexes for just that purpose. This section contains tables for 162 U.S. and Canadian cities, based on a 30 major city

INTERIOR CONSTR.	06.4-146	Doors, Interior Flush, Metal

This page illustrates and describes interior metal door systems including a metal door, metal frame and hardware. Lines 06.4-146-04 thru 07 give the unit price and total price on a cost each for this system. Prices for alternate interior metal door systems are on Line Items 06.4-146-11 thru 21. Both material quantities and labor costs have been adjusted for the system listed.

Factors: To adjust for job conditions other than normal working situations use Lines 06.5-146-29 thru 40.

Example: You are to install the system while protecting existing construction. Go to Line 06.4-146-37 and apply these percentages to the appropriate MAT. and INST. costs.

LINE NO.	DESCRIPTION	QUANTITY	COST EACH		
			MAT.	INST.	TOTAL
01	**Single metal door, including frame and hardware.**				
02					
03					
04	Hollow metal door, 1-⅜" thick, 2'-6" x 6'-8", painted	1 Ea.	151.24	64.76	216
05	Metal frame, 5-¾" deep	1 Set	71.50	33.50	105
06	Hinges and passage lockset	1 Set	64.90	17.10	82
07	TOTAL	Ea.	287.64	115.36	403
08					
09					
10	For alternate systems:				
11	Hollow metal doors, 1-⅜" thick, 2'-8" x 6'-8"	Ea.	295.34	117.66	413
12	3'-0" x 7'-0"	"	303.04	124.96	428
13					
14	Interior fire door, 1-⅜" thick, 2'-6" x 6'-8"	Ea.	329.44	118.56	448
15	2'-8" x 6'-8"	↓	337.14	120.86	458
16	3'-0" x 7'-0"	↓	350.34	122.66	473
17					
18	Add to fire doors:				
19	Baked enamel finish	Ea.	30%	90%	
20	Galvanizing	↓	9.50%		
21	Porcelain finish	↓	100%	150%	
22					
23					
24					
25					
26					
27					
28					
29	Cut & patch to match existing construction, add, minimum		2%	3%	
30	Maximum		5%	9%	
31	Dust protection, add, minimum		1%	2%	
32	Maximum		4%	11%	
33	Equipment usage curtailment, add, minimum		1%	1%	
34	Maximum		3%	10%	
35	Material handling & storage limitation, add, minimum		1%	1%	
36	Maximum		6%	7%	
37	Protection of existing work, add, minimum		2%	2%	
38	Maximum		5%	7%	
39	Shift work requirements, add, minimum			5%	
40	Maximum			30%	
41					
42					

285

Figure 25.5

CIRCLE REFERENCE NUMBERS

(14) Steel Tubular Scaffolding (Div. 015-256)

On new construction, tubular scaffolding is efficient up to 60' high or five stories. Above this it is usually better to use a hung scaffolding if construction permits.

In repairing or cleaning the front of an existing building the cost of tubular scaffolding per S.F. of building front increases as the height increases above the first tier. The first tier cost is relatively high due to leveling and alignment. Swing scaffolding operations may interfere with tenants. In this case the tubular is more practical at all heights.

The minimum efficient crew for erection is three men. For heights over 50', a four-man crew is more efficient. Use two or more on top and two at the bottom for handing up or hoisting. Four men can erect and dismantle about nine frames per hour up to five stories. From five to eight stories they will average six frames per hour. With 7' horizontal spacing this will run about 300 S.F. and 200 S.F. of wall surface, respectively. Time for placing planks must be added to the above. On heights above 50', five planks can be placed per man-hour.

The cost per 1,000 S.F. of building front in the table below was developed by pricing the materials required for a typical tubular scaffolding system eleven frames long and two frames high. Planks were figured five wide for standing plus two wide for materials.

Frames are 2', 4' and 5' wide and usually spaced 7' O.C. horizontally. Sidewalk frames are 6' wide. Rental rates will be lower for jobs over three months duration.

For jobs under twenty-five frames, figure rental at $6.00 per frame. For jobs over one hundred frames, rental can go as low as $2.65 per frame. These figures do not include accessories which are listed separately below. Large quantities for long periods can reduce rental rates by 20%.

Item	Unit	Purchase, Each		Monthly Rent, Each		Per 1,000 S.F. of Building Front	
		Regular	Heavy Duty	Regular	Heavy Duty	No. of Frames	Rental per Mo.
5' Wide Frames, 3' High	Ea.	$55	$ —	$3.65	$ —	—	—
*5'-0" High		70	—	3.65	—	—	—
*6'-6" High		85	—	3.65	—	24	$ 87.60
2' & 4' Wide, 5' High		—	75	—	3.75	—	—
6'-0" High		—	85	—	3.75	—	—
6' Wide Frame, 7'-6" High		130	155	7.95	10	—	—
Sidewalk Bracket, 20"		20	—	1.60	—	12	19.20
Guardrail Post		15	—	1.10	—	12	13.20
Guardrail, 7' section		7	—	.80	—	11	8.80
Cross Braces		15	17	.75	.75	44	33.00
Screw Jacks & Plates		20	30	2.00	2.50	24	48.00
8" Casters		50	—	5.75	—	—	—
16' Plank, 2" x 10"		22	—	5.10	—	35	178.50
8' Plank, 2" x 10"	↓	11	—	3.75	—	7	26.25
1' to 6' Extension Tube		—	70	—	2.50	—	—
Shoring Stringers, steel, 10' to 12' long	L.F.	—	7	—	.40	—	—
Aluminum, 12' to 16' long		—	16	—	.60	—	—
Aluminum joists with nailers, 10' to 22' long	↓	—	12.50	—	.50	—	—
Flying Truss System, Aluminum	S.F.C.A.	—	10	—	.60	—	—
						Total	$414.55
						2 Use/Mo.	$207.28

*Most commonly used

Scaffolding is often used as falsework over 15' high during construction of cast-in-place concrete beams and slabs. Two ft. wide scaffolding is generally used for heavy beam construction. The span between frames depends upon the load to be carried with a maximum span of 5'.

Heavy duty scaffolding with a capacity of 10,000#/leg can be spaced up to 10' O.C. depending upon form support design and loading.

Scaffolding used as horizontal shoring requires less than half the material required with conventional shoring.

On new construction, erection is done by carpenters.

Rolling towers supporting horizontal shores can reduce labor and speed the job. For maintenance work, catwalks with spans up to 70' can be supported by the rolling towers.

(15) Concrete Pipe (Div. 027-162)

Prices given are for inside 20 mile delivery zone. Add $1.40 per ton of pipe for each additional 10 miles. Minimum truckload is 10 tons. The non-reinforced pipe listed in the front of the book is designation ASTM C14-59 extra strength. The reinforced pipe listed is ASTM C76-65T class 3, no gaskets. The installation cost given includes shaping bottom of the trench, placing the pipe, and backfilling and tamping to the top of the pipe only.

Figure 25.6

average of 100. The figures are broken down into material and installation for all CSI Divisions, as shown in Figure 25.7. Please note that for each city there is a weighted average based on total project costs. This average is based on the relative contribution of each division to the construction process as a whole.

In addition to adjusting the figures in *Means Repair & Remodeling Cost Data* for particular locations, the City Cost Index can also be used to adjust costs from one city to another. For example, costs for a particular building type are known for City A. In order to budget the costs of the same building type in City B, the following calculation can be made:

$$\frac{\text{City B Index}}{\text{City A Index}} \times \text{City A Cost} = \text{City B Cost}$$

While City Cost Indexes provide a means to adjust prices for location, the Historical Cost Index, (also included in *Means Repair & Remodeling Cost Data* and shown in Figure 25.8) provides a method for adjusting for time. Using the same principle as the City Cost Index, a time adjustment factor can be calculated as follows.

$$\frac{\text{Index for Year X}}{\text{Index for Year Y}} = \text{Time Adjustment Factor}$$

This time adjustment factor can be used to determine the budget costs for a particular building type in Year X, based on the cost for a similar building type known from Year Y. Used together, the two indexes allow for cost adjustments from one city during a given year to another city in another year (the present or otherwise). For example, an office building built in San Francisco in 1974 originally cost $1,000,000. How much will a similar building cost in Phoenix in 1989? Adjustment factors are developed as shown above using data from Figures 25.8 and 25.9:

$$\frac{\text{Phoenix Index}}{\text{San Francisco Index}} = \frac{92.8}{124.5} = 0.75$$

$$\frac{\text{1989 Index}}{\text{1974 Index}} = \frac{207.1}{94.7} = 2.19$$

Original cost × location adjustment × time adjustment = Proposed new cost,

$$\$1{,}000{,}000 \times 0.75 \times 2.19 = \$1{,}642{,}500$$

CIRCLE REFERENCE NUMBERS

(86) Thirty City Lumber Prices (Jan. 1st, 1989) (Div. 061)

Prices for boards are for #2 or better or sterling, whichever is in best supply. Dimension lumber is "Standard or Better" either Southern Yellow Pine (S.Y.P.), Spruce-Pine-Fir (S.P.F.), Hem-Fir (H.F.) or Douglas Fir (D.F.). The species of lumber used in a geographic area is listed by city. Rough Sawn lumber is Douglas Fir, Hem-Fir, or a variety of hardwood, sheathing or lagging grade. Plyform is 3/4" BB oil sealed fir or S.Y.P. whichever prevails locally, 5/8" CDX is S.Y.P. or Fir.

For 10 MBF lots add 5%; for retail add 10% to prices.

These are prices at the time of publication and should be checked against the current market price. Relative differences between cities will stay approximately constant.

City	Species	Carload Lots per M.B.F.								Carload Lots per M.S.F.	
		S4S					Rough Sawn Lumber				
		Dimensions			Boards						
		2"x4"	2"x6"	2"x10"	1"x6"	1"x12"	3"x12"	6"x12"	12"x12"	3/4" Ext. Plyform	5/8" Thick CDX
Atlanta	S.Y.P.	$320	$335	$410	$685	$885	$415	$485	$480	$715	$345
Baltimore	S.P.F.	330	330	395	815	915	400	450	440	730	480
Boston	S.P.F.	360	365	400	830	930	420	500	480	765	470
Buffalo	S.P.F.	355	370	435	815	945	450	505	490	775	450
Chicago	S.P.F.	370	380	450	815	910	470	500	490	775	460
Cincinnati	S.Y.P.	340	365	440	715	915	470	510	475	715	400
Cleveland	S.P.F.	330	345	410	800	930	430	520	500	760	450
Columbus	S.P.F.	350	360	415	815	900	435	530	505	765	465
Dallas	S.Y.P.	355	380	450	750	915	475	500	500	725	400
Denver	H.F.	335	335	395	765	900	410	510	505	790	430
Detroit	S.P.F.	360	360	410	815	920	420	535	510	770	450
Houston	S.Y.P.	355	380	450	755	925	470	510	490	730	390
Indianapolis	S.P.F.	340	345	420	820	940	435	515	495	765	435
Kansas City	D.F.	380	380	430	765	905	440	520	510	825	450
Los Angeles	D.F.	340	345	400	735	895	410	500	480	795	410
Memphis	S.Y.P.	345	370	430	700	890	435	515	500	730	395
Milwaukee	S.P.F.	325	330	400	810	905	415	495	475	760	450
Minneapolis	S.P.F.	325	330	410	820	920	425	500	480	775	460
Nashville	S.Y.P.	355	380	450	750	890	475	525	495	730	400
New Orleans	S.Y.P.	325	345	420	720	900	440	510	500	740	410
New York City	H.F.	380	380	405	715	925	430	555	520	795	445
Philadelphia	H.F.	355	360	390	775	915	415	490	480	770	450
Phoenix	S.Y.P.	360	380	470	740	920	490	510	500	730	400
Pittsburgh	S.P.F.	330	335	415	815	900	425	550	540	750	440
St. Louis	S.Y.P.	355	380	450	700	905	470	495	490	825	395
San Antonio	S.Y.P.	360	385	460	720	955	480	515	510	730	390
San Diego	D.F.	335	335	390	730	875	410	510	500	750	405
San Francisco	D.F.	330	330	395	710	860	415	510	505	770	420
Seattle	D.F.	310	315	350	700	830	400	500	490	770	400
Washington, DC	H.F.	340	340	390	755	905	415	480	470	735	450
Average		$345	$355	$420	$760	$910	$435	$510	$495	$760	$425

To convert square feet of surface to board feet, 4% waste included

S4S Size	Multiply S.F. by	T & G Size	Multiply S.F. by	Flooring Size	Multiply S.F. by
1 x 4	1.18	1 x 4	1.27	25/32" x 2-1/4"	1.37
1 x 6	1.13	1 x 6	1.18	25/32" x 3-1/4"	1.29
1 x 8	1.11	1 x 8	1.14	15/32" x 1-1/2"	1.54
1 x 10	1.09	2 x 6	2.36	1" x 3"	1.28
				1" x 4"	1.24

Figure 25.7

CITY COST INDEXES

DIVISION	ALABAMA												ALASKA			ARIZONA		
	BIRMINGHAM			HUNTSVILLE			MOBILE			MONTGOMERY			ANCHORAGE			PHOENIX		
	MAT.	INST.	TOTAL	MAT.	INST.	TOTAL	MAT.	INST.	TOTAL	MAT.	INST.	TOTAL	MAT.	INST.	TOTAL	MAT.	INST.	TOTAL
2 SITE WORK	100.1	89.9	95.6	119.7	91.1	107.1	122.7	87.2	107.0	91.9	88.9	90.6	159.2	127.1	145.0	92.9	95.6	94.1
3.1 FORMWORK	97.4	71.2	77.1	103.5	72.3	79.3	106.6	74.4	81.7	112.5	70.9	80.3	124.4	140.2	136.7	108.6	88.1	92.7
3.2 REINFORCING	94.5	72.5	85.4	95.8	72.0	86.0	82.9	72.6	78.7	82.9	72.5	78.6	117.8	131.7	123.5	111.0	90.7	102.6
3.3 CAST IN PLACE CONC.	89.2	91.5	90.6	101.9	93.0	96.5	99.9	93.2	95.9	101.2	92.0	95.6	225.7	111.3	156.3	107.5	93.2	98.8
3 CONCRETE	92.0	81.8	85.5	100.9	83.0	89.5	97.5	84.0	88.9	99.5	81.9	88.3	181.8	124.6	145.5	108.5	91.0	97.4
4 MASONRY	81.7	71.0	73.5	88.5	72.1	75.9	93.9	78.0	81.7	86.6	70.4	74.2	150.2	138.0	140.9	93.3	79.7	82.9
5 METALS	95.6	79.1	89.8	100.0	79.0	92.5	93.4	79.9	88.6	95.8	79.1	89.9	116.3	124.4	119.2	99.1	91.9	96.5
6 WOOD & PLASTICS	91.8	72.4	80.9	107.3	72.7	88.0	92.2	76.5	83.4	101.5	71.4	84.7	118.1	139.8	130.2	99.6	86.4	92.2
7 MOISTURE PROTECTION	84.5	72.3	80.6	92.1	72.1	85.8	87.2	77.6	84.2	88.5	69.9	82.7	102.5	141.2	114.8	92.7	84.9	90.2
8 DOORS, WINDOWS, GLASS	90.7	71.8	80.9	101.0	72.1	86.0	98.5	73.9	85.7	98.0	70.7	83.8	128.5	139.4	134.2	103.0	83.8	93.0
9.2 LATH & PLASTER	96.0	70.4	76.6	91.2	72.4	76.9	91.8	80.8	83.4	108.3	72.3	81.0	120.4	140.6	135.7	93.6	92.6	92.9
9.2 DRYWALL	100.4	71.5	86.8	108.5	72.1	91.4	92.3	76.8	85.0	100.5	70.9	86.6	122.1	141.3	131.1	90.6	86.6	88.7
9.5 ACOUSTICAL WORK	97.7	71.7	83.5	100.1	72.1	84.8	93.1	75.7	83.6	93.1	70.4	80.7	124.1	141.3	133.4	103.7	85.4	93.8
9.6 FLOORING	112.0	73.6	101.7	97.5	72.1	90.6	114.0	79.5	104.7	100.8	70.4	92.6	117.3	141.3	123.7	93.1	89.1	92.0
9.9 PAINTING	104.3	74.3	80.2	110.7	72.1	79.7	121.5	78.6	87.1	119.7	70.4	80.1	123.2	141.3	137.7	96.2	82.0	84.9
9 FINISHES	103.0	72.5	86.7	105.2	72.1	87.4	100.1	77.7	88.1	102.0	70.7	85.3	121.2	141.2	131.9	92.8	85.5	88.9
10-14 TOTAL DIV. 10-14	100.0	74.4	92.3	100.0	73.6	92.1	100.0	79.5	93.8	100.0	72.9	91.9	100.0	141.3	112.3	100.0	91.1	97.3
15 MECHANICAL	96.4	72.7	84.4	99.3	72.6	85.8	97.3	76.6	86.8	98.9	69.5	84.1	107.3	141.1	124.3	98.4	87.9	93.1
16 ELECTRICAL	94.9	72.8	79.4	92.5	71.8	78.0	90.4	76.5	80.7	91.6	70.4	76.7	107.6	141.3	131.1	105.4	85.6	91.6
1-16 WEIGHTED AVERAGE	94.8	75.4	84.4	100.2	75.6	87.0	97.8	79.1	87.7	96.8	74.0	84.5	125.9	135.7	131.2	99.3	87.2	92.8

DIVISION	ARIZONA			ARKANSAS						CALIFORNIA								
	TUCSON			FORT SMITH			LITTLE ROCK			ANAHEIM			BAKERSFIELD			FRESNO		
	MAT.	INST.	TOTAL	MAT.	INST.	TOTAL	MAT.	INST.	TOTAL	MAT.	INST.	TOTAL	MAT.	INST.	TOTAL	MAT.	INST.	TOTAL
2 SITE WORK	110.4	97.4	104.7	100.0	91.8	96.4	107.0	94.2	101.3	104.7	112.0	107.9	97.1	110.4	103.0	94.7	122.8	107.1
3.1 FORMWORK	109.1	88.0	92.7	111.4	70.5	79.7	103.4	72.1	79.1	104.6	123.7	119.4	124.7	123.7	123.9	110.1	124.6	121.4
3.2 REINFORCING	95.1	90.7	93.3	124.5	69.9	102.0	117.8	71.5	98.7	99.3	129.8	111.9	96.1	129.8	110.0	106.5	129.8	116.1
3.3 CAST IN PLACE CONC.	105.5	97.7	100.8	90.5	92.0	91.4	98.4	92.8	95.0	109.3	109.5	109.5	103.2	109.5	107.0	93.0	108.7	102.5
3 CONCRETE	103.9	93.2	97.2	102.1	81.5	89.0	103.6	82.7	90.4	106.2	116.9	113.0	105.9	116.9	112.9	99.4	116.9	110.5
4 MASONRY	92.2	79.7	82.7	95.4	70.0	75.9	88.8	71.6	75.6	108.6	130.4	125.3	100.8	120.6	116.0	119.9	122.0	121.5
5 METALS	90.9	93.5	91.8	96.6	77.7	89.9	106.2	78.9	96.5	99.1	121.5	107.1	99.4	120.9	107.0	95.0	124.3	105.4
6 WOOD & PLASTICS	105.8	86.0	94.7	107.0	71.0	86.9	94.6	72.6	82.3	96.0	119.4	109.1	95.3	119.4	108.8	97.1	121.2	110.6
7 MOISTURE PROTECTION	105.5	76.2	96.2	84.8	70.0	80.1	84.3	71.6	80.3	108.0	133.4	116.0	84.9	119.3	95.7	107.5	115.1	109.9
8 DOORS, WINDOWS, GLASS	88.2	83.8	85.9	92.8	70.0	80.9	95.2	71.6	82.9	93.4	123.4	108.9	99.9	122.0	111.3	101.2	123.4	112.7
9.2 LATH & PLASTER	109.1	87.2	92.5	93.1	70.8	76.1	98.5	72.3	78.6	97.2	127.8	120.4	92.3	119.6	113.0	102.1	118.5	114.5
9.2 DRYWALL	82.2	86.6	84.2	95.4	70.0	83.4	114.8	71.6	94.5	97.6	124.1	110.0	98.2	120.3	108.6	98.7	121.2	109.3
9.5 ACOUSTICAL WORK	113.6	85.4	98.3	83.7	70.0	76.2	83.7	71.6	77.1	81.3	120.1	102.5	93.2	120.1	107.8	96.6	122.1	110.5
9.6 FLOORING	110.0	85.9	103.5	89.5	70.0	84.2	88.7	71.6	84.1	117.1	126.0	119.5	111.8	120.6	114.2	88.7	122.1	97.6
9.9 PAINTING	98.5	86.0	88.4	111.1	70.0	78.1	104.7	71.6	78.1	108.3	127.4	123.6	120.1	120.7	120.5	107.9	122.1	119.3
9 FINISHES	93.1	86.2	89.4	94.6	70.0	81.4	105.1	71.6	87.2	101.7	125.2	114.3	102.9	120.4	112.3	97.3	121.5	110.2
10-14 TOTAL DIV. 10-14	100.0	90.3	97.1	100.0	70.0	91.0	100.0	71.6	91.5	100.0	126.4	107.9	100.0	123.9	107.1	100.0	145.2	113.5
15 MECHANICAL	98.6	88.3	93.4	97.3	70.1	83.5	96.8	69.9	83.2	96.7	125.5	111.2	94.8	101.2	98.0	92.6	118.6	105.7
16 ELECTRICAL	103.2	86.0	91.2	100.2	71.3	80.0	94.2	74.7	80.6	99.5	127.4	118.9	107.2	120.7	116.6	110.7	122.1	118.6
1-16 WEIGHTED AVERAGE	99.0	87.7	92.9	97.2	73.9	84.7	98.9	75.3	86.2	101.1	123.9	113.4	99.1	116.0	108.2	99.6	121.3	111.3

DIVISION	CALIFORNIA																	
	LOS ANGELES			OXNARD			RIVERSIDE			SACRAMENTO			SAN DIEGO			SAN FRANCISCO		
	MAT.	INST.	TOTAL	MAT.	INST.	TOTAL	MAT.	INST.	TOTAL	MAT.	INST.	TOTAL	MAT.	INST.	TOTAL	MAT.	INST.	TOTAL
2 SITE WORK	98.2	115.4	105.8	102.0	103.5	102.7	98.8	110.7	104.1	86.5	108.5	96.3	95.3	108.2	101.0	102.2	117.3	108.9
3.1 FORMWORK	112.0	124.2	121.5	98.4	124.2	118.4	114.0	123.7	121.6	109.9	128.1	124.0	105.1	125.5	120.9	103.6	138.5	130.7
3.2 REINFORCING	87.2	129.8	104.8	99.3	129.8	111.9	124.5	129.8	126.7	99.3	129.8	111.9	118.4	129.8	123.1	123.4	129.8	126.1
3.3 CAST IN PLACE CONC.	98.9	112.6	107.2	102.4	110.2	107.1	102.4	109.9	106.9	115.9	107.8	111.0	99.9	105.3	103.2	101.6	118.0	111.6
3 CONCRETE	98.9	118.7	111.5	100.9	117.5	111.4	109.6	117.1	114.4	111.1	117.8	115.3	105.0	115.5	111.7	106.8	127.2	119.7
4 MASONRY	108.9	130.4	125.3	100.8	126.7	120.6	105.3	118.7	115.5	103.2	126.5	121.1	110.6	109.8	110.0	127.0	143.6	139.7
5 METALS	101.6	123.0	109.2	105.2	121.7	111.1	99.2	121.5	107.1	111.1	124.2	115.7	99.0	120.8	106.7	104.0	127.2	112.2
6 WOOD & PLASTICS	99.8	120.5	111.4	92.8	120.3	108.2	94.7	119.4	108.5	78.5	125.6	104.9	96.5	121.6	110.5	93.6	137.1	118.0
7 MOISTURE PROTECTION	104.0	133.4	113.3	90.0	131.5	103.1	90.5	127.7	102.3	85.2	122.2	96.9	94.5	111.9	100.0	100.5	133.4	110.9
8 DOORS, WINDOWS, GLASS	102.9	123.4	113.5	102.6	123.4	113.4	103.1	123.4	113.7	91.9	127.1	110.2	107.4	123.3	115.7	113.6	134.3	124.4
9.2 LATH & PLASTER	96.1	127.8	120.2	97.7	125.8	119.0	97.7	127.3	120.2	99.0	126.0	119.5	102.3	113.8	111.1	101.3	147.9	136.7
9.2 DRYWALL	89.6	124.1	105.7	98.9	122.3	109.8	94.8	124.1	108.5	97.2	126.7	111.0	100.1	122.2	110.5	81.2	139.0	108.3
9.5 ACOUSTICAL WORK	98.8	120.1	110.4	87.5	120.1	105.3	87.5	120.1	105.3	85.6	126.6	107.9	100.8	122.5	112.6	100.8	138.9	121.5
9.6 FLOORING	96.3	126.0	104.3	95.8	126.0	103.9	95.8	126.0	103.9	85.9	126.6	96.8	98.4	131.6	107.3	107.2	136.7	115.1
9.9 PAINTING	83.9	126.7	118.2	92.2	127.1	120.1	100.7	126.8	121.6	112.3	126.6	123.7	91.5	128.7	121.3	102.1	145.0	136.5
9 FINISHES	91.4	125.0	109.4	96.6	124.2	111.4	95.1	125.0	111.1	95.2	126.6	112.0	99.0	124.6	112.7	91.1	141.4	118.1
10-14 TOTAL DIV. 10-14	100.0	126.7	107.9	100.0	126.3	107.8	100.0	126.2	107.8	100.0	146.8	114.0	100.0	124.3	107.2	100.0	152.8	115.8
15 MECHANICAL	97.6	125.5	111.7	98.5	123.0	110.9	96.4	126.3	111.5	97.9	126.1	112.2	102.9	123.8	113.5	101.0	166.4	134.0
16 ELECTRICAL	102.0	126.5	119.1	99.5	111.9	108.2	99.0	126.8	118.4	110.7	126.6	121.8	105.8	107.2	106.8	108.1	149.8	137.2
1-16 WEIGHTED AVERAGE	99.6	124.4	113.0	99.4	120.6	110.8	99.6	122.0	111.7	99.8	124.6	113.2	101.8	116.8	109.9	103.5	142.5	124.5

Figure 25.8

CITY COST INDEXES

Historical Cost Indexes

The table below lists both the Means City Cost Index based on Jan. 1, 1975 = 100 as well as the computed value of an index based on January 1, 1989 costs. Since the Jan. 1, 1989 figure is estimated, space is left to write in the actual index figures as they become available thru either the quarterly "Means Construction Cost Indexes" or as printed in the "Engineering News-Record". To compute the actual index based on Jan. 1, 1989 = 100, divide the Quarterly City Cost Index for a particular year by the actual Jan. 1, 1989 Quarterly City Cost Index. Space has been left to advance the index figures as the year progresses.

Year	"Quarterly City Cost Index" Jan. 1, 1975 = 100		Current Index Based on Jan. 1, 1989 = 100	
	Est.	Actual	Est.	Actual
Oct. 1989				
July 1989				
April 1989				
Jan. 1989	207.1		100.0	100.0
July 1988		205.7	99.3	
1987		200.7	96.9	
1986		192.8	93.1	
1985		189.1	91.3	
1984		187.6	90.6	
1983		183.5	88.6	
1982		174.3	84.2	
1981		160.2	77.4	
1980		144.0	69.5	
1979		132.3	63.9	
1978		122.4	59.1	
1977		113.3	54.7	

Year	"Quarterly City Cost Index" Jan. 1, 1975 = 100	Current Index Based on Jan. 1, 1989 = 100	
	Actual	Est.	Actual
July 1976	107.3	51.8	
1975	102.6	49.5	
1974	94.7	45.7	
1973	86.3	41.7	
1972	79.7	38.5	
1971	73.5	35.5	
1970	65.8	31.8	
1969	61.6	29.7	
1968	56.9	27.5	
1967	53.9	26.0	
1966	51.9	25.1	
1965	49.7	24.0	
1964	48.6	23.5	
1963	47.3	22.8	
1962	46.2	22.3	
1961	45.4	21.9	

Year	"Quarterly City Cost Index" Jan. 1, 1975 = 100	Current Index Based on Jan. 1, 1989 = 100	
	Actual	Est.	Actual
July 1960	45.0	21.7	
1959	44.2	21.3	
1958	43.0	20.8	
1957	42.2	20.4	
1956	40.4	19.5	
1955	38.1	18.4	
1954	36.7	17.7	
1953	36.2	17.5	
1952	35.3	17.0	
1951	34.4	16.6	
1950	31.4	15.2	
1949	30.4	14.7	
1948	30.4	14.7	
1947	27.6	13.3	
1946	23.2	11.2	
1945	20.2	9.8	

City Cost Indexes

Tabulated on the following pages are average construction cost indexes for 162 major U.S. and Canadian cities. Index figures for both material and installation are based on the 30 major city average of 100 and represent the cost relationship as of July 1, 1988. The index for each division is computed from representative material and labor quantities for that division. The weighted average for each city is a weighted total of the components listed above it, but does not include relative productivity between trades or cities.

The material index for the weighted average includes about 100 basic construction materials with appropriate quantities of each material to represent typical "average" building construction projects.

The installation index for the weighted average includes the contribution of about 30 construction trades with their representative man-days in proportion to the material items installed. Also included in the installation costs are the representative equipment costs for those items requiring equipment.

Since each division of the book contains many different items, any particular item multiplied by the particular city index may give incorrect results. However, when all the book costs for a particular division are summarized and then factored, the result should be very close to the actual costs for that particular division for that city.

If a project has a preponderance of materials from any particular division (say structural steel), then the weighted average index should be adjusted in proportion to the value of the factor for that division.

Adjustments to Costs

Time Adjustment using the Historical Cost Indexes:

$$\frac{\text{Index for Year A}}{\text{Index for Year B}} \times \text{Cost in Year B} = \text{Cost in Year A}$$

Location Adjustment using the City Cost Indexes:

$$\frac{\text{Index for City A}}{\text{Index for City B}} \times \text{Cost in City B} = \text{Cost in City A}$$

Adjustment from the National Average:

$$\text{National Average Cost} \times \frac{\text{Index for City A}}{100} = \text{Cost in City A}$$

Note: The City Cost Indexes for Canada can be used to convert U.S. national averages to local costs in Canadian dollars.

Figure 25.9

Part III

ESTIMATING EXAMPLES

Chapter 26

UNIT PRICE ESTIMATING

The Unit Price Estimate is the most detailed and most accurate type of estimate. The estimator should have working drawings and specifications to provide sufficient information to complete this estimate effectively. In commercial renovation, however, it is not enough to have the plans and specifications. The estimator must also perform a thorough evaluation of the site to understand how the existing conditions will affect the work.

The following example is a Unit Price estimate for a hypothetical commercial renovation project. A description of the project is included at the beginning. The site evaluation and discussion of the existing conditions is provided for each division. The individual items in the estimate may not represent every item that will be found in a renovation project. But the example will provide a basis for understanding, evaluating, and estimating commercial renovation as a whole. If the reader uses this example as a guideline for actual projects, consideration must be given to all building, fire, health, and safety codes and regulations effective in a given locality.

The example assumes that working drawings and specifications have been provided, and refers to them.

Project Description

The sample project is the renovation of a turn-of-the-century mill building into retail and commercial office space. The building is located in the downtown area of a small city. The exterior walls are brick and the floor systems are cast steel columns and wood beams with heavy wood decking. The roof structure is heavy timber trusses.

The building was originally used for manufacturing and has recently been used primarily for warehousing. The building has not been well maintained and is in a general state of disrepair. A retail tenant currently occupying the premises is to remain in operation throughout the renovation.

Figures 26.1 through 26.7 are plans, a section, and an elevation of the existing conditions of the building. Figures 26.8 through 26.13 provide the same information for the proposed renovations. The owner has secured one office tenant for half of the third floor. Because the owner must know the costs for pricing future tenant renovations, the costs for the tenant improvements are estimated separately in each appropriate division. All costs in the sample estimate are from *Means Repair and Remodeling Cost Data*, 1989.

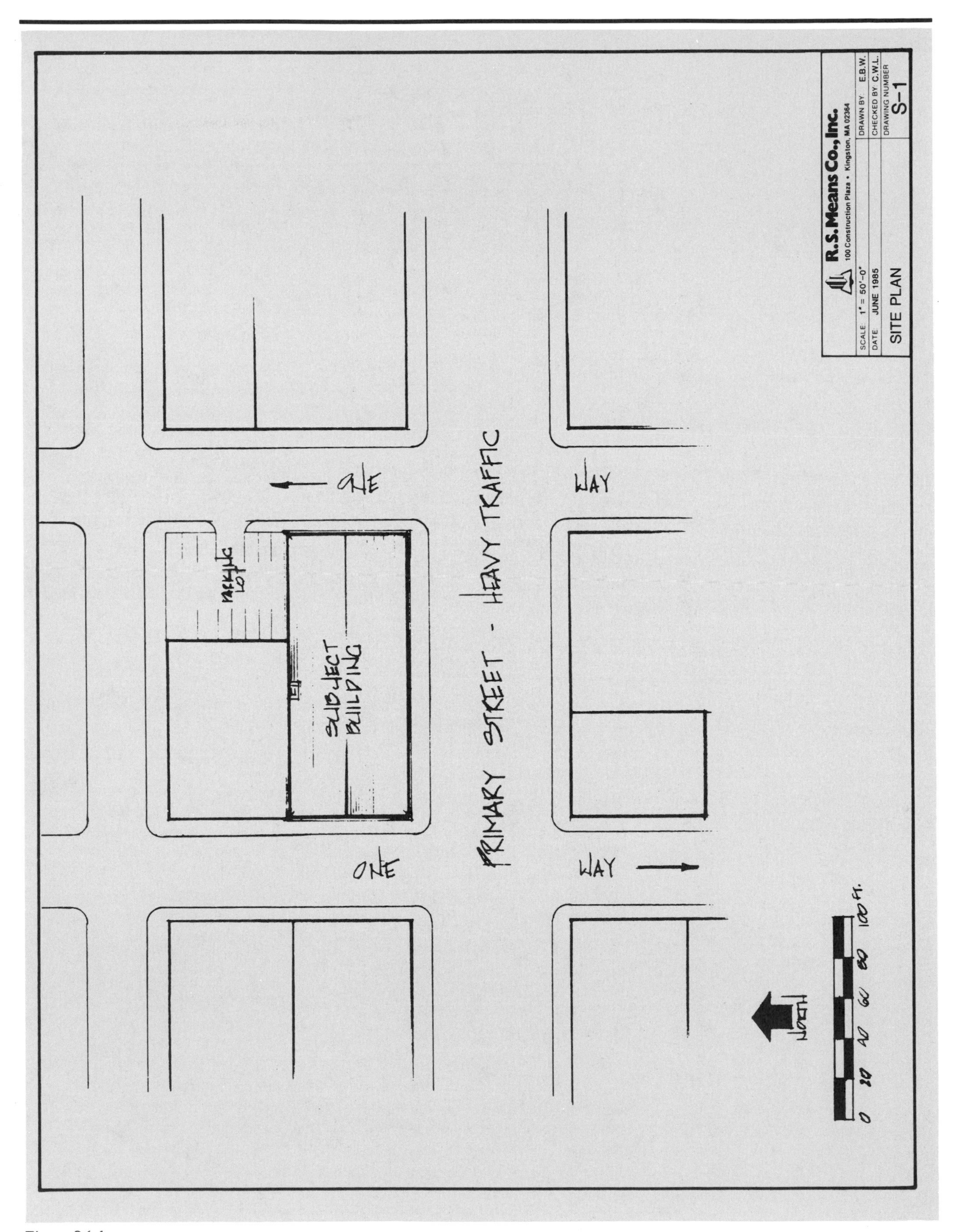
PRIMARY STREET - HEAVY TRAFFIC
ONE WAY
ONE WAY
SUBJECT BUILDING
PARKING LOT
NORTH
0 20 40 60 80 100 FT.
R.S. Means Co., Inc.
100 Construction Plaza • Kingston, MA 02364
SCALE 1" = 50'-0"
DATE JUNE 1985
SITE PLAN
DRAWN BY E.B.W.
CHECKED BY C.W.L.
DRAWING NUMBER
S-1

Figure 26.1

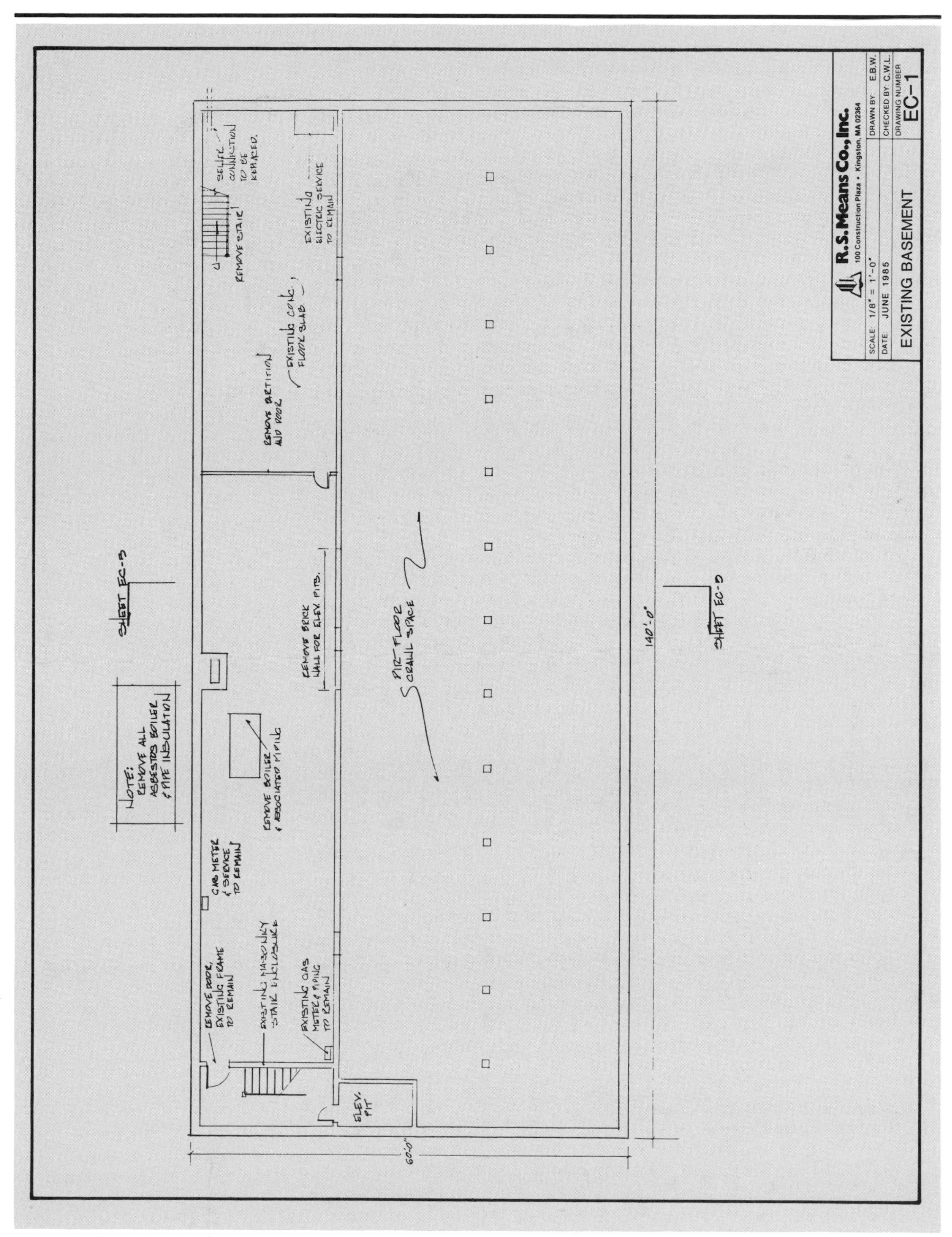
R.S. Means Co., Inc.
100 Construction Plaza • Kingston, MA 02364
SCALE: 1/8" = 1'-0"
DATE: JUNE 1985
DRAWN BY: E.B.W.
CHECKED BY: C.W.L.
DRAWING NUMBER
EC-1
EXISTING BASEMENT
NOTE: REMOVE ALL ASBESTOS BOILER & PIPE INSULATION
ELEV. PIT
140'-0"
60'-0"

Figure 26.2

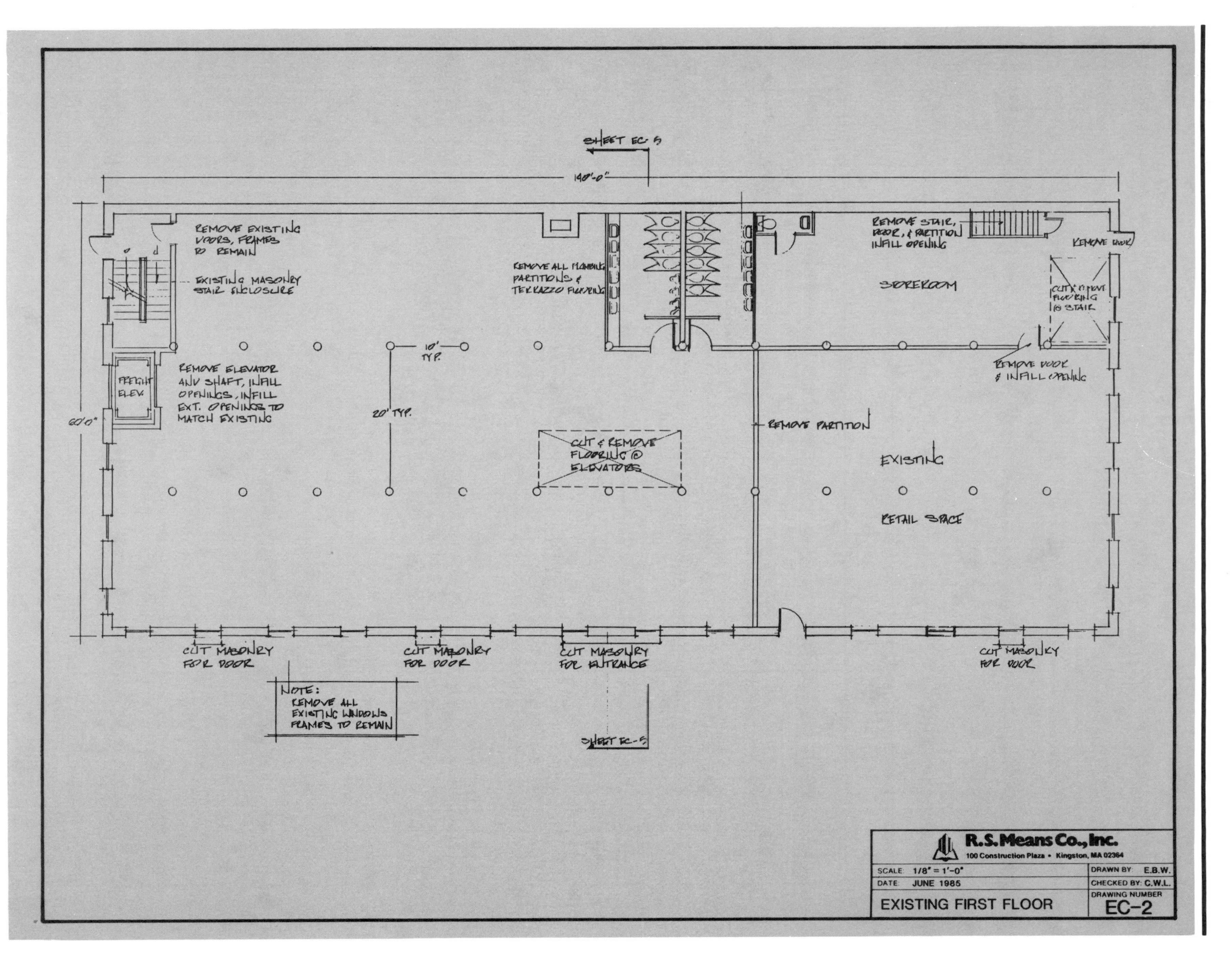

SHEET EC-5
140'-0"
60'-0"
REMOVE EXISTING DOORS, FRAMES TO REMAIN
EXISTING MASONRY STAIR ENCLOSURE
FREIGHT ELEV.
REMOVE ELEVATOR AND SHAFT, INFILL OPENINGS, INFILL EXT. OPENINGS TO MATCH EXISTING
10' TYP.
20' TYP.
REMOVE ALL PLUMBING PARTITIONS & TERRAZZO FLOORING
CUT & REMOVE FLOORING @ ELEVATORS
REMOVE STAIR, DOOR, & PARTITION INFILL OPENING
REMOVE DOOR
STOREROOM
CUT & REMOVE FLOORING @ STAIR
REMOVE DOOR & INFILL OPENING
REMOVE PARTITION
EXISTING
RETAIL SPACE
CUT MASONRY FOR DOOR
CUT MASONRY FOR DOOR
CUT MASONRY FOR ENTRANCE
CUT MASONRY FOR DOOR
NOTE: REMOVE ALL EXISTING WINDOWS FRAMES TO REMAIN
SHEET EC-5
R.S. Means Co., Inc.
100 Construction Plaza • Kingston, MA 02364
SCALE: 1/8" = 1'-0"
DATE: JUNE 1985
DRAWN BY: E.B.W.
CHECKED BY: C.W.L.
DRAWING NUMBER
EC-2
EXISTING FIRST FLOOR

Figure 26.3

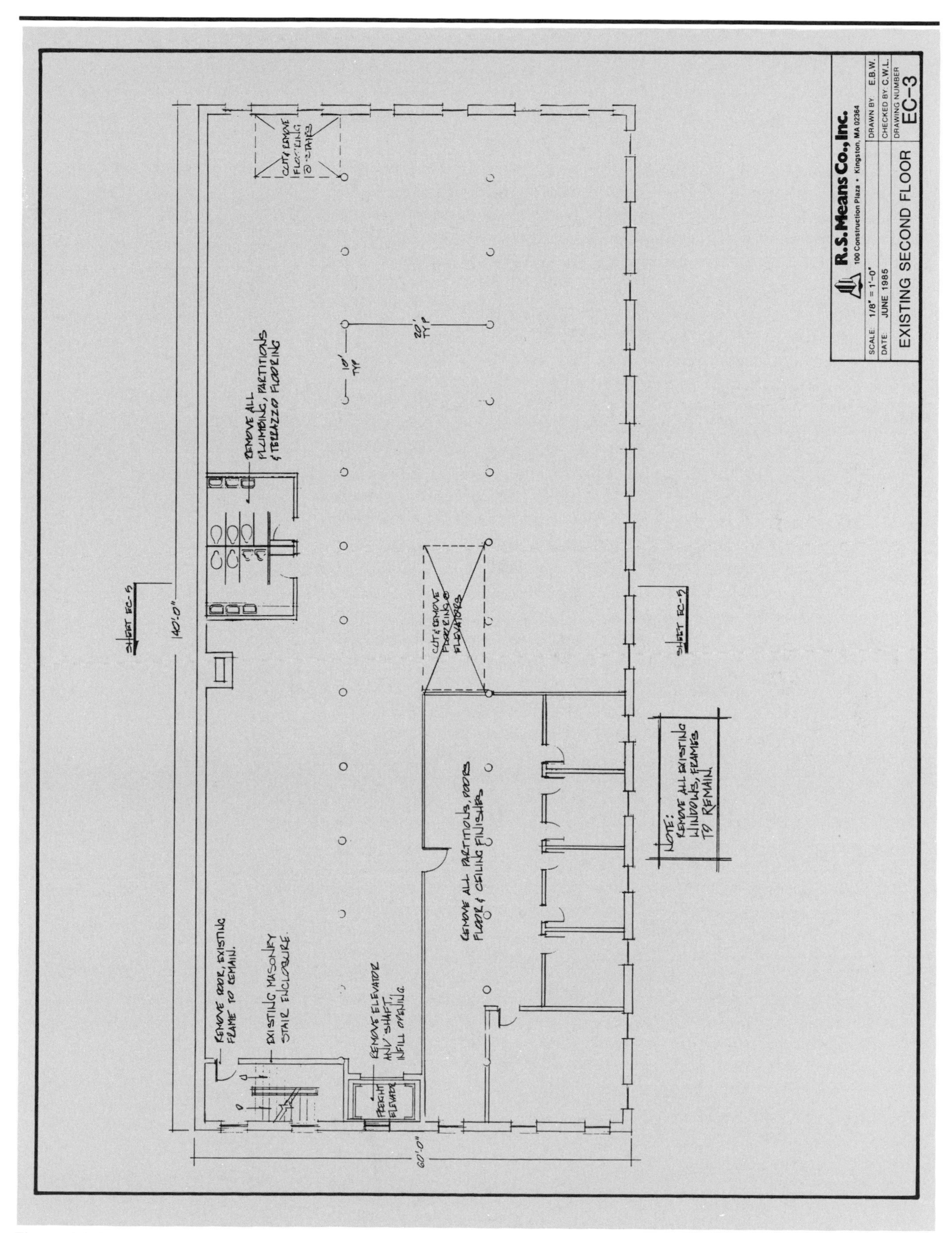

R.S. Means Co., Inc.
100 Construction Plaza • Kingston, MA 02364
SCALE: 1/8" = 1'-0"
DATE: JUNE 1985
DRAWN BY E.B.W.
CHECKED BY: C.W.L.
DRAWING NUMBER
EC-3
EXISTING SECOND FLOOR
140'-0"
60'-0"
NOTE: REMOVE ALL EXISTING WINDOWS, FRAMES TO REMAIN.
REMOVE ALL PARTITIONS, DOORS FLOOR & CEILING FINISHES
REMOVE ALL PLUMBING, PARTITIONS & TERRAZZO FLOORING
REMOVE DOOR, EXISTING FRAME TO REMAIN.
EXISTING MASONRY STAIR ENCLOSURE.
REMOVE ELEVATOR AND SHAFT, INFILL OPENING.
FREIGHT ELEVATOR

Figure 26.4

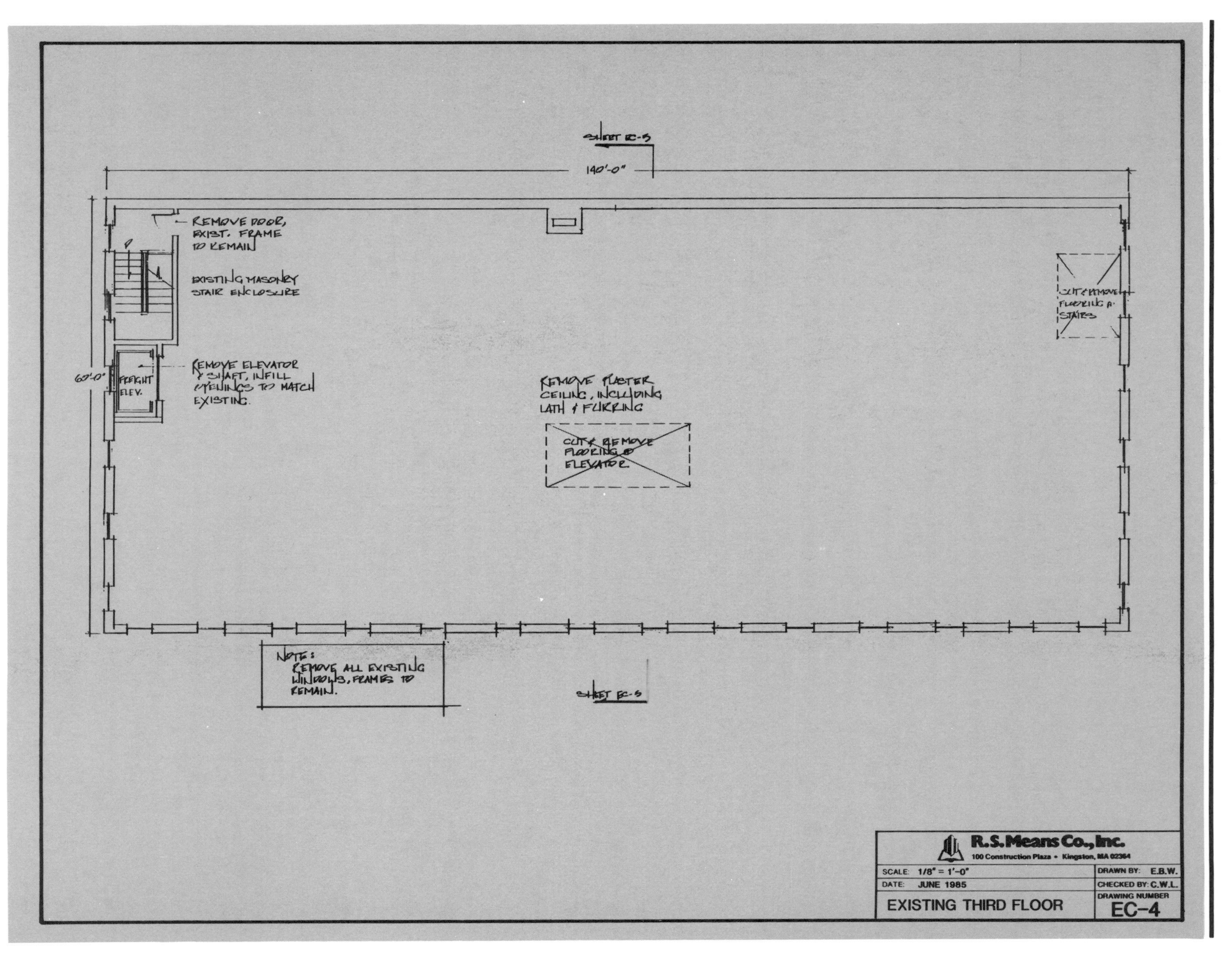

SHEET EC-5
140'-0"
REMOVE DOOR, EXIST. FRAME TO REMAIN
EXISTING MASONRY STAIR ENCLOSURE
REMOVE ELEVATOR & SHAFT, INFILL OPENINGS TO MATCH EXISTING.
60'-0"
FREIGHT ELEV.
REMOVE PLASTER CEILING, INCLUDING LATH & FURRING
CUT & REMOVE FLOORING @ ELEVATOR
CUT & REMOVE FLOORING @ STAIRS
NOTE: REMOVE ALL EXISTING WINDOWS, FRAMES TO REMAIN.
SHEET EC-5
R.S. Means Co., Inc.
100 Construction Plaza • Kingston, MA 02364
SCALE: 1/8" = 1'-0"
DATE: JUNE 1985
DRAWN BY: E.B.W.
CHECKED BY: C.W.L.
DRAWING NUMBER
EC-4
EXISTING THIRD FLOOR

Figure 26.5

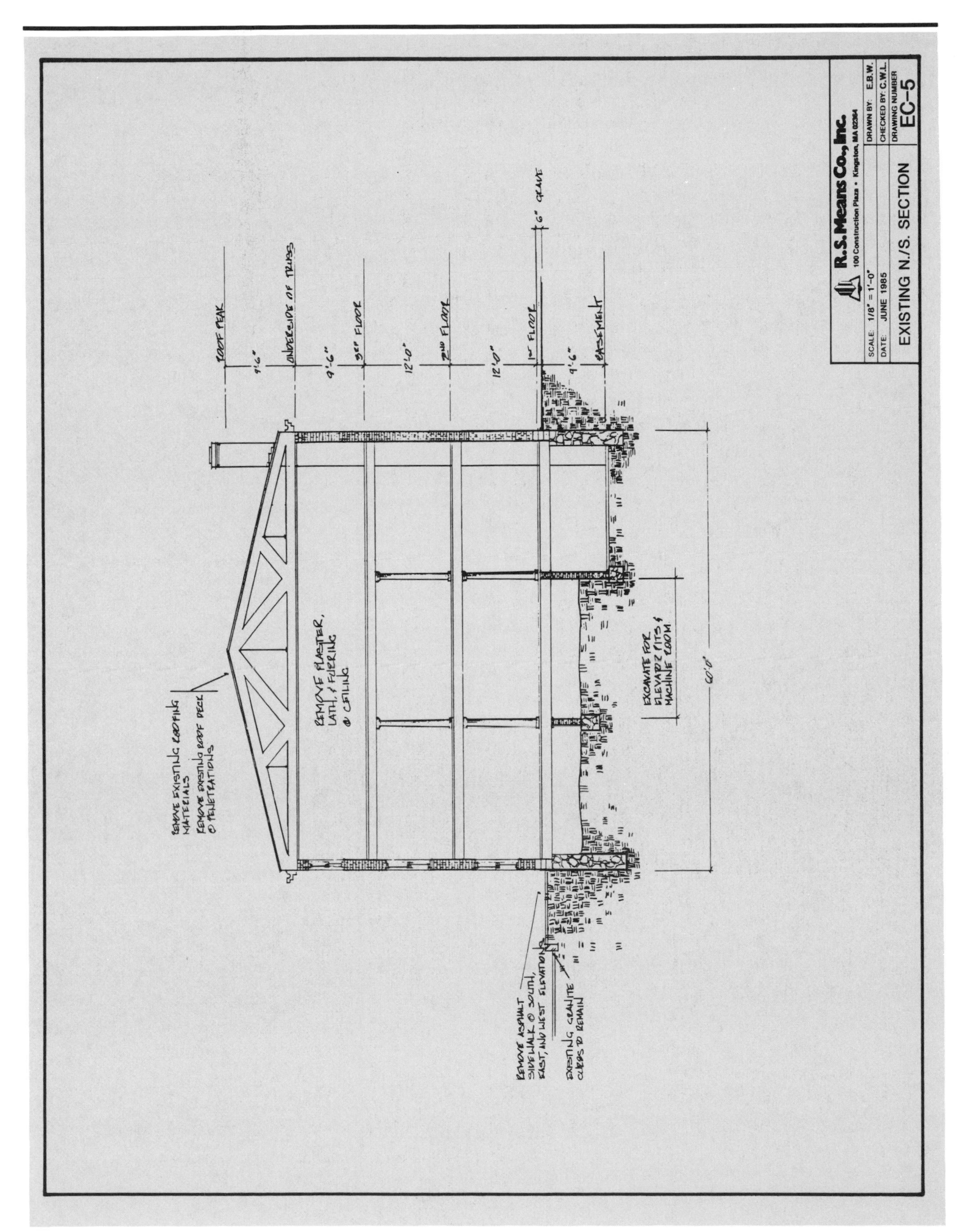

R.S. Means Co., Inc.
100 Construction Plaza • Kingston, MA 02364
SCALE: 1/8" = 1'-0"
DATE: JUNE 1985
DRAWN BY: E.B.W.
CHECKED BY: C.W.L.
DRAWING NUMBER
EC-5
EXISTING N./S. SECTION
ROOF PEAK
UNDERSIDE OF TRUSS
3RD FLOOR
2ND FLOOR
1ST FLOOR
BASEMENT
6" GRADE
9'-6"
9'-6"
12'-0"
12'-0"
9'-6"
60'-0"
REMOVE EXISTING ROOFING MATERIALS
REMOVE EXISTING ROOF DECK @ PENETRATIONS
REMOVE PLASTER, LATH, & FURRING @ CEILING
EXCAVATE FOR ELEVATOR PITS & MACHINE ROOM
REMOVE ASPHALT SIDEWALK @ SOUTH, EAST, AND WEST ELEVATION
EXISTING GRANITE CURBS TO REMAIN

Figure 26.6

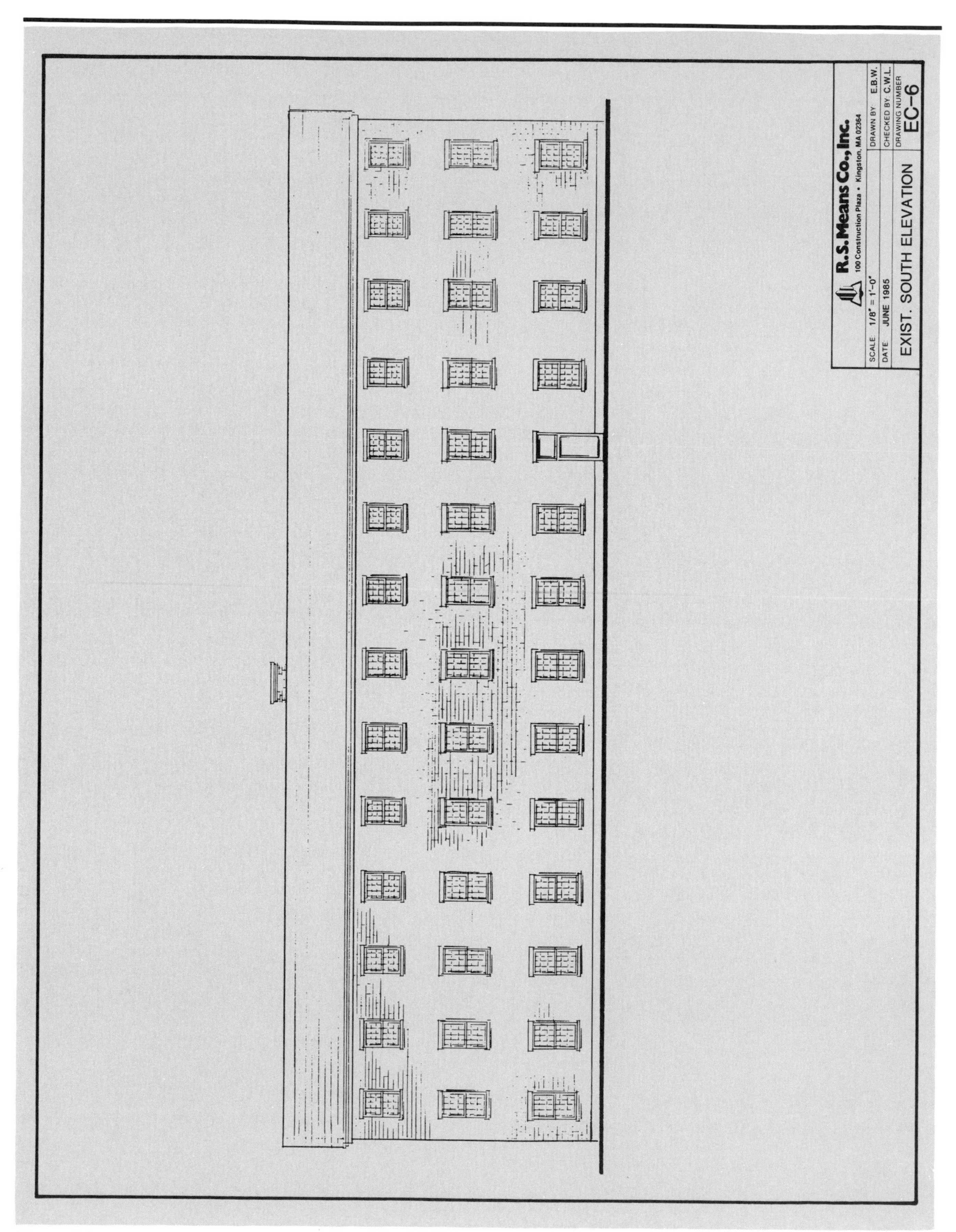
R.S. Means Co., Inc.
100 Construction Plaza • Kingston, MA 02364
SCALE 1/8" = 1'-0"
DATE JUNE 1985
DRAWN BY E.B.W.
CHECKED BY C.W.L.
DRAWING NUMBER
EC-6
EXIST. SOUTH ELEVATION

Figure 26.7

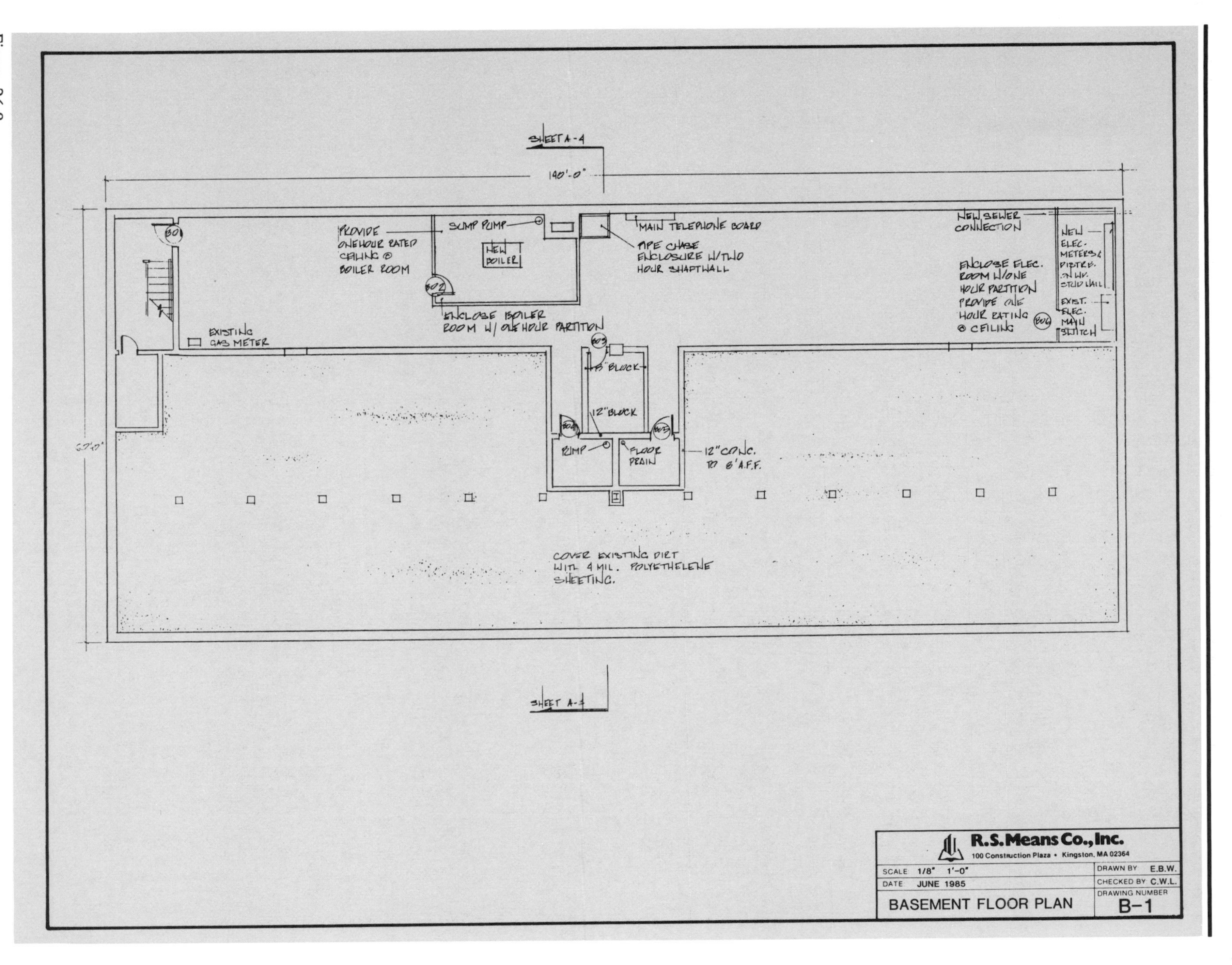
SHEET A-4
140'-0"
PROVIDE ONE HOUR RATED CEILING @ BOILER ROOM
SUMP PUMP
NEW BOILER
MAIN TELEPHONE BOARD
PIPE CHASE ENCLOSURE W/TWO HOUR SHAFTWALL
NEW SEWER CONNECTION
ENCLOSE ELEC. ROOM W/ONE HOUR PARTITION PROVIDE ONE HOUR RATING @ CEILING
EXIST. ELEC. MAIN SWITCH
ENCLOSE BOILER ROOM W/ ONE HOUR PARTITION
EXISTING GAS METER
12" BLOCK
PUMP
FLOOR DRAIN
12" CONC. TO 8' A.F.F.
COVER EXISTING DIRT WITH 4 MIL. POLYETHELENE SHEETING.
R.S. Means Co., Inc.
100 Construction Plaza • Kingston, MA 02364
SCALE 1/8" 1'-0"
DATE JUNE 1985
DRAWN BY E.B.W.
CHECKED BY C.W.L.
DRAWING NUMBER
B-1
BASEMENT FLOOR PLAN

Figure 26.8

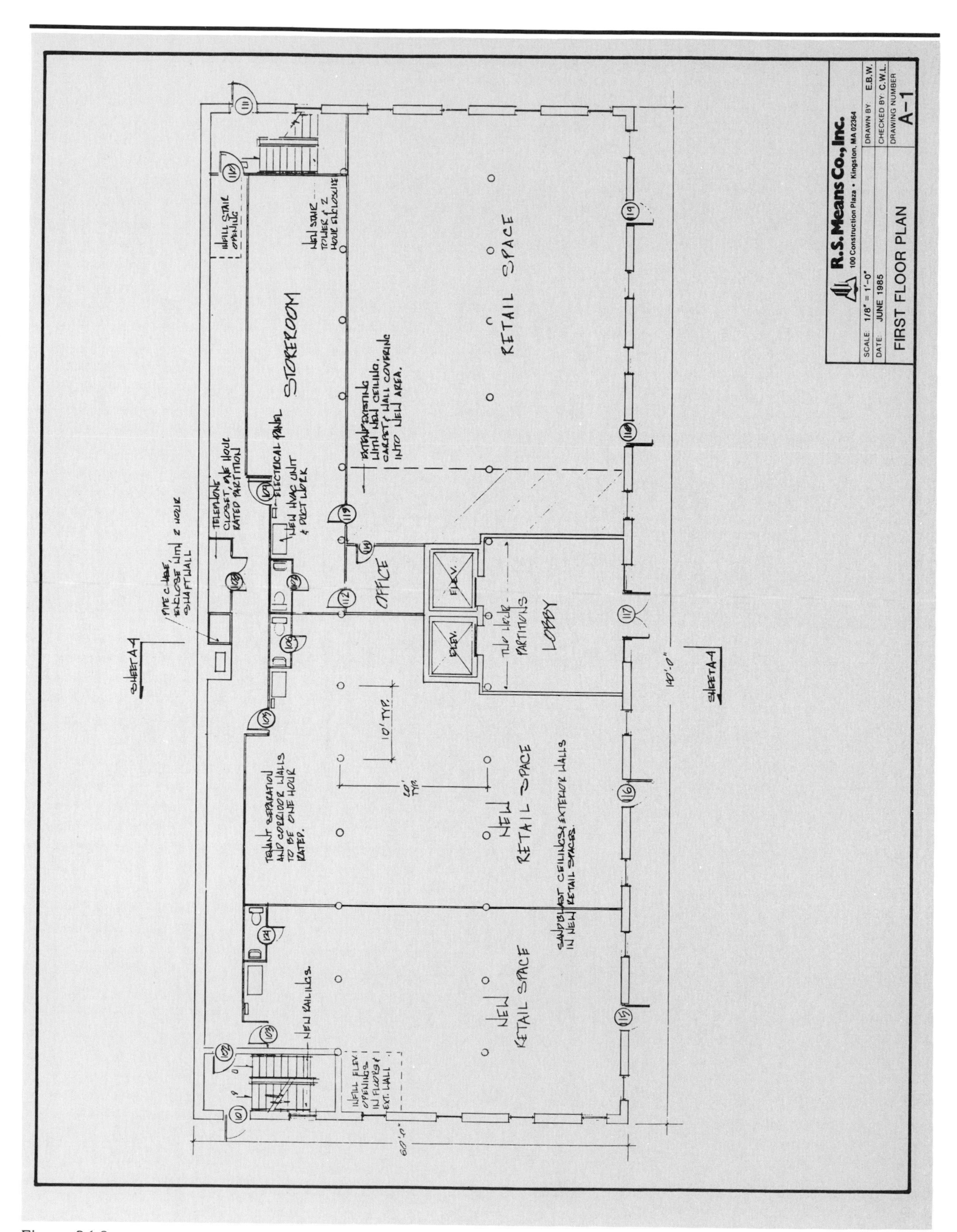
R.S. Means Co., Inc.
100 Construction Plaza • Kingston, MA 02364
SCALE: 1/8" = 1'-0"
DATE: JUNE 1985
DRAWN BY: E.B.W.
CHECKED BY: C.W.L.
DRAWING NUMBER
A-1
FIRST FLOOR PLAN
STOREROOM
RETAIL SPACE
NEW RETAIL SPACE
NEW RETAIL SPACE
OFFICE
LOBBY
ELEV.
ELEV.
TWO HOUR PARTITIONS
10' TYP.
140'-0"
60'-0"
SHEET A-4
SHEET A-4
NEW RAILINGS
ELECTRICAL PANEL
NEW HVAC UNIT & DUCT WORK

Figure 26.9

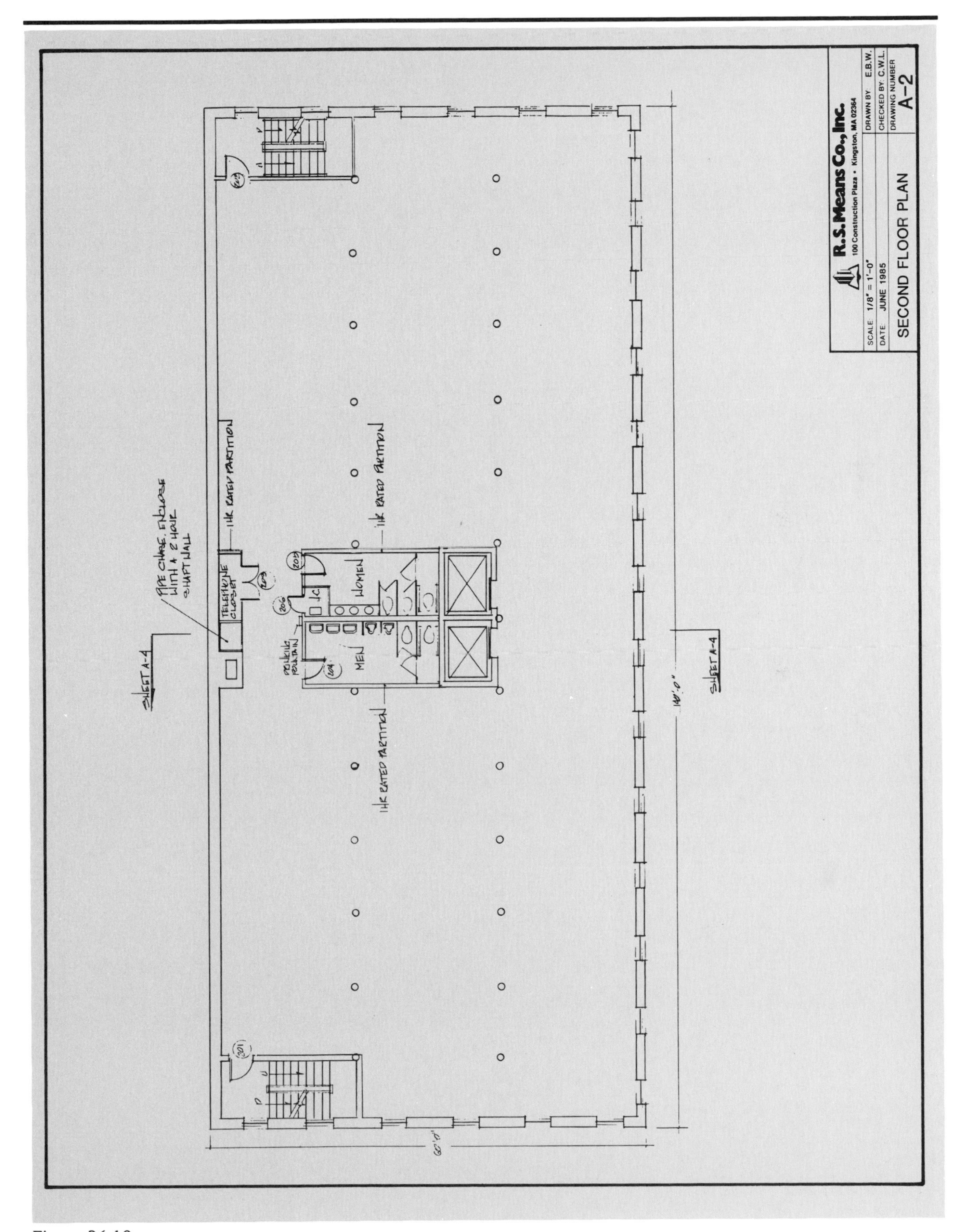
R.S. Means Co., Inc.
100 Construction Plaza • Kingston, MA 02364
SCALE: 1/8" = 1'-0"
DATE: JUNE 1985
DRAWN BY: E.B.W.
CHECKED BY: C.W.L.
DRAWING NUMBER
A-2
SECOND FLOOR PLAN
PIPE CHASE. ENCLOSE WITH A 2 HOUR SHAFT WALL
1HR RATED PARTITION
TELEPHONE CLOSET
WOMEN
MEN
DRINKING FOUNTAIN
SHEET A-4
140'-0"
60'-0"

Figure 26.10

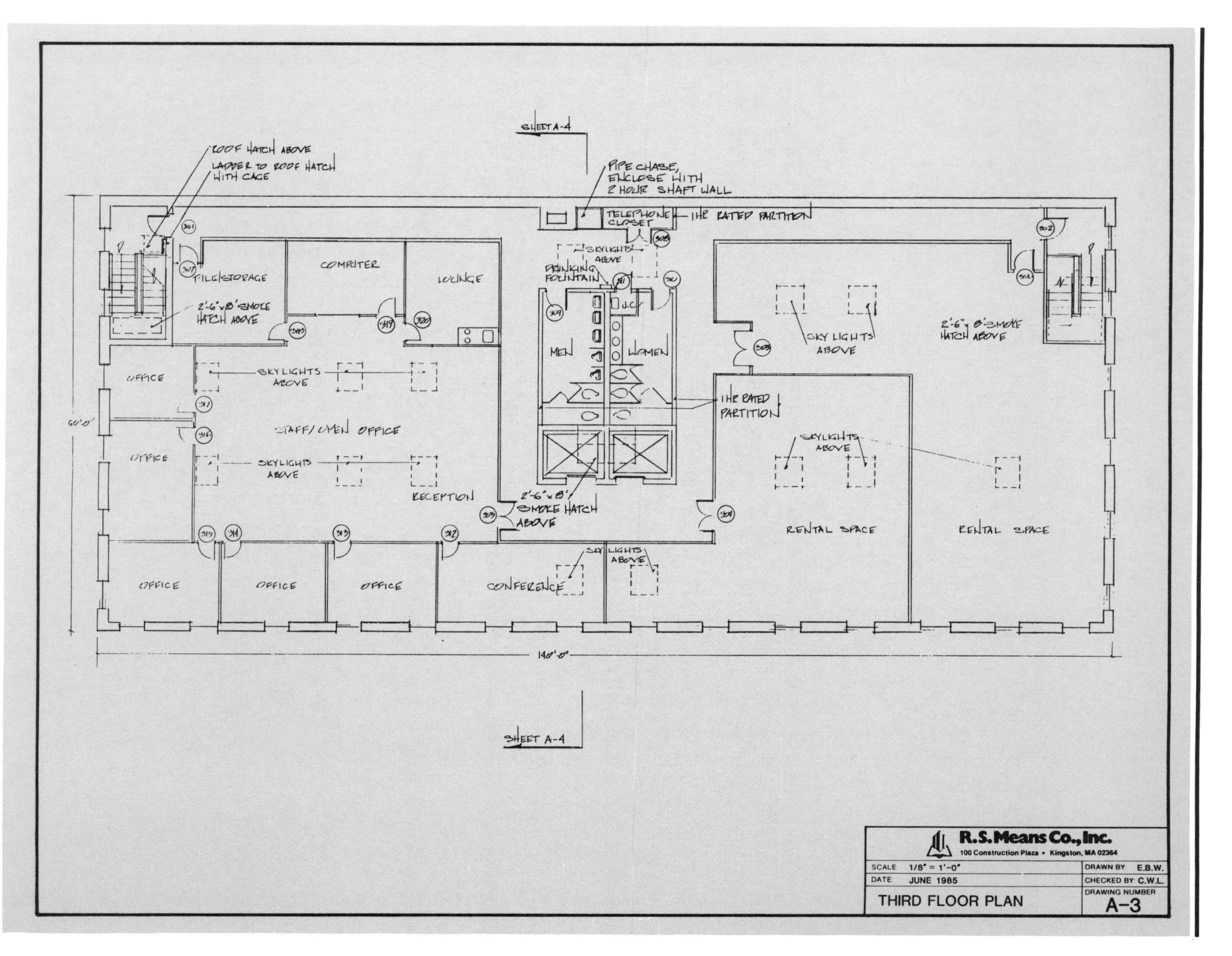
SHEET A-4
ROOF HATCH ABOVE
LADDER TO ROOF HATCH WITH CAGE
PIPE CHASE, ENCLOSE WITH 2 HOUR SHAFT WALL
TELEPHONE CLOSET
1HR RATED PARTITION
FILE/STORAGE
COMPUTER
LOUNGE
SKYLIGHTS ABOVE
DRINKING FOUNTAIN
J.C.
2'-6" x 8' SMOKE HATCH ABOVE
MEN
WOMEN
SKY LIGHTS ABOVE
2'-6" x 8' SMOKE HATCH ABOVE
OFFICE
SKYLIGHTS ABOVE
1HR RATED PARTITION
60'-0"
STAFF/OPEN OFFICE
SKYLIGHTS ABOVE
OFFICE
SKYLIGHTS ABOVE
RECEPTION
2'-6" x 8' SMOKE HATCH ABOVE
RENTAL SPACE
RENTAL SPACE
SKYLIGHTS ABOVE
OFFICE
OFFICE
OFFICE
CONFERENCE
140'-0"
SHEET A-4
R.S. Means Co., Inc.
100 Construction Plaza • Kingston, MA 02364
SCALE 1/8" = 1'-0"
DATE: JUNE 1985
DRAWN BY: E.B.W.
CHECKED BY: C.W.L.
DRAWING NUMBER
A-3
THIRD FLOOR PLAN

Figure 26.11

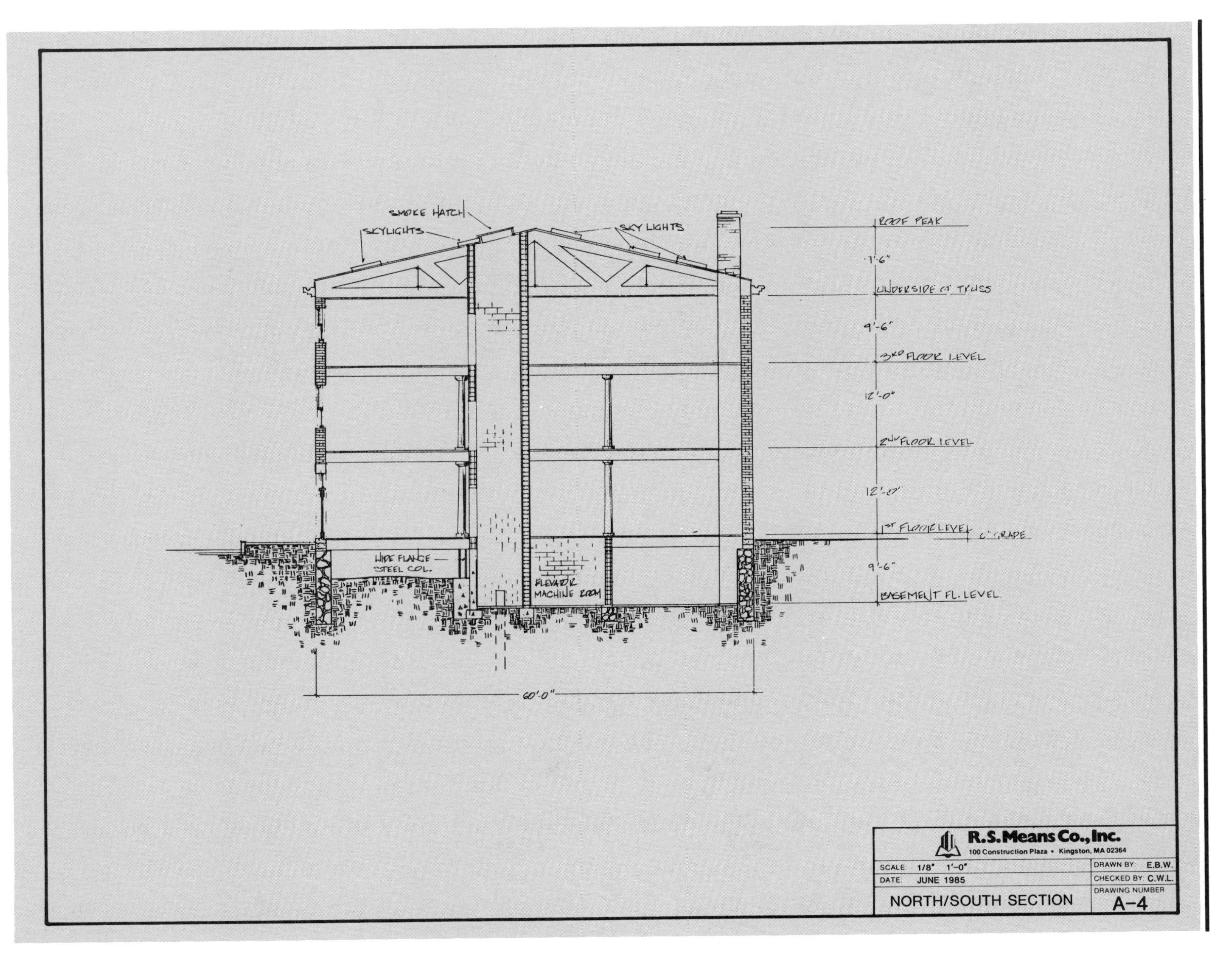

SMOKE HATCH
SKYLIGHTS
SKY LIGHTS
ROOF PEAK
7'-6"
UNDERSIDE OF TRUSS
9'-6"
3RD FLOOR LEVEL
12'-0"
2ND FLOOR LEVEL
12'-0"
1ST FLOOR LEVEL
6" GRADE
9'-6"
BASEMENT FL. LEVEL
WIDE FLANGE STEEL COL.
ELEVATOR MACHINE ROOM
60'-0"
R.S. Means Co., Inc.
100 Construction Plaza • Kingston, MA 02364
SCALE: 1/8" 1'-0"
DATE: JUNE 1985
DRAWN BY: E.B.W.
CHECKED BY: C.W.L.
DRAWING NUMBER
A-4
NORTH/SOUTH SECTION

Figure 26.12

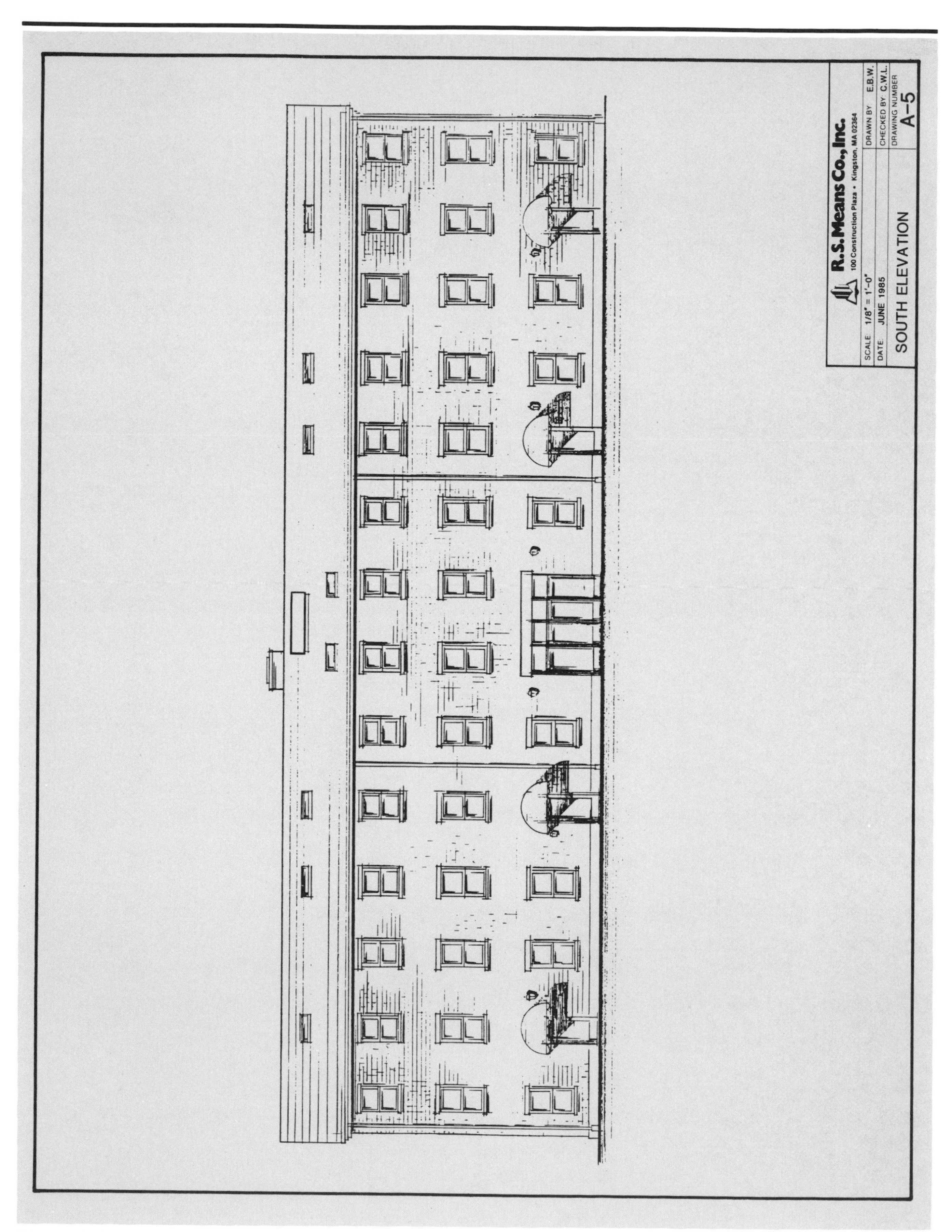
R.S. Means Co., Inc.
100 Construction Plaza • Kingston, MA 02364
SCALE 1/8" = 1'-0"
DATE JUNE 1985
SOUTH ELEVATION
DRAWN BY E.B.W.
CHECKED BY C.W.L.
DRAWING NUMBER
A-5

Figure 26.13

Division 1: General Requirements

The General Requirements of a renovation estimate should include the costs for all items that are not directly part of the physical construction (permits, insurance, bonds) as well as those direct costs that cannot be allocated to a particular division (clean-up, temporary construction, scaffolding) used by many trades.

Many of the items to be included in the General Requirements are dependent upon the total cost and/or the time duration of the project, and therefore cannot be quantified or priced until the Estimate Summary.

Figures 26.14a and 26.14b are a Project Overhead Summary form, which lists most items to be included in General Requirements. This form will be used throughout the appropriate divisions of the sample estimate wherever such items are to be included.

Please note that "Main Office Expense" and "Contingencies" are among the final items of Figure 26.14b. Some contractors include pro-rated costs for these items as indirect job costs within the estimate. Most contractors, however, include these costs as part of the overhead and profit percentages added at the Estimate Summary. The latter method is used in this sample estimate.

Included in one column are the costs for equipment, fees, rentals, and other items that are not labor or purchased material costs. All such costs receive the same mark-up in the estimate summary, usually 10% for supervision and handling. The material costs also receive a 10% mark-up as well as added sales tax. These percentages may vary depending upon local practice.

"Factors" that affect the costs are included in each division and are discussed in Chapter 6 of this book. The application of these factors is crucial to effective estimating for remodeling and renovation. They help to determine added costs and restrictions caused by existing conditions. These factors are illustrated in Figure 26.15. The notations in the highlighted area are those that will be used throughout the sample estimate.

Division 2: Site Work and Demolition

After a quick glance at the floor plans of existing conditions, the estimator at first concludes that site work and demolition costs are minimal. But upon investigation of the proposed plans and specifications, and upon evaluation of the site, it is determined that these costs are significant. Please refer to Figures 26.16 to 26.19 throughout the following discussion.

Asbestos removal has become a job only for licensed and experienced companies. A quotation is required and is shown in Figure 26.20. When the quotation is by telephone, the estimator must be sure to obtain all pertinent information. In this case, the estimator must determine what work required by the general contractor is not included in the quote. A factor for dust protection must be added to the costs. Figure 26.16 shows how the costs are included in the estimate. It is also important for the estimator to determine how long the work will take, since other workers will not be allowed in the area during the asbestos removal. This will affect the Project Schedule.

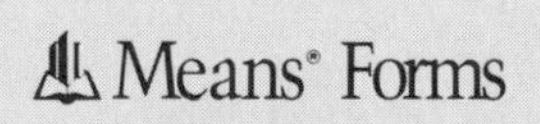

PROJECT
OVERHEAD SUMMARY

SHEET NO.

PROJECT

ESTIMATE NO.

LOCATION

ARCHITECT

DATE

QUANTITIES BY:

PRICES BY:

EXTENSIONS BY:

CHECKED BY:

DESCRIPTION	QUANTITY	UNIT	MATERIAL/EQUIPMENT		LABOR		TOTAL COST	
			UNIT	TOTAL	UNIT	TOTAL	UNIT	TOTAL
Job Organization: Superintendent								
Project Manager								
Timekeeper & Material Clerk								
Clerical								
Safety, Watchman & First Aid								
Travel Expense: Superintendent								
Project Manager								
Engineering: Layout								
Inspection/Quantities								
Drawings								
CPM Schedule								
Testing: Soil								
Materials								
Structural								
Equipment: Cranes								
Concrete Pump, Conveyor, Etc.								
Elevators, Hoists								
Freight & Hauling								
Loading, Unloading, Erecting, Etc.								
Maintenance								
Pumping								
Scaffolding								
Small Power Equipment/Tools								
Field Offices: Job Office								
Architect/Owner's Office								
Temporary Telephones								
Utilities								
Temporary Toilets								
Storage Areas & Sheds								
Temporary Utilities: Heat								
Light & Power								
Water								
PAGE TOTALS								

Page 1 of 2

Figure 26.14a

DESCRIPTION	QUANTITY	UNIT	MATERIAL/EQUIPMENT		LABOR		TOTAL COST	
			UNIT	TOTAL	UNIT	TOTAL	UNIT	TOTAL
Total Brought Forward								
Winter Protection: Temp. Heat/Protection								
Snow Plowing								
Thawing Materials								
Temporary Roads								
Signs & Barricades: Site Sign								
Temporary Fences								
Temporary Stairs, Ladders & Floors								
Photographs								
Clean Up								
Dumpster								
Final Clean Up								
Punch List								
Permits: Building								
Misc.								
Insurance: Builders Risk								
Owner's Protective Liability								
Umbrella								
Unemployment Ins. & Social Security								
Taxes								
City Sales Tax								
State Sales Tax								
Bonds								
Performance								
Material & Equipment								
Main Office Expense								
Special Items								
TOTALS:								

Page 2 of 2

Figure 26.14b

010 | Overhead

		010 000 | Overhead	CREW	DAILY OUTPUT	MAN-HOURS	UNIT	BARE COSTS MAT.	BARE COSTS LABOR	BARE COSTS EQUIP.	BARE COSTS TOTAL	TOTAL INCL O&P	
004	0011	ARCHITECTURAL FEES (10)										004
	0020	For work to $10,000				Project					15%	
	0040	To $25,000									13%	
	0060	To $100,000									10%	
	0080	To $500,000									8%	
	0090	To $1,000,000									7%	
016	0011	CONSTRUCTION MANAGEMENT FEES										016
	0060	For work to $10,000				Project					10%	
	0070	To $25,000									9%	
	0090	To $100,000									6%	
	0100	To $500,000									5%	
	0110	To $1,000,000									4%	
020	0010	**CONTINGENCIES** Allowance to add at conceptual stage									20%	020
	0050	Schematic stage									15%	
	0100	Preliminary working drawing stage									10%	
	0150	Final working drawing stage									2%	
028	0010	**ENGINEERING FEES** Educational planning consultant, minimum				Contrct					4.10%	028
	0100	Maximum									10.10%	
	0400	Elevator & conveying systems, minimum (11)									2.50%	
	0500	Maximum									5%	
	1000	Mechanical (plumbing & HVAC), minimum									4.10%	
	1100	Maximum									10.10%	
	1200	Structural, minimum				Project					1%	
	1300	Maximum				"					2.50%	
032	0010	**FACTORS** To be added to construction costs for particular job (143)										032
	0200											
	0500	Cut & patch to match existing construction, add, minimum [1]	−			Costs	2%	3%				
	0550	Maximum	+				5%	9%				
	0800	Dust protection, add, minimum [2]	−				1%	2%				
	0850	Maximum	+				4%	11%				
	1100	Equipment usage curtailment, add, minimum [3]	−				1%	1%				
	1150	Maximum	+				3%	10%				
	1400	Material handling & storage limitation, add, minimum [4]	−				1%	1%				
	1450	Maximum	+				6%	7%				
	1700	Protection of existing work, add, minimum [5]	−				2%	2%				
	1750	Maximum	+				5%	7%				
	2000	Shift work requirements, add, minimum [6]	−					5%				
	2050	Maximum	+					30%				
	2300	Temporary shoring and bracing, add, minimum [7]	−				2%	5%				
	2350	Maximum	+				5%	12%				
036	0011	FIELD PERSONNEL										036
	0020											
	0180	Project manager, minimum				Week		925		925	1,405	
	0200	Average						1,030		1,030	1,555	
	0220	Maximum						1,170		1,170	1,790	
	0240	Superintendent, minimum						875		875	1,325	
	0260	Average						975		975	1,480	
	0280	Maximum						1,095		1,095	1,655	
040	0010	**INSURANCE** Builders risk, standard, minimum				Job					.22%	040
	0050	Maximum									.59%	
	0200	All-risk type, minimum (2)									.25%	
	0250	Maximum									.62%	
	0400	Contractor's equipment floater, minimum				Value					.90%	
	0450	Maximum				"					1.60%	
	0600	Public liability, average				Job					1.55%	
	0810	Workers compensation & employer's liability										
	2000	Range of 36 trades in 50 states, excl. wrecking, minimum				Payroll		1.80%				
	2100	Average				"		12.50%				

For expanded coverage of these items see *Means Building Construction Cost Data 1989* 1

Figure 26.15

Means® Forms
COST ANALYSIS

Division 2

SHEET NO. 1 of 4
PROJECT Commercial Renovation — CLASSIFICATION — ESTIMATE NO. 89-1
LOCATION — ARCHITECT — DATE 1989
TAKE OFF BY EBW — QUANTITIES BY EBW — PRICES BY RSM — EXTENSIONS BY — CHECKED BY

DESCRIPTION	SOURCE/DIMENSIONS	QUANTITY	UNIT	MATERIAL		LABOR		EQUIPMENT		SUBCONTRACT			
				UNIT COST	TOTAL	UNIT COST	TOTAL	UNIT COST	TOTAL	UNIT COST	TOTAL	UNIT COST	TOTAL
Division 2: Site Work Demolition													
Asbestos Removal	Telephone Quote			SUBCONTRACT							3400		48
Factor [2] +										11%	374	11%	5
Remove wood deck @ Elev, Stair & Skylights	Crew F-2	2	Day			342.40	685	15	30			16	32
Brick Wall @ Pits	028-544-1200	180	CF			1.36	245	.81	146			.071	13
Brick Walls @ Ext. Openings	020-544-1200	314	CF			1.36	427	.81	254			.071	22
Factors [5] [7]	19% Labor					19%	81					19%	4
Remove Elev. & Machinery	Written Quote			SUBCONTRACT							800		80
Elevator Shaft	020-544-1200	1040	CF			1.36	1414	.81	842			.071	74
[2] + [5] +	18% Labor					18%	255					18%	13
Roofing (Compensate SF for slope)	020-726-3000	8660	SF			.43	3724					.025	217
Page Subtotals							6831		1272		4574		508

Figure 26.16

Means® Forms
COST ANALYSIS

Division 2

SHEET NO. 2 of 4

ESTIMATE NO. 89-1

DATE 1989

PROJECT Commercial Renovation

CLASSIFICATION

LOCATION

ARCHITECT

TAKE OFF BY EBW

QUANTITIES BY EBW

PRICES BY RSM

EXTENSIONS BY

CHECKED BY

DESCRIPTION	SOURCE/DIMENSIONS	QUANTITY	UNIT	MATERIAL		LABOR		EQUIPMENT		SUBCONTRACT			
				UNIT COST	TOTAL	UNIT COST	TOTAL	UNIT COST	TOTAL	UNIT COST	TOTAL	UNIT COST	TOTAL
Division 2: Sitework													
Demolition (Cont'd)													
Ceilings: Suspended	020-702-1580	2440	SF			.35	854					.021	51
Plaster	020-702-1000	8400	SF			.39	3276					.023	193
Doors:													
Exterior	0 0-706-0200	3	EA.			8.45	25					.5	2
Interior	020-706-0500	15	EA.			6.75	101					.4	6
Stair	020-706-0200	4	EA.			8.45	34					.5	2
Walls:													
Drywall	020-732-1000	400	SF			.13	52					.008	3
Studs	020-714-6600	680	LF			.13	88					.008	5
2 - 5 - 6 +	43% Labor					43%	38					13%	1
Drywall	020-732-1000	1920	SF			.13	250					.008	15
Plaster	020-732-3000	1560	SF			.34	530					.02	31
Studs	020-714-6600	3306	LF			.13	430					.008	26
Carpet	020-712-0400	1800	SF			.13	234					.008	14
Terrazzo	020-712-2640	640	SF			.77	493	.29	186			.046	29
Stair	020-714-6200	15	RIS			6.75	101					.4	6
Page Subtotals							6506		186				384

Figure 26.17

Means® Forms
COST ANALYSIS

Division 2

PROJECT Commercial Renovation CLASSIFICATION SHEET NO. 3 of 4

LOCATION ARCHITECT ESTIMATE NO. 89-1

DATE 1989

TAKE OFF BY EBN QUANTITIES BY EBN PRICES BY RSM EXTENSIONS BY CHECKED BY

DESCRIPTION	SOURCE/DIMENSIONS	QUANTITY	UNIT	MATERIAL UNIT COST	MATERIAL TOTAL	LABOR UNIT COST	LABOR TOTAL	EQUIPMENT UNIT COST	EQUIPMENT TOTAL	SUBCONTRACT UNIT COST	SUBCONTRACT TOTAL	UNIT COST	TOTAL
Division 2: Sitework Demolition (Cont'd)													
Windows	020-734-2020	75	EA.			11.25	844					.667	50
Plumbing:													
Water Closet	020-724-1400	16	EA.			24	384					1	16
Lavatories	020-724-1200	19	EA.			19.55	371					.8	15
Urinals	020-724-1520	6	EA.			28	168					1.14	7
Toilet Partitions	020-732-3800	19	EA.			27	513					1.6	30
Trash Chute	020-620-0440	30	LF	18.70	561	14	420					.8	24
Load & Haul (6 Loads)	020-620-2040	180	CY			16.35	2943					.97	175
Asphalt Sidewalk	020-554-4000	249	SY			1.34	334	.54	134			.074	18
HVAC: Boiler, Radiators, Piping; Plumb: Piping; Elec	Removed by Salvage Co. at no cost												
Page Subtotals					561		5977		134				335

Figure 26.18

Means® Forms
COST ANALYSIS

Division 2

SHEET NO. 4 of 4

PROJECT Commercial Renovation | CLASSIFICATION | ESTIMATE NO. 89-1

LOCATION | ARCHITECT | DATE 1989

TAKE OFF BY EBW | QUANTITIES BY EBW | PRICES BY RSM | EXTENSIONS BY | CHECKED BY

DESCRIPTION	SOURCE/DIMENSIONS	QUANTITY	UNIT	MATERIAL UNIT COST	MATERIAL TOTAL	LABOR UNIT COST	LABOR TOTAL	EQUIPMENT UNIT COST	EQUIPMENT TOTAL	SUBCONTRACT UNIT COST	SUBCONTRACT TOTAL	UNIT COST	TOTAL
Division 2: Sitework													
Sidewalk	025-128-0310	2241	SF	.95	2129	.79	1770					.04	90
Sewer Connection													
Excav.	022-254-1500	4	CY			34	136					2	8
Backfill	022-204-0100/0800	4	CY			16.21	65					.962	4
Pipe	027-168-2080	10	LF	1.75	18	1.38	14					.072	1
Factors [6] +	30% Labor					30%	4						
Excavation @ Elev. shaft	022-250-0010	60	CY			16.85	1011					1	60
[7] +	12% Labor					12%	121					12%	7
Page Subtotals					(2147)		(3121)						(170)
Sheet 1							6831		1272		4574		508
Sheet 2							6506		186				384
Sheet 3					561		5977		134				335
Demolition Subtotal					561		19314		1592		4574		1227
Sheet 4					2147		3121						170
Division 2 Totals					2708		22435		1592				1397

Figure 26.19

Means® Forms

TELEPHONE QUOTATION

PROJECT	Office Renovation	DATE	
FIRM QUOTING	Asbestos Removal Co.	TIME	
ADDRESS		PHONE ( )	
ITEM QUOTED		BY	
		RECEIVED BY	EBN

WORK INCLUDED	AMOUNT OF QUOTATION
Removal of asbestos pipe & boiler insulation in basement (includes required permits)	
Lump Sum	3400
Job requires 3 workers / 2 days	
TOTAL BID	3400

DELIVERY TIME

DOES QUOTATION INCLUDE THE FOLLOWING:			If ☐ NO is checked, determine the following:
STATE & LOCAL SALES TAXES	☐ YES	☐ NO N/A	MATERIAL VALUE
DELIVERY TO THE JOB SITE	☐ YES	☐ NO N/A	WEIGHT
COMPLETE INSTALLATION	☐ YES	☐ NO N/A	QUANTITY
COMPLETE SECTION AS PER PLANS & SPECIFICATIONS	☐ YES	☒ NO	DESCRIBE BELOW

EXCLUSIONS AND QUALIFICATIONS

Does not include dust protection of existing Tenant

Add factor to quote [2] + 11% on Labor

ADDENDA ACKNOWLEDGEMENT	**TOTAL ADJUSTMENTS**
	ADJUSTED TOTAL BID

ALTERNATES

ALTERNATE NO.

ALTERNATE NO.

ALTERNATE NO.

ALTERNATE NO.

ALTERNATE NO.

ALTERNATE NO.

ALTERNATE NO.

Figure 26.20

Even when specific items are not listed as individual line items in *Means Repair and Remodeling Cost Data*, the costs can be derived from the information that is available. In Figure 26.16, the costs for cutting the mill-type wood flooring for the elevator, stair, and roof openings have been calculated by estimating the time required and determining the appropriate crew (F-2) to perform the work. Referring to the Crew Lists in Repair and Remodeling Cost Data provides the labor and equipment costs to be used. The estimator should make a note that the flooring that is removed should be saved to fill in the openings where the freight elevator is to be removed.

The estimator must use experience and good judgment when applying the "Factors" to the estimate. The project includes openings to be cut in the exterior masonry walls for the lobby entrance and retail doorways, as shown in Figure 26.3. Figure 26.21 contains the appropriate line item. The estimator knows that this masonry removal will have to be performed carefully because the edges of the openings will have to be rebuilt as finished surfaces. The estimator chooses the appropriate factors as shown in Figure 26.22, and records them in Figure 26.16.

The existing freight elevator is to be used for hauling materials and is to remain in operation for as long as possible. This forces a delay of the demolition of the shaft and elevator as well as the construction of the third floor office. Since the new windows and other work will be in place when the shaft is demolished, factors are applied for the protection of existing work and for dust protection. The estimator should also note for scheduling purposes that workers should place all materials on the upper floors before removal of the elevator.

The removal of the wall at the existing retail space will also entail extra labor expense to protect existing work, and the work must be performed after business hours. Note on Figure 26.17 that the total factors (43% for labor) are added only to the labor cost. The 30% added for overtime affects only the cost and is not used as a factor in determining the man-hours.

Other work, from other divisions, is required when the masonry is removed at the openings. For example, needling will be necessary at the wall above the entrance opening, and all opening jambs require toothing and rebuilding, using existing brick that has been removed. When items such as these are encountered the estimator should make notes on separate sheets, as in Figure 26.23, so that items will not be omitted.

Salvaged materials can help to reduce the costs of demolition. In this case, a wrecking subcontractor is to remove all piping, wiring, radiators, and the boiler at no cost. Ingenuity and legwork by the estimator can result in lower costs for the client and lower prices for competitive bids.

When all items in Division 2 have been entered, the estimator should review the division to note and include all related items to be entered on the Project Overhead Summary, as shown in Figures 26.24a and 26.24b. The costs for a dumpster and the scaffolding, for example, are included in Division 1 in this estimate because they are to be used by different trades throughout the job, not just for

020 | Subsurface Investigation and Demolition

	020 120 Std Penetration Tests		CREW	DAILY OUTPUT	MAN-HOURS	UNIT	BARE COSTS MAT.	LABOR	EQUIP.	TOTAL	TOTAL INCL O&P	
125	1840	6" diameter core	B-89A	46	.348	Ea.	4.79	6.75	.99	12.53	17.10	125
	1850	Each added inch thick, add		175	.091		.80	1.77	.26	2.83	4	
	1860	8" diameter core		31	.516		6.55	10	1.46	18.01	25	
	1870	Each added inch thick, add		135	.119		1.08	2.30	.34	3.72	5.25	
	1880	10" diameter core	A-1	28	.286		8.80	4.81	1.84	15.45	19.45	
	1890	Each added inch thick, add		115	.070		1.08	1.17	.45	2.70	3.56	
	1900	12" diameter core		25	.320		10.65	5.40	2.06	18.11	23	
	1910	Each added inch thick, add		95	.084		1.78	1.42	.54	3.74	4.83	
	1950	Minimum charge for above, 3" diameter core	B-89A	6.35	2.520	Total		49	7.15	56.15	86	
	2000	4" diameter core		6.15	2.600			50	7.40	57.40	89	
	2050	6" diameter core		5.50	2.910			56	8.25	64.25	99	
	2100	8" diameter core		5.10	3.140			61	8.90	69.90	105	
	2150	10" diameter core		4.50	3.560			69	10.10	79.10	120	
	2200	12" diameter core		3.70	4.320			84	12.25	96.25	145	
	2250	14" diameter core		3.20	5			97	14.20	111.20	170	
	2300	18" diameter core		3	5.330			105	15.15	120.15	180	

	020 550 Site Demolition		CREW	DAILY OUTPUT	MAN-HOURS	UNIT	BARE COSTS MAT.	LABOR	EQUIP.	TOTAL	TOTAL INCL O&P	
554	0010	**SITE DEMOLITION** No hauling, abandon catch basin or manhole	B-6	7	3.430	Ea.		62	25	87	125	554
	0020	Remove existing catch basin or manhole		4	6			110	44	154	220	
	0030	Catch basin or manhole frames and covers stored		13	1.850			33	13.40	46.40	68	
	0040	Remove and reset		7	3.430			62	25	87	125	
	0045	Minimum labor/equipment charge		4	6	Job		110	44	154	220	
	0600	Fencing, barbed wire, 3 strand	2 Clab	430	.037	L.F.		.63		.63	1.01	
	0650	5 strand		280	.057			.96		.96	1.55	
	0700	Chain link, remove only, 8' to 10' high		310	.052			.87		.87	1.40	
	0800	Guide rail, remove only		85	.188			3.17		3.17	5.10	
	0850	Remove and reset		35	.457			7.70		7.70	12.35	
	0890	Minimum labor/equipment charge		4	4	Job		67		67	110	
	0900	Hydrants, fire, remove only	2 Plum	4.70	3.400	Ea.		83		83	125	
	0950	Remove and reset		1.40	11.430	"		280		280	425	
	0990	Minimum labor/equipment charge		2	8	Job		195		195	300	
	1000	Remove masonry walls, block or tile, solid	B-5	1,800	.036	C.F.		.68	.41	1.09	1.52	
	1100	Cavity		2,200	.029			.56	.33	.89	1.24	
	1200	Brick, solid		900	.071			1.36	.81	2.17	3.04	
	1300	With block		1,130	.057			1.08	.65	1.73	2.42	
	1400	Stone, with mortar		900	.071			1.36	.81	2.17	3.04	
	1500	Dry set		1,500	.043			.81	.49	1.30	1.82	
	1590	Minimum labor/equipment charge	A-1	4	2	Job		34	12.85	46.85	68	
	1710	Pavement removal, bituminous, 3" thick	B-38	690	.058	S.Y.		1.10	1.50	2.60	3.39	
	1750	4" to 6" thick		420	.095			1.81	2.46	4.27	5.55	
	1800	Bituminous driveways		680	.059			1.12	1.52	2.64	3.44	
	1900	Concrete to 6" thick, mesh reinforced		255	.157			2.97	4.06	7.03	9.15	
	2000	Rod reinforced		200	.200			3.79	5.15	8.94	11.70	
	2100	Concrete 7" to 24" thick, plain		13.10	3.050	C.Y.		58	79	137	180	
	2200	Reinforced		9.50	4.210	"		80	110	190	245	
	2250	Minimum labor/equipment charge		6	6.670	Job		125	170	295	390	
	2300	With hand held air equipment, bituminous	B-39	1,900	.025	S.F.		.45	.06	.51	.79	
	2320	Concrete to 6" thick, no reinforcing		1,200	.040			.71	.10	.81	1.24	
	2340	Mesh reinforced		830	.058			1.03	.14	1.17	1.80	
	2360	Rod reinforced		765	.063			1.12	.15	1.27	1.95	
	2390	Minimum labor/equipment charge	B-38	6	6.670	Job		125	170	295	390	
	2400	Curbs, concrete, plain	B-6	325	.074	L.F.		1.34	.54	1.88	2.71	
	2500	Reinforced		220	.109			1.97	.79	2.76	4	
	2600	Granite curbs		355	.068			1.22	.49	1.71	2.48	
	2700	Bituminous curbs		830	.029			.52	.21	.73	1.06	
	2790	Minimum labor/equipment charge		6	4	Job		72	29	101	145	
	2900	Pipe removal, concrete, no excavation, 12" diameter		175	.137	L.F.		2.48	.99	3.47	5.05	

For expanded coverage of these items see *Means Site Work Cost Data 1989*

11

Figure 26.21

010 | Overhead

		010 000 Overhead	CREW	DAILY OUTPUT	MAN-HOURS	UNIT	BARE COSTS MAT.	BARE COSTS LABOR	BARE COSTS EQUIP.	BARE COSTS TOTAL	TOTAL INCL O&P	
004	0011	**ARCHITECTURAL FEES** (10)										004
	0020	For work to $10,000				Project					15%	
	0040	To $25,000									13%	
	0060	To $100,000									10%	
	0080	To $500,000									8%	
	0090	To $1,000,000									7%	
016	0011	**CONSTRUCTION MANAGEMENT FEES**										016
	0060	For work to $10,000				Project					10%	
	0070	To $25,000									9%	
	0090	To $100,000									6%	
	0100	To $500,000									5%	
	0110	To $1,000,000									4%	
020	0010	**CONTINGENCIES** Allowance to add at conceptual stage									20%	020
	0050	Schematic stage									15%	
	0100	Preliminary working drawing stage									10%	
	0150	Final working drawing stage									2%	
028	0010	**ENGINEERING FEES** Educational planning consultant, minimum				Contrct					4.10%	028
	0100	Maximum									10.10%	
	0400	Elevator & conveying systems, minimum									2.50%	
	0500	Maximum (11)									5%	
	1000	Mechanical (plumbing & HVAC), minimum									4.10%	
	1100	Maximum									10.10%	
	1200	Structural, minimum				Project					1%	
	1300	Maximum				"					2.50%	
032	0010	**FACTORS** To be added to construction costs for particular job (143)										032
	0200											
	0500	Cut & patch to match existing construction, add, minimum [1] −				Costs	2%	3%				
	0550	Maximum +					5%	9%				
	0800	Dust protection, add, minimum [2] −					1%	2%				
	0850	Maximum +					4%	11%				
	1100	Equipment usage curtailment, add, minimum [3] −					1%	1%				
	1150	Maximum +					3%	10%				
	1400	Material handling & storage limitation, add, minimum [4] −					1%	1%				
	1450	Maximum +					6%	7%				
	1700	Protection of existing work, add, minimum [5] −					2%	2%				
	1750	Maximum +					5%	7%				
	2000	Shift work requirements, add, minimum [6] −						5%				
	2050	Maximum +						30%				
	2300	Temporary shoring and bracing, add, minimum [7] −					2%	5%				
	2350	Maximum +					5%	12%				
036	0011	**FIELD PERSONNEL**										036
	0020											
	0180	Project manager, minimum				Week		925		925	1,405	
	0200	Average						1,030		1,030	1,555	
	0220	Maximum						1,170		1,170	1,790	
	0240	Superintendent, minimum						875		875	1,325	
	0260	Average						975		975	1,480	
	0280	Maximum						1,095		1,095	1,655	
040	0010	**INSURANCE** Builders risk, standard, minimum				Job					.22%	040
	0050	Maximum									.59%	
	0200	All-risk type, minimum									.25%	
	0250	Maximum (2)									.62%	
	0400	Contractor's equipment floater, minimum				Value					.90%	
	0450	Maximum				"					1.60%	
	0600	Public liability, average				Job					1.55%	
	0810	Workers compensation & employer's liability										
	2000	Range of 36 trades in 50 states, excl. wrecking, minimum				Payroll		1.80%				
	2100	Average				"		12.50%				

For expanded coverage of these items see *Means Building Construction Cost Data 1989* 1

Figure 26.22

Means® Forms
COST ANALYSIS

Division 4

PROJECT Commercial Renovation　　CLASSIFICATION　　SHEET NO. 1 of 1

LOCATION　　ARCHITECT　　ESTIMATE NO. 89-1

TAKE OFF BY EBW　　QUANTITIES BY EBW　　PRICES BY RSM　　EXTENSIONS BY　　DATE 1989

CHECKED BY

DESCRIPTION	SOURCE/DIMENSIONS			QUANTITY	UNIT	MATERIAL		LABOR		EQUIPMENT		SUBCONTRACT			
						UNIT COST	TOTAL	UNIT COST	TOTAL	UNIT COST	TOTAL	UNIT COST	TOTAL	UNIT COST	TOTAL
Division 4: Masonry															
Needling															
Extra Floors	045	240	1080												
@Entrance	045	240	2000												
Toothing															
@ Ext. Openings	045	290	0520												

Figure 26.23

Means® Forms

PROJECT OVERHEAD SUMMARY

Division I

SHEET NO. 1 of 2

PROJECT Commercial Renovation

ESTIMATE NO. 89-1

LOCATION

ARCHITECT

DATE 1989

QUANTITIES BY: EBW

PRICES BY: RSM

EXTENSIONS BY:

CHECKED BY:

DESCRIPTION	MAN-HOURS	QUANTITY	UNIT	MATERIAL		LABOR		EQUIP. FEES, RENTAL	
				UNIT	TOTAL	UNIT	TOTAL	UNIT	TOTAL
Job Organization: Superintendent									
Project Manager									
Timekeeper & Material Clerk									
Clerical									
Safety, Watchman & First Aid									
Travel Expense: Superintendent									
Project Manager									
Engineering: Layout									
Inspection/Quantities									
Drawings									
CPM Schedule									
Testing: Soil									
Materials									
Structural									
Equipment: Cranes									
Concrete Pump, Conveyor, Etc.									
Elevators, Hoists									
Freight & Hauling									
Loading, Unloading, Erecting, Etc.									
Maintenance									
Pumping									
Scaffolding 015-254-0090	170	119	CSF			31	3689		
~~Small Power Equipment/Tools~~ Rental - 381 Frames			Mo.						
@ $3.65/MO. - Circle 14									
Field Offices: Job Office									
Architect/Owner's Office									
Temporary Telephones			Mo.					100	
Utilities (Power)		28	MSF					93	2604
Temporary Toilets									
Storage Areas & Sheds									
Temporary Utilities: Heat									
Light ~~& Power~~ 015-104-0350	66	280	CSF	2.06	577	5.70	1596		
Water									
PAGE TOTALS									

Page 1 of 2

Figure 26.24a

Means® Forms

DIVISION 1 SHEET 2 OF 2 DESCRIPTION	MAN-HOURS	QUANTITY	UNIT	MATERIAL UNIT	MATERIAL TOTAL	LABOR UNIT	LABOR TOTAL	EQUIP. FEES, RENTAL UNIT	EQUIP. FEES, RENTAL TOTAL
Total Brought Forward									
Winter Protection: Temp. Heat/Protection									
Snow Plowing									
Thawing Materials									
Temporary Roads									
RENT 6 PARKING SPACES			MO.					300	
Signs & Barricades: Site Sign		1	EA.		250				
Temporary Fences									
Temporary Stairs, Ladders & Floors									
RAILS @ STAIRS & ELEV. 015-302-1000	22	270	LF	1.35	365	1.71	462		
Photographs									
Clean Up									
Dumpster 020-620-0800			WK					265	
Final Clean Up									
Punch List									
Permits: Building									
Misc. STREET USE		1	EA.					50	50
Insurance: Builders Risk									
Owner's Protective Liability									
Umbrella									
Unemployment Ins. & Social Security									
Taxes									
City Sales Tax									
State Sales Tax									
Bonds									
Performance									
Material & Equipment									
Main Office Expense									
Special Items									
TOTALS:									

Page 2 of 2

Figure 26.24b

demolition. Note that quantities only have been listed for these and other items. The total costs are dependent on the time extent of the project and will be entered when the Project Schedule has been completed.

Division 3: Concrete

The cast-in-place concrete for the project is limited to the footings, walls, and slab in the basement for the new elevator shafts. These items can be estimated in two ways using *Means Repair and Remodeling Cost Data*.

The individual components, forms, reinforcing, labor, and concrete can be priced separately. The costs for the floor slab are shown in Figure 26.25. No forms are required because the new floor slab is to be placed flush with the existing basement floor. The "out" for the sump pump is too small to be deducted. At this point, the sump pumps and associated floor drain should be listed on a sheet for Division 15: Mechanical. (Figure 26.25) For small pours, the estimator should be aware of minimum concrete costs.

The second method for pricing the cast-in-place concrete is shown in Figure 26.27. *Means Repair and Remodeling Cost Data* provides costs for complete concrete installations, including forms, reinforcing, concrete and placement. This method is used for the footings and walls in Figure 26.25.

The precast concrete beam that will serve as the lintel at the lobby entrance cannot be lowered into place with a crane. The beam must be slid into place by hand and winch. A factor for the added labor is included in the estimate. The specifications call for the beam to receive a special stucco finish to match the existing limestone window lintels. This is included in Division 9: Finishes.

During the estimate, the owner asks for an alternate price to lower the basement dirt floor to create an area for tenant storage. While the estimator knows from experience that such a project would not be cost effective, the owner needs to know the costs. In a short period of time, the estimator can calculate fairly accurate costs, without plans and specifications, by visualizing the work to be performed. The costs for the alternate are shown in Figure 26.28. This price does not include the costs for continuing the new stairway to the basement (required by code for basement occupancy) or the resulting costs for relocating the electrical service. After a phone call to the owner, the estimator does not have to waste any more time.

Division 4: Masonry

In commercial renovation, new masonry work is usually limited compared to the extent of repair and restoration of existing masonry, which takes extensive time and labor. Normal methods of work must be altered to accommodate existing conditions. Figure 26.16 in Division 2 lists costs for removing the brick at the new exterior openings. The demolition of the brick for the lobby entrance cannot occur until the structure above is supported by needles as shown in Figure 26.29. Such an operation should always be planned and supervised by a structural engineer. The costs are included in Figure 26.31.

Included in the costs for removing the exterior brick is a factor for protection of existing work (Figure 26.15). This factor is applied not

Means® Forms
COST ANALYSIS

Division 3

SHEET NO. 1 of 1

PROJECT Commercial Renovation | CLASSIFICATION | ESTIMATE NO. 89-1

LOCATION | ARCHITECT | DATE 1989

TAKE OFF BY EBW | QUANTITIES BY EBW | PRICES BY RSW | EXTENSIONS BY | CHECKED BY

DESCRIPTION	SOURCE/DIMENSIONS	QUANTITY	UNIT	MATERIAL		LABOR		EQUIPMENT		SUBCONTRACT			
				UNIT COST	TOTAL	UNIT COST	TOTAL	UNIT COST	TOTAL	UNIT COST	TOTAL	UNIT COST	TOTAL
Division 3: Concrete													
Footings	033-130-3850	12	CY	76	912	25	300	5.40	65			1.08	13
Walls	033-130-4260	18	CY	134	2412	135	2430	10.15	183			6	108
3 Both Above +	10% Labor					10%	273					10%	12
Slab:													
Reinforcing	032-207-0010	4	CSF	8	32	10.60	42					.457	2
Labor	033-172-4300	5	CY			7.80	39	.53	3			.436	2
Concrete (8CY. Min)	033-126-0150	8	CY	55.80	446								
Wheeled	033-172-9000	5	CY			Min.	67	Min.	26			Min.	4
Precast Beam @ Entrance	034-104-0050	16	LF	1.45	2320	6.85	110	5.25	84			.3	5
3 +	10% Labor					10%	11					10%	1
Division 3: Totals					6122		3272		361				147

Figure 26.25

Means® Forms
COST ANALYSIS

Division 15

SHEET NO. 1 of 1

PROJECT Commercial Renovation

CLASSIFICATION

ESTIMATE NO. 89-1

LOCATION

ARCHITECT

DATE 1989

TAKE OFF BY EBW

QUANTITIES BY EBW

PRICES BY RSM

EXTENSIONS BY

CHECKED BY

DESCRIPTION	SOURCE/DIMENSIONS	QUANTITY	UNIT	MATERIAL		LABOR		EQUIPMENT		SUBCONTRACT			
				UNIT COST	TOTAL	UNIT COST	TOTAL	UNIT COST	TOTAL	UNIT COST	TOTAL	UNIT COST	TOTAL
Division 15: Mechanical													
Plumbing													
Sump Pump @ Elev. Pits	152-480-7100	1	EA.		SUBCONTRACT					285		1.33	
Floor Drain	151-125-2040	1	EA.							76		1.33	

Figure 26.26

033 | Cast-In-Place Concrete

	033 100 Structural Concrete		CREW	DAILY OUTPUT	MAN-HOURS	UNIT	BARE COSTS MAT.	LABOR	EQUIP.	TOTAL	TOTAL INCL O&P	
118	2120	Maximum				Lb.	1.50			1.50	1.65	118
	2200	Water reducing admixture, average				Gal.	7.75			7.75	8.55	
126	0010	**CONCRETE, READY MIX** Regular weight, 2000 psi (42)				C.Y.	52.30			52.30	58	126
	0100	2500 psi					54			54	59	
	0150	3000 psi (43)					55.80			55.80	61	
	0200	3500 psi					57.50			57.50	63	
	0250	3750 psi					57.90			57.90	64	
	0300	4000 psi					58.25			58.25	64	
	0350	4500 psi					60			60	66	
	0400	5000 psi					62.35			62.35	69	
	1000	For high early strength cement, add					10%					
	2000	For all lightweight aggregate, add					45%					
	3000	For integral colors, 2500 psi, 5 bag mix										
	3100	Red, yellow or brown, 1.8 lb. per bag, add				C.Y.	12.60			12.60	13.85	
	3200	9.4 lb. per bag, add					66			66	73	
	3400	Black, 1.8 lb. per bag, add					13.10			13.10	14.40	
	3500	7.5 lb. per bag, add (41)					55			55	61	
	3700	Green, 1.8 lb. per bag, add					27			27	30	
	3800	7.5 lb. per bag, add					115			115	125	
130	0010	**CONCRETE IN PLACE** Including forms (4 uses), reinforcing										130
	0050	steel, including finishing unless otherwise indicated										
	0100	Average for concrete framed building, (35)										
	0110	including finishing	C-17B	15.75	5.210	C.Y.	105	115	16.70	236.70	320	
	0130	Average for substructure only, simple design, incl. finishing		29.07	2.820		75	63	9.05	147.05	195	
	0150	Average for superstructure only, including finishing		13.42	6.110		113	135	19.60	267.60	365	
	0200	Base, granolithic, 1" x 5" high, straight (50)	C-10	175	.137	L.F.	.13	2.67	.35	3.15	4.71	
	0220	Cove	"	140	.171	"	.13	3.34	.44	3.91	5.85	
	0300	Beams, 5 kip per L.F., 10' span	C-17A	6.28	12.900	C.Y.	185	290	22	497	685	
	0350	25' span		7.40	10.950		145	245	18.50	408.50	570	
	0500	Chimney foundations, minimum		26.70	3.030		98	68	5.10	171.10	220	
	0510	Maximum		19.70	4.110		113	92	6.95	211.95	280	
	0700	Columns, square, 12" x 12", minimum reinforcing (49)		4.60	17.610		200	395	30	625	880	
	0720	Average reinforcing		4.10	19.760		285	440	33	758	1,050	
	0740	Maximum reinforcing	C-17B	3.84	21.350		420	475	69	964	1,300	
	0800	16" x 16", minimum reinforcing	C-17A	6	13.500		175	300	23	498	695	
	0820	Average reinforcing (47)	"	4.97	16.300		280	365	28	673	915	
	0840	Maximum reinforcing	C-17B	8.34	9.830		460	220	32	712	890	
	1200	Columns, round, tied, 16" diameter, minimum reinforcing		13.02	6.300		245	140	20	405	515	
	1220	Average reinforcing		8.30	9.880		350	220	32	602	770	
	1240	Maximum reinforcing (49)		6.05	13.550		560	300	43	903	1,150	
	1300	20" diameter, minimum reinforcing		17.35	4.730		230	105	15.15	350.15	435	
	1320	Average reinforcing		10.43	7.860		370	175	25	570	715	
	1340	Maximum reinforcing		7.47	10.980		500	245	35	780	980	
	1700	Curbs, formed in place, 6" x 18", straight,	C-15	400	.180	L.F.	3.20	3.63	.12	6.95	9.45	
	1750	Curb and gutter	"	170	.424	"	5.05	8.55	.27	13.87	19.55	
	3800	Footings, spread under 1 C.Y.	C-17B	35.95	2.280	C.Y.	80	51	7.30	138.30	175	
	3850	Over 5 C.Y.	C-17C	73.91	1.120		76	25	5.40	106.40	130	
	3900	Footings, strip, 18" x 9", plain	C-17B	29.24	2.800		69	63	9	141	185	
	3950	36" x 12", reinforced		51.42	1.590		75	36	5.10	116.10	145	
	4000	Foundation mat, under 10 C.Y.		32.32	2.540		124	57	8.15	189.15	235	
	4050	Over 20 C.Y.		47.37	1.730		113	39	5.55	157.55	190	
	4200	Grade walls, 8" thick, 8' high	C-17A	9.50	8.530		115	190	14.40	319.40	445	
	4250	14' high	C-20	7.30	8.770		167	160	71	398	515	
	4260	12" thick, 8' high	C-17A	13.50	6		134	135	10.15	279.15	370	
	4270	14' high	C-20	11.60	5.520		124	100	45	269	345	
	4300	15" thick, 8' high	C-17B	18.40	4.460		91	99	14.30	204.30	275	
	4350	12' high	C-20	14.80	4.320		106	79	35	220	280	

For expanded coverage of these items see *Means Concrete Cost Data 1989* 43

Figure 26.27

Means® Forms
COST ANALYSIS

Basement Alternate

SHEET NO. 1 of 1
ESTIMATE NO. 89-1
DATE 1989

PROJECT Office Renovation
CLASSIFICATION
LOCATION
ARCHITECT
TAKE OFF BY EBW
QUANTITIES BY EBW
PRICES BY RSM
EXTENSIONS BY
CHECKED BY

DESCRIPTION	SOURCE/DIMENSIONS	QUANTITY	UNIT	MATERIAL		LABOR		EQUIPMENT		SUBCONTRACT			
				UNIT COST	TOTAL	UNIT COST	TOTAL	UNIT COST	TOTAL	UNIT COST	TOTAL	UNIT COST	TOTAL
Underpinning @ Columns	021-564-0100	13	CY	150	1950	480	6240	135	1755			24.35	317
Under 50 CY	021-564-0900			10%	195	40%	2496					40%	127
Factors [3] [7]				8%	172	22%	1922					22%	98
Excavating	022-250-0010	700	CY			16.85	11795					1	700
Factors [3] [7]						22%	2595					22%	154
Conveyor	016-406-0900	2	WK.					395	790				
Reinforcing	032-207-0010	56	CSF	8	448	10.60	594					.457	26
Floor Slab	033-172-4300	69	CY			7.80	538	.53	37			.436	30
Wheeled	033-172-5610	69	CY			4.85	300	1.66	115			.258	18
[1] [3]	4% Labor						34					4%	2
Concrete	033-126-0150	69	CY	55.80	3850								
Subtotals (Bare Costs)					6615		26514		2697				1472
Overhead	10% Mat. + Equip. 59.2 Labor				662		15696		270				
					7277		42210		2967				
Profit 10%					728		4221		297				
Total							57700						

Figure 26.28

only for protection of the remaining adjacent surfaces, but also because the existing brick must be removed carefully to be reused to construct the jambs at the new exterior openings. Brick in older buildings usually cannot be matched with new brick. Since the old brick is to be used, there are no material costs for the specified brick masonry. There is no specific line item for rebuilding jambs to match existing conditions at new openings, using existing materials. The estimator must use good judgment to determine the most accurate costs. The work is similar to constructing brick columns, so line 042-108-0300 in Figure 26.30 is chosen. Since the work requires care to match existing conditions, a factor is included as shown in Figure 26.31. Figures 26.31 to 26.33 are the estimate sheets for Division 4: Masonry.

Unless the contractor has a great deal of experience in masonry restoration, it is recommended that a knowledgeable architect or a

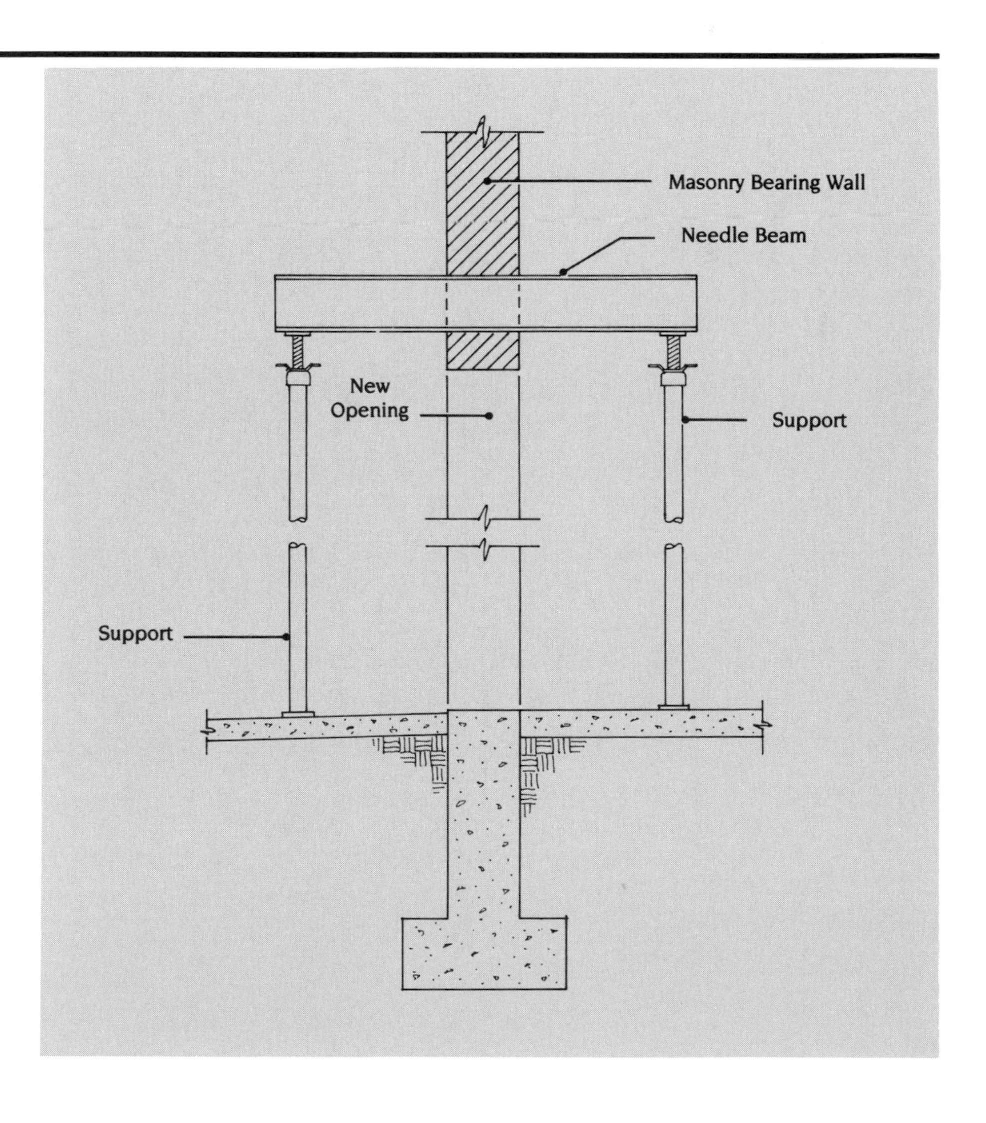

Figure 26.29

041 | Mortar and Masonry Accessories

		041 000 | Mortar	CREW	DAILY OUTPUT	MAN-HOURS	UNIT	BARE COSTS MAT.	BARE COSTS LABOR	BARE COSTS EQUIP.	BARE COSTS TOTAL	TOTAL INCL O&P	
016	0800	Door frames, 3' x 7' opening, 2.5 C.F. per opening	D-4	60	.533	Opng.	7.90	10.25	2.22	20.37	27	016
	0850	6' x 7' opening, 3.5 C.F. per opening	"	45	.711	"	11.05	13.65	2.96	27.66	37	
	9000	Minimum labor/equipment charge	1 Bric	2	4	Job		88		88	140	
020	0010	LIME Masons, hydrated, 50 lb. bag, T.L. lots				Bag	4.15			4.15	4.57	020
	0050	L.T.L. lots					4.50			4.50	4.95	
	0200	Finish, double hydrated, 50 lb. bag, T.L. lots					5.40			5.40	5.95	
	0250	L.T.L. lots				↓	6.15			6.15	6.75	
		041 500 | Masonry Accessories										
504	0010	**ANCHOR BOLTS** Hooked type with nut, 1/2" diam., 8" long	1 Bric	200	.040	Ea.	.34	.88		1.22	1.77	504
	0030	12" long		190	.042		.44	.93		1.37	1.95	
	0060	3/4" diameter, 8" long		160	.050		1.39	1.11		2.50	3.27	
	0070	12" long		150	.053	↓	1.71	1.18		2.89	3.74	
512	0010	**JOINT REINFORCING** Steel bars, placed horizontal, #3 & #4 bars		450	.018	Lb.	.30	.39		.69	.95	512
	0020	#5 & #6 bars		800	.010		.30	.22		.52	.68	
	0050	Placed vertical, #3 & #4 bars		350	.023		.30	.51		.81	1.13	
	0060	#5 & #6 bars	↓	650	.012	↓	.30	.27		.57	.76	
	0500	Wire strips, ladder type, galvanized										
	0600	8 ga. sides, 9 ga. ties, 4" wall	1 Bric	30	.267	C.L.F.	9.90	5.90		15.80	20	
	0750	10" wall	"	20	.400	"	11.65	8.85		20.50	27	
	1000	Wire strips, truss type, galvanized										
	1010	To 8" wide	1 Bric	3,000	.003	L.F.	.18	.06		.24	.29	
	1030	12" wide		2,000	.004	"	.19	.09		.28	.35	
	1100	9 ga. sides, 9 ga. ties, 4" wall		30	.267	C.L.F.	12.50	5.90		18.40	23	
	1250	10" wall	↓	20	.400	"	15.50	8.85		24.35	31	
	3500	For hot dip galvanizing, add					65%					
520	0010	**WALL TIES** To brick veneer, galv., corrugated, 7/8" x 7", 24 gauge	1 Bric	10.50	.762	C	3.50	16.85		20.35	30	520
	0050	16 gauge		10.50	.762		9.60	16.85		26.45	37	
	0600	Cavity wall, 6" long, Z type, galvanized, 1/8" diameter		10.50	.762		16.50	16.85		33.35	45	
	0650	3/16" diameter		10.50	.762		7.20	16.85		24.05	35	
	1500	Stone anchors, galv., U or Z shaped, 6" long, 1/8" x 1"		10.50	.762		65	16.85		81.85	98	
	1550	1/4" x 1"	↓	10.50	.762	↓	100	16.85		116.85	135	

042 | Unit Masonry

		042 050 | Chimneys	CREW	DAILY OUTPUT	MAN-HOURS	UNIT	BARE COSTS MAT.	BARE COSTS LABOR	BARE COSTS EQUIP.	BARE COSTS TOTAL	TOTAL INCL O&P	
054	0010	**CHIMNEY** For foundation, add to prices below, see div 033-130-0500										054
	0100	Brick @ $240/M, 16" x 16", 8" flue, scaff. not incl.	D-1	18.20	.879	V.L.F.	11.30	17.25		28.55	40	
	0150	16" x 20" with one 8" x 12" flue (65)		16	1		14.10	19.60		33.70	46	
	0200	16" x 24" with two 8" x 8" flues		14	1.140		16.65	22		38.65	54	
	0250	20" x 20" with one 12" x 12" flue		13.70	1.170		16.65	23		39.65	54	
	0300	20" x 24" with two 8" x 12" flues		12	1.330		22.20	26		48.20	66	
	0350	20" x 32" with two 12" x 12" flues	↓	10	1.600	↓	28.15	31		59.15	80	
		042 100 | Brick Masonry										
108	0010	**COLUMNS** Brick @ $240 per M, 8" x 8", 9 brick, scaff. not incl.	D-1	56	.286	V.L.F.	2.45	5.60		8.05	11.55	108
	0100	12" x 8", 13.5 brick		37	.432		3.70	8.50		12.20	17.45	
	0200	12" x 12", 20.3 brick		25	.640		5.55	12.55		18.10	26	
	0300	16" x 12", 27 brick	↓	19	.842	↓	7.40	16.50		23.90	34	

48

For expanded coverage of these items see *Means Concrete Cost Data 1989*

Figure 26.30

Means® Forms
COST ANALYSIS

Division 4

SHEET NO. 1 of 3
PROJECT Commercial Renovation
CLASSIFICATION
ESTIMATE NO. 89-1
LOCATION
ARCHITECT
DATE 1989
TAKE OFF BY EBW
QUANTITIES BY EBW
PFICES BY RSM
EXTENSIONS BY
CHECKED BY

DESCRIPTION	SOURCE/DIMENSIONS	QUANTITY	UNIT	MATERIAL		LABOR		EQUIPMENT		SUBCCNTRACT			
				UNIT COST	TOTAL	UNIT COST	TOTAL	UNIT COST	TOTAL	UNIT COST	TOTAL	UNIT COST	TOTAL
Division 4 : Masonry													
Needling	045-240-1080	4	EA.	50	200	155	620	26	104			8.89	36
Extra Floors	045-240-2000	(2x4) 8	EA.	38	304	70	560					4	32
@ Entrance													
Toothing	045-290-0520	80	VLF			4.49	359					.267	21
@ Ext. Openings													
Stone Lintel &	044-604-2100	5	CF	51	255	41	205	15.40	77			2	10
Sill @ Elevator													
Brick @ Door	042-108-0300	80	VLF			16.50	1320					.842	67
Jambs (No Material - Use exist'g)													
[illegible] +	9% Labor					9%	119					9%	6
Brick Wall @	042-184-1050	42	SF			10.70	449					.533	22
Ext. Elevator (No Material - Use Exist'g)													
[illegible] +	9% Labor					9%	40					9%	2
Page Subtotals					759		3672		181				196

Figure 26.31

historic preservationist be consulted regarding masonry cleaning and repair in older buildings. It is very easy to damage existing masonry if the wrong methods and materials are used. In this example the work is very clearly specified. High pressure water with no chemical cleaners is required (see Figure 26.32). The cleaning is to be performed during normal working hours. Pedestrian protection and tarpaulins are included (although not directly specified) to prevent the high volumes of water and mist from causing complaints or damage. Also, the type of mortar should be defined in detail. Modern mortars have different strengths and expansion/contraction properties from former types and can cause damage to old brick.

Sandblasting is often rejected as a method of cleaning exterior masonry because it destroys the weatherproof integrity of the masonry surface. For interior surfaces, however, it is widely used. In the sample project, the perimeter walls and wood ceiling of the new retail spaces, the entire second floor ceiling, and the perimeter walls of the third floor tenant space are to be sandblasted. A subcontractor will perform the work, but since the estimator has not yet received quotations, the costs must be calculated. Because the building is occupied by the existing retail tenant and is located in an active urban area, workers must take precautionary measures and perform the job after business hours. The appropriate factors are added to the costs. The factors (43%) are applied to labor costs only (see Figures 26.15 and 26.33). Since the work is to be subcontracted, the increase must be figured to include overhead and profit for the subcontractor. Note that a factor of only 13% is added to the man-hours. This is because the overtime factor (30%) does not entail more time expended, only greater expense.

Division 5: Metals

The specified wide flange steel column at the elevator pit is 4′ high with special bearing plates. This is a unique item and requires a quotation from a steel fabricator for the material cost. The quotation would most likely come from the subcontractor who will be installing the new stairs. Since this is a single, odd item, the installation must be visualized in order to estimate it properly. Figures 26.34 and 26.35 are the estimate sheets for Division 5.

Figure 26.36 lists costs for structural steel. Note on line 051-260-9000 that the minimum labor and equipment charges are $635 and $505, respectively. If the estimator does not use common sense, the total installation costs for this single, small column could be $1140. When using *Means Repair and Remodeling Cost Data* or any other data source, the estimator must be aware of the crew and equipment necessary for installation. The crew listed for the minimum labor and equipment charges in Figure 26.36 is E-2, shown in Figure 26.37. The minimum charge is for a half day. Obviously this kind of crew and equipment is not required for the column installation in the sample estimate. The specifications call for expansion bolts into the concrete footing and lag bolts into the wood beam above. These are easily priced, as shown in Figure 26.34. Note that drilling the holes is a separate item. The crew used for these prices is one carpenter. There is a minimum charge of $43 for the expansion anchors, substantially less than

Means® Forms
COST ANALYSIS

Division 4

SHEET NO. 2 of 3

PROJECT Commercial Renovation | CLASSIFICATION | ESTIMATE NO. 89-1

LOCATION | ARCHITECT | DATE 1989

TAKE OFF BY EBW | QUANTITIES BY EBW | PRICES BY RSM | EXTENSIONS BY | CHECKED BY

DESCRIPTION	SOURCE/DIMENSIONS	QUANTITY	UNIT	MATERIAL		LABOR		EQUIPMENT		SUBCONTRACT			
				UNIT COST	TOTAL	UNIT COST	TOTAL	UNIT COST	TOTAL	UNIT COST	TOTAL	UNIT COST	TOTAL
Division 4: Masonry													
High Pressure H_2O													
Cleaning Ext.	045-102-0440	11,900	SF			.69	8211	.12	1428			.04	476
Pedestrian Protection	045-102-4200	10%	Job			10%	821	10%	143			10%	48
Tarpaulins	015-602-0300	1000	SF	.06	60								
move 6 Times	2 Laborers	6	HR			34	204					2	12
Cut & Repoint @ Exterior Walls	045-270-0320	6000	SF	.19	1140	2.30	13800					.104	624
Cut & Repoint @ Stone Foundation	045-270-0700	370	LF	.27	100	1.26	466					.057	21
Caulk @ Sills & Lintels	045-210-1100	780	LF	.45	351	1.41	1100					.064	50
Cut & Repoint @ Chimney	045-270-0320	140	SF	.19	27	2.30	322					.104	15
Page Subtotals					1678		24924		1571				1246

Figure 26.32

Means® Forms
COST ANALYSIS

Division 4

SHEET NO. 3 of 3
ESTIMATE NO. 89-1
DATE 1989

PROJECT Commercial Renovation — CLASSIFICATION

LOCATION — ARCHITECT

TAKE OFF BY EBN — QUANTITIES BY EBN — PRICES BY RSM — EXTENSIONS BY — CHECKED BY

DESCRIPTION	SOURCE/DIMENSIONS	QUANTITY	UNIT	MATERIAL UNIT COST	MATERIAL TOTAL	LABOR UNIT COST	LABOR TOTAL	EQUIPMENT UNIT COST	EQUIPMENT TOTAL	SUBCONTRACT UNIT COST	SUBCONTRACT TOTAL	UNIT COST	TOTAL
Division 4: Masonry													
Concrete Block 8"	042-232-4200	288	SF	1.20	346	2.14	616					.107	31
Lintels @ 3°x6⁸ Doors	051-232-2100	8	EA.	10.70	86	3.93	31					.178	1
Concrete Blocks 12" (Elev. Shafts)	042-232-4300	2696	SF	1.84	4961	2.77	7468					.141	380
Lintels @ Elev. Doors	051-232-2600	12	EA.	13.60	163	4.42	53					.2	2
Sandblast Int. Brick	045-102-1420	2100	SF	TO BE SUBCONTRACTED						.94	1974	.023	48
2 5 6 (+ − +)	43% Labor	(Add to Sub's Total)								.27	567	13%	6
Sandblast Int. Wood	045-102-1420	11,800	SF							.94	11092	.023	271
Corn Chips	045-102-1820	11,800	SF							.33	3894		
2 5 6 (+ − +)	43% Labor	(Add to Sub's Total)								.27	3186	13%	35
Page Subtotals					5556		8168		—		20713		774
Sheet 1					759		3672		181				196
Sheet 2 (Restoration)					1678		24924		1571				1246
Sheet 3					5556		8168				20713		774
Division 4 Totals					7993		36764		1752		20713		2216

Figure 26.33

Means® Forms
COST ANALYSIS

Division 5

SHEET NO. 1 of 2

PROJECT Commercial Renovation | CLASSIFICATION | ESTIMATE NO. 89-1

LOCATION | ARCHITECT | DATE 1989

TAKE OFF BY EBW | QUANTITIES BY EBW | PRICES BY RSM | EXTENSIONS BY | CHECKED BY

DESCRIPTION	SOURCE/DIMENSIONS	QUANTITY	UNIT	MATERIAL		LABOR		EQUIPMENT		SUBCONTRACT			
				UNIT COST	TOTAL	UNIT COST	TOTAL	UNIT COST	TOTAL	UNIT COST	TOTAL	UNIT COST	TOTAL
Division 5: Metals													
Steel Col.	Quote (Mat. Only)	1	EA.	250	250								
Expansion Anchors	050-520-1100	4	EA.	2.60	10	Min.	43					.123	1
Lag Bolts	050-530-0300	4	EA.	1.02	4	2.01	8					.094	1
Drill-Concrete	050-515-0500	4	EA.	.06	1	7.45	30					.348	1
Drill-Timber	050-515-0800	12	EA.	.09	1	9.50	114					.444	5
Ladder to Roof	055-158-0010	12	VLF		SUBCONTRACT					74	888	.64	8
Wall Railings:													
Exist'g stair	055-203-0945	78	LF							12.50	975	.15	12
New Stair	055-203-0945	52	LF							12.50	650	.15	8
New Metal Stair 4'-0" Wide	055-104-0300	36	Ris.							110	3960	1.07	39
Landings	055-104-1500	72	SF							35	2520	.20	14
Subcontract Total											8993		
Page Subtotals					266		195				8993		89

Figure 26.34

Means® Forms
COST ANALYSIS

Division 5

SHEET NO. 2 of 2

PROJECT Commercial Rennovation — CLASSIFICATION — ESTIMATE NO. 89-1

LOCATION — ARCHITECT — DATE 1989

TAKE OFF BY EBN — QUANTITIES BY EBN — PRICES BY RSM — EXTENSIONS BY — CHECKED BY

DESCRIPTION	SOURCE/DIMENSIONS	QUANTITY	UNIT	MATERIAL UNIT COST	MATERIAL TOTAL	LABOR UNIT COST	LABOR TOTAL	EQUIPMENT UNIT COST	EQUIPMENT TOTAL	SUBCONTRACT UNIT COST	SUBCONTRACT TOTAL	UNIT COST	TOTAL
Division 5: Metals													
Beam Hangers	060-512-1800	76	EA.	1.13	86	110	84					.052	4
@ Roof Openings	(19 Openings-4 EA.)												
Stud Driver:													
Chargers	050-550-0300	3	C	12.15	36								
Studs	050-550-0600	3	C	34	102								
Page Subtotals					224		84						4
Sheet 1					266		195				8993		89
Sheet 2					224		84						4
Division 5 Totals					490		279				8993		93

Figure 26.35

051 | Structural Metal Framing

051 200 | Structural Steel

		051 200 Structural Steel	CREW	DAILY OUTPUT	MAN-HOURS	UNIT	BARE COSTS MAT.	LABOR	EQUIP.	TOTAL	TOTAL INCL O&P	
215	2150	Hung from pre-set inserts	E-4	12	2.670	Ea.	67	63	5.20	135.20	190	215
	2400	Toilet partition support		36	.889	L.F.	29	21	1.73	51.73	70	
	2500	X-ray travel gantry support		12	2.670	"	85	63	5.20	153.20	210	
220	0010	**COLUMNS** Aluminum, extruded, stock units, 6" diameter	E-4	240	.133	L.F.	8.15	3.17	.26	11.58	14.75	220
	0100	8" diameter	"	170	.188		11	4.47	.37	15.84	20	
	0500	For square columns, add to column prices above					50%					
	0800	Steel, concrete filled, extra strong pipe, 3-½" diameter	E-2	660	.085		9.20	1.93	1.53	12.66	15.05	
	0830	4" diameter		780	.072		11	1.63	1.30	13.93	16.30	
	0890	5" diameter		1,020	.055		15	1.25	.99	17.24	19.70	
	1500	Steel pipe, extra strong, no concrete, 3" to 5" O.D.		12,960	.004	Lb.	.78	.10	.08	.96	1.11	
	3300	Square structural tubing, 4" to 6" square, light section		11,270	.005	"	.40	.11	.09	.60	.73	
	9000	Minimum labor/equipment charge	1 Sswk	1	8	Job		185		185	325	
230	0010	**LIGHTWEIGHT FRAMING**										230
	0200	For steel studs see division 092-612										
	0400	Angle framing, 4" and larger	E-4	3,000	.011	Lb.	.69	.25	.02	.96	1.22	
	0450	Less than 4" angles	"	1,800	.018		.75	.42	.03	1.20	1.60	
	1400	Tie rod, not upset, 1-½" to 4" diameter, with turnbuckle	2 Sswk	800	.020		.70	.47		1.17	1.58	
	1420	No turnbuckle		700	.023		.64	.53		1.17	1.63	
	9000	Minimum labor/equipment charge		2	8	Job		185		185	325	
232	0010	**LINTELS** Plain steel angles, under 500 lb.	1 Bric	500	.016	Lb.	.47	.35		.82	1.08	232
	0100	500 to 1000 lb.		600	.013	"	.40	.29		.69	.91	
	2000	Steel angles, 3-½" x 3", ¼" thick, 2'-6" long		50	.160	Ea.	6	3.54		9.54	12.20	
	2100	4'-6" long		45	.178		10.70	3.93		14.63	18	
	2600	4" x 3-½", ¼" thick, 5'-0" long		40	.200		13.60	4.42		18.02	22	
	2700	9'-0" long		35	.229		24	5.05		29.05	34	
	3500	For precast concrete lintels, see div. 034-802										
	9000	Minimum labor/equipment charge	1 Bric	4	2	Job		44		44	70	
240	0010	**STEEL CUTTING** Hand burning, including preparation,										240
	0020	torch cutting, and grinding, no staging										
	0100	Steel to ½" thick	A-1	70	.114	L.F.	.38	1.93	.73	3.04	4.32	
	0150	¾" thick		50	.160		.53	2.70	1.03	4.26	6.05	
	0200	1" thick		45	.178		.59	3	1.14	4.73	6.70	
	9000	Minimum labor/equipment charge		2	4	Job		67	26	93	135	
255	0010	**STRUCTURAL STEEL PROJECTS** Bolted, unless mentioned otherwise										255
	1300	Industrial bldgs., 1 story, beams & girders, steel bearing	E-5	12.90	6.200	Ton	965	145	88	1,198	1,400	
	1400	Masonry bearing		10	8		965	185	115	1,265	1,500	
	1600	1 story with roof trusses, steel bearing		10.60	7.550		1,135	175	105	1,415	1,675	
	1700	Masonry bearing		8.30	9.640		1,135	225	135	1,495	1,775	
260	0010	**STRUCTURAL STEEL** Bolted, incl. fabrication, not incl. trucking										260
	0050	Beams, 6 WF 9	E-2	720	.078	L.F.	8.05	1.77	1.40	11.22	13.40	
	0100	8 WF 10		720	.078		8.35	1.77	1.40	11.52	13.70	
	0150	10 WF 15		720	.078		12.75	1.77	1.40	15.92	18.55	
	0200	Columns, 6 WF 15.5		540	.104		13.20	2.36	1.87	17.43	21	
	0250	8 WF 31		540	.104		19.05	2.36	1.87	23.28	27	
	0500	Girders, 12 WF 22		900	.062		14.60	1.41	1.12	17.13	19.70	
	0550	14 WF 26		900	.062		16.80	1.41	1.12	19.33	22	
	0600	16 WF 31		900	.062		17.65	1.41	1.12	20.18	23	
	0700	Joists (bar joists, H series), span to 30'	E-7	30,000	.003	Lb.	.34	.06	.04	.44	.52	
	0750	Span to 50'	"	20,000	.004	"	.31	.09	.06	.46	.56	
	9000	Minimum labor/equipment charge	E-2	2	28	Job		635	505	1,140	1,625	

(76)

Figure 26.36

CREWS

Crew No.	Bare Costs		Incl. Subs O & P		Cost Per Man-hour	
Crew E-2	**Hr.**	**Daily**	**Hr.**	**Daily**	**Bare Costs**	**Incl. O&P**
1 Struc. Steel Foreman	$25.25	$202.00	$43.80	$350.40	$22.72	$38.39
4 Struc. Steel Workers	23.25	744.00	40.35	1291.20		
1 Equip. Oper. (crane)	22.40	179.20	34.85	278.80		
1 Equip. Oper. Oiler	18.45	147.60	28.70	229.60		
1 Crane, 90 Ton		1011.00		1112.10	18.05	19.85
56 M.H., Daily Totals		$2283.80		$3262.10	$40.77	$58.24
Crew E-3	**Hr.**	**Daily**	**Hr.**	**Daily**	**Bare Costs**	**Incl. O&P**
1 Struc. Steel Foreman	$25.25	$202.00	$43.80	$350.40	$23.91	$41.50
1 Struc. Steel Worker	23.25	186.00	40.35	322.80		
1 Welder	23.25	186.00	40.35	322.80		
1 Gas Welding Machine		62.20		68.40		
1 Torch, Gas & Air		57.20		62.90	4.97	5.47
24 M.H., Daily Totals		$693.40		$1127.30	$28.88	$46.97
Crew E-4	**Hr.**	**Daily**	**Hr.**	**Daily**	**Bare Costs**	**Incl. O&P**
1 Struc. Steel Foreman	$25.25	$202.00	$43.80	$350.40	$23.75	$41.21
3 Struc. Steel Workers	23.25	558.00	40.35	968.40		
1 Gas Welding Machine		62.20		68.40	1.94	2.13
32 M.H., Daily Totals		$822.20		$1387.20	$25.69	$43.34
Crew E-5	**Hr.**	**Daily**	**Hr.**	**Daily**	**Bare Costs**	**Incl. O&P**
2 Struc. Steel Foremen	$25.25	$404.00	$43.80	$700.80	$23.08	$39.32
5 Struc. Steel Workers	23.25	930.00	40.35	1614.00		
1 Equip. Oper. (crane)	22.40	179.20	34.85	278.80		
1 Welder	23.25	186.00	40.35	322.80		
1 Equip. Oper. Oiler	18.45	147.60	28.70	229.60		
1 Crane, 90 Ton		1011.00		1112.10		
1 Gas Welding Machine		62.20		68.40		
1 Torch, Gas & Air		57.20		62.90	14.13	15.54
80 M.H., Daily Totals		$2977.20		$4389.40	$37.21	$54.86
Crew E-6	**Hr.**	**Daily**	**Hr.**	**Daily**	**Bare Costs**	**Incl. O&P**
3 Struc. Steel Foreman	$25.25	$606.00	$43.80	$1051.20	$23.10	$39.40
9 Struc. Steel Workers	23.25	1674.00	40.35	2905.20		
1 Equip. Oper. (crane)	22.40	179.20	34.85	278.80		
1 Welder	23.25	186.00	40.35	322.80		
1 Equip. Oper. Oiler	18.45	147.60	28.70	229.60		
1 Equip. Oper. (light)	20.60	164.80	32.05	256.40		
1 Crane, 90 Ton		1011.00		1112.10		
1 Gas Welding Machine		62.20		68.40		
1 Torch, Gas & air		57.20		62.90		
1 Air Compr., 160 C.F.M.		77.60		85.35		
2 Impact Wrenches		52.80		58.10	9.85	10.83
128 M.H., Daily Totals		$4218.40		$6430.85	$32.95	$50.23
Crew E-7	**Hr.**	**Daily**	**Hr.**	**Daily**	**Bare Costs**	**Incl. O&P**
1 Struc. Steel Foreman	$25.25	$202.00	$43.80	$350.40	$23.08	$39.32
4 Struc. Steel Workers	23.25	744.00	40.35	1291.20		
1 Equip. Oper. (crane)	22.40	179.20	34.85	278.80		
1 Equip. Oper. Oiler	18.45	147.60	28.70	229.60		
1 Welder Foreman	25.25	202.00	43.80	350.40		
2 Welders	23.25	372.00	40.35	645.60		
1 Crane, 90 Ton		1011.00		1112.10		
2 Gas Welding Machines		124.40		136.85	14.19	15.61
80 M.H., Daily Totals		$2982.20		$4394.95	$37.27	$54.93

Crew No.	Bare Costs		Incl. Subs O & P		Cost Per Man-hour	
Crew E-8	**Hr.**	**Daily**	**Hr.**	**Daily**	**Bare Costs**	**Incl. O&P**
1 Struc. Steel Foreman	$25.25	$202.00	$43.80	$350.40	$22.91	$38.92
4 Struc. Steel Workers	23.25	744.00	40.35	1291.20		
1 Welder Foreman	25.25	202.00	43.80	350.40		
4 Welders	23.25	744.00	40.35	1291.20		
1 Equip. Oper. (crane)	22.40	179.20	34.85	278.80		
1 Equip. Oper. Oiler	18.45	147.60	28.70	229.60		
1 Equip. Oper. (light)	20.60	164.80	32.05	256.40		
1 Crane, 90 Ton		1011.00		1112.10		
4 Gas Welding Machines		248.80		273.70	12.11	13.32
104 M.H. Daily Totals		$3643.40		$5433.80	$35.02	$52.24
Crew E-9	**Hr.**	**Daily**	**Hr.**	**Daily**	**Bare Costs**	**Incl. O&P**
2 Struc. Steel Foremen	$25.25	$404.00	$43.80	$700.80	$23.10	$39.40
5 Struc. Steel Workers	23.25	930.00	40.35	1614.00		
1 Welder Foreman	25.25	202.00	43.80	350.40		
5 Welders	23.25	930.00	40.35	1614.00		
1 Equip. Oper. (crane)	22.40	179.20	34.85	278.80		
1 Equip. Oper. Oiler	18.45	147.60	28.70	229.60		
1 Equip. Oper. (light)	20.60	164.80	32.05	256.40		
1 Crane, 90 Ton		1011.00		1112.10		
5 Gas Welding Machines		311.00		342.10		
1 Torch, Gas & Air		57.20		62.90	10.77	11.85
128 M.H., Daily Totals		$4336.80		$6561.10	$33.87	$51.25
Crew E-10	**Hr.**	**Daily**	**Hr.**	**Daily**	**Bare Costs**	**Incl. O&P**
1 Welder Foreman	$25.25	$202.00	$43.80	$350.40	$24.25	$42.07
1 Welder	23.25	186.00	40.35	322.80		
4 Gas Welding Machines		248.80		273.70		
1 Truck, 3 Ton		111.60		122.75	22.52	24.77
16 M.H., Daily Totals		$748.40		$1069.65	$46.77	$66.84
Crew E-11	**Hr.**	**Daily**	**Hr.**	**Daily**	**Bare Costs**	**Incl. O&P**
2 Painters, Struc. Steel	$20.85	$333.60	$37.85	$605.60	$19.78	$33.70
1 Building Laborer	16.85	134.80	27.05	216.40		
1 Equip. Oper. (light)	20.60	164.80	32.05	256.40		
1 Air Compressor 250 C.F.M.		80.80		88.90		
1 Sand Blaster		43.20		47.50		
1 Sand Blasting Accessories		9.40		10.35	4.16	4.58
32 M.H., Daily Totals		$766.60		$1225.15	$23.94	$38.28
Crew E-12	**Hr.**	**Daily**	**Hr.**	**Daily**	**Bare Costs**	**Incl. O&P**
1 Welder Foreman	$25.25	$202.00	$43.80	$350.40	$22.92	$37.92
1 Equip. Oper. (light)	20.60	164.80	32.05	256.40		
1 Gas Welding Machine		62.20		68.40	3.88	4.27
16 M.H., Daily Totals		$429.00		$675.20	$26.80	$42.19
Crew E-13	**Hr.**	**Daily**	**Hr.**	**Daily**	**Bare Costs**	**Incl. O&P**
1 Welder Foreman	$25.25	$202.00	$43.80	$350.40	$23.70	$39.88
.5 Equip. Oper. (light)	20.60	82.40	32.05	128.20		
1 Gas Welding Machine		62.20		68.40	5.18	5.70
12 M.H., Daily Totals		$346.60		$547.00	$28.88	$45.58
Crew E-14	**Hr.**	**Daily**	**Hr.**	**Daily**	**Bare Costs**	**Incl. O&P**
1 Welder Foreman	$25.25	$202.00	$43.80	$350.40	$25.25	$43.80
1 Gas Welding Machine		62.20		68.40	7.77	8.55
8 M.H., Daily Totals		$264.20		$418.80	$33.02	$52.35

Figure 26.37

$1140. A little extra time and sound judgment can help to prevent expensive mistakes. The temporary shoring and bracing is included in Division 2, Figure 26.19.

It should be noted that there is relatively little structural metal work in this sample project or in any efficiently designed commercial renovation. This keeps the costs down and makes renovation a favorable alternative, in terms of cost, to new construction. Whether the work is planned by an architect or by the contractor, the designer should try to work with and around the existing structural elements.

The new stairway, ladder, and railings are to be subcontracted to a steel fabricator. The estimator has received quotations but feels that they are not quite right, either too high or too low, based upon past experience. The costs are included in the estimate as a crosscheck in Figure 26.34. These costs are subtotaled so that the quotation can easily be substituted in the Estimate Summary.

The specifications (and building codes) require that the contractor place fire extinguishers at every level when welding. Since the extinguishers will be in place throughout the project for all trades, they are included in Division 1. Figures 26.38a and 26.38b show the Project Overhead Summary as items are added throughout the estimate. The purchase of a powder-activated stud driver is also listed as project overhead because it will be used for future jobs. The charges and studs to be used only for this job are included in Division 6 of the estimate.

Division 6: Wood and Plastics

The use of wood in commercial renovation is often dependent upon two factors: the type of existing construction and the local building code requirements for fire-resistance. When the use of wood (whether structural, rough, or finish) is permitted in buildings of fire-resistant construction, fire-retardant treatment is usually specified. Because the interior structure of the sample project is wood, such treatment is not required. Light gauge metal studs are becoming more common in commercial renovation because they are lightweight and easy to install. It is faster and less costly for a carpenter, with a pair of tin snips and a screw gun, to cut and install metal studs to conform to the variety of existing conditions in older buildings than to do the same with wood. Also, with the wide use of materials such as metal door frames and vinyl base, wood is becoming less prevalent as a primary material in commercial renovation as well as new construction. On some projects, it is not unusual for a carpenter never to work with wood at all.

In this example, metal studs are specified for all interior, non-load bearing partitions. Wood is specified only to match existing work, and for blocking and framing for fixtures and equipment. Figures 26.39, 26.40 and 26.41 are the estimate sheets for Division 6.

To frame the floor openings at the freight elevator, contractors use planking that has been removed from other areas, so no material costs are included. The planks must be cut to match existing conditions. This work will occur in the late stages of the renovation, so a factor for protection of existing work must be applied.

The roof openings for hatches and skylights are also framed using existing materials. In Division 2, the costs for cutting the openings

Means® Forms

PROJECT OVERHEAD SUMMARY

Division 1

SHEET NO. 1 of 2

PROJECT Commercial Renovation

ESTIMATE NO. 89-1

LOCATION

ARCHITECT

DATE 1989

QUANTITIES BY: EBN

PRICES BY: RSM

EXTENSIONS BY:

CHECKED BY:

DESCRIPTION	MAN HOURS	QUANTITY	UNIT	MATERIAL UNIT	MATERIAL TOTAL	LABOR UNIT	LABOR TOTAL	EQUIP. FEES, RENTAL UNIT	EQUIP. FEES, RENTAL TOTAL
Job Organization: Superintendent									
Project Manager									
Timekeeper & Material Clerk									
Clerical									
Safety, Watchman & First Aid									
010-036-0260			WK			975			
Travel Expense: Superintendent									
Project Manager									
Engineering: Layout									
Inspection/Quantities									
Drawings									
CPM Schedule									
Testing: Soil									
Materials									
Structural									
Equipment: Cranes									
Concrete Pump, Conveyor, Etc.									
Elevators, Hoists									
Freight & Hauling									
Loading, Unloading, Erecting, Etc.									
Maintenance									
Pumping									
Scaffolding 015-254-0090	170	119	CSF			31	3689		
~~Small Power Equipment/Tools~~ Rental - 381 Frames @ $3.65/MO. - circle	14		Mo.						
Stud Driver 050-550-0010		1	EA	230	230				
Field Offices: Job Office									
Architect/Owner's Office									
Temporary Telephones			MO.					100	
Utilities (Power)		28	MSF					93	2604
Temporary Toilets									
Storage Areas & Sheds									
Temporary Utilities: Heat									
Light ~~& Power~~ 015-104-0350	66	280	CSF	2.06	577	5.70	1596		
Water									
PAGE TOTALS									

Page 1 of 2

Figure 26.38a

Means® Forms

DIVISION 1 SHEET 2 OF 2 — DESCRIPTION	QUANTITY	UNIT	MATERIALS UNIT	MATERIALS TOTAL	LABOR UNIT	LABOR TOTAL	EQUIP. FEES & RENTAL UNIT	EQUIP. FEES & RENTAL TOTAL
Total Brought Forward								
Winter Protection: Temp. Heat/Protection								
Snow Plowing								
Thawing Materials								
Temporary Roads								
Rent 6 Parking Spaces		MO					300	
Signs & Barricades: Site Sign	1	EA		250				
Temporary Fences								
Temporary Stairs, Ladders & Floors								
RAILS @ STAIRS & ELEV. (22)	270	LF	1.35	365	1.71	462		
Photographs 015-302-1000								
Clean Up								
Dumpster 020-620-0800		WK					265	
Final Clean Up 017-104-0100 (44)	28	MSF	1.60	45	26	728	2.37	66
Punch List								
Permits: Building								
Misc. Street Use	1	EA.					50	50
Sewer Connection	1	EA.					300	300
Insurance: Builders Risk								
Owner's Protective Liability								
Umbrella								
Unemployment Ins. & Social Security								
Taxes								
City Sales Tax								
State Sales Tax								
Bonds								
Performance								
Material & Equipment								
Main Office Expense								
Special Items								
Fire Exting. 154-125-2080	4	EA.	42.50	170				
Legal: Contract Review							100	100
TOTALS:								

Page 2 of 2

Figure 26.38b

Means® Forms
COST ANALYSIS

Division 6

SHEET NO. 1 of 3

PROJECT Commercial Renovation | CLASSIFICATION | ESTIMATE NO. 89-1

LOCATION | ARCHITECT | DATE 1989

TAKE OFF BY EBW | QUANTITIES BY EBW | PRICES BY RSM | EXTENSIONS BY | CHECKED BY

DESCRIPTION	SOURCE/DIMENSIONS	QUANTITY	UNIT	MATERIAL		LABOR		EQUIPMENT		SUBCONTRACT			
				UNIT COST	TOTAL	UNIT COST	TOTAL	UNIT COST	TOTAL	UNIT COST	TOTAL	UNIT COST	TOTAL
Division 6: Wood & Plastics													
Rough Carpentry													
Blocking - 2X8	061-102-2660	.125	MBF	425	53	635	79	28	4			29.63	4
Blocking · 2X4 @ Door Frames	061-102-2620	.314	MBF	390	122	1,000	314	44	14			47.06	15
Stud Wall @ Elec. Meters	061-138-0380	10	LF	3.62	36	4.28	43	.19	2			.20	2
Plywood @ Elec. & Telephone	061-154-0700	196	SF	.44	86	Min.	86	.01	2			.015	3
Infill Elev.	061-304-1100	639	BF			.33	211	.01	6			.015	10
Openings (Use Existing Material)													
[1] [5]	10% Labor					10%	21					10%	1
Frame Roof Openings	Crew F-2	1.5	DAY			342.40	514	15	23			16	24
[1]	9% Labor					9%	46					9%	2
Underlayment	051-168-0010	390	SF	.41	160	.23	90	.01	4			.011	4
	061-168-0010	2080	SF	.41	853	.23	478	.01	21			.011	23
Page Subtotals					1310		1882		76				88

Figure 26.39

Means® Forms
COST ANALYSIS

Division 6

SHEET NO. 2 of 3

PROJECT Commercial Renovation | CLASSIFICATION | ESTIMATE NO. 89-1

LOCATION | ARCHITECT | DATE 1989

TAKE OFF BY EBW | QUANTITIES BY EBW | PRICES BY RSM | EXTENSIONS BY | CHECKED BY

DESCRIPTION	SOURCE/DIMENSIONS	QUANTITY	UNIT	MATERIAL		LABOR		EQUIPMENT		SUBCONTRACT			
				UNIT COST	TOTAL	UNIT COST	TOTAL	UNIT COST	TOTAL	UNIT COST	TOTAL	UNIT COST	TOTAL
Division 6: Wood & Plastics Rough carpentry (cont'd)													
Framedeck @ AC Units (2x8)	061-120-7060	220	LF	.57	125	.36	79	.02	4			.017	4
Complexity	061-120-7790					50%	40					50%	2
[1]+ [4]-	6% Mat. 10% Labor			6%	8	10%	12					10%	1
Plywood Deck	061-154-0300	150	SF	.52	78	Min	86	.01	2			.013	2
[4]+	6% Mat. 7% Labor			6%	5	7%	6					7%	1
Rough Hardware	061-140-0200	1526		1.5%	23								
Material Subtotal → (to Quantity 1526)													
Page Subtotals					239		223		6				10

Figure 26.40

Means® Forms
COST ANALYSIS

Division 6

SHEET NO. 3 of 3

PROJECT Commercial Renovation — CLASSIFICATION — ESTIMATE NO. 89-1

LOCATION — ARCHITECT — DATE 1989

TAKE OFF BY EBN — QUANTITIES BY EBN — PRICES BY RSM — EXTENSIONS BY — CHECKED BY

DESCRIPTION	SOURCE/DIMENSIONS	QUANTITY	UNIT	MATERIAL		LABOR		EQUIPMENT		SUBCONTRACT			
				UNIT COST	TOTAL	UNIT COST	TOTAL	UNIT COST	TOTAL	UNIT COST	TOTAL	UNIT COST	TOTAL
Division 6: Wood & Plastics													
Finish Carpentry													
Shelves @	062-304-0600	30	LF	.98	29	2.28	68	.10	3			.107	3
Jan. Closets													
Vanities: Base	064-140-8100	4	EA.	185	740	43	172					2	8
Top	062-408-1000	12	LF	25	300	5.70	68					.267	4
Cutouts	062-408-1900	6	EA.	2.50	15	5.35	32					.25	2
Window Sills													
Material	Quotation (Incl. Milling)	116	LF	1.80	209								
Installation	062-228-5100	116	LF			1.14	132					.053	6
[1]	3% Labor					3%	5					3%	1
Grounds (@ sills, headers, & jambs)	061-132-0100	606	LF	.08	48	.60	364					.028	17
Page Subtotals					1341		841		3				41
Rough: Sheet 1					1310		1882		76				88
Sheet 2					239		223		6				10
Subtotal					1549		2105		82				98
Finish: Sheet 3					1341		841		3				41
Division 6 Totals					2890		2946		85				139

Figure 26.41

were calculated by determining the crew size and time required. The same method and crew are used to price the framing because the two will be done together.

The exterior walls of older masonry buildings are usually very thick. Oversized window sills are required. The plans specify solid oak for the sills, as shown in Figure 26.42. A material price is obtained from a local supplier. Note that 1 x 10 (nominal) boards are needed, cut to 8″ wide and surfaced on three sides. Installation is similar to that of normal window stools, so line item 062-228-5100, in Figure 26.43, is used for labor only. A factor is added for cutting to conform to the variations in wall thickness and window installation.

In Figure 26.41, the quantities for the sills and grounds are 116 L.F. and 606 L.F., respectively. It is important to record how these, and all quantities, are derived for cross-checking and future reference. A quantity sheet, as shown in Figure 26.44, is useful to keep track of this information. Sketches are helpful to show how dimensions are obtained.

The first floor windows receive no sills or grounds because the windows at the existing retail space are already finished and trimmed, and the exterior walls in the new retail spaces will be sandblasted. The existing stone sills will remain exposed.

All windows on the second floor, except in the stairways, will receive grounds and sills. On the third floor, the walls in the tenant space will be sandblasted, so only the remaining eleven windows are included. Walls in the stairways will also be sandblasted.

In order to price rough hardware for the project, the material costs for rough hardware are shown in Figure 26.45 as a minimum and a maximum. The maximum percentage is used because material costs are slightly low due to the use of existing materials for some of the work.

Division 7: Thermal and Moisture Protection

Four types of insulation are specified in the project. The urethane roof insulation is included in the roofing subcontract. The polystyrene foundation insulation is installed on the interior of the foundation walls at the crawl space. The takeoff for these types is straightforward. The exterior wall fiberglass (for thermal protection) and the interior wall fiberglass (for reduced sound transmission) are an integral part of the drywall partition systems. As the quantities are calculated, the estimator should note and record the wall types and dimensions. This information will save time when estimating Division 9: Finishes. Figures 26.46 to 26.48 are the estimate sheets for Division 7.

The polyethylene vapor barrier in the basement is an item that might easily be overlooked. Many such items in a complex package of construction documents may be shown only once on the plans and not even mentioned directly in the specifications. The estimator must be careful and thorough in order to avoid omissions.

The roofing and roof insulation are estimated for the complete roof surface with no deductions for the hatches or skylights. This is because these accessories will require extra labor for cutting and flashing. When bids are submitted, the estimator must be sure that

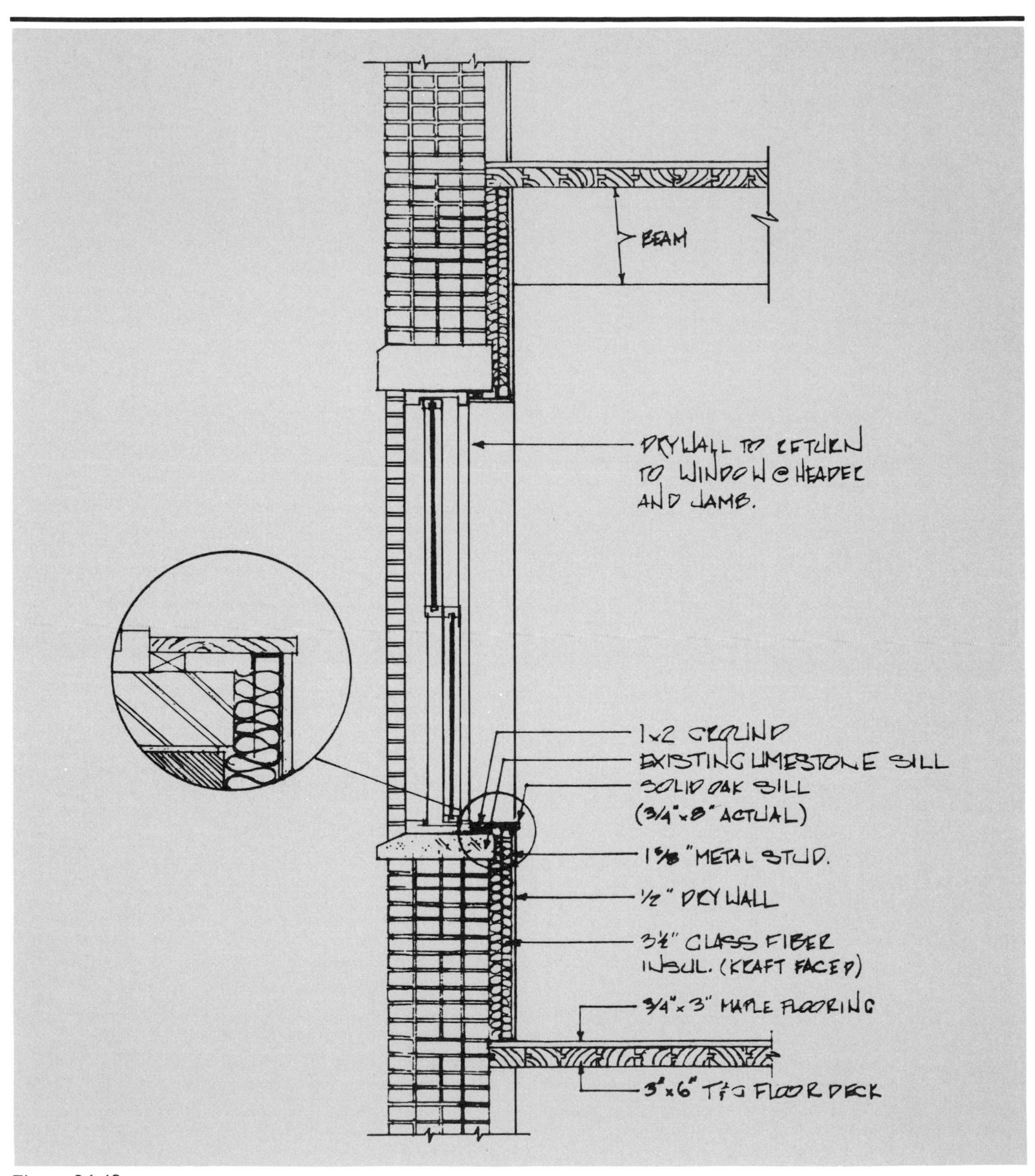
BEAM
DRYWALL TO RETURN
TO WINDOW @ HEADER
AND JAMB.
1x2 GROUND
EXISTING LIMESTONE SILL
SOLID OAK SILL
(3/4"x8" ACTUAL)
1⅝" METAL STUD.
½" DRYWALL
3½" GLASS FIBER
INSUL. (KRAFT FACED)
3/4"x3" MAPLE FLOORING
3"x6" T&G FLOOR DECK

Figure 26.42

062 | Finish Carpentry

		062 200 | Millwork Moldings	CREW	DAILY OUTPUT	MAN-HOURS	UNIT	BARE COSTS MAT.	BARE COSTS LABOR	BARE COSTS EQUIP.	BARE COSTS TOTAL	TOTAL INCL O&P	
224	3800	Miscellaneous, custom, pine or cedar, 1" x 1"	1 Carp	270	.030	L.F.	.14	.63		.77	1.17	224
	3900	Nominal 1" x 3"		240	.033		.33	.71		1.04	1.51	
	4100	Birch or oak, custom, nominal 1" x 1"		240	.033		.20	.71		.91	1.37	
	4200	Nominal 1" x 3"		215	.037		.55	.80		1.35	1.88	
	4400	Walnut, custom, nominal 1" x 1"		215	.037		.30	.80		1.10	1.61	
	4500	Nominal 1" x 3"		200	.040		.85	.86		1.71	2.31	
	4700	Teak, custom, nominal 1" x 1"		215	.037		.75	.80		1.55	2.10	
	4800	Nominal 1" x 3"		200	.040		2	.86		2.86	3.57	
	4900	Quarter round, stock pine, ¼" x ¼"		275	.029		.10	.62		.72	1.11	
	4950	¾" x ¾"		255	.031		.25	.67		.92	1.35	
	5600	Wainscot moldings, 1-⅛" x 9/16", 2' high, minimum		76	.105	S.F.	5.30	2.25		7.55	9.45	
	5700	Maximum		65	.123	"	10.50	2.63		13.13	15.80	
	9000	Minimum labor/equipment charge		4	2	Job		43		43	69	
228	0010	**MOLDINGS, WINDOW AND DOOR**										228
	0020											
	2800	Door moldings, stock, decorative, 1-⅛" wide, plain	1 Carp	17	.471	Set	20	10.05		30.05	38	
	2900	Detailed	"	17	.471	"	45	10.05		55.05	66	
	3100	Door trim, interior, including headers,										
	3150	stops and casings, 2 sides, pine, 2-½" wide	1 Carp	5.90	1.360	Opng.	20	29		49	69	
	3170	4-½" wide		5.30	1.510	"	26	32		58	80	
	3200	Glass beads, stock pine, ¼" x 11/16"		285	.028	L.F.	.17	.60		.77	1.15	
	3250	⅜" x ½"		275	.029		.24	.62		.86	1.26	
	3270	⅜" x ⅞"		270	.030		.30	.63		.93	1.35	
	4850	Parting bead, stock pine, ⅜" x ¾"		275	.029		.19	.62		.81	1.21	
	4870	½" x ¾"		255	.031		.23	.67		.90	1.33	
	5000	Stool caps, stock pine, 11/16" x 3-½"		200	.040		.89	.86		1.75	2.35	
	5100	1-1/16" x 3-¼"		150	.053		1.98	1.14		3.12	4.01	
	5300	Threshold, oak, 3' long, inside, ⅝" x 3-⅝"		32	.250	Ea.	5.10	5.35		10.45	14.20	
	5400	Outside, 1-½" x 7-⅝"		16	.500	"	20	10.70		30.70	39	
	5900	Window trim sets, including casings, header, stops,										
	5910	stool and apron, 2-½" wide, minimum	1 Carp	13	.615	Opng.	10.50	13.15		23.65	33	
	5950	Average		10	.800		15	17.10		32.10	44	
	6000	Maximum		6	1.330		20	29		49	68	
	9000	Minimum labor/equipment charge		4	2	Job		43		43	69	

		062 300 | Shelving	CREW	DAILY OUTPUT	MAN-HOURS	UNIT	MAT.	LABOR	EQUIP.	TOTAL	TOTAL INCL O&P	
304	0010	**SHELVING** Pine, clear grade, no edge band, 1" x 8"	F-1	115	.070	L.F.	.92	1.49	.07	2.48	3.47	304
	0100	1" x 10"		110	.073		1.16	1.56	.07	2.79	3.85	
	0200	1" x 12"		105	.076		1.40	1.63	.07	3.10	4.24	
	0300											
	0400	For lumber edge band, by hand, add				L.F.	1.25			1.25	1.38	
	0420	By machine, add					.80			.80	.88	
	0600	Plywood, ¾" thick with lumber edge, 12" wide	F-1	75	.107		.98	2.28	.10	3.36	4.85	
	0700	24" wide		70	.114		1.84	2.45	.11	4.40	6.05	
	0900	Bookcase, pine, clear grade, 8" shelves, 12" O.C.		70	.114	S.F.	3.65	2.45	.11	6.21	8.05	
	1000	12" wide shelves		65	.123	"	4.30	2.63	.12	7.05	9.10	
	1200	Adjustable closet rod and shelf, 12" wide, 3' long		20	.400	Ea.	7.30	8.55	.38	16.23	22	
	1300	8' long		15	.533	"	13.25	11.40	.50	25.15	33	
	1500	Prefinished shelves with supports, stock, 8" wide		75	.107	L.F.	5.15	2.28	.10	7.53	9.45	
	1600	10" wide		70	.114	"	5.75	2.45	.11	8.31	10.35	
	1800	Custom, high quality dadoed pine shelving units, minimum				S.F.				23	28	
	1900	Maximum				"				31.50	40	
	9000	Minimum labor/equipment charge	F-1	4	2	Job		43	1.88	44.88	71	

For expanded coverage of these items see *Means Interior Cost Data 1989*

77

Figure 26.43

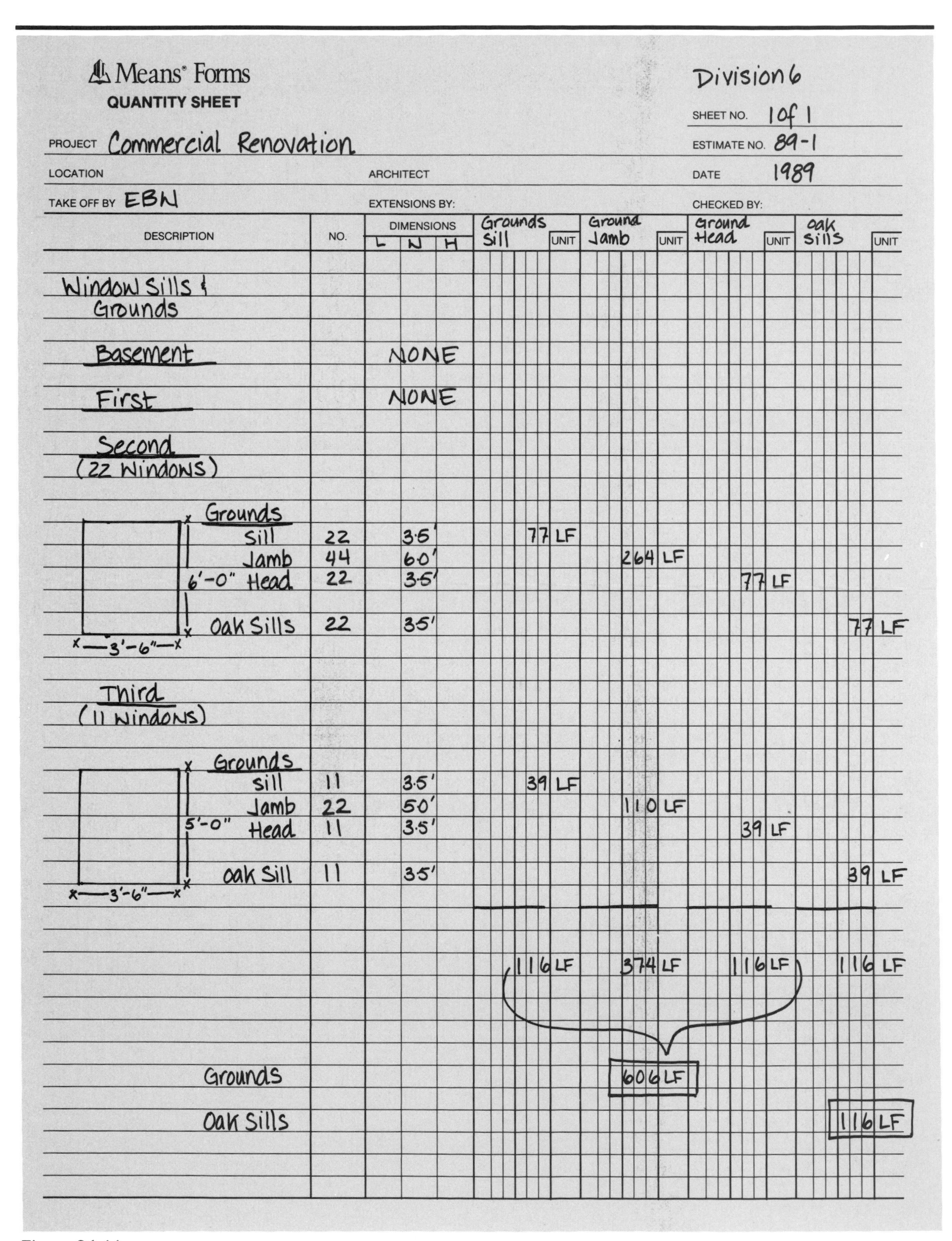

Means® Forms
QUANTITY SHEET

Division 6

SHEET NO. 1 of 1
PROJECT Commercial Renovation
ESTIMATE NO. 89-1
LOCATION
ARCHITECT
DATE 1989
TAKE OFF BY EBN
EXTENSIONS BY:
CHECKED BY:

DESCRIPTION	NO.	DIMENSIONS L	W	H	Grounds Sill	UNIT	Ground Jamb	UNIT	Ground Head	UNIT	Oak Sills	UNIT
Window Sills & Grounds												
Basement		NONE										
First		NONE										
Second (22 Windows)												
Grounds												
Sill	22	3.5'			77	LF						
Jamb	44	6.0'					264	LF				
6'-0" Head	22	3.5'							77	LF		
Oak Sills	22	3.5'									77	LF
3'-6"												
Third (11 Windows)												
Grounds												
Sill	11	3.5'			39	LF						
Jamb	22	5.0'					110	LF				
5'-0" Head	11	3.5'							39	LF		
Oak Sill	11	3.5'									39	LF
3'-6"												
					116	LF	374	LF	116	LF	116	LF
Grounds							606	LF				
Oak Sills											116	LF

Figure 26.44

061 | Rough Carpentry

		061 100 Wood Framing	CREW	DAILY OUTPUT	MAN-HOURS	UNIT	BARE COSTS MAT.	BARE COSTS LABOR	BARE COSTS EQUIP.	BARE COSTS TOTAL	TOTAL INCL O&P
132	0500	On masonry	1 Carp	225	.036	L.F.	.08	.76		.84	1.31
	0600	On concrete		175	.046		.08	.98		1.06	1.66
	0700	On metal lath		200	.040	↓	.08	.86		.94	1.46
	9000	Minimum labor/equipment charge	↓	4	2	Job		43		43	69
134	0010	**INSULATION** See division 072									
136	0010	**LAMINATED** See division 061-804									
138	0010	**PARTITIONS** Wood stud with single bottom plate and									
	0020	double top plate, no waste, std. & better lumber									
	0180	2" x 4" studs, 8' high, studs 12" O.C.	F-2	80	.200	L.F.	2.95	4.28	.19	7.42	10.35
	0200	16" O.C.		100	.160		2.56	3.42	.15	6.13	8.50
	0300	24" O.C.		125	.128		2.32	2.74	.12	5.18	7.10
	0380	10' high, studs 12" O.C.		80	.200		3.62	4.28	.19	8.09	11.05
	0400	16" O.C.		100	.160		3.04	3.42	.15	6.61	9
	0500	24" O.C.		125	.128		2.59	2.74	.12	5.45	7.40
	0580	12' high, studs 12" O.C.		65	.246		4.02	5.25	.23	9.50	13.15
	0600	16" O.C.		80	.200		3.49	4.28	.19	7.96	10.90
	0700	24" O.C.	↓	100	.160	↓	2.95	3.42	.15	6.52	8.90
	0702										
	0780	2" x 6" studs, 8' high, studs 12" O.C.	F-2	70	.229	L.F.	4.42	4.89	.21	9.52	12.95
	0800	16" O.C.		90	.178		3.88	3.80	.17	7.85	10.55
	0900	24" O.C.		115	.139		3.36	2.98	.13	6.47	8.60
	0980	10' high, studs 12" O.C.		70	.229		5.23	4.89	.21	10.33	13.85
	1000	16" O.C.		90	.178		4.55	3.80	.17	8.52	11.30
	1100	24" O.C.		115	.139		3.88	2.98	.13	6.99	9.20
	1180	12' high, studs 12" O.C.		55	.291		6.04	6.25	.27	12.56	16.95
	1200	16" O.C.		70	.229		5.23	4.89	.21	10.33	13.85
	1300	24" O.C.		90	.178		4.42	3.80	.17	8.39	11.15
	1400	For horizontal blocking, 2" x 4", add		600	.027		.26	.57	.03	.86	1.23
	1500	2" x 6", add		600	.027		.39	.57	.03	.99	1.37
	1600	For openings, add	↓	250	.064	↓		1.37	.06	1.43	2.26
	1700	Headers for above openings, material only, add				B.F.	.45			.45	.50
	9000	Minimum labor/equipment charge	1 Carp	4	2	Job		43		43	69
140	0010	**ROUGH HARDWARE** Average % of carpentry material, minimum					.50%				
	0200	Maximum					1.50%				
		061 150 Sheathing									
154	0010	**SHEATHING** Plywood on roof, CDX									
	0030	5/16" thick	F-2	1,600	.010	S.F.	.29	.21	.01	.51	.67
	0050	3/8" thick (83)		1,525	.010		.33	.22	.01	.56	.73
	0100	1/2" thick		1,400	.011		.41	.24	.01	.66	.85
	0200	5/8" thick		1,300	.012		.44	.26	.01	.71	.92
	0300	3/4" thick		1,200	.013		.52	.29	.01	.82	1.04
	0500	Plywood on walls with exterior CDX, 3/8" thick		1,200	.013		.33	.29	.01	.63	.83
	0600	1/2" thick		1,125	.014		.41	.30	.01	.72	.95
	0700	5/8" thick		1,050	.015		.44	.33	.01	.78	1.02
	0800	3/4" thick	↓	975	.016		.52	.35	.02	.89	1.15
	1000	For shear wall construction, add						20%			
	1200	For structural 1 exterior plywood, add					10%				
	1400	With boards, on roof 1" x 6" boards, laid horizontal	F-2	725	.022		.78	.47	.02	1.27	1.64
	1500	Laid diagonal		650	.025		.78	.53	.02	1.33	1.73
	1700	1" x 8" boards, laid horizontal		875	.018		.83	.39	.02	1.24	1.56
	1800	Laid diagonal	↓	725	.022		.83	.47	.02	1.32	1.69
	2000	For steep roofs, add						40%			
	2200	For dormers, hips and valleys, add					5%	50%			
	2400	Boards on walls, 1" x 6" boards, laid regular	F-2	650	.025		.78	.53	.02	1.33	1.73
	2500	Laid diagonal	"	585	.027	↓	.78	.59	.03	1.40	1.83

For expanded coverage of these items see *Means Interior Cost Data 1989*

71

Figure 26.45

Means® Forms
COST ANALYSIS

Division 7

PROJECT	Commercial Renovation
CLASSIFICATION	
LOCATION	
ARCHITECT	
TAKE OFF BY	EBN
QUANTITIES BY	EBN
PRICES BY	RSM
EXTENSIONS BY	
SHEET NO.	1 of 3
ESTIMATE NO.	89-1
DATE	1989
CHECKED BY	

DESCRIPTION	SOURCE/DIMENSIONS	QUANTITY	UNIT	MATERIAL		LABOR		EQUIPMENT		SUBCONTRACT		TOTAL	
				UNIT COST	TOTAL	UNIT COST	TOTAL	UNIT COST	TOTAL	UNIT COST	TOTAL	UNIT COST	TOTAL
Division 7: Thermal & Moisture Protect.													
Insulation:													
Roof	072-203-2240	8660	SF		SUBCONTRACT					1.22	10565	.010	87
Exterior Wall	072-118-0080	4799	SF	.22	1056	.11	528					.005	24
Interior Wall (Sound Proofing)	072-118-0820	10,131	SF	.20	2026	.13	1317					.006	61
Foundation @ Crawl Space	072-109-0700	880	SF	.34	299	.25	220					.012	11
[1]	2% Mat. 3% Labor			2%	6	3%	7					3%	1
Vapor Barrier	071-922-0700	52	SQ	1.15	60	4.63	241					.216	11
Roofing	075-302-7100	8660	SF		SUBCONTRACT					2.25	19485	.015	130
Copper Down Spouts	076-201-2500	204	LF	5.70	1163	1.33	271					.055	11
Strainers	076-201-2800	6	EA.	3.80	23	1.33	8					.055	1
Page Subtotals					4633		2592				30050		337

Figure 26.46

Means® Forms
COST ANALYSIS

Division 7

SHEET NO. 2 of 3

PROJECT Commercial Renovation | CLASSIFICATION | ESTIMATE NO. 89-1

LOCATION | ARCHITECT | DATE 1989

TAKE OFF BY EBN | QUANTITIES BY EBN | PFICES BY RSM | EXTENSIONS BY | CHECKED BY

DESCRIPTION	SOURCE/DIMENSIONS	QUANTITY	UNIT	MATERIAL		LABOR		EQUIPMENT		SUBCCNTRACT			
				UNIT COST	TOTAL	UNIT COST	TOTAL	UNIT COST	TOTAL	UNIT COST	TOTAL	UNIT COST	TOTAL
Division 7: (cont'd)													
Accessories													
Roof Hatch	077-206-0500	1	EA.	310	310	65	65	2.36	2			3.2	3
Skylights (14)	078-101-0600	196	SF	15.40	3018	2.08	408	.07	14			.102	20
Curbs @ Skylts.	078-101-1800			30%	905								
Smoke Hatches	077-206-1400	3	EA.	730	2190	99	297	3.57	11			4.85	15
Add				10%	219	5%	15					5%	1
Page Subtotals					6642		785		27				39

Figure 26.47

Means® Forms
COST ANALYSIS

Division 7

SHEET NO. 3 of 3

PROJECT Commercial Renovation — CLASSIFICATION — ESTIMATE NO. 89-1

LOCATION — ARCHITECT — DATE 1989

TAKE OFF BY EBW — QUANTITIES BY EBW — PRICES BY RSM — EXTENSIONS BY — CHECKED BY

DESCRIPTION	SOURCE/DIMENSIONS	QUANTITY	UNIT	MATERIAL		LABOR		EQUIPMENT		SUBCONTRACT			
				UNIT COST	TOTAL	UNIT COST	TOTAL	UNIT COST	TOTAL	UNIT COST	TOTAL	UNIT COST	TOTAL
Division 7: (Cont'd)													
Tenant:													
Insulation (sound proofing)	072-118-0820	3015	SF	.20	603	.13	392					.006	18
Skylight	078-101-0600	14	SF	15.40	216	2.08	29	.07	1			.102	1
Curb	078-101-1800		30%		65								
Page Subtotals					884		421		1				19
Sheet 1					4633		2592				30050		337
Sheet 2					6642		785		27				39
Subtotal					11275		3377		27		30050		376
Sheet 3 Tenant					884		421		1				19
Division 7: Totals					12159		3798		28		30050		395

Figure 26.48

the quotes are for the roofing and materials exactly as specified. There are many materials and methods of installation, especially for single-ply roofs.

Division 7 is the first instance where work for the third floor tenant is included in the estimate. In Figure 26.48, the items are priced separately so that the total tenant costs can easily be determined at the end of the estimate. The tenant has requested that an additional skylight be installed. Note that curbs must be added to the cost of the skylights.

During the estimating process, the estimator should begin to think about scheduling and the progression of the work. Figure 26.49 is a preliminary schedule for Division 2 through Division 7. The schedule, at this point, is used only as a reference and will undoubtedly be changed as the estimate is completed. The estimator should note such items as the demolition of the freight elevator and shaft, which will occur out of a normal sequence. The sidewalk is scheduled near the end of the project because the scaffolding must be removed, and all exterior work completed, before installation. (See Chapter 8, "Pre-bid Scheduling.") Proper preparation of an accurate Project Schedule is very important when determining the allocation of workers for a project. It is preferable to keep the work force as constant as possible throughout the project. The schedule also provides a basis for determining costs of time-related items in Division 1.

Division 8: Doors, Windows and Glass

A proper set of plans and specifications will have door, window and hardware schedules. These schedules should provide all information that is necessary for the quantity takeoff. Figure 26.50 shows portions of the door schedule for the sample project. There are two common types of door schedules. The first, as in Figure 26.50, lists each door separately. The second lists each type of door (usually accompanied by elevations) that have common characteristics. Hardware is usually only designated by "set" on the schedules. These "sets" are described elsewhere on the plans or in the specifications. The estimator should crosscheck quantities obtained from the schedules by counting and marking each item on the plans.

Specifications often state that all doors, frames, and hardware be as specified, or an "approved equal." This means that if the estimator can find similar quality products for less cost, the architect may accept these products as alternatives. Specifications usually require that the contractor submit shop drawings, schedules (as prepared by the contractor or supplier) or "cuts" (product literature). Most door, frame, and hardware suppliers provide these services to the contractor. It is recommended that the estimator obtain material prices from a supplier, because prices vary and fluctuate widely.

Installation costs, also, can vary due to restrictive existing conditions in commercial renovation. During the site evaluation the estimator should note any areas where such restrictions require the addition of factors. In the sample project, the metal frames with transoms at the stairway exits must be installed in existing masonry openings. These openings most likely will not conform exactly to the sizes of new metal door frames. A factor is added to match existing conditions.

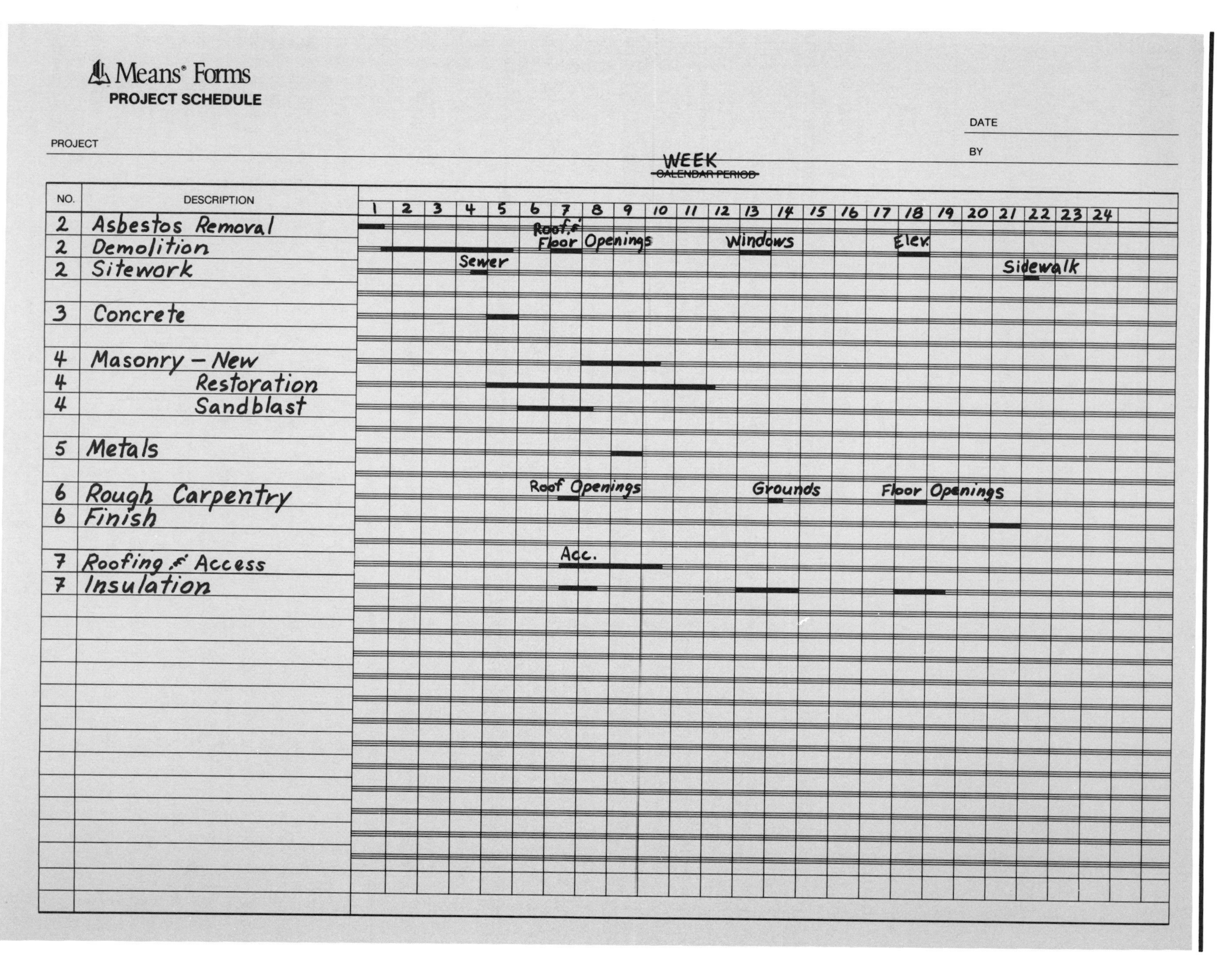

Means® Forms
PROJECT SCHEDULE

PROJECT
DATE
BY

WEEK ~~CALENDAR PERIOD~~

NO.	DESCRIPTION
2	Asbestos Removal
2	Demolition
2	Sitework
3	Concrete
4	Masonry – New
4	Restoration
4	Sandblast
5	Metals
6	Rough Carpentry
6	Finish
7	Roofing & Access
7	Insulation

Weeks: 1 2 3 4 5 6 7 8 9 10 11 12 13 14 15 16 17 18 19 20 21 22 23 24

Figure 26.49

The existing interior door frames at the west stairway are to remain and be reused. Again, a factor must be added to the appropriate door installations to match existing conditions. (See Figure 26.52.) Figures 26.51 to 26.56 are the estimate sheets for Division 8.

Note that the tenant entrance door costs in Figure 26.52 have a deduction for 6′-8″ high doors. This is obtained from Figure 26.57.

Door Schedule

Door	Size	Type	Rating	Frame	Depth	Rating	Set	Remarks
B01	3^0 x 6^8	Flush Steel 18 ga.	"B" 1-1/2 Hr.	Exist'g.	—	—	HW-1	10" x 10" Vision Lite Shop-Primed
B02	3^0 x 6^8	Flush Steel 18 ga.	"B" 1 Hr.	HMKD 16 ga.	4-7/8"	"B" 1 Hr.	HW-2	Shop-Primed
B03	3^0 x 6^8	Flush Steel 18 ga.	"B" 1 Hr.	H.M. Welded 16 ga.	8"	"B" 1 Hr.	HW-3	w/Masonry Anchors Shop-Primed
B04	3^0 x 6^8	Flush Steel 18 ga.	"B" 1-1/2 Hr.	H.M. Welded 16 ga.	8"	"B" 1-1/2 Hr.	HW-3	w/Masonry Anchors Shop-Primed
B05	3^0 x 6^8	Flush Steel 18 ga.	"B" 1-1/2 Hr.	H.M. Welded 16 ga.	8"	"B" 1-1/2 Hr.	HW-3	w/Masonry Anchors Shop-Primed
B06	3^0 x 6^8	Flush Steel 18 Ga.	"B" 1 Hr.	HMKD 16 ga.	4-7/8"	"B" 1 Hr.	HW-2	Shop-Primed
101	3^0 x 6^8	Flush Steel 18 ga.	—	HMKD 16 Ga.	4-7/8"	—	HW-4	Transom Frame Above w/Masonry Anchors
102	3^0 x 6^8	Flush Steel 18 ga.	"B" 1-1/2 Hr.	Exist'g.	—	—	HW-1	Shop-Primed
103	3^0 x 6^8	Flush Oak Face	"B" 1 Hr.	HMKD 16 ga.	4-7/8"	"B" 1 Hr.	HW-5	
104	2^6 x 6^8	Flush Oak Face SC	—	HMKD 16 ga.	4-5/8"	—	HW-6	
105	3^0 x 6^8	Flush Oak Face	"B" 1 Hr.	HMKD 16 ga.	4-7/8"	"B" 1 Hr.	HW-5	

Figure 26.50

Means® Forms
COST ANALYSIS

Division 8

SHEET NO. 1 of 6

PROJECT Commercial Rennovation | CLASSIFICATION | ESTIMATE NO. 89-1

LOCATION | ARCHITECT | DATE 1989

TAKE OFF BY EBW | QUANTITIES BY EBW | PRICES BY RSM | EXTENSIONS BY | CHECKED BY

DESCRIPTION	SOURCE/DIMENSIONS			QUANTITY	UNIT	MATERIAL		LABOR		EQUIPMENT		SUBCONTRACT			
						UNIT COST	TOTAL	UNIT COST	TOTAL	UNIT COST	TOTAL	UNIT COST	TOTAL	UNIT COST	TOTAL
Division 8: Door, Windows, Glass															
Metal Frames															
16ga. to 5 3/4"	081	118	3600	6	EA.	64	384	23	138	1.	6			1.07	6
16ga. "B" Label	081	118	5400	13	EA.	80	1040	23	299	1.	13			1.07	14
Welded	081	118	4900	3	EA.	25	75								
16ga. "B" Label Double	081	118	5440	6	EA.	104	624	29	174	1.25	8			1.33	8
16ga. "B" Label to 6 3/4"	081	118	5800	3	EA.	86	258	23	69	1.	3			1.07	3
16ga. - Stair Exit "B" Label	081	118	5400	2	EA.	80	160	23	46	1.	2			1.07	2
Transoms	081	118	7900	9	SF	12	108	2.21	20	.10	1			.103	1
16ga. "B" Label to 8 3/4"	081	118	6200	3	EA.	83	279	23	69	1.	3			1.07	3
Page Subtotals							2928		815		36				37

Figure 26.51

Means® Forms
COST ANALYSIS

Division 8

SHEET NO. 2 of 6

PROJECT Commercial Renovation | CLASSIFICATION | ESTIMATE NO. 89-1

LOCATION | ARCHITECT | DATE 1989

TAKE OFF BY EBW | QUANTITIES BY EBW | PRICES BY RSM | EXTENSIONS BY | CHECKED BY

DESCRIPTION	SOURCE/DIMENSIONS	QUANTITY	UNIT	MATERIAL		LABOR		EQUIPMENT		SUBCONTRACT			
				UNIT COST	TOTAL	UNIT COST	TOTAL	UNIT COST	TOTAL	UNIT COST	TOTAL	UNIT COST	TOTAL
Division 8: (Cont'd)													
Doors													
Exterior Exit	081-103-1140	2	EA.	171	342	21	42	.94	2			1	2
Stair (Exist'g Fr.)	081-110-0140	4	EA.	190	760	21	84	.94	4			1	4
+	5% Mat. 9% Labor			5%	38	9%	8					9%	1
Stair & Basement (New Frames)	081-110-0140	8	EA.	190	1520	21	168	.94	8			1	8
Vision Lites	081-110-0240	7	EA.	43	301								
Telephone Cl.	082-070-0140	6	EA.	210	1260	24	144	1.07	6			1.14	7
Tenant Entrance Doors @ Corridor	082-070-0190	17	EA.	237 213.30	3626	29	493	1.25	21			1.33	23
6'-8" Deduct	082-070-2460	(Deduct)		↑ (10%)									
Retail - Interior	082-062-2240	3	EA.	82	246	24	72	1.07	3			1.14	3
Retail - Bath	082-062-2220	3	EA.	73	219	23	69	1	3			1.07	3
Page Subtotals					8312		1080		47				51

Figure 26.52

Means® Forms
COST ANALYSIS

Division 8

SHEET NO. 3 of 6

PROJECT Commercial Renovation | CLASSIFICATION | ESTIMATE NO. 89-1

LOCATION | ARCHITECT | DATE 1989

TAKE OFF BY EBW | QUANTITIES BY EBW | PRICES BY RSM | EXTENSIONS BY | CHECKED BY

DESCRIPTION	SOURCE/DIMENSIONS	QUANTITY	UNIT	MATERIAL UNIT COST	MATERIAL TOTAL	LABOR UNIT COST	LABOR TOTAL	EQUIPMENT UNIT COST	EQUIPMENT TOTAL	SUBCONTRACT UNIT COST	SUBCONTRACT TOTAL	UNIT COST	TOTAL
Division 8: (Cont'd)													
Windows:				Incl. O&P									
Subcontractor Quote		70	EA.	512		30				542	37940	2	140
Hardware:													
Closers	087-206-2400	28	EA.	63	1764	29	812					1.33	37
Floor Bumpers	087-110-1600	21	EA.	1.40	29	7.15	150					.333	7
Hinges	087-116-1400	23	PR.	75	2475								
Kickplates	087-118-0500	4	EA.	35	140	11.40	46					.533	2
Locksets:													
Privacy	087-120-0100	3	EA.	33	99	14.25	43					.667	2
Keyed	087-120-0400	13	EA.	58	754	17.10	222					.80	10
Mortise	087-125-2130	8	EA.	316	2528	24	192					1.14	9
Exterior Panic	087-127-0210	2	EA.	400	800	43	86					2	4
Stair Panic	087-127-0210	7	EA.	268	1876	34	238					1.6	11
Push/Pull	087-129-0500	4	EA.	58	232	14.25	57					.667	3
Ext. Thresholds	087-304-0100	2	EA.	65	130	14.25	29					.667	1
Page Subtotals					10827		1875				37940		226

Figure 26.53

Means® Forms
COST ANALYSIS

Division 8

SHEET NO. 4 of 6

PROJECT Commercial Renovation | CLASSIFICATION | ESTIMATE NO. 89-1

LOCATION | ARCHITECT | DATE 1989

TAKE OFF BY EBW | QUANTITIES BY EBW | PFICES BY RSM | EXTENSIONS BY | CHECKED BY

DESCRIPTION	SOURCE/DIMENSIONS	QUANTITY	UNIT	MATERIAL UNIT COST	MATERIAL TOTAL	LABOR UNIT COST	LABOR TOTAL	EQUIPMENT UNIT COST	EQUIPMENT TOTAL	SUBCCNTRACT UNIT COST	SUBCCNTRACT TOTAL	UNIT COST	TOTAL
Division 8: (Cont'd)													
Aluminum Doors													
Retail Entrances	081-212-0200	4	EA.									8.89	36
Black Finish	081-212-1500			35%			SUBCONTRACTED		}				
Concealed Closer	081-212-1900	4	EA.							2463	9852	12.31	49
1 + 5 −	7% Mat. 11% Labor			7%		11%						11%	9
Concealed Panic	081-212-1600	4	EA.							440	1760		
Glass: Doors	088-118-0800	84	SF							9.95	836	.133	11
Transoms	088-118-0600	36	SF							7.35	265	.133	5
Stair Exit Transoms	088-1 8-0800	18	SF							9.95	179	.133	2
Lobby Entrance													
13'-6" x 10'-0"	084-105-1100	135	SF									.139	19
Black Finish	084-105-1600			25%						29.39	3968		
1 + 5 −	7% Mat. 11% Labor			7%		11%						11%	2
Page Subtotals											16860		133

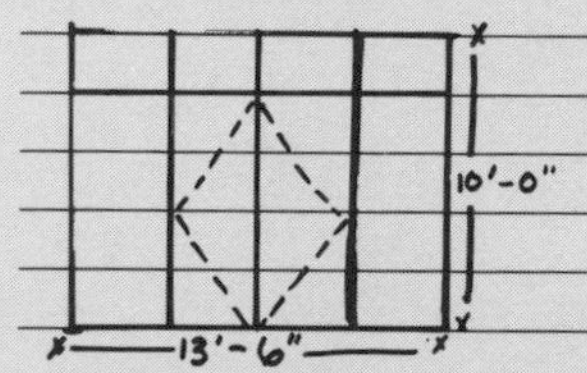

Figure 26.54

Means® Forms
COST ANALYSIS

Division 8

SHEET NO. 5 of 6

PROJECT Commercial Renovation | CLASSIFICATION | ESTIMATE NO. 89-1

LOCATION | ARCHITECT | DATE 1989

TAKE OFF BY EBN | QUANTITIES BY EBN | PRICES BY RSM | EXTENSIONS BY | CHECKED BY

DESCRIPTION	SOURCE/DIMENSIONS	QUANTITY	UNIT	MATERIAL		LABOR		EQUIPMENT		SUBCONTRACT			
				UNIT COST	TOTAL	UNIT COST	TOTAL	UNIT COST	TOTAL	UNIT COST	TOTAL	UNIT COST	TOTAL
Division 8 : (Cont'd)													
Tenant :													
Doors	082-062-2240	9	EA.	82	738	24	216	1.07	10			1.14	10
Frames	081-118-3600	9	EA.	64	576	23	207	1	9			1.07	10
Hinges	087-116-1400	12	PR.	75	900								
Floor Bumpers	087-110-1600	9	EA.	1.40	13	7.15	64					.333	3
Locksets:													
Passage	087-120-1000	7	EA.	87	609	14.35	100					.667	5
Keyed	087-120-1400	2	EA.	128	256	17.10	34					.80	2
Computer Window													
Frame	089-204-0200	20	LF		SUBCONTRACT							.178	6
Glass Bead	089-204-0450	20	LF							20.88	418	.067	1
Black Finish	089-204-8020			27%									
Glass	088-118-0800	24	SF							10.43	250	1.33	3
Page Subtotals					3092		621		19		668		40

Figure 26.55

Means® Forms
COST ANALYSIS

Division 8

SHEET NO. 6 of 6

PROJECT Commercial Renovation | CLASSIFICATION | ESTIMATE NO. 89-1

LOCATION | ARCHITECT | DATE 1989

TAKE OFF BY EBW | QUANTITIES BY EBW | PRICES BY RSM | EXTENSIONS BY | CHECKED BY

DESCRIPTION	SOURCE/DIMENSIONS	QUANTITY	UNIT	MATERIAL		LABOR		EQUIPMENT		SUBCONTRACT			
				UNIT COST	TOTAL	UNIT COST	TOTAL	UNIT COST	TOTAL	UNIT COST	TOTAL	UNIT COST	TOTAL
Division 8: (Cont'd)													
Totals													
Sheet 1 Frames					2928		815		36				37
Sheet 2 Doors					8312		1080		47				51
Sheet 3 Hardware					10827		1875						86
Subtotal					22067		3770		83				174
Windows											37940		140
Sheet 4 Glazing													133
Sheet 5 Tenant					3092		621		19		668		40
Division 8: Totals					25159		4391		102		55468		487

Figure 26.56

082 | Wood and Plastic Doors

		082 050 Wood And Plastic Doors	CREW	DAILY OUTPUT	MAN-HOURS	UNIT	BARE COSTS MAT.	BARE COSTS LABOR	BARE COSTS EQUIP.	BARE COSTS TOTAL	TOTAL INCL O&P	
062	4300	For 6'-8" high door, deduct from 7'-0" door				Ea.	10%					062
	9000	Minimum labor/equipment charge	F-1	4	2	Job		43	1.88	44.88	71	
066	0010	**WOOD DOORS, DECORATOR**										066
	4000	Hand carved door, mahogany, simple design										
	4020	1-3/4" x 7'-0" x 3'-0" wide	F-2	14	1.140	Ea.	412	24	1.07	437.07	495	
	4040	3'-6" wide		13	1.230		505	26	1.15	532.15	600	
	4200	Rosewood, 1-3/4" x 7'-0" x 3'-0" wide		14	1.140		585	24	1.07	610.07	685	
	4220	3'-6" wide		13	1.230		675	26	1.15	702.15	785	
	4280	For 6'-8" high door, deduct from 7'-0" door					10%					
	4320	For detailed design, add					50%					
	4340	For hand carved back, add					20%					
	4360	For ornate mahogany door, 2-1/4" thick, add					20%					
	4380	For ornate rosewood door, 2-1/4" thick, add					20%					
	4400	For custom finish, add					95			95	105	
	4600	Side panel, mahogany, simple design, 7'-0" x 1'-0" wide	F-2	21	.762		111	16.30	.71	128.01	150	
	4620	1'-2" wide		20	.800		125	17.10	.75	142.85	165	
	4640	1'-4" wide		19	.842		135	18	.79	153.79	180	
	4800	Rosewood, simple design 7'-0" x 1'-0" wide		21	.762		165	16.30	.71	182.01	210	
	4820	1'-2" wide		20	.800		184	17.10	.75	201.85	230	
	4840	1'-4" wide		19	.842		205	18	.79	223.79	255	
	4900	For detailed design, add					50%					
	4920	For hand carved back, add					20%					
	6520	Interior cafe doors, 2'-6" opening, stock, panel pine	F-2	16	1		95	21	.94	116.94	140	
	6540	3'-0" opening	"	16	1		108	21	.94	129.94	155	
	6550	Custom hardwood or louvered pine										
	6560	2'-6" opening	F-2	16	1	Ea.	84	21	.94	105.94	130	
	8000	3'-0" opening	"	16	1	"	93	21	.94	114.94	140	
	8800	Pre-hung doors, see division 082-082										
	9000	Minimum labor/equipment charge	F-1	4	2	Job		43	1.88	44.88	71	
070	0010	**WOOD FIRE DOORS** Mineral core, 3 ply stile, "B" label,										070
	0040	1 hour, birch face, 2'-6" x 6'-8"	F-2	14	1.140	Ea.	182	24	1.07	207.07	240	
	0090	3'-0" x 7'-0"	"	12	1.330	"	213	29	1.25	243.25	280	
	0110											
	0140	Oak face, 2'-6" x 6'-8"	F-2	14	1.140	Ea.	210	24	1.07	235.07	270	
	0190	3'-0" x 7'-0"		12	1.330		237	29	1.25	267.25	310	
	0240	Walnut face, 2'-6" x 6'-8"		14	1.140		235	24	1.07	260.07	300	
	0290	3'-0" x 7'-0"		12	1.330		269	29	1.25	299.25	345	
	0440	M.D. overlay on hardboard, 2'-6" x 6'-8"		15	1.070		180	23	1	204	235	
	0490	3'-0" x 7'-0"		13	1.230		208	26	1.15	235.15	270	
	0540	H.P. plastic laminate, 2'-6" x 6'-8"		13	1.230		135	26	1.15	162.15	190	
	0590	3'-0" x 7'-0"		11	1.450		145	31	1.36	177.36	210	
	2460	For 6'-8" high door, deduct from 7'-0" door					10%					
	2480	For oak veneer, add					50%					
	2500	For walnut veneer, add					75%					
	9000	Minimum labor/equipment charge	F-1	4	2	Job		43	1.88	44.88	71	
074	0010	**WOOD DOORS, PANELED** Interior, six panel, hollow core, 1-3/8" thick										074
	0040	Molded hardboard, 2'-0" x 6'-8"	F-2	17	.941	Ea.	42	20	.88	62.88	80	
	0060	2'-6" x 6'-8"		17	.941		47	20	.88	67.88	85	
	0080	3'-0" x 6'-8"		17	.941		52	20	.88	72.88	91	
	0140	Embossed print, molded hardboard, 2'-0" x 6'-8"		17	.941		48	20	.88	68.88	86	
	0160	2'-6" x 6'-8"		17	.941		53	20	.88	73.88	92	
	0180	3'-0" x 6'-8"		17	.941		58	20	.88	78.88	97	
	0540	Six panel, solid, 1-3/8" thick, pine, 2'-0" x 6'-8"		15	1.070		90	23	1	114	135	
	0560	2'-6" x 6'-8"		14	1.140		99	24	1.07	124.07	150	
	0580	3'-0" x 6'-8"		13	1.230		115	26	1.15	142.15	170	
	1020	Two panel, bored rail, solid, 1-3/8" thick, pine, 1'-6" x 6'-8"		16	1		119	21	.94	140.94	165	
	1040	2'-0" x 6'-8"		15	1.070		176	23	1	200	230	

114

For expanded coverage of these items see *Means Interior Cost Data 1989*

Figure 26.57

This deduction should be subtracted from the material price before extension of the costs. Otherwise, a minus figure in the "TOTAL" column would be confusing and inconsistent.

The new windows for the project are not standard sizes and require a special black finish. A quotation from a subcontractor is obtained and is included in Figure 26.53. There are no factors added separately because any special installation or material costs should be included in the subcontract price. It is important, therefore, that the subcontractor examine the existing conditions thoroughly before submitting the quote. The estimator must be sure that no "surprises" will be encountered during installation.

As with the windows, the retail entrance doors and lobby entrance require a black finish. These items will be furnished and installed by a subcontractor. The added costs for the black finish are provided as a percentage of the bare material cost as shown in Figure 26.58. Also, the factors for matching existing conditions and protecting existing work (applied to bare costs) must be added to the total including the installing subcontractor's overhead and profit. The calculations are shown in Figure 26.59.

The factors do not apply to the costs for the concealed panic devices. These items are installed in the door and frame at the supplier's shop and affect only the material cost of the door units. The installation of the concealed closers, however, is affected by the existing conditions, so the factors do apply. The black finish cost is also included before the factors are calculated, because any required shims or spacers to conform to the existing opening will have the same black finish. Similar calculations are required for the lobby entrance storefront.

Storefront systems can be estimated in two ways, using *Means Repair and Remodeling Cost Data*. The first method is to separate the system into components, tube framing, glass, doors, and hardware, and to price each item individually. The second method is to use the data provided for complete systems as shown in Figure 26.60. The latter method is used in this example. When pricing is used for a system such as this, the estimator must be sure that all specified items are included in the system or added if necessary.

Figure 26.56 illustrates how the different components of Division 8 are totalled. The other estimate sheets (Figures 26.51 to 26.55) have been organized so that groups of similar items are priced together. The estimator is able to use this information to determine the percentage of total job costs, or the costs per square foot of the different components of commercial renovation. These percentages and square foot costs may be helpful when estimating future jobs.

As in Division 7, the tenant work for Division 8 is priced separately. The estimator must remember, for scheduling purposes, that all materials for the second and third floors should be in place before the freight elevator is removed.

Division 9: Finishes

In commercial renovation, Division 9 usually represents a major portion of the work. The seven estimate sheets for the finishes are shown in Figures 26.61 to 26.67.

081 | Metal Doors and Frames

		081 100 | Steel Doors And Frames	CREW	DAILY OUTPUT	MAN-HOURS	UNIT	BARE COSTS MAT.	BARE COSTS LABOR	BARE COSTS EQUIP.	BARE COSTS TOTAL	TOTAL INCL O&P	
118	6500	For galvanizing, add					9.50%					118
	6600	For porcelain enamel finish, add					100%	150%				
	7900	Transom lite frames, fixed, add	F-2	155	.103	S.F.	12	2.21	.10	14.31	16.85	
	8000	Movable, add	"	130	.123	"	15	2.63	.12	17.75	21	
	9000	Minimum labor/equipment charge	F-1	4	2	Job		43	1.88	44.88	71	

		081 200 | Alum Doors And Frames	CREW	DAILY OUTPUT	MAN-HOURS	UNIT	MAT.	LABOR	EQUIP.	TOTAL	TOTAL INCL O&P	
204	0010	ALUMINUM FRAMES Entrance, 3' x 7' opening, clear finish	2 Sswk	7	2.290	Opng.	210	53		263	325	204
	0100	Bronze finish		7	2.290		240	53		293	355	
	0500	6' x 7' opening, clear finish		6	2.670		260	62		322	395	
	0520	Bronze finish		6	2.670		305	62		367	445	
	1000	With 3' high transoms, 3' x 10' opening, clear finish		6.50	2.460		310	57		367	440	
	1050	Bronze finish		6.50	2.460		365	57		422	500	
	1100	Black finish		6.50	2.460		425	57		482	565	
	1500	With 3' high transoms, 6' x 10' opening, clear finish		5.50	2.910		395	68		463	550	
	1550	Bronze finish		5.50	2.910		460	68		528	625	
	1600	Black finish		5.50	2.910		535	68		603	705	
	9000	Minimum labor/equipment charge		4	4	Job		93		93	160	
212	0010	ALUMINUM DOORS & FRAMES Entrance, narrow stile, including										212
	0020	hardware & closer, clear finish, not incl. glass, 3' x 7' opening	2 Sswk	2	8	Ea.	650	185		835	1,050	
	0100	3' x 10' opening, 3' high transom		1.80	8.890		745	205		950	1,175	
	0200	3'-6" x 10' opening, 3' high transom		1.80	8.890		775	205		980	1,200	
	0300	6' x 7' oponing		1.30	12.310	Pr.	1,125	285		1,410	1,725	
	0400	6' x 10' opening, 3' high transom		1.10	14.550	"	1,250	340		1,590	1,950	
	1000	Add to above for wide stile doors				Leaf	45%					
	1100	Full vision doors, with ½" glass, add					55%					
	1200	Non-standard size, add					35%					
	1300	Light bronze finish, add					20%					
	1400	Dark bronze finish, add					18%					
	1500	Black finish, add					35%					
	1600	Concealed panic device, add					400			400	440	
	1700	Electric striker release, add				Opng.	450			450	495	
	1800	Floor check, add				Leaf	570			570	625	
	1900	Concealed closer, add				"	710			710	780	
	9000	Minimum labor/equipment charge	F-2	4	4	Job		86	3.75	89.75	140	

082 | Wood and Plastic Doors

		082 050 | Wood And Plastic Doors	CREW	DAILY OUTPUT	MAN-HOURS	UNIT	BARE COSTS MAT.	BARE COSTS LABOR	BARE COSTS EQUIP.	BARE COSTS TOTAL	TOTAL INCL O&P	
054	0010	WOOD FRAMES										054
	0400	Exterior frame, incl. ext. trim, pine, 5/4 x 4-9/16" deep	F-2	375	.043	L.F.	2.75	.91	.04	3.70	4.54	
	0420	5-3/16" deep		375	.043		3	.91	.04	3.95	4.81	
	0440	6-9/16" deep		375	.043		3.50	.91	.04	4.45	5.35	
	0600	Oak, 5/4 x 4-9/16" deep		350	.046		3.45	.98	.04	4.47	5.40	
	0620	5-3/16" deep		350	.046		3.75	.98	.04	4.77	5.75	
	0640	6-9/16" deep		350	.046		4.45	.98	.04	5.47	6.50	
	0800	Walnut, 5/4 x 4-9/16" deep		350	.046		5.05	.98	.04	6.07	7.20	
	0820	5-3/16" deep		350	.046		5.85	.98	.04	6.87	8.05	
	0840	6-9/16" deep		350	.046		6.65	.98	.04	7.67	8.95	

For expanded coverage of these items see *Means Interior Cost Data 1989*

Figure 26.58

The estimator must be careful when taking off quantities for metal stud and drywall partitions. It would be simple if each type and size of stud had a different drywall application. But different size studs will have the same drywall treatment and vice versa. Thoroughly complete plans and specifications will often have a schedule of the wall types as well as a Room Finish Schedule. In either case, the different wall types should be marked on the plans, with different colors, during the quantity takeoff. The estimator should determine quantities for these "assembled" wall types before attempting to figure quantities of individual components. The estimator should also make notes during the takeoff on existing conditions that will require added factors. In the sample project, the wood beams at the ceilings occur every 10′. The studs and drywall will have to be cut to fit around the beams. No deduction for material is made for this type of condition, but extra labor expense is included.

When the quantities of the different wall types have been determined, the estimator should list each individual component and the appropriate quantities. Then the total quantities of each component for all wall types are added and entered on the estimate sheet. As a crosscheck, the resulting totals for studs and drywall can be compared. Remember, when comparing, that the drywall quantities will be twice those for some of the studs because of application on two sides. Figure 26.68 is the comparison for the sample project. Refer to Figures 26.61 and 26.62. This method will help to ensure that all items have been included.

In Figure 26.62, the 5/8″ fire-resistant drywall is listed in two ways, taped and with no finish. The drywall that requires no finish is the

Aluminum Entrance Doors				
	Material	Labor	Total	Total Including O & P
Entrance Door	$ 775	$205	$ 980	
Black Finish (35% of $775)	271			
Concealed Closer	710		710	
	1,756	205	$1,690	
Factors [1] [5]	(7%) 123	(11%) 23		
+ −	1,879	228		
Overhead and Profit 10%	(10%) 188			
Material 73.5% Labor		(73.5%) 168		
Totals Include Overhead and Profit	$2,067	$396		$2,463

Figure 26.59

083 | Special Doors

083 750 | Swing Doors

		083 750 Swing Doors	CREW	DAILY OUTPUT	MAN-HOURS	UNIT	BARE COSTS MAT.	BARE COSTS LABOR	BARE COSTS EQUIP.	BARE COSTS TOTAL	TOTAL INCL O&P	
752	2050	7′ wide	2 Carp	3.80	4.210	Pr.	1,030	90		1,120	1,275	752
	9000	Minimum labor/equipment charge	"	2	8	Job		170		170	275	
754	0010	**GLASS, SWING** Tempered, ½″ thick, incl. hardware, 3′ x 7′ opening	2 Glaz	2	8	Opng.	1,810	175		1,985	2,275	754
	0100	6′ x 7′ opening		1.40	11.430	"	3,495	250		3,745	4,250	
	9000	Minimum labor/equipment charge	↓	2	8	Job		175		175	275	

083 800 | Sound Retardant Doors

		083 800 Sound Retardant Doors	CREW	DAILY OUTPUT	MAN-HOURS	UNIT	MAT.	LABOR	EQUIP.	TOTAL	TOTAL INCL O&P	
804	0010	**ACOUSTICAL** Incl. framed seals, 3′ x 7′, wood, 27 STC rating	F-2	1.50	10.670	Ea.	425	230	10	665	845	804
	0100	Steel, 40 STC rating		1.50	10.670		2,190	230	10	2,430	2,775	
	0200	45 STC rating		1.50	10.670		2,350	230	10	2,590	2,950	
	0300	48 STC rating		1.50	10.670		2,465	230	10	2,705	3,100	
	0400	52 STC rating	↓	1.50	10.670	↓	2,640	230	10	2,880	3,275	
	9000	Minimum labor/equipment charge	F-1	4	2	Job		43	1.88	44.88	71	

083 900 | Screen And Storm Doors

		083 900 Screen And Storm Doors	CREW	DAILY OUTPUT	MAN-HOURS	UNIT	MAT.	LABOR	EQUIP.	TOTAL	TOTAL INCL O&P	
904	0010	**STORM DOORS & FRAMES** Aluminum, residential,										904
	0020	combination storm and screen										
	0400	Clear anodic coating, 6′-8″ x 2′-6″ wide	F-2	15	1.070	Ea.	145	23	1	169	195	
	0420	2′-8″ wide		14	1.140		150	24	1.07	175.07	205	
	0440	3′-0″ wide	↓	14	1.140		160	24	1.07	185.07	215	
	0500	For 7′-0″ door, add					5%					
	1000	Mill finish, 6′-8″ x 2′-6″ wide	F-2	15	1.070		135	23	1	159	185	
	1020	2′-8″ wide		14	1.140		135	24	1.07	160.07	190	
	1040	3′-0″ wide	↓	14	1.140		140	24	1.07	165.07	195	
	1100	For 7′-0″ door, add					5%					
	1500	White painted, 6′-8″ x 2′-6″ wide	F-2	15	1.070		148	23	1	172	200	
	1520	2′-8″ wide		14	1.140		152	24	1.07	177.07	210	
	1540	3′-0″ wide	↓	14	1.140		154	24	1.07	179.07	210	
	1600	For 7′-0″ door, add				↓	5%					
	2000	Wood door & screen, see division 082-078										
	2020											
	9000	Minimum labor/equipment charge	F-1	4	2	Job		43	1.88	44.88	71	

084 | Entrances and Storefronts

084 100 | Aluminum

		084 100 Aluminum	CREW	DAILY OUTPUT	MAN-HOURS	UNIT	BARE COSTS MAT.	BARE COSTS LABOR	BARE COSTS EQUIP.	BARE COSTS TOTAL	TOTAL INCL O&P	
103	0010	**BALANCED DOORS** Hdwre & frame, alum. & glass, 3′ x 7′, econ.	2 Sswk	.90	17.780	Ea.	2,375	415		2,790	3,325	103
	0150	Premium		.70	22.860	"	4,150	530		4,680	5,475	
	9000	Minimum labor/equipment charge	↓	1	16	Job		370		370	645	
105	0010	**STOREFRONT SYSTEMS** Aluminum frame, clear ⅜″ plate glass,										105
	0020	incl. 3′ x 7′ door with hardware (400 sq. ft. max. wall)										
	0500	Wall height to 12′ high, commercial grade	2 Glaz	150	.107	S.F.	11.55	2.35		13.90	16.40	
	0600	Institutional grade		130	.123		14.95	2.71		17.66	21	
	0700	Monumental grade		115	.139		21.50	3.07		24.57	28	
	1000	6′ x 7′ door with hardware, commercial grade		135	.119		10.65	2.61		13.26	15.80	
	1100	Institutional grade		115	.139		15.95	3.07		19.02	22	
	1200	Monumental grade	↓	100	.160		22.25	3.53		25.78	30	
	1500	For bronze anodized finish, add					17%					
	1600	For black anodized finish, add				↓	25%					

122 For expanded coverage of these items see *Means Interior Cost Data 1989*

Figure 26.60

Means® Forms
COST ANALYSIS

Division 9

SHEET NO. 1 of 7

PROJECT Commercial Renovation | CLASSIFICATION | ESTIMATE NO. 89-1

LOCATION | ARCHITECT | DATE 1989

TAKE OFF BY EBN | QUANTITIES BY EBN | PRICES BY RSM | EXTENSIONS BY | CHECKED BY

DESCRIPTION	SOURCE/DIMENSIONS	QUANTITY	UNIT	MATERIAL		LABOR		EQUIPMENT		SUBCONTRACT			
				UNIT COST	TOTAL	UNIT COST	TOTAL	UNIT COST	TOTAL	UNIT COST	TOTAL	UNIT COST	TOTAL
Division 9: Finishes													
Metal Studs													
Non-load Bearing 25ga. 16" O.C.													
1 5/8"	092-612-2000	7144	SF	.24	1715	.38	2715					.018	129
3 5/8"	092-612-2300	11,775	SF	.32	3768	.40	4710					.019	224
6"	092-612-2500	560	SF	.47	263	.42	235					.020	11
1	2% Mat. 3% Labor			2%	115	3%	230					3%	11
Furring @ Stair & Elev.	092-804-0900	26.38	CLF	15.50	409	66	1741					3.08	81
Shaft Wall @ Pipe Chase	092-624-0060	552	SF	1.75	966	1.56	861					.073	40
Tape & Finish	092-624-0900	552	SF	.05	28	.16	88					.008	4
Page Subtotals					7264		10580						500

Figure 26.61

Means® Forms
COST ANALYSIS

Division 9

PROJECT Commercial Renovation — CLASSIFICATION — SHEET NO. 2 of 7

LOCATION — ARCHITECT — ESTIMATE NO. 89-1

DATE 1989

TAKE OFF BY EBN — QUANTITIES BY EBN — PRICES BY RSM — EXTENSIONS BY — CHECKED BY

DESCRIPTION	SOURCE/DIMENSIONS	QUANTITY	UNIT	MATERIAL UNIT COST	MATERIAL TOTAL	LABOR UNIT COST	LABOR TOTAL	EQUIPMENT UNIT COST	EQUIPMENT TOTAL	SUBCONTRACT UNIT COST	SUBCONTRACT TOTAL	UNIT COST	TOTAL
Division 9: (Cont'd)													
Drywall:													
1/2" Standard Taped	092-608-0350	13,292	SF	.25	3323	.35	4652					.017	226
5/8" FR No Finish	092-608-2100	3288	SF	.26	855	.17	559					.008	26
5/8" FR Taped	092-608-2150	22,014	SF	.31	6824	.35	7705					.017	374
5/8" FR Ceil. Taped	092-608-3150	490	SF	.31	152	.45	220					.021	10
High Ceilings	092-608-5200	39,084	SF	.10	3908	.11	4299					.005	195
1 4	3% Mat. 4% Labor			3%	452	4%	697					4%	33
Corner Bead	092-804-0800	14.4	CLF	7.50	108	59	850					2.76	40
Finish Corners	092-608-5350	2588	LF	.06	155	.31	802					.015	39
@ Windows													
1/2" Standard No Finish	092-608-0300	404	SF	.20	81	.17	69					.008	3
1 4	3% Mat. 4% Labor			3%	2	4%	3					4%	0
Corner Bead	092-804-0300	6.06	CLF	7.50	45	59	358					2.76	17
Finish Corners	092-608-5350	1344	LF	.06	81	.31	417					.015	20
Page Subtotals					15986		20631						983

Figure 26.62

Means® Forms
COST ANALYSIS

Division 9

SHEET NO. 3 of 7

PROJECT Commercial Renovation | CLASSIFICATION | ESTIMATE NO. 89-1

LOCATION | ARCHITECT | DATE 1989

TAKE OFF BY EBN | QUANTITIES BY EBN | PRICES BY RSM | EXTENSIONS BY | CHECKED BY

DESCRIPTION	SOURCE/DIMENSIONS	QUANTITY	UNIT	MATERIAL		LABOR		EQUIPMENT		SUBCONTRACT			
				UNIT COST	TOTAL	UNIT COST	TOTAL	UNIT COST	TOTAL	UNIT COST	TOTAL	UNIT COST	TOTAL
Division 9: (cont'd)													
Ceramic Tile:													
Base	093-102-1300	252	LF		SUBCONTRACT					7.10	1789		32
Floor	093-102-3400	684	SF							6.30	4309		59
Wall	093-102-5400	1008	SF							7.35	7409		90
Bullnose	093-102-2500	252	LF							5.55	1399		31
Carpet:	096-852-3200	129	SY		SUBCONTRACT					28	3612		21
Resilient:													
VCT	096-601-7350	153	SF							1.72	263		4
Vinyl Base	096-601-1150	993	LF							1.39	1380		25
Corners	096-601-1630	60	EA.							1.55	93		2
Page Subtotals											20254		264

Figure 26.63

Means® Forms
COST ANALYSIS

Division 9

PROJECT: Commercial Renovation | CLASSIFICATION: | SHEET NO. 4 of 7

LOCATION: | ARCHITECT: | ESTIMATE NO. 89-1

DATE: 1989

TAKE OFF BY: EBW | QUANTITIES BY: EBW | PRICES BY: RSM | EXTENSIONS BY: | CHECKED BY:

DESCRIPTION	SOURCE/DIMENSIONS	QUANTITY	UNIT	MATERIAL UNIT COST	MATERIAL TOTAL	LABOR UNIT COST	LABOR TOTAL	EQUIPMENT UNIT COST	EQUIPMENT TOTAL	SUBCONTRACT UNIT COST	SUBCONTRACT TOTAL	UNIT COST	TOTAL
Division 9: (Cont'd)													
Ceiling:													
Tile	095 104-3740	2837	SF	1.10	3121	.30	851					.014	40
Grid	091-304-0300	2837	SF	.42	1192	.26	738					.012	34
Lobby: Walls	Written Quote	876	SF										
Custom Floors		441	SF								17184		
Millwork Ceiling		441	SF										
Stair Treads	096-781-1000	192	LF	4.80	922	1.50	288					.070	13
Risers	096-781-1900	192	LF	1.30	250	.69	132					.032	6
Landings	086-781-1300	108	SF	2.80	302	1.43	154					.067	7
1	2% Mat. 3% Labor			2%	29	3%	17					3%	1
Refinish Wood Floor	095-604-7600	3190	SF	.75	2392	1.32	4211					.062	198
Stucco @ Entrance Lintel	092-304-0100	4	SY	1.84	7								
Min. Charge	092-304-1550		Job			170	170					8	8
Special Finish	092-304-0700	4	SY	1	4								
Page Subtotals					8219		6561				17184		307

Figure 26.64

first layer of a two layer application, at the new stairway enclosure. The estimator should visualize the work in order to include the proper costs.

The drywall returns at the window jambs and headers could easily be overlooked (Figure 26.42). For quantities, refer back to Figure 26.44, the Quantity Sheet, for the sills and grounds. the measurements have already been calculated and can be used for the drywall takeoff. The drywall installation at the window is almost all corners, and the appropriate finish work is included in the estimate (Figure 26.62).

It is important, when estimating drywall, to be aware of units. Studs, drywall, and taping are priced by the square foot, furring and corner bead per one hundred linear feet, and finishing corners by the linear foot. Confusion of units can result in expensive mistakes.

The estimator must be constantly aware of how the existing conditions and factors will affect the costs. For example, the wallcovering, ceiling, and carpet in the existing retail space must be installed after business hours, and precautions are necessary. The appropriate factors and the costs are included in Figure 26.65. The work, without factors, would cost $3,242. With the factors, however, the cost is $4,132, an increase of 27%.

Division 10: Specialties

Specialties are items of construction that do not fall within other divisions and are permanently attached or built into the work. The materials are usually pre-finished items installed at or near the end of the project. The estimate sheet for Division 10, Figure 26.69, does not include any factors. This is because previous work has "accounted for" or "corrected" discrepancies caused by existing conditions, and the items are installed at areas of new work. This is not to say that factors will never be applicable. For example, bathroom accessories may be specified at existing walls. If there is no backing, the wall must be cut and patched for proper installation.

The estimator must be very careful when examining the plans and specifications to be sure to include all required items that may be shown or described only once. Sometimes these items are included in a Room Finish Schedule. If there are no plans and specifications, the estimator should refer to past estimates or to a checklist.

Division 11: Architectural Equipment

This division includes pre-fabricated items or items that may be built and installed by specialty subcontractors. Often, the architect arranges to have the owner purchase architectural equipment. The estimator must include installation costs if necessary. Figure 26.70 is the estimate sheet for Divisions 11, 12, and 13. The only work in Division 11 is the installation of the kitchen unit in the third floor lounge.

Division 12: Furnishings

Furnishings are usually purchased by the owner. In the sample project, the blinds are included in the project specifications because all windows receive the same treatment. A construction contract will most likely not include furniture, but the estimator may be asked to provide budget prices.

Means® Forms
COST ANALYSIS

Division 9

PROJECT Commercial Renovation | CLASSIFICATION | SHEET NO. 5 of 7

LOCATION | ARCHITECT | ESTIMATE NO. 89-1

DATE 1989

TAKE OFF BY EBN | QUANTITIES BY EBN | PRICES BY RSM | EXTENSIONS BY | CHECKED BY

DESCRIPTION	SOURCE/DIMENSIONS	QUANTITY	UNIT	MATERIAL UNIT COST	MATERIAL TOTAL	LABOR UNIT COST	LABOR TOTAL	EQUIPMENT UNIT COST	EQUIPMENT TOTAL	SUBCONTRACT UNIT COST	SUBCONTRACT TOTAL	UNIT COST	TOTAL
Division 9: (Cont'd)													
Paint:													
Doors - Oak 7 Each	099-216-1800	58	EA.	3.20	186	32	1856					1.6	93
Metal 3 Side	099-216-1000	28	EA.	1.40	39	12.35	346					.615	17
Walls	099-224-0240	6680	SF	.04	267	.07	468					.004	27
	099-224-0840	6680	SF	.09	601	.14	935					.007	47
Sills	099-228-0010	77	SF	.07	5	.18	14					.009	1
Wall Covering	099-701-3300	7950	SF	.62	4929	.34	2703					.017	135
Spray Sandblast:													
Wood (Inc. 3rd Ceiling)	099-224-4100	20,460	SF	.06	1228	.18	3683					.009	184
Brick	099-224-4100	2100	SF	.06	126	.18	378					.009	19
Existing Retail													
Wall Covering	099-701-3300	580	SF	.62	360	.34	197					.017	10
Ceiling - Tile	095-104-3740	486	SF	1.10	535	.30	146					.014	7
Grid	091-304-0300	486	SF	.42	204	.26	126					.012	6
Carpet	096-852-1100	54	SY		SUBCONTRACT					31	1674	.14	8
1 + 2 + 5 + 6	14% Mat. 57% Labor			14%	154	57%	267			28%	469	27%	8
Page Subtotals					8634		11119				2143		562

Figure 26.65

Means® Forms
COST ANALYSIS

Division 9

SHEET NO. 6 of 7
ESTIMATE NO. 89-1
DATE 1989

PROJECT Commercial Renovation
CLASSIFICATION
LOCATION
ARCHITECT
TAKE OFF BY EBW
QUANTITIES BY EBW
PRICES BY RSM
EXTENSIONS BY
CHECKED BY

DESCRIPTION	SOURCE/DIMENSIONS	QUANTITY	UNIT	MATERIAL UNIT COST	MATERIAL TOTAL	LABOR UNIT COST	LABOR TOTAL	EQUIPMENT UNIT COST	EQUIPMENT TOTAL	SUBCONTRACT UNIT COST	SUBCONTRACT TOTAL	UNIT COST	TOTAL
Division 9:(Cont'd)													
Tenant:													
3 5/8" 25ga. 16" O.C.	092-612-2300	3015	SF	.32	965	.40	1206					.019	57
1/2" Drywall-Taped	092-608-0350	6030	SF	.25	1508	.35	2110					.017	103
High Ceilings	092-608-5200	6030	SF	.10	603	.11	663					.005	30
1 5	4% Mat. 5% Labor			4%	123	5%	199					5%	10
Corner Bead	092-804-0300	.30	CLF	7.50	2	59	18					2.76	1
Finish Corners	092-608-5350	615	LF	.06	37	.31	191					.015	9
Paint: Doors (Ea. Side)	099-216-1800	16	EA.	3.20	51	32	512					1.6	26
Walls	099-224-0240	2610	SF	.04	104	.07	183					.004	10
	099-224-0840	2610	SF	.09	235	.14	365					.007	18
Wall Covering	099-701-3080	4320	SF	.46	1987	.25	1080					.013	56
Ceiling	095-104-3740	607	SF	1.10	668	.30	182					.014	8
	091-304-0300	607	SF	.42	255	.26	158					.012	7
VCT	096-104-3750	607	SF		SUBCONTRACT	↓				1.72	1044	.025	15
Vinyl Base	096-601-1150	482	LF							1.39	670	.025	12
Carpe	096-852-4500	222	SY							37	8214	.14	31
Refinish Wood Floor	095-604-7600	418	SF	.75	313	1.32	552					.062	26
Page Subtotals					6851		7419				9928		419

Figure 26.66

Means® Forms
COST ANALYSIS

Division 9

SHEET NO. 7 of 7

PROJECT Commercial Renovation — CLASSIFICATION — ESTIMATE NO. 89-1

LOCATION — ARCHITECT — DATE 1989

TAKE OFF BY EBN — QUANTITIES BY EBN — PRICES BY RSM — EXTENSIONS BY — CHECKED BY

DESCRIPTION	SOURCE/DIMENSIONS	QUANTITY	UNIT	MATERIAL		LABOR		EQUIPMENT		SUBCONTRACT			
				UNIT COST	TOTAL	UNIT COST	TOTAL	UNIT COST	TOTAL	UNIT COST	TOTAL	UNIT COST	TOTAL
Division 9 : (Cont'd)													
Sheet 1					7264		10580						500
Sheet 2					15986		20631						983
Subtotal : Drywall Partitions					23250		31211						1483
Sheet 3 : Floor Covering											20254		264
Sheet 4					8219		6561				17184		307
Sheet 5					8634		11119				2143		562
Subtotal					40103		48891				39581		2616
Sheet 6 : Tenant					6851		7419				9928		419
Division 9 : Totals					46954		56310				49509		3035

Figure 26.67

Division 13: Special Construction

The estimator must carefully examine and analyze work in Division 13 on an item by item basis, obtaining prices from appropriate specialty subcontractors. Special construction often requires preparation work, excavation, and unloading, which may not be included in the subcontract. All such requirements must be included in the estimate.

Division 14: Conveying Systems

Installation of elevators in commercial renovation is often difficult at best. The estimator and the installing subcontractor must be thoroughly aware of the existing conditions and be familiar with the complete installation process. Hydraulic elevators are used most often in commercial renovation, unless the buildings are tall.

In the sample project, the price included in Figure 26.71 is a budget price only. A firm subcontract price must be obtained. Such a bid from the elevator installer will itemize what work is included and what is excluded. For the exclusions, the estimator must either determine and estimate the appropriate costs or define the exclusions in the final project bid. For example, ledge or rock encountered in drilling for a hydraulic elevator piston is very expensive. The potential extra costs should be included as a possible addendum in the prime contract.

The costs for the elevator budget price are derived from Figure 26.72. Note that there is a deduction for fewer than five stops. Factors are applied to the budget price because access for drilling is very restricted and the handling of the drilling waste requires special consideration. The calculations for the budget price are shown in Figure 26.73. However ridiculous it may seem, a common error would be to forget to double the elevator cost to include the two elevators.

Division 15: Mechanical

When mechanical plans and specifications are provided, and prepared by an engineer, takeoff and pricing can be relatively simple.

Quantity Comparison of Studs vs. Drywall

Metal Studs		Drywall	
1-5/8"	7,144 S.F.	1/2"	13,292 S.F.
3-5/8" (x2)	23,550	5/8"	22,014
6" (x2)	1,120	(disregard one layer on two lower walls and ceiling)	
Furring	3,492		
(convert to S.F.)			
	35,306 S.F.		35,306 S.F.

Figure 26.68

Means® Forms
COST ANALYSIS

Division 10

SHEET NO. 1 of 1

PROJECT Commercial Renovation | CLASSIFICATION | ESTIMATE NO. 89-1

LOCATION | ARCHITECT | DATE 1989

TAKE OFF BY EBW | QUANTITIES BY EBW | PRICES BY RSM | EXTENSIONS BY | CHECKED BY

DESCRIPTION	SOURCE/DIMENSIONS	QUANTITY	UNIT	MATERIAL		LABOR		EQUIPMENT		SUBCONTRACT			
				UNIT COST	TOTAL	UNIT COST	TOTAL	UNIT COST	TOTAL	UNIT COST	TOTAL	UNIT COST	TOTAL
Division 10: Specialties													
Grab Bars	108-204-1100	8	EA.	17	136	8.55	68					.4	3
1 1/4" Dia.	108-204-2000			15%	20								
Mirrors	108-204-3800	4	EA.	368	1472	29	116					1.33	5
Napkin Dispenser	108-204-4200	2	EA.	285	570	11.40	23					.533	1
Soap Dispenser	108-204-4600	11	EA.	52	572	8.55	94					.40	4
T.P. Holder	108-204-6100	13	EA.	19	247	5.70	74					.267	3
Towel Dispenser	108-204-6700	7	EA	52	364	10.70	75					.50	3
Waste Receptacle	108-204-8000	4	EA.	110	440	17.10	68					.80	3
Canvas Awnings	Written Quote	4	EA.		SUBCONTRACTOR					725	2900	8	32
Medicine Cabinets	108-208-0010	3	EA.	40	120	12.25	37					.571	2
Toilet Partitions	101-602-2500	10	EA.	235	2350	57	570					2.67	27
Handicapped	101-602-2900			70	700								
Urinal Screens	101-602-5300	4	EA	120	480	43	172					2	8
Division 10: Totals					7471		1297				2900		91

Figure 26.69

Means® Forms
COST ANALYSIS

Division 11,12,13

SHEET NO. 1of1

PROJECT Commercial Renovation — CLASSIFICATION — ESTIMATE NO. 89-1

LOCATION — ARCHITECT — DATE 1989

TAKE OFF BY EBW — QUANTITIES BY EBW — PRICES BY RSM — EXTENSIONS BY — CHECKED BY

DESCRIPTION	SOURCE/DIMENSIONS	QUANTITY	UNIT	MATERIAL		LABOR		EQUIPMENT		SUBCONTRACT			
				UNIT COST	TOTAL	UNIT COST	TOTAL	UNIT COST	TOTAL	UNIT COST	TOTAL	UNIT COST	TOTAL
Division 11: Architectural Equipment													
Kitchen Unit Labor Only Tenant	114-002-1660	1	EA.			43	43					1.78	2
Division 12: Furnishings													
Blinds (44)	125-103-0010	924	SF	2.20	2033.29		268					.014	13
(26)	125-103-0010	385	SF	2.20	847.29		112					.014	5
Division 12: Totals					2880		380						18
Division 13: Special Construction													
No Work													

Figure 26.70

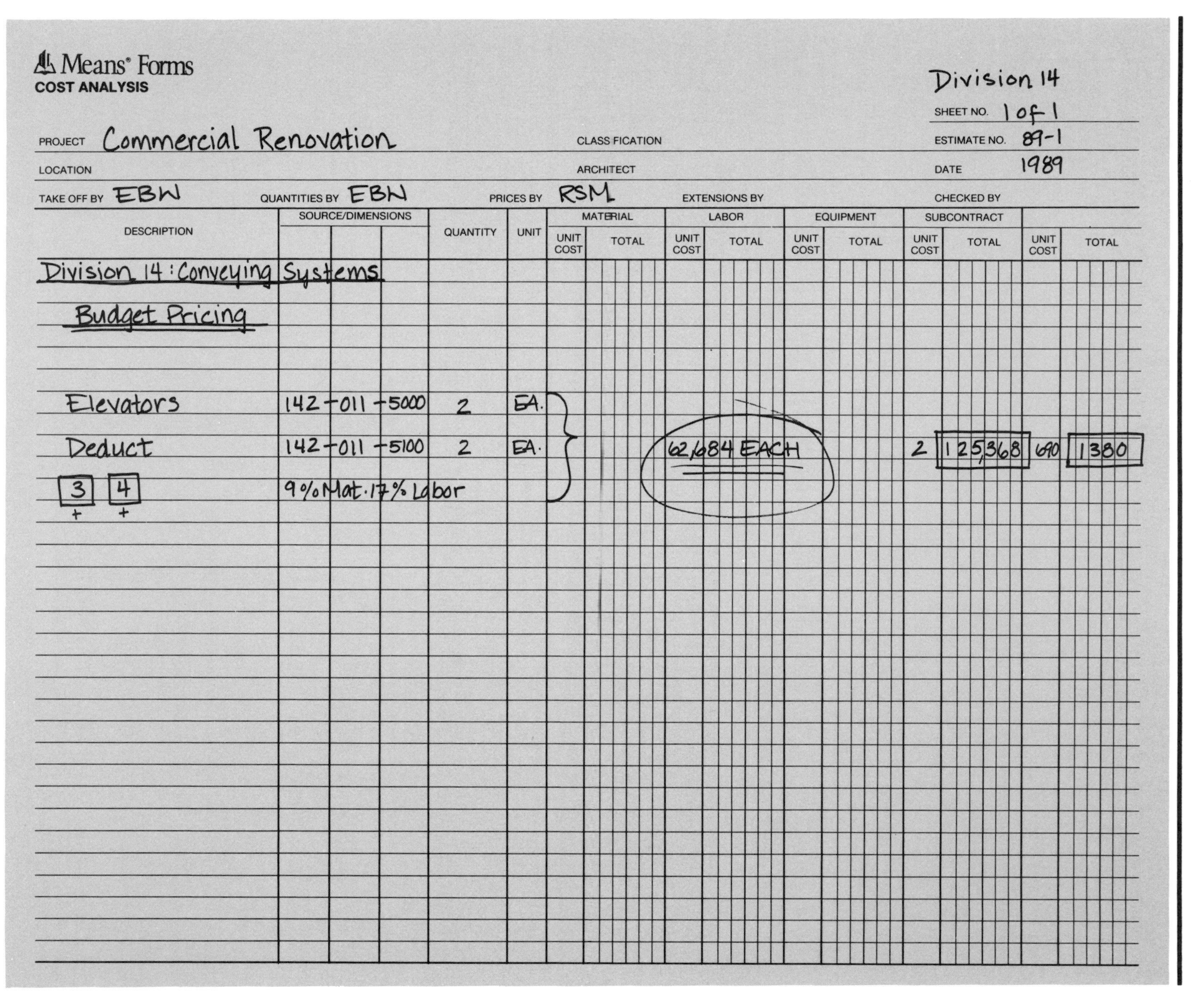

Means® Forms
COST ANALYSIS

Division 14

SHEET NO. 1 of 1

PROJECT Commercial Renovation　CLASSIFICATION　ESTIMATE NO. 89-1

LOCATION　ARCHITECT　DATE 1989

TAKE OFF BY EBW　QUANTITIES BY EBW　PRICES BY RSM　EXTENSIONS BY　CHECKED BY

DESCRIPTION	SOURCE/DIMENSIONS	QUANTITY	UNIT	MATERIAL UNIT COST	MATERIAL TOTAL	LABOR UNIT COST	LABOR TOTAL	EQUIPMENT UNIT COST	EQUIPMENT TOTAL	SUBCONTRACT UNIT COST	SUBCONTRACT TOTAL	UNIT COST	TOTAL
Division 14: Conveying Systems													
Budget Pricing													
Elevators	142-011-5000	2	EA.										
Deduct	142-011-5100	2	EA.			62,684 EACH				2	125,368	690	1380
3 + 4 +	9% Mat. 17% Labor												

Figure 26.71

141 | Dumbwaiters

	141 200 Electric Dumbwaiters		CREW	DAILY OUTPUT	MAN-HOURS	UNIT	BARE COSTS MAT.	BARE COSTS LABOR	BARE COSTS EQUIP.	BARE COSTS TOTAL	TOTAL INCL O&P	
201	0010	**DUMBWAITERS** 2 stop, electric, minimum	2 Elev	.13	123	Ea.	3,575	2,975		6,550	8,500	201
	0100	Maximum		.11	145	"	8,500	3,525		12,025	14,700	
	0600	For each additional stop, add		.54	29.630	Stop	850	715		1,565	2,025	

142 | Elevators

	142 010 Elevators		CREW	DAILY OUTPUT	MAN-HOURS	UNIT	BARE COSTS MAT.	BARE COSTS LABOR	BARE COSTS EQUIP.	BARE COSTS TOTAL	TOTAL INCL O&P	
011	0012	**ELEVATORS**										011
	5000	Passenger, pre-engineered, 5 story, hydraulic, 2,500 lb. cap.	M-1	.04	800	Ea.	42,700	18,400	1,725	62,825	77,000	
	5100	For less than 5 stops, deduct	"	.29	110	Stop	5,780	2,525	240	8,545	10,500	
	5200	For 4,000 lb. capacity, general purpose, add				Ea.	7,400			7,400	8,150	
	5400	10 story, geared traction, 200 FPM, 2,500 lb. capacity	M-1	.02	600	"	41,400	36,700	3,450	81,550	105,500	
	5500	For less than 10 stops, deduct		.34	94.120	Stop	3,150	2,150	205	5,505	7,000	
	5600	For 4,500 lb. capacity, general purpose		.02	600	Ea.	44,500	36,700	3,450	84,650	109,000	
	5800											
	7000	Residential, cab type, 1 floor, 2 stop, minimum	2 Elev	.20	80	Ea.	5,500	1,925		7,425	9,025	
	7100	Maximum		.10	160		9,800	3,875		13,675	16,700	
	7200	2 floor, 3 stop, minimum		.12	133		6,800	3,225		10,025	12,400	
	7300	Maximum		.06	267		16,500	6,450		22,950	28,000	
	7700	Stair climber (chair lift), single seat, minimum		1	16		3,550	385		3,935	4,500	
	7800	Maximum		.20	80		5,000	1,925		6,925	8,475	
	8000	Wheelchair, porch lift, minimum		1	16		3,200	385		3,585	4,125	
	8500	Maximum		.50	32		7,000	775		7,775	8,875	
	8700	Stair lift, minimum		1	16		5,800	385		6,185	6,975	
	8900	Maximum		.20	80		8,800	1,925		10,725	12,600	

144 | Lifts

	144 010 Lifts		CREW	DAILY OUTPUT	MAN-HOURS	UNIT	BARE COSTS MAT.	BARE COSTS LABOR	BARE COSTS EQUIP.	BARE COSTS TOTAL	TOTAL INCL O&P	
011	0010	**CORRESPONDENCE LIFT** 1 floor 2 stop, 25 lb. capacity, electric	2 Elev	.20	80	Ea.	3,575	1,925		5,500	6,900	011
	0100	Hand, 5 lb. capacity	"	.20	80	"	1,475	1,925		3,400	4,600	

145 | Material Handling Systems

	145 600 Chutes		CREW	DAILY OUTPUT	MAN-HOURS	UNIT	BARE COSTS MAT.	BARE COSTS LABOR	BARE COSTS EQUIP.	BARE COSTS TOTAL	TOTAL INCL O&P	
601	0010	**CHUTES** Linen or refuse, incl. sprinklers										601
	0020											
	0050	Aluminized steel, 16 ga., 18" diameter	2 Shee	3.50	4.570	Floor	525	110		635	750	
	0100	24" diameter	"	3.20	5	"	625	120		745	875	

176

Figure 26.72

If no plans are available, the work should be estimated by an experienced subcontractor who will be sure to include all requirements.

In the sample project, the plumbing work is itemized and estimated. A subcontractor will perform the work. Costs, including overhead and profit, are used. The estimator must visualize the whole system and follow the path of the piping to be sure that existing conditions will not restrict installation. If no plans are available, a quick riser diagram will be helpful.

Figures 26.74 to 26.77 are the estimate sheets for Division 15. In order to include all components, the estimator should make a list of all fixtures and equipment, complete with required faucets, fittings, and hanging hardware. All required backing should be included in Division 6.

Costs for water supply risers and waste vent and stack, only, are estimated because rough-in piping is included with the fixtures. A factor for working around existing conditions is applied to the entire piping cost.

Because of the scope and size of the sample project, a subcontractor bids the cost of heating, ventilation, and air conditioning. The estimator must be sure that all necessary work is included in the bids. For example, since the third floor ceiling is open to the trusses and roof deck, ductwork will be exposed. The corridors must be

Calculations for Elevator Budget Price

	Material	Labor	Equipment	Total Including O & P
Elevator 142-011-5000	$42,700	$18,400	$1,725	
Deduct (2 Stops) 142-011-5100	(11,560)	(5,050)	(480)	
	31,140	13,350	1,245	
Factors	(9%) 2,803	(17%) 2,270		
	33,943	15,620		
Overhead and Profit	(10%) 3,394	(53.5%) 8,357	(10%) 125	
	$37,337	$23,977	$1,370	$62,684
$62,684 × 2 Elevators = $125,368				

Figure 26.73

Means® Forms
COST ANALYSIS

Division 15

SHEET NO. 1 of 4

PROJECT Commercial Renovation | CLASSIFICATION | ESTIMATE NO. 89-1

LOCATION | ARCHITECT | DATE 1989

TAKE OFF BY EBW | QUANTITIES BY EBW | PRICES BY RSM | EXTENSIONS BY | CHECKED BY

DESCRIPTION	SOURCE/DIMENSIONS	QUANTITY	UNIT	MATERIAL		LABOR		EQUIPMENT		SUBCONTRACT			
				UNIT COST	TOTAL	UNIT COST	TOTAL	UNIT COST	TOTAL	UNIT COST	TOTAL	UNIT COST	TOTAL
Division 15: Mechanical													
Plumbing:													
Backflow Preventor	151-105-5120	1	EA.		SUBCONTRACT					5900	5900	8	8
Faucets:													
Lavatory	151-141-2100	15	EA.							53	795	.80	12
W/Drain	151-141-2200	15	EA.	9.40		15%				14.82	222	15%	2
Service	151-141-3000	2	EA.							69	138	.571	1
Fixtures:													
Drink Fountain	152-116-2820	2	EA.							765	1530	2.5	5
Rough	152-116-3980	2	EA.							190	380	4.37	9
Lavatories													
Vanity	152-136-0600	6	EA.							210	1260	2.5	15
Wall	152-136-4180	9	EA.							210	1890	2	18
Rough	152-136-6960	15	EA.							445	6675	9.64	145
Jan. Sink	152-140-0100	2	EA.							385	770	2.67	5
Rough	152-140-9600	2	EA.							365	730	8.7	17
Page Subtotals											20290		237

Figure 26.74

Means® Forms
COST ANALYSIS

Division 15

SHEET NO. 2 of 4

PROJECT Commercial Renovation   CLASSIFICATION   ESTIMATE NO. 89-1

LOCATION   ARCHITECT   DATE 1989

TAKE OFF BY EBW   QUANTITIES BY EBW   PRICES BY RSM   EXTENSIONS BY   CHECKED BY

DESCRIPTION	SOURCE/DIMENSIONS	QUANTITY	UNIT	MATERIAL		LABOR		EQUIPMENT		SUBCONTRACT			
				UNIT COST	TOTAL	UNIT COST	TOTAL	UNIT COST	TOTAL	UNIT COST	TOTAL	UNIT COST	TOTAL
Division 15 : Mechanical (Cont'd)													
Fixtures (Cont'd)													
Urinals	152-168-3000	4	EA.	SUBCONTRACT						505	2020	5.33	21
Rough	152-168-3300	4	EA.							330	1320	8.04	32
Water Cl.													
Flush	152-180-3100	10	EA.							365	3650	2.76	28
Rough	152-180-3200	10	EA.							410	4100	7.8	78
Tank	152-180-1100	3	EA.							230	690	3.02	9
Rough	152-180-1980	3	EA.							370	1110	8.25	25
Water Heater	152-110-6040	1	EA.							1025	1025	5.71	6
Sump Pumps	152-480-7100	2	EA.							285	570	1.33	3
Floor Drains	152-125-2040	1	EA.							76	76	1.33	1
Page Subtotals											14561		203

Figure 26.75

Means® Forms
COST ANALYSIS

Division 15

SHEET NO. 3 of 4

PROJECT Commercial Renovation

CLASSIFICATION

ESTIMATE NO. 89-1

LOCATION

ARCHITECT

DATE 1989

TAKE OFF BY EBW

QUANTITIES BY EBW

PRICES BY RSM

EXTENSIONS BY

CHECKED BY

DESCRIPTION	SOURCE/DIMENSIONS	QUANTITY	UNIT	MATERIAL UNIT COST	MATERIAL TOTAL	LABOR UNIT COST	LABOR TOTAL	EQUIPMENT UNIT COST	EQUIPMENT TOTAL	SUBCONTRACT UNIT COST	SUBCONTRACT TOTAL	UNIT COST	TOTAL
Division 15: Mechanical (Cont'd)													
Piping:													
6" PVC	151-551-4490	132	LF		SUBCONTRACT					20	2640	.41	54
90° Elbows	151-558-0620	16	EA.							96	1536	2	32
Tees	151-558-0890	12	EA.							155	1860	3.2	38
Copper													
2 1/2"	151-401-2280	112	LF							15.65	1753	.258	29
1"	151-401-2200	86	LF							6.30	542	.118	10
2 1/2" Elbows	151-430-0350	6	EA.							50	300	1.23	7
1" Elbows	151-430-0310	12	EA.							20	240	.5	6
2 1/2" Tees	151-430-0650	4	EA.							82	328	2	8
1" Tees	151-430-0510	7	EA.							32	224	.80	6
2 1/2" Coupl.	151-430-0750	8	EA.							40	320	1.07	9
1" Coupl.	151-430-0710	12	EA.							17.20	206	.444	5
Factors													
1 +	5% Mat. 9% Labor			5%		9%				7%	696	9%	20
Page Subtotals											10645		224

Figure 26.76

Means® Forms
COST ANALYSIS

Division 15

SHEET NO. 4 of 4
PROJECT Commercial Renovation
CLASSIFICATION
ESTIMATE NO. 89-1
LOCATION
ARCHITECT
DATE 1989
TAKE OFF BY EBW
QUANTITIES BY EBW
PRICES BY RSM
EXTENSIONS BY
CHECKED BY

DESCRIPTION	SOURCE/DIMENSIONS	QUANTITY	UNIT	MATERIAL		LABOR		EQUIPMENT		SUBCONTRACT			
				UNIT COST	TOTAL	UNIT COST	TOTAL	UNIT COST	TOTAL	UNIT COST	TOTAL	UNIT COST	TOTAL
Division 15: Mechanical-(Cont'd)													
Heating & Air Conditioning	Written Quote										225375		1650
Tenant-HVAC											17476		
Page Subtotals											242851		1650
Sheet 1											20290		237
Sheet 2											14561		203
Sheet 3											10645		224
Plumbing Subtotal											45496		664
Sheet 4											242851		1650
Division 15: Totals											288347		2314

Figure 26.77

heated and cooled, and the bathrooms require ventilation. The subcontractor should supply shop drawings of all work for clarification. At the owner's request, tenant costs are priced separately.

Division 16: Electrical

As with the heating, ventilation, and air conditioning, the electrical work is bid by a subcontractor (Figure 26.77). For purposes of this example, the itemized estimate for the tenant electrical work is included in Figure 26.78.

When estimating electrical work for commercial renovation, the estimator should visualize the installation and follow the proposed paths of the wiring during the site visit. Wrapping conduit around beams and columns can become very expensive. Such restrictive conditions exist in the tenant space. The ceiling is open to the roof deck. Light fixture wiring is exposed and carried in EMT. The conduit must be bent and neatly installed over, along, and around the exposed trusses. The exterior walls are sandblasted brick. Wiremold is specified. Factors must be applied to this work for conforming to existing conditions and for the protection of existing work, as shown in Figure 26.78 and 26.79.

The wiring for switches and receptacles in the perimeter offices must be concealed in the walls since there is no suspended ceiling. Much more wire is needed because straight runs are not possible. The estimator must check to be sure that the subcontractor bids the work to conform to all code requirements. Even if the plans are prepared by an electrical engineer, approval by local authorities is usually necessary, especially for fire alarms.

Estimate Summary

When the work for all divisions is priced, the estimator should complete the Project Schedule so that time-related costs in the Project Overhead Summary can be determined. When preparing the schedule, the estimator must visualize the entire construction process so that the correct sequence of work is determined. Certain tasks must be completed before others are begun. Different trades will work simultaneously. Material deliveries will affect scheduling. All such variables must be incorporated into the Project Schedule, as shown in Figure 26.80. The man-hour figures, which have been calculated for each division, are used to assist with scheduling. The estimator must be careful not to use the man-hours for each division independently. Each division must be coordinated with related work.

The schedule shows that the project will last approximately six months. Time-dependent items, such as equipment rental and superintendent costs, can be included in the Project Overhead Summary. Some items are dependent on total job costs, such as permits and insurance. The total direct costs for the project must be determined as shown in the Condensed Estimate Summary, Figure 26.81. All costs can now be included and totalled on the Project Overhead Summary, Figures 26.82a and 26.82b.

The estimator is now able to complete the Estimate Summary as shown in Figure 26.83. Appropriate contingency, sales tax, and overhead and profit costs must be added to direct costs of the project. Ten percent is added to material, equipment, and

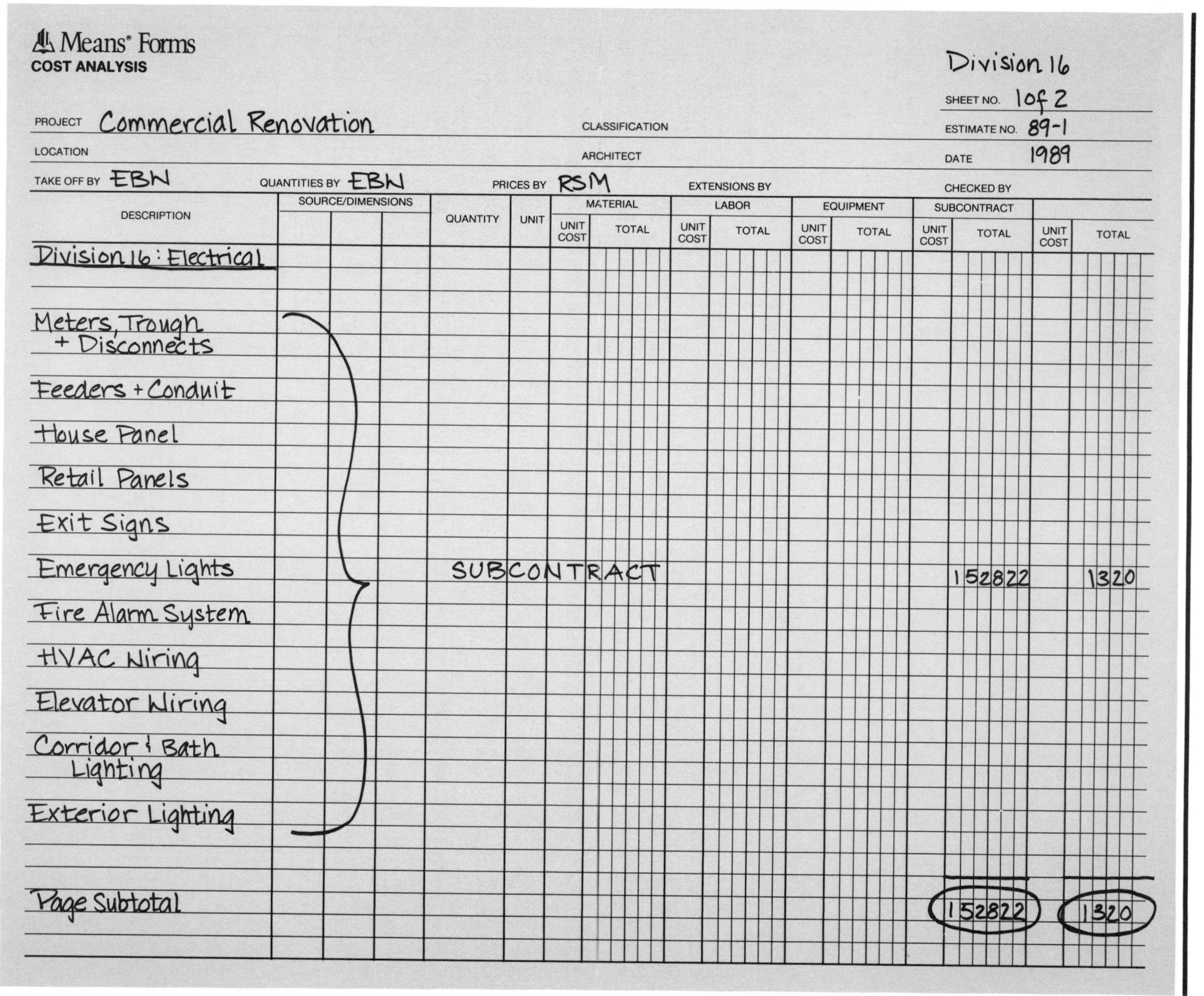
Means® Forms
COST ANALYSIS

Division 16

PROJECT Commercial Renovation | CLASSIFICATION | SHEET NO. 1 of 2
LOCATION | ARCHITECT | ESTIMATE NO. 89-1
TAKE OFF BY EBW | QUANTITIES BY EBW | PRICES BY RSM | EXTENSIONS BY | CHECKED BY | DATE 1989

DESCRIPTION	SOURCE/DIMENSIONS	QUANTITY	UNIT	MATERIAL UNIT COST	MATERIAL TOTAL	LABOR UNIT COST	LABOR TOTAL	EQUIPMENT UNIT COST	EQUIPMENT TOTAL	SUBCONTRACT UNIT COST	SUBCONTRACT TOTAL	UNIT COST	TOTAL
Division 16: Electrical													
Meters, Trough, + Disconnects													
Feeders + Conduit													
House Panel													
Retail Panels													
Exit Signs													
Emergency Lights		SUBCONTRACT									152822		1320
Fire Alarm System													
HVAC Wiring													
Elevator Wiring													
Corridor & Bath Lighting													
Exterior Lighting													
Page Subtotal											152822		1320

Figure 26.78

Means® Forms
COST ANALYSIS

Division 16

PROJECT	Commercial Renovation
CLASSIFICATION	
SHEET NO.	2 of 2
ESTIMATE NO.	89-1
LOCATION	
ARCHITECT	
DATE	1989
TAKE OFF BY	EBN
QUANTITIES BY	EBN
PRICES BY	RSM
EXTENSIONS BY	
CHECKED BY	

DESCRIPTION	SOURCE/DIMENSIONS	QUANTITY	UNIT	MATERIAL		LABOR		EQUIPMENT		SUBCONTRACT		TOTAL	
				UNIT COST	TOTAL	UNIT COST	TOTAL	UNIT COST	TOTAL	UNIT COST	TOTAL	UNIT COST	TOTAL
Division 16: Electrical (Cont'd)													
Tenant Work:													
EMT 3/4"	160-205-5020	285	LF		SUBCONTRACT					2.75	784	.062	18
Bends	160-205-5220	92	EA.							3.68	339	.10	9
Wiremold	160-290-0100	112	LF.							3.45	386	.08	9
Boxes	160-290-3000	16	EA.							23	368	.50	8
1 + 5 +	10% Mat. 16% Labor			10%		16%				13%	244	16%	7
Panel	163-245-0850	1	EA.							1325	1325	17.78	18
BX	161-105-0150	4.6	CLF							170	782	3.81	18
Boxes	162-110-0150	57	EA.							17.65	1006	.444	25
Rings	162-110-0300	57	EA.							5.35	305	.125	7
Switches	162-320-0500	14	EA.							15.45	216	.296	4
Recept.	162-320-2300	59	EA.							18.95	1118	.296	17
Lighting Recessed	166-130-0600	10	EA.							125	1250	1.7	17
Track Lighting	166-150-0100	24	EA.							110	2640	1.51	36
Fixtures	Quote	48	EA.							103	4944		
Page Subtotals											15707		193
Sheet 1											152822		1320
Sheet 2 (Tenant)											15707		193
Division 16 Totals											168529		1513

Figure 26.79

Means® Forms
PROJECT SCHEDULE

PROJECT Commercial Renovation

DATE

BY

WEEK ~~CALENDAR PERIOD~~

NO.	DESCRIPTION	1	2	3	4	5	6	7	8	9	10	11	12	13	14	15	16	17	18	19	20	21	22	23	24	25
	Asbestos Removal																									
	Demolition							Roof						Windows					Elev							
	Site Work				Sewer																		Sidewalk			
	Concrete																									
	Masonry - New																									
	Restoration																									
	Sandblast																									
	Metals																									
	Rough Carpentry							Roof Openings							Grounds				Floor Openings							
	Finish																									
	Insulation																									
	Roof																									
	Doors																									
	Windows																									
	Drywall																									
	Paint & Wall Cov.																									
	Floor																									
	Specialities																									
	Elevators																									
	Plumbing											Rough									Finish					
	HVAC										Boiler															
	Electrical		Temp.							Rough									Finish							
	Punch List																									

Figure 26.80

Means® Forms

CONDENSED ESTIMATE SUMMARY

PROJECT	Commercial Renovation	SHEET NO.	1 of 1
		ESTIMATE NO.	89-1
LOCATION		TOTAL AREA/VOLUME 25,200 SF (Rentable)	DATE 1989
ARCHITECT		COST PER S.F./C.F.	NO. OF STORIES 3
PRICES BY: EBW		EXTENSIONS BY: RSM	CHECKED BY:

DIV.	DESCRIPTION	MAN HOURS	MATERIAL	LABOR	EQUIPMENT	SUBCONTRACT	TOTAL
1.0	General Requirements						
2.0	Site Work	170	2147	3121			5268
	Demolition	1227	561	19314	1592	4574	26041
3.0	Concrete	147	6122	3272	361		9755
4.0	Masonry	970	6315	11840	181	20713	39049
	Restoration	1246	1678	24924	1571		28173
5.0	Metals	93	490	279		8993	9762
6.0	Carpentry Rough	98	1549	2105	82		3736
	Finish	41	1341	841	3		2185
7.0	Moisture & Thermal Protection	376	11275	3377	27	30050	44729
	Tenant	19	884	421	1		1306
8.0	Doors, Windows, Glass	447	22067	3770	83	54800	80720
	Tenant	40	3092	621	19	668	4400
9.0	Finishes	2616	40103	48890		39581	128574
	Tenant	419	6851	7419		9928	24198
10.0	Specialties	91	7471	1297		2900	11668
11.0	Equipment Tenant	2		43			43
12.0	Furnishings	18	2880	380			3260
13.0	Special Construction						
14.0	Conveying Systems	1380				125368	125368
	Budget Price						
15.0	Mechanical	2314				270871	270871
	Tenant					17476	17476
16.0	Electrical	1320				152822	152822
	Tenant	193				15707	15707
	Subtotals – Total Direct Costs		114826	131914	3920	754451	1,005,111
	Sales Tax 6 %						
	Overhead (Project) %						
	Subtotal						
	Profit & Overhead 10/59.2/10/10						
	Contingency 2 %						
	Adjustments						
	TOTAL BID						1,005,111

Figure 26.81

Means® Forms

PROJECT OVERHEAD SUMMARY

Division 1

SHEET NO. 1 of 2

PROJECT Commercial Renovation

ESTIMATE NO. 89-1

LOCATION

ARCHITECT

DATE 1989

QUANTITIES BY: EBW PRICES BY: RSM EXTENSIONS BY: CHECKED BY:

DESCRIPTION	MAN-HOURS	QUANTITY	UNIT	MATERIAL UNIT	MATERIAL TOTAL	LABOR UNIT	LABOR TOTAL	EQUIPMENT UNIT	EQUIPMENT TOTAL
Job Organization: Superintendent									
Project Manager									
Timekeeper & Material Clerk									
Clerical									
Safety, Watchman & First Aid									
010-036-0260	960	24	WK			975	23400		
Travel Expense: Superintendent									
Project Manager									
Engineering: Layout									
Inspection/Quantities									
Drawings									
CPM Schedule									
Testing: Soil									
Materials									
Structural									
Equipment: Cranes									
Concrete Pump, Conveyor, Etc.									
Elevators, Hoists									
Freight & Hauling									
Loading, Unloading, Erecting, Etc.									
Maintenance									
Pumping									
Scaffolding 015-254-0090 381 Frames @ $3.65/MO.	170	119	CSF			31	3689	1391	8346
Small Power Equipment/Tools Circle (14)									
STUD DRIVER 050-550-0010		1	EA.	230	230				
Field Offices: Job Office									
Architect/Owner's Office									
Temporary Telephones		6	MO.					100	600
Utilities (Power)		28	MSF					93	2604
Temporary Toilets									
Storage Areas & Sheds									
Temporary Utilities: Heat									
Light ~~& Power~~ 015-104-0350	66	280	CSF	2.06	577	5.70	1596		
Water									
PAGE TOTALS					807		28685		11550

Page 1 of 2

Figure 26.82a

Means® Forms

DIVISION 1 SHEET 2 OF 2 DESCRIPTION	QUANTITY	UNIT	MATERIAL UNIT	MATERIAL TOTAL	LABOR UNIT	LABOR TOTAL	EQUIP. FEES & RENTAL UNIT	EQUIP. FEES & RENTAL TOTAL
Total Brought Forward				807		28685		11550
Winter Protection: Temp. Heat/Protection								
Snow Plowing								
Thawing Materials								
Temporary Roads								
Rent 6 Parking Spaces	6	MO.					300	1800
Signs & Barricades: Site Sign	1	EA.		250				
Temporary Fences								
Temporary Stairs, Ladders & Floors								
Rails @ stair & Elev. (22)	270	LF	1.35	365	1.71	462		
Photographs 015-302-1000								
Clean Up								
Dumpster 020-620-0800	22	WK					265	5830
Final Clean Up 017-104-0100 (44)	28	MSF	1.60	45	26	728	2.37	66
Punch List								
Permits: Building 010-070-0010	0.5%							5026
Misc. Street Use	1	EA.					50	50
Sewer Connection	1	EA.					300	300
Insurance: Builders Risk								
Owner's Protective Liability	25%	JOB						2513
Umbrella								
Unemployment Ins. & Social Security								
Taxes								
City Sales Tax								
State Sales Tax								
Bonds								
Performance								
Material & Equipment								
Main Office Expense								
Special Items								
Fire Exting. 154-125-2080	4	EA.	42.50	170				
Legal: Contract Review							100	100
TOTALS: 1262 MAN-HOURS				1637		29875		27235

Page 2 of 2

Figure 26.82b

Means® Forms

CONDENSED ESTIMATE SUMMARY

SHEET NO. 1 of 1

PROJECT Commercial Renovation — ESTIMATE NO. 89-1

LOCATION — TOTAL AREA/VOLUME 25,200 SF (Rentable) — DATE 1989

ARCHITECT — COST PER S.F./C.F. — NO. OF STORIES 3

PRICES BY: EBW — EXTENSIONS BY: RSM — CHECKED BY:

DIV.	DESCRIPTION	MAN HOURS	MATERIAL	LABOR	EQUIPMENT	SUBCONTRACT	TOTAL
1.0	General Requirements						
2.0	Site Work	170	2147	3121			5268
	Demolition	1227	561	19314	1592	4574	26041
3.0	Concrete	147	6122	3272	361		9755
4.0	Masonry	970	6315	11840	181	20713	39049
	Restoration	1246	1678	24924	1571		28173
5.0	Metals	93	490	279		8993	9762
6.0	Carpentry Rough	98	1549	2105	82		3736
	Finish	41	1341	841	3		2185
7.0	Moisture & Thermal Protection	376	11275	3377	27	30050	44729
	Tenant	19	884	421	1		1306
8.0	Doors, Windows, Glass	447	22067	3770	83	54800	80720
	Tenant	40	3092	621	19	668	4400
9.0	Finishes	2616	40103	48890		39581	128574
	Tenant	419	6851	7419		9928	24198
10.0	Specialties	91	7471	1297		2900	11668
11.0	Equipment Tenant	2		43			43
12.0	Furnishings	18	2880	380			3260
13.0	Special Construction						
14.0	Conveying Systems	1380				125368	125368
	Budget Price						
15.0	Mechanical	2314				270871	270871
	Tenant					17476	17476
16.0	Electrical	1320				152822	152822
	Tenant	193				15707	15707
	Subtotals - Total Direct Costs		114826	131914	3920	754451	1,005,111
	Sales Tax 6%		6890				6890
	Overhead (Project)	1262	1637	29875	27235		58747
	Subtotal		123353	161789	31155	754451	1070748
	Profit & Overhead 10/59.2/10/10		12335	95779	3116	75445	186675
	Contingency 2%		2714	5151	685	16598	25148
	Adjustments						
	TOTAL BID		138402	262719	34956	846494	1,282,571

Figure 26.83

subcontractor costs for handling and supervision. The overhead and profit percentage of 59.2% for labor is obtained from Figure 26.84, as the average mark-up for skilled workers. Contractors should determine appropriate mark-ups for their own companies as discussed in Chapter 6, Indirect Costs.

As part of the requirements for the estimate, the tenant improvement costs have been separated within each appropriate Division and are included and totalled on a separate estimate sheet (Figure 26.85). Cost per square foot is shown at the top of the Estimate Summary. The owner can use this figure to budget costs for future tenant improvements in the building. This square foot cost figure should be used with discretion. Even in the same building, different areas will require special considerations because of existing conditions.

Note in Figures 26.83 and 26.85 that totals are recorded horizontally for each division, and vertically for the different cost categories. This method of recording the numbers provides a way for crosschecking the final calculations, an important feature since a mistake at this final stage of the estimate could be very costly.

Installing Contractor's Overhead & Profit

Below are the **average** installing contractor's percentage mark-ups applied to base labor rates to arrive at typical billing rates.

Column A: Labor rates are based on union wages averaged for 30 major U.S. cities. Base rates including fringe benefits are listed hourly and daily. These figures are the sum of the wage rate and employer-paid fringe benefits such as vacation pay, employer-paid health and welfare costs, pension costs, plus appropriate training and industry advancement funds costs.

Column B: Workers' Compensation rates are the national average of state rates established for each trade.

Column C: Column C lists average fixed overhead figures for all trades. Included are Federal and State Unemployment costs set at 6.2%; Social Security Taxes (FICA) set at 7.65%; Builder's Risk Insurance costs set at 0.34%; and Public Liability costs set at 1.55%. All the percentages except those for Social Security Taxes vary from state to state as well as from company to company.

Column D and E: Percentages in Columns D and E are based on the presumption that the installing contractor has annual billing of $500,000 and up. Overhead percentages may increase with smaller annual billing. The overhead percentages for any given contractor may vary greatly and depend on a number of factors, such as the contractor's annual volume, engineering and logistical support costs, and staff requirements. The figures for overhead and profit will also vary depending on the type of job, the job location, and the prevailing economic conditions. All factors should be examined very carefully for each job.

Column F: Column F lists the total of columns B, C, D, and E.

Column G: Column G is Column A (hourly base labor rate) multiplied by the percentage in Column F (O&P percentage).

Column H: Column H is the total of Column A (hourly base labor rate) plus Column G (Total O&P).

Column I: Column I is Column H multiplied by eight hours.

		A		B	C	D	E	F	G	H	I
Abbr.	Trade	Base Rate Incl. Fringes		Workers' Comp. Ins.	Average Fixed Overhead	Overhead	Profit	Total Overhead & Profit		Rate with O & P	
		Hourly	Daily					%	Amount	Hourly	Daily
Skwk	Skilled Workers Average (35 trades)	$21.90	$175.20	12.5%	15.7%	16.0%	15.0%	59.2%	$12.95	$34.85	$278.80
	Helpers Average (5 trades)	16.60	132.80	13.8				60.5	10.05	26.65	213.20
	Foremen Average, Inside (50¢ over trade)	22.40	179.20	12.5				59.2	13.25	35.65	285.20
	Foremen Average, Outside ($2.00 over trade)	23.90	191.20	12.5				59.2	14.15	38.05	304.40
Clab	Common Building Laborers	16.85	134.80	13.9				60.6	10.20	27.05	216.40
Asbe	Asbestos Workers	23.85	190.80	11.6				58.3	13.90	37.75	302.00
Boil	Boilermakers	23.90	191.20	7.9				54.6	13.05	36.95	295.60
Bric	Bricklayers	22.10	176.80	11.3				58.0	12.80	34.90	279.20
Brhe	Bricklayer Helpers	17.10	136.80	11.3				58.0	9.90	27.00	216.00
Carp	Carpenters	21.40	171.20	13.9				60.6	12.95	34.35	274.80
Cefi	Cement Finishers	20.80	166.40	7.9				54.6	11.35	32.15	257.20
Elec	Electricians	24.25	194.00	5.2				51.9	12.60	36.85	294.80
Elev	Elevator Constructors	24.15	193.20	6.8				53.5	12.90	37.05	296.40
Eqhv	Equipment Operators, Crane or Shovel	22.40	179.20	8.8				55.5	12.45	34.85	278.80
Eqmd	Equipment Operators, Medium Equipment	21.65	173.20	8.8				55.5	12.00	33.65	269.20
Eqlt	Equipment Operators, Light Equipment	20.60	164.80	8.8				55.5	11.45	32.05	256.40
Eqol	Equipment Operators, Oilers	18.45	147.60	8.8				55.5	10.25	28.70	229.60
Eqmm	Equipment Operators, Master Mechanics	23.10	184.80	8.8				55.5	12.80	35.90	287.20
Glaz	Glaziers	22.05	176.40	10.4				57.1	12.60	34.65	277.20
Lath	Lathers	21.60	172.80	8.4				55.1	11.90	33.50	268.00
Marb	Marble Setters	21.95	175.60	11.3				58.0	12.75	34.70	277.60
Mill	Millwrights	22.15	177.20	8.4				55.1	12.20	34.35	274.80
Mstz	Mosaic and Terrazzo Workers	21.45	171.60	7.0				53.7	11.50	32.95	263.60
Pord	Painters, Ordinary	20.10	160.80	10.4				57.1	11.50	31.60	252.80
Psst	Painters, Structural Steel	20.85	166.80	34.9				81.6	17.00	37.85	302.80
Pape	Paper Hangers	20.25	162.00	10.4				57.1	11.55	31.80	254.40
Pile	Pile Drivers	21.50	172.00	21.9				68.6	14.75	36.25	290.00
Plas	Plasterers	21.10	168.80	11.0				57.7	12.15	33.25	266.00
Plah	Plasterer Helpers	17.35	138.80	11.0				57.7	10.00	27.35	218.80
Plum	Plumbers	24.45	195.60	6.2				52.9	12.95	37.40	299.20
Rodm	Rodmen (Reinforcing)	23.15	185.20	22.0				68.7	15.90	39.05	312.40
Rofc	Roofers, Composition	19.75	158.00	25.7				72.4	14.30	34.05	272.40
Rots	Roofers, Tile & Slate	19.85	158.80	25.7				72.4	14.35	34.20	273.60
Rohe	Roofer Helpers (Composition)	14.65	117.20	25.7				72.4	10.60	25.25	202.00
Shee	Sheet Metal Workers	24.05	192.40	8.8				55.5	13.35	37.40	299.20
Spri	Sprinkler Installers	25.25	202.00	6.7				53.4	13.50	38.75	310.00
Stpi	Steamfitters or Pipefitters	24.55	196.40	6.2				52.9	13.00	37.55	300.40
Ston	Stone Masons	22.00	176.00	11.3				58.0	12.75	34.75	278.00
Sswk	Structural Steel Workers	23.25	186.00	26.8				73.5	17.10	40.35	322.80
Tilf	Tile Layers (Floor)	21.50	172.00	7.0				53.7	11.55	33.05	264.40
Tilh	Tile Layer Helpers	17.05	136.40	7.0				53.7	9.15	26.20	209.60
Trlt	Truck Drivers, Light	17.45	139.60	11.8				58.5	10.20	27.65	221.20
Trhv	Truck Drivers, Heavy	17.60	140.80	11.8				58.5	10.30	27.90	223.20
Sswl	Welders, Structural Steel	23.25	186.00	26.8				73.5	17.10	40.35	322.80
Wrck	*Wrecking	16.80	134.40	28.1				74.8	12.55	29.35	234.80

*Not included in Averages.

Figure 26.84

Means® Forms

CONDENSED ESTIMATE SUMMARY

PROJECT Tenant Improvements		SHEET NO. 1 of 1	
		ESTIMATE NO. 89-1	
LOCATION Commercial	TOTAL AREA/VOLUME 2916 SF	DATE 1989	
ARCHITECT	COST PER S.F./C.F. $ 23.56	NO. OF STORIES	
PRICES BY: EBW	EXTENSIONS BY:	CHECKED BY:	

DIV.	DESCRIPTION		MATERIAL	LABOR	EQUIPMENT	SUBCONTRACT	TOTAL
1.0	General Requirements						
2.0	Site Work						
3.0	Concrete						
4.0	Masonry						
5.0	Metals						
6.0	Carpentry						
7.0	Moisture & Thermal Protection		884	421	1		1306
8.0	Doors, Windows, Glass		3092	621	19	668	4400
9.0	Finishes		6851	7419		9928	24198
10.0	Specialties						
11.0	Equipment		43				43
12.0	Furnishings						
13.0	Special Construction						
14.0	Conveying Systems						
15.0	Mechanical						
	HVAC					17476	17476
16.0	Electrical					15707	15707
	Subtotals		10827	8504	20	43779	63130
	Sales Tax	%					
	Overhead	4 %	433	340	1	1751	2525
	Subtotal						
	OVERHEAD & Profit	10/59.2/10/10 %	1217	5340	2	4644	11203
	Contingency	2 %	225	177		911	1313
	TAX	6 %	689				689
	Adjustments						
	TOTAL BID		13391	14361	23	51085	78860

Figure 26.85

Chapter 27

ASSEMBLIES ESTIMATING EXAMPLE

The Assemblies, or Systems, estimate is useful during the design development stage of a project. The estimator needs only certain parameters and perhaps a preliminary floor plan to complete the estimate effectively. The advantage of using an Assemblies estimate is the ability to develop costs quickly and to establish a budget before preparation of working drawings and specifications. The estimator can easily substitute one assembly for another to determine the most cost effective approach. The Assemblies estimate can be completed in much less time than the Unit Price estimate. Some accuracy is sacrificed, however, and the Assemblies estimate should be used only for budgetary purposes.

In remodeling and renovation, costs vary greatly from project to project because of different requirements and the restrictions caused by existing conditions. Budgets and cost control are becoming increasingly important before the project enters the final design process and owners take on the expense of working drawings and specifications. It is crucial that the estimator combine a thorough evaluation of the existing conditions with the design parameters in order to properly complete the Assemblies estimate. The estimator must use experience to be sure to include all requirements since little information is provided. Applicable building and fire codes and local regulations must also be considered.

Prices used in the following example are from *Means Repair and Remodeling Cost Data*. A description of the use of the book is included in Chapter 25 of this book. The sample project below will vary in detail from actual projects. Every project must be treated individually.

Project Description

The sample project is the renovation of a twenty-five year old, two story suburban office building to eight apartments. The owner feels that the building is not profitable as office space and wants to know how much it will cost to convert to apartments. An architect has prepared a preliminary floor plan for only the ground level floor. The only information available to the estimator is the findings of the site visit, items passed on in a few discussions, and the floor plan, as shown in Figure 27.1.

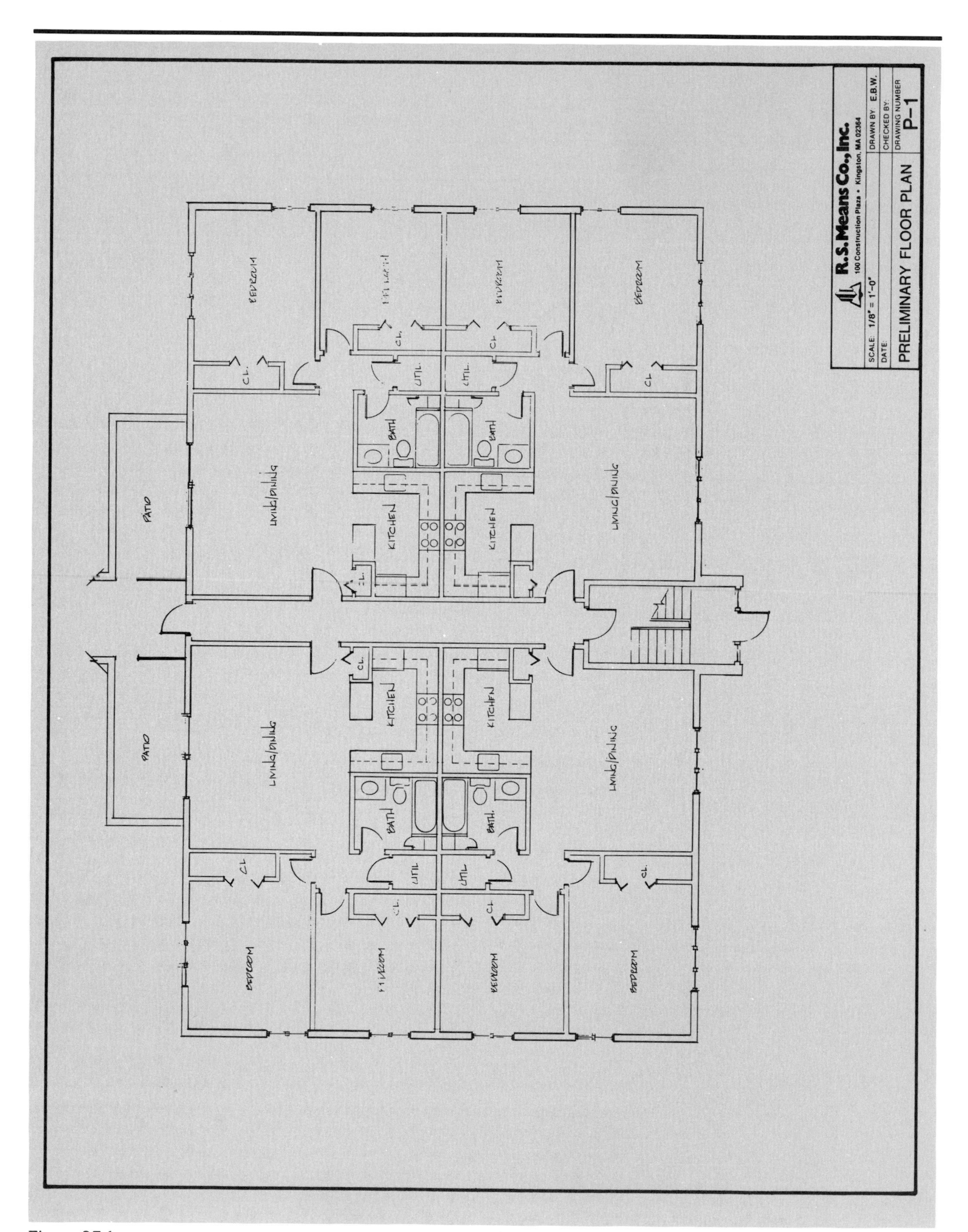
R.S. Means Co., Inc.
100 Construction Plaza • Kingston, MA 02364
SCALE: 1/8" = 1'-0"
DATE:
DRAWN BY: E.B.W.
CHECKED BY:
DRAWING NUMBER
P-1
PRELIMINARY FLOOR PLAN
PATIO
LIVING/DINING
KITCHEN
BATH
UTIL
CL.
BEDROOM

Figure 27.1

The exterior of the building is concrete block, with single pane, steel frame windows. The roof is wood trusses with old, curling asphalt shingles. The ground floor is partially below grade. The ground floor structure is a concrete slab and the second floor is wood joist. The requirements for the project and existing conditions will be discussed throughout the appropriate divisions.

Division 1: Foundations

Preliminary floor plans often do not designate what is new work and what is existing. The estimator must determine the scope of work during the site visit. Footings are required for the retaining walls at the patios. Costs are determined from Figure 27.2. The footing system as shown is appropriate. Access for trucks, however, is restricted, so the concrete must be placed by hand. The appropriate factor for equipment usage curtailment is added. The quantities and costs are entered on the cost analysis in Figure 27.3.

Costs for the retaining wall are determined from Figure 27.4. The system, as shown, must be modified to meet the requirements of the project. The waterproofing, insulation, and anchor bolts are not needed. The costs for these items must be deducted from the complete system. Also the wall is 6′ high and must be attached to the existing building foundation. The calculations, including the appropriate factors are shown in Figure 27.5. The costs are entered on the cost analysis (Figure 27.3).

Division 2: Substructure

Sections of the ground floor slab must be cut and removed to install plumbing pipes. The estimator must be sure to include the demolition costs in Division 12 of the Assemblies portion of *Means Repair and Remodeling Cost Data*. The system used to replace the concrete is shown in Figure 27.6. The forms are not required and are deducted, and the appropriate factors are added as shown in Figure 27.7. The total costs are entered in Figure 27.8.

Division 3: Superstructure

While no complete floor systems are specified for the project, note that the superstructure systems in *Means Repair and Remodeling Cost Data* (Figure 27.9) include floor and ceiling finishes. The local building code requires that apartment ceilings have a one-hour fire rating. The Floor and Ceiling Selective Price Sheet in Figure 27.10 is used to determine the costs for the rated ceiling. Painting costs are taken from Figure 27.9. The floor finishes will be included in Division 6, Interior Construction, of the Assemblies Section of *Means Repair and Remodeling Cost Data*.

During the site visit, the estimator must also determine the requirements for an exterior exit stairway from the second floor corridor. There is no indication of this item on the plan, so it is left to the estimator to itemize the requirements and calculate costs. Figure 27.11 is used to determine the costs for the stairway. The costs are entered on the cost analysis for Division 3 in Figure 27.12.

In Systems Estimating, the estimator must often make choices of methods and materials. Experience and a thorough evaluation of the existing conditions are important for making the correct choices.

FOUNDATIONS	**01.1-144**	**Strip Footing**

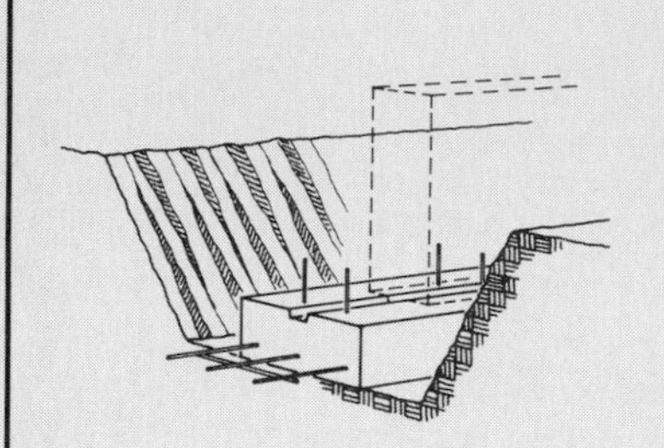

This page illustrates and describes a strip footing system including concrete, forms, reinforcing, keyway and dowels. Lines 01.1-144-04 thru 10 give the unit price and total price per linear foot for this system. Prices for alternate strip footing systems are on Line Items 01.1-144-15 thru 25. Both material quantities and labor costs have been adjusted for the system listed.

Factors: To adjust for job conditions other than normal working situations use Lines 01.1-144-29 thru 40.

Example: You are to install this footing, and due to a lack of accessibility, only hand tools can be used. Material handling is also a problem. Go to Lines 01.1-144-34 and 36 and apply these percentages to the appropriate MAT. and INST. costs.

LINE NO.	DESCRIPTION	QUANTITY	COST PER L.F.		
			MAT.	INST.	TOTAL
01	**Strip footing, 2'-0" wide x 1'-0" thick, 2000 psi concrete including forms**				
02	**Reinforcing, keyway, and dowels.**				
03					
04	Concrete, 2000 psi	.074 C.Y.	4.29		4.29
05	Placing concrete	.074 C.Y.		.88	.88
06	Forms, footing, 4 uses	2 S.F.	.64	4.40	5.04
07	Reinforcing	3.17 Lb.	.94	.93	1.87
08	Keyway, 2" x 4", 4 uses	1 L.F.	.08	.52	.60
09	Dowels, #4 bars, 2' long, 24" O.C.	.5 Ea.	.52	2.51	3.03
10	TOTAL	L.F.	6.47	9.24	15.71
11			3% .19	10% .92	
12			6.66	10.16	
13					
14	Above system with the following:				
15	2'-0" wide x 1' thick, 3000 psi concrete	L.F.	6.69	9.24	15.93
16	4000 psi concrete	"	6.92	9.24	16.16
17					
18	For alternate footing systems:				
19	2'-6" wide x 1' thick, 2000 psi concrete	L.F.	7.82	9.71	17.53
20	3000 psi concrete		8.10	9.71	17.81
21	4000 psi concrete	↓	8.38	9.71	18.09
22					
23	3'-0" wide x 1' thick, 2000 psi concrete	L.F.	9.04	10.15	19.19
24	3000 psi concrete		9.37	10.15	19.52
25	4000 psi concrete	↓	9.70	10.15	19.85
26					
27					
28					
29	Cut & patch to match existing construction, add, minimum		2%	3%	
30	Maximum		5%	9%	
31	Dust protection, add, minimum		1%	2%	
32	Maximum		4%	11%	
33	Equipment usage curtailment, add, minimum		1%	1%	
34	Maximum		3%	10%	
35	Material handling & storage limitation, add, minimum		1%	1%	
36	Maximum		6%	7%	
37	Protection of existing work, add, minimum		2%	2%	
38	Maximum		5%	7%	
39	Shift work requirements, add, minimum			5%	
40	Maximum			30%	
41					
42					

238

Figure 27.2

Means® Forms

COST ANALYSIS

SHEET NO. 1 of 11

PROJECT Apartment Renovation

ESTIMATE NO. 89-2

ARCHITECT

DATE 1989

TAKE OFF BY: QUANTITIES BY: EBW PRICES BY: RSM EXTENSIONS BY: CHECKED BY:

DESCRIPTION	SOURCE/DIMENSIONS			QUANTITY	UNIT	MATERIAL		LABOR		EQ./TOTAL	
						UNIT COST	TOTAL	UNIT COST	TOTAL	UNIT COST	TOTAL
Division 1: Foundations											
Footings				64	LF	6.66	426	10.16	650		
Retaining Wall				64	LF	20.73	1327	48.50	3104		
Division 1: Totals							1753		3754		

Figure 27.3

FOUNDATIONS	01.1-214	Concrete Wall

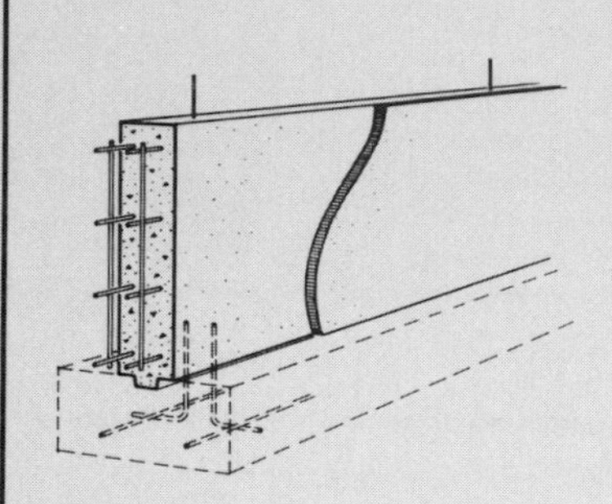

This page illustrates and describes a concrete wall system including concrete, placing concrete, forms, reinforcing, insulation, waterproofing and anchor bolts. Lines 01.1-214-04 thru 11 give the unit price and total price per linear foot for this system. Prices for alternate concrete wall systems are on Line Items 01.1-214-15 thru 26. Both material quantities and labor costs have been adjusted for the system listed.

Factors: To adjust for job conditions other than normal working situations use Lines 01.1-214-29 thru 40.

Example: You are to install this wall system where delivery of material is difficult. Go to Line 01.1-214-36 and apply these percentages to the appropriate MAT. and INST. costs.

LINE NO.	DESCRIPTION	QUANTITY	COST PER L.F. MAT.	INST.	TOTAL
01	**Cast in place concrete foundation wall, 8" thick, 3' high, 2500 psi**				
02	**Concrete including forms, reinforcing, waterproofing, and anchor bolts.**				
03					
04	Concrete, 2500 psi, 8" thick, 3' high	.07 C.Y.	4.13		4.13
05	Forms, wall, 4 uses	6 S.F.	3.70	19.64	23.34
06	Reinforcing	6 Lb.	1.85	1.27	3.12
07	Placing concrete	.07 C.Y.		2.03	2.03
08	Waterproofing (Deduct)	3 S.F.	.50	1.63	2.13
09	Rigid insulaton, 1" polystyrono (Deduct)	3 S.F.	.66	1.20	1.86
10	Anchor bolts, ½" diameter, 4' O.C. (Deduct)	.25 Ea.	.12	.37	.49
11	TOTAL	L.F.	10.96	26.14	37.10
12					
13					
14	For alternate wall systems:				
15	8" thick, 2500 psi concrete, 4' high	L.F.	14.95	34.94	49.89
16	6' high		22.37	52.22	74.59
17	8' high		29.79	69.50	99.29
18	3500 psi concrete, 4' high		15.35	34.94	50.29
19	6' high		22.97	52.22	75.19
20	8' high		30.59	69.50	100.09
21	12" thick, 2500 psi concrete, 4' high		19.14	37.23	56.37
22	6' high		27.73	55.10	82.83
23	8' high		36.92	73.25	110.17
24	3500 psi concrete, 4' high		19.74	37.23	56.97
25	8' high		38.12	73.25	111.37
26	10' high		47.30	91.33	138.63
27					
28					
29	Cut & patch to match existing construction, add, minimum		2%	3%	
30	Maximum		5%	9%	
31	Dust protection, add, minimum		1%	2%	
32	Maximum		4%	11%	
33	Equipment usage curtailment, add, minimum		1%	1%	
34	Maximum		3%	10%	
35	Material handling & storage limitation, add, minimum		1%	1%	
36	Maximum		6%	7%	
37	Protection of existing work, add, minimum		2%	2%	
38	Maximum		5%	7%	
39	Shift work requirements, add, minimum			5%	
40	Maximum			30%	
41					
42					

239

Figure 27.4

Division 4: Exterior Closure

The requirements for the exterior of the building are determined through discussions with the owner.

1. Stucco over the existing concrete block
2. Insulation
3. New casement windows
4. New entrance door
5. New corridor exit doors
6. Patio doors at ground floor

These items can be easily priced, but the existing conditions will have a great effect on the work. The quantity of the stucco finish must be determined at the site. New door and window openings must be deducted. The costs for stucco are found in Figure 27.13, and are included in the estimate in Figure 27.14.

While the systems as described in *Means Repair and Remodeling Cost Data* may not conform exactly to job requirements, portions of the systems may be used as needed. The owner has requested that the building be well insulated, but has not specified the type of insulation. That choice is left to the estimator. Figures 27.15 and 27.16 illustrate two types of exterior wall systems that have different interior insulation and finish treatments. By comparing the costs of the two treatments (and also the R-Values) the estimator is able to determine which is better for the particular application. This cost comparison is demonstrated in Figure 27.17. Note that the drywall in the second calculation is 1/2'' substituted for 5/8'' as shown, in order to compare the systems equally. The wood stud/fiberglass insulation system is chosen because of lower cost and higher R-value.

Calculations for Retaining Wall

	Material	Installation	Total
Concrete Wall 01.1-214-16	$22.37	$52.22	$74.59
Deducts			
Waterproofing	(1.00)	(3.26)	(4.26)
Insulation	(1.32)	(2.40)	(3.72)
Anchor Bolts	(.12)	(.37)	(0.49)
Subtotal	19.93	46.19	66.12
Factors 4% Material 5% Installation	.80	2.31	3.11
Total Costs per L.F.	$20.73	$48.50	$69.23

Figure 27.5

SUBSTRUCTURE	02.1-104	Interior Slab On Grade

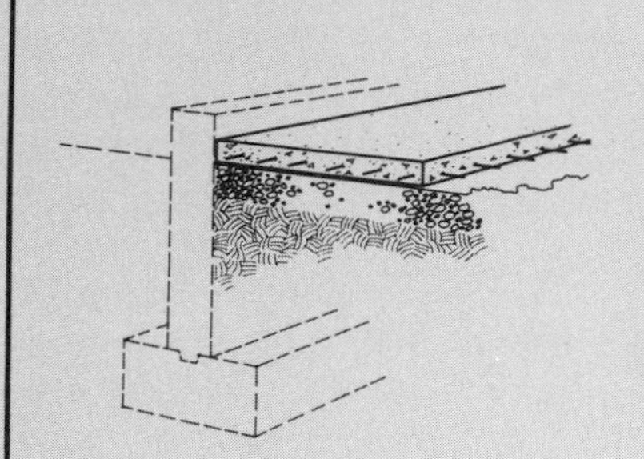

This page illustrates and describes a slab on grade system including slab, bank run gravel, bulkhead forms, placing concrete, welded wire fabric, vapor barrier, steel trowel finish and curing paper. Lines 02.1-104-04 thru 11 give the unit price and total price per square foot for this system. Prices for alternate slab on grade systems are on Line Items 02.1-104-15 thru 26. Both material quantities and labor costs have been adjusted for the system listed.

Factors: To adjust for job conditions other than normal working situations use Line 02.1-104-29 thru 40.

Example: You are to install the system at a site where protection of the existing building is required. Go to Line 02.1-104-38 and apply these percentages to the appropriate MAT. and INST. costs.

LINE NO.	DESCRIPTION	QUANTITY	COST PER S.F. MAT.	COST PER S.F. INST.	COST PER S.F. TOTAL
01	**Ground slab, 4" thick, 3000 psi concrete, 4" granular base, vapor barrier**				
02	**Welded wire fabric, screed and steel trowel finish.**				
03					
04	Concrete, 4" thick, 3000 psi concrete	.012 C.Y.	.73		.73
05	Bank run gravel, 4" deep	.074 C.Y.	.11	.04	.15
06	Polyethylene vapor barrier, 10 mil.	.011 C.S.F.	.03	.09	.12
07	Bulkhead forms, expansion material *Deduct*	.1 L.F.	.02	.18	.20
08	Welded wire fabric, 6 x 6 - #10/10	.011 C.S.F.	.09	.21	.30
09	Place concrete	.012 C.Y.		.16	.16
10	Screed & steel trowel finish	1 S.F.		.53	.53
11	TOTAL	S.F.	.98	1.21	2.19
12					
13					
14	Above system with the following:				
15	4" thick slab, 3000 psi concrete, 6" deep bank run gravel	S.F.	1.04	1.22	2.26
16	12" deep bank run gravel	"	1.21	1.25	2.46
17					
18					
19					
20	For alternate slab systems:				
21	5" thick slab, 3000 psi concrete, 6" deep bank run gravel	S.F.	1.23	1.26	2.49
22	12" deep bank run gravel	"	1.40	1.29	2.69
23					
24					
25	6" thick slab, 3000 psi concrete, 6" deep bank run gravel	S.F.	1.47	1.31	2.78
26	12" deep bank run gravel	"	1.64	1.34	2.98
27					
28					
29	Cut & patch to match existing construction, add, minimum		2%	3%	
30	Maximum		5%	9%	
31	Dust protection, add, minimum		1%	2%	
32	Maximum		4%	11%	
33	Equipment usage curtailment, add, minimum		1%	1%	
34	Maximum		3%	10%	
35	Material handling & storage limitation, add, minimum		1%	1%	
36	Maximum		6%	7%	
37	Protection of existing work, add, minimum		2%	2%	
38	Maximum		5%	7%	
39	Shift work requirements, add, minimum			5%	
40	Maximum			30%	
41					
42			*14%*	*26%*	

241

Figure 27.6

By thinking ahead, the estimator should realize that the stud system will allow for easier installation of electrical receptacles and switches in the exterior walls. The furring system would require that the concrete block be chipped away at every box. This is a good example of the advantage of being able to visualize the whole job.

The doors and windows, with appropriate factors added, are taken from Figures 27.18 to 27.20. The entrance door and corridor exit door systems include panic hardware. The estimator must be sure that the hardware is as required. Figure 27.21, Hardware Selective Price Sheet, shows that the panic device is rim-type, exit only. A mortise lock for exit and entrance is required at the entrance and is substituted as shown in Figure 27.22.

Division 5: Roofing

The existing roof truss system is to remain, with some modifications. New shingles and roof trim are specified, and the owner wants 9" of fiberglass insulation installed. The costs for these items are obtained from Figures 27.23 and 27.24. Remember that the drywall ceiling was included in Division 3. The costs for new aluminum gutters and downspouts are found in the Roof Accessory Selective Price Sheet, Figures 27.25 and 27.26. The calculations for adding the appropriate factors are shown in Figure 27.27, and the total costs for Division 5 of the Assemblies Section of *Means Repair and Remodeling Cost Data* are entered on Figure 27.28.

Already, the advantage of speed in Systems Estimating can be seen. In a relatively short period of time, a large portion of the renovation has been estimated.

Substructure			
	Material	Installation	Total
Interior Slab 02.1-104	$ .98	$1.21	$2.19
Deducts Forms	(.02)	(.18)	(.20)
Subtotal	0.96	1.03	1.99
Factors 14% Material 26% Installation	0.13	.27	.40
Total/S.F.	$1.09	$1.30	$2.39

Figure 27.7

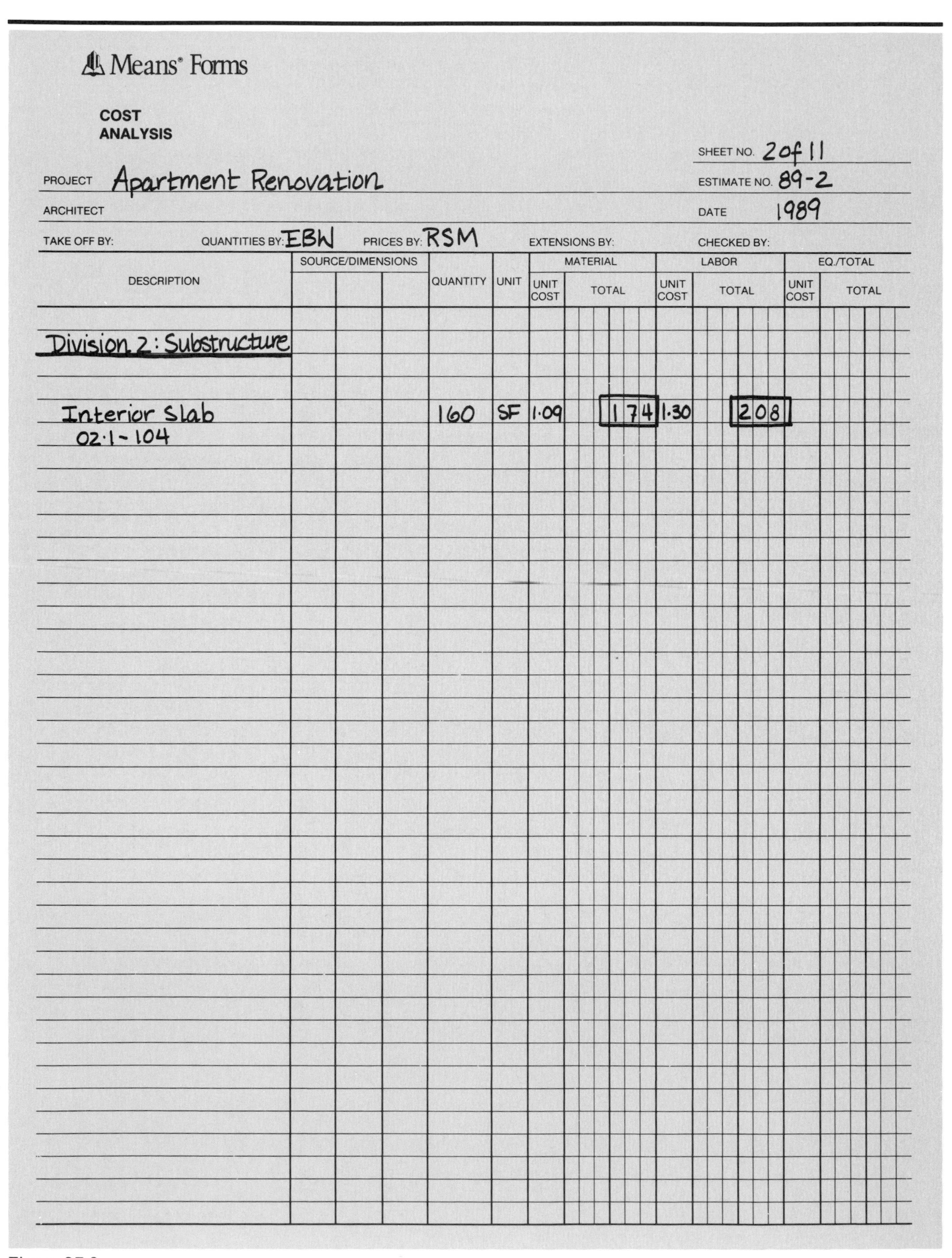

Means® Forms

COST ANALYSIS

SHEET NO. 2 of 11

PROJECT Apartment Renovation

ESTIMATE NO. 89-2

ARCHITECT

DATE 1989

TAKE OFF BY: QUANTITIES BY: EBW PRICES BY: RSM EXTENSIONS BY: CHECKED BY:

DESCRIPTION	SOURCE/DIMENSIONS			QUANTITY	UNIT	MATERIAL		LABOR		EQ./TOTAL	
						UNIT COST	TOTAL	UNIT COST	TOTAL	UNIT COST	TOTAL
Division 2: Substructure											
Interior Slab 02.1-104				160	SF	1.09	174	1.30	208		

Figure 27.8

SUPERSTRUCTURE	03.5-714	Floor-Ceiling, Wood Joists

This page illustrates and describes a wood joist floor system including wood joist, oak floor, sub-floor, bridging, sand and finish floor, furring, plasterboard, taped, finished and painted ceiling. Lines 03.5-714-04 thru 13 give the unit price and total price per square foot for this system. Prices for alternate wood joist floor systems are on Line Items 03.5-714-17 thru 23. Both material quantities and labor costs have been adjusted for the system listed.

Factors: To adjust for job conditions other than normal working situations use Lines 03.5-714-27 thru 40.

Example: You are to install the system during off peak hours, 6 P.M. to 2 A.M. Go to Line 03.5-714-38 and apply this percentage to the appropriate INST. costs

LINE NO.	DESCRIPTION	QUANTITY	COST PER S.F.		
			MAT.	INST.	TOTAL
01	**Wood joists, 2" x 8", 16" O.C.,oak floor (sanded & finished),½" sub**				
02	**Floor, 1"x 3" bridging, furring, ⅝" drywall, taped, finished and painted.**				
03					
04	Wood joists, 2" x 8", 16" O.C.	1 L.F.	.63	.51	1.14
05	C.D.X. plywood sub floor, ½" thick	1 S.F.	.45	.38	.83
06	Bridging 1" x 3"	.15 Pr.	.04	.33	.37
07	Oak flooring, No. 1 common	1 S.F.	2.97	1.49	4.46
08	Sand and finish floor	1 S.F.	.55	.93	1.48
09	Wood furring, 1" x 3", 16" O.C.	1 L.F.	.11	.79	.90
10	Gypsum drywall, ⅝" thick	1 S.F.	.26	.31	.57
11	Tape and finishing	1 S.F.	.06	.27	.33
12	Paint ceiling	1 S.F.	.10	.27	.37
13	TOTAL	S.F.	5.17	5.28	10.45
14					
15					
16	For alternate wood joist systems:				
17	16" on center, 2" x 6" joists	S.F.	4.97	5.22	10.19
18	2" x 10"		5.40	5.40	10.80
19	2" x 12"		5.54	5.42	10.96
20	2" x 14"		5.72	5.51	11.23
21	12" on center, 2" x 10" joists		5.61	5.56	11.17
22	2" x 12"		5.78	5.59	11.37
23	2" x 14"		6	5.71	11.71
24					
25					
26					
27	Cut & patch to match existing construction, add, minimum		2%	3%	
28	Maximum		5%	9%	
29	Dust protection, add, minimum		1%	2%	
30	Maximum		4%	11%	
31	Equipment usage curtailment, add, minimum		1%	1%	
32	Maximum		3%	10%	
33	Material handling & storage limitation, add, minimum		1%	1%	
34	Maximum		6%	7%	
35	Protection of existing work, add, minimum		2%	2%	
36	Maximum		5%	7%	
37	Shift work requirements, add, minimum			5%	
38	Maximum			30%	
39	Temporary shoring and bracing, add, minimum		2%	5%	
40	Maximum		5%	12%	
41					
42					

247

Figure 27.9

03.9-900	Floor & Ceiling Selective Price Sheet			
LINE NO.	DESCRIPTION	COST PER S.F.		
		MAT.	INST.	TOTAL
01	Flooring, carpet, nylon, level loop, 26 oz. light traffic	1.89	.55	2.44
02	40 oz. heavy traffic	2.87	.57	3.44
03	Nylon, plush, 20 oz. light traffic	1.16	.52	1.68
04	24 oz. medium traffic	1.86	.47	2.33
05	26 oz. heavy traffic	2.56	.55	3.11
06	28 oz. heavy traffic	3.11	.55	3.66
07	Tile, foamed back, needle punch	2.20	.41	2.61
08	Tufted loop	1.93	.40	2.33
09	Wool, 36 oz. medium traffic, level loop	3.20	.57	3.77
10	48 oz. heavy traffic, patterned	4.48	.63	5.11
11	Composition, epoxy, with colored chips, minimum	1.43	2.12	3.55
12	Maximum	2.40	2.90	5.30
13	Trowelled, minimum	1.93	2.55	4.48
14	Maximum	2.64	2.96	5.60
15	Terrazzo, ¼" thick, chemical resistant, minimum	3.47	3.83	7.30
16	Maximum	8.47	5.08	13.55
17	Resilient, asphalt tile, ⅛" thick	.77	.66	1.43
18	Conductive flrg, rubber, ⅛" thick	1.98	.84	2.82
19	Cork tile, ⅛" thick, standard finish	1.27	.83	2.10
20	Urethane finish	1.71	.83	2.54
21	PVC sheet goods for gyms, ¼" thick	2.97	3.28	6.25
22	⅜" thick	3.30	4.40	7.70
23	Vinyl composition 12" x 12" tile, plain, 1/16" thick	.55	.84	1.39
24	⅛" thick	1.49	.83	2.32
25	Vinyl tile, 12" x 12" x ⅛" thick, minimum	1.65	.84	2.49
26	Maximum	7.98	.82	8.80
27	Vinyl sheet goods, backed, .093" thick	1.38	1.14	2.52
28	.250" thick	2.48	1.14	3.62
29	Wood, maple strip 25/32" x 2-¼", finished, select	3.14	2.54	5.68
30	2nd and better	2.37	2.54	4.91
31	Oak, 25/32" x 2-¼" finished, clear	6.93	2.55	9.48
32	No. 1 common	3.52	2.42	5.94
33	Parquet, standard 5/16" thick finished, minimum	1.93	2.64	4.57
34	Maximum	5.50	3.68	9.18
35	Custom finished, minimum	11.55	2.75	14.30
36	Maximum	16.23	5.77	22
37	Prefinished, oak, 2-¼" wide	5.50	1.60	7.10
38	Ranch plank	6.60	1.90	8.50
39	Sleepers on concrete, treated, 24" O.C., 1" x 2"	.08	.19	.27
40	1" x 3"	.12	.22	.34
41	2" x 4"	.30	.30	.60
42	2" x 6"	.44	.36	.80
43	Ceiling, plaster, gypsum, 2 coats	.34	1.56	1.90
44	3 coats	.47	1.86	2.33
45	Perlite or vermiculite, 2 coats	.39	1.81	2.20
46	3 coats	.62	2.27	2.89
47	Gypsum lath, plain, ⅜" thick	.39	.35	.74
48	½" thick	.42	.37	.79
49	Firestop, ½" thick	.42	.42	.84
50	⅝" thick	.44	.46	.90
51	Metal lath, rib, 2.75 lb.	.19	.40	.59
52	3.40 lb.	.22	.43	.65
53	Diamond, 2.50 lb.	.13	.41	.54
54	3.40 lb.	.14	.51	.65
55	Drywall, taped and finished, standard, ½" thick	.28	.71	.99
56	⅝" thick	.32	.72	1.04
57	Fire resistant, ½" thick	.32	.72	1.04
58	⅝" thick	.34	.72	1.06

249

Figure 27.10

SUPERSTRUCTURE	03.9-104	Stairs

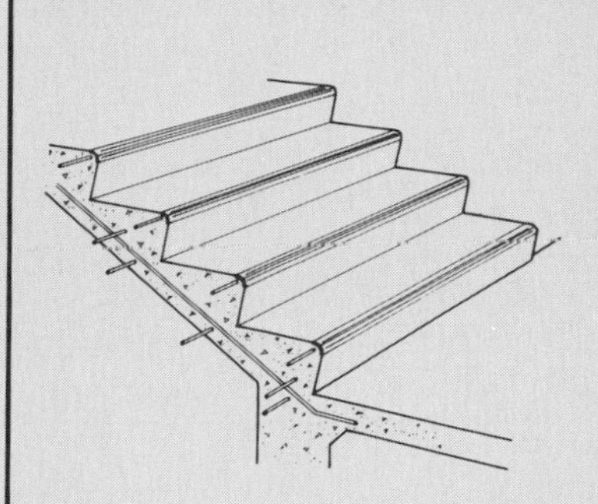

This page illustrates and describes a stair system based on a cost per flight price. Prices for various stair systems are on Line Items 03.9-104-07 thru 32. Both material quantities and labor costs have been adjusted for the system listed.

Factors: To adjust for job conditions other than normal working situations use Lines 03.9-104-35 thru 42.

Example: You are to install the system during evenings only. Go to Line 03.9-104-42 and apply this percentage to the appropriate MAT. and INST. costs.

LINE NO.	DESCRIPTION	QUANTITY	COST PER FLIGHT		
			MAT.	INST.	TOTAL
01	**Below are various stair systems based on cost per flight of stairs, no side**				
02	**Walls. Stairs are 4'-0" wide, railings are included unless otherwise noted**				
03					
04					
05					
06					
07	Concrete, cast in place, no nosings, no railings, 12 risers	Flight	261.36	938.64	1200
08	24 risers		522.72	1877.28	2400
09	Add for 1 intermediate landing		72.16	267.04	339.20
10	Concrete, cast in place, with nosings, no railings, 12 risers		617.76	1103.04	1720.80
11	24 risers		1235.52	2206.08	3441.60
12	Add for 1 intermediate landing		101.86	280.74	382.60
13	Steel, grating tread, safety nosing, 12 risers		871.20	514.80	1386
14	24 risers		1742.40	1029.60	2772
15	Add for intermediate landing		422.40	137.60	560
16	Steel, cement fill pan tread, 12 risers		740.52	526.68	1267.20
17	24 risers		1481.04	1053.36	2534.40
18	Add for intermediate landing		422.40	137.60	560
19	Spiral, industrial, 4' - 6" diameter, 12 risers		1056	384	1440
20	24 risers		2112	768	2880
21	Wood, box stairs, oak treads, 12 risers		1520.75	425.75	1946.50
22	24 risers		3041.50	851.50	3893
23	Add for 1 intermediate landing		242.40	70.96	313.36
24	Wood, basement stairs, no risers, 12 steps		96.80	150.70	247.50
25	24 steps		193.60	301.40	495
26	Add for 1 intermediate landing		17.82	13.94	31.76
27	Wood, open, rough sawn cedar, 12 steps		1560.90	265.10	1826
28	24 steps		3121.80	530.20	3652
29	Add for 1 intermediate landing		14.31	14.37	28.68
30	Wood, residential, oak treads, 12 risers		907.50	1375	2282.50
31	24 risers		1815	2750	4565
32	Add for 1 intermediate landing		233.49	64.03	297.52
33			1575.21	279.47	
34					
35	Dust protection, add, minimum		1%	2%	
36	Maximum		4%	11%	
37	Material handling & storage limitation, add, minimum		1%	1%	
38	Maximum		6%	7%	
39	Protection of existing work, add, minimum		2%	2%	
40	Maximum		5%	7%	
41	Shift work requirements, add, minimum			5%	
42	Maximum			30%	

248

Figure 27.11

Means® Forms

COST ANALYSIS

PROJECT Apartment Renovation	SHEET NO. 3 of 11
ARCHITECT	ESTIMATE NO. 89-2
	DATE 1989
TAKE OFF BY: QUANTITIES BY: EBW PRICES BY: RSM EXTENSIONS BY:	CHECKED BY:

DESCRIPTION	SOURCE/DIMENSIONS			QUANTITY	UNIT	MATERIAL UNIT COST	MATERIAL TOTAL	LABOR UNIT COST	LABOR TOTAL	EQ./TOTAL UNIT COST	EQ./TOTAL TOTAL
Division 3: Superstructure											
Ceiling Drywall				5,808	SF	.34	1975	.72	4182		
Paint				5,808	SF	.10	581	.27	1568		
Stairs				1	Flight	1,575	1575	279	279		
Division 3: Totals							4131		6029		

Figure 27.12

04.9-200	Exterior Wall Selective Price Sheet			
LINE NO.	DESCRIPTION	COST PER S.F.		
		MAT.	INST.	TOTAL
01	Exterior surface, masonry, concrete block, standard 4″ thick	.85	2.95	3.80
02	6″ thick	1.11	3.17	4.28
03	8″ thick	1.32	3.39	4.71
04	12″ thick	2.02	4.38	6.40
05	Split rib, 4″ thick	1.68	3.67	5.35
06	8″ thick	2.99	4.16	7.15
07	Brick running bond, standard size, 6.75/S.F.	2.28	5.77	8.05
08	Buff, 6.75/S.F.	2.51	5.79	8.30
09	Stucco, on frame	.64	3.36	4
10	On masonry	.22	2.56	2.78
11	Metal, aluminum, horizontal, plain	1.10	1.08	2.18
12	Insulated	1.30	1.08	2.38
13	Vertical, plain	1.43	1.08	2.51
14	Insulated	1.49	1.13	2.62
15	Wood, beveled siding, "A" grade cedar, ½″ x 6″	1.79	1.10	2.89
16	½″ x 8″	2.06	1	3.06
17	Shingles, 16″ #1 red, 7-½″ exposure	.88	1.32	2.20
18	18″ perfections, 7-½″ exposure	.72	1.13	1.85
19	Handsplit, 10″ exposure	.94	1.11	2.05
20	White cedar, 7-½″ exposure	.88	1.37	2.25
21	Vertical, board & batten, redwood	1.44	1.38	2.82
22	White pine	.57	1	1.57
23	T. & G. boards, redwood, 1″ x 4″	2.68	1.89	4.57
24	1′ x 8″	2.41	1.51	3.92
25				
26	Interior surface, drywall, taped & finished, standard, ½″	.28	.56	.84
27	⅝″ thick	.32	.57	.89
28	Fire resistant, ½″ thick	.32	.57	.89
29	⅝″ thick	.34	.57	.91
30	Moisture resistant, ½″ thick	.39	.56	.95
31	⅝″ thick	.43	.57	1
32	Core board, 1″ thick	.88	1.15	2.03
33	Plaster, gypsum, 2 coats	.34	1.36	1.70
34	3 coats	.47	1.65	2.12
35	Perlite or vermiculite, 2 coats	.39	1.55	1.94
36	3 coats	.62	1.93	2.55
37	Gypsum lath, standard, ⅜″ thick	.39	.35	.74
38	½″ thick	.42	.37	.79
39	Fire resistant, ½″ thick	.42	.42	.84
40	⅝″ thick	.44	.46	.90
41	Metal lath, diamond, 2.5 lb.	.13	.36	.49
42	Rib, 3.4 lb.	.22	.43	.65
43	Framing metal studs including top and bottom			
44	Runners, walls 10′ high			
45	24″ O.C., non load bearing 20 gauge, 2-½″ wide	.42	.54	.96
46	3-⅝″ wide	.51	.55	1.06
47	4″ wide	.58	.56	1.14
48	6″ wide	.67	.57	1.24
49	Load bearing 18 gauge, 2-½″ wide	.90	1.15	2.05
50	3-⅝″ wide	.96	1.19	2.15
51	4″ wide	1.02	1.25	2.27
52	6″ wide	1.20	1.31	2.51
53	16″ O.C., non load bearing 20 gauge, 2-½″ wide	.51	.62	1.13
54	3-⅝″ wide	.61	.63	1.24
55	4″ wide	.70	.66	1.36
56	6″ wide	.80	.67	1.47
57	Load bearing 18 gauge, 2-½″ wide	1.08	1.37	2.45
58	3-⅝″ wide	1.16	1.44	2.60

265

Figure 27.13

Means® Forms

COST ANALYSIS

SHEET NO. 4 of 11

PROJECT Apartment Renovation

ESTIMATE NO. 89-2

ARCHITECT

DATE 1989

TAKE OFF BY: QUANTITIES BY: EBW PRICES BY: RSM EXTENSIONS BY: CHECKED BY:

DESCRIPTION	SOURCE/DIMENSIONS			QUANTITY	UNIT	MATERIAL		LABOR		EQ./TOTAL	
						UNIT COST	TOTAL	UNIT COST	TOTAL	UNIT COST	TOTAL
Division 4: Exterior Closure											
Stucco				3,120	SF	.22	686	2.56	7987		
Interior Treatment				4,420	SF	1.11	4906	1.79	7912		
Patio Doors				2	EA.	839	1678	277	554		
Windows-Type I				16	EA.	454	7264	105	1680		
Type II				14	EA.	950	13300	160	2240		
Entrance Door				1	EA.	1665	1665	657	657		
Corridor Doors				2	EA.	723	1446	296	592		
Division 4: Totals							30945		21622		

Figure 27.14

EXTERIOR CLOSURE	04.1-258	Masonry Wall, Brick - Stone

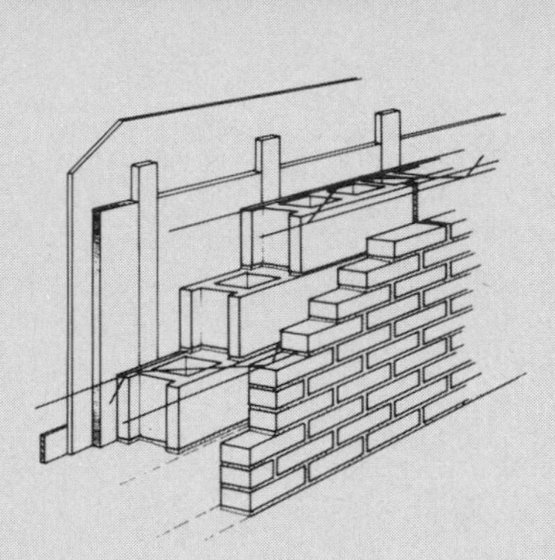

This page illustrates and describes a masonry wall, brick-stone system including brick, concrete block, durawall, insulation, plasterboard, taped and finished, furring, baseboard and painting interior. Lines 04.1-258-04 thru 12 give the unit price and total price per square foot for this system. Prices for alternate masonry wall, brick-stone systems are on Line Item 04.1-258-15 thru 25. Both material quantities and labor costs have been adjusted for the system listed.

Factors: To adjust for job conditions other than normal working situations use Lines 04.1-258-31 thru 42.

Example: You are to install the system without damaging the existing work. Go to Line 04.1-258-39 and apply these percentages to the appropriate MAT. and INST. costs.

LINE NO.	DESCRIPTION	QUANTITY	COST PER S.F. MAT.	INST.	TOTAL
01	**Face brick, 4"thick, concrete block back-up, reinforce every second course,**				
02	**¾"insulation, furring, ½"drywall, taped, finish, and painted, baseboard**				
03					
04	Face brick, 4" brick $270 per M	1 S.F.	2.28	5.77	8.05
05	Concrete back-up block, reinforced 8" thick	1 S.F.	1.51	3.21	4.72
06	¾" rigid polystyrene insulation	1 S.F.	.45	.34	.79
07	Furring, 1" x 3", wood, 16" O.C.	1 L.F.	.12	.56	.68
08	Drywall, ½" thick	1 S.F.	.22	.27	.49
09	Taping & finishing	1 S.F.	.06	.27	.33
10	Painting, 2 coats	1 S.F.	.10	.27	.37
11	Baseboard, wood, 9/16" x 2-5/8"	.1 L.F.	.08	.19	.27
12	TOTAL	S.F.	4.82	10.88	15.70
13					
14	For alternate exterior wall systems:				
15	Face brick, Norman, 4" x 2-⅔" x 12" (4.5 per S.F.) $450 per M	S.F.	5.03	9.07	14.10
16	Roman, 4" x 2" x 12" (6.0 per S.F.) $545 per M		6.49	10.21	16.70
17	Engineer, 4" x 3-⅙" x 8" (5.63 per S.F.) $300 per M		4.63	9.97	14.60
18	S.C.R., 6" x 2-⅔" x 12" (4.5 per S.F.) $625 per M		6.02	9.23	15.25
19	Jumbo, 6" x 4" x 12" (3.0 per S.F.) $1,015 per M		6.20	8.05	14.25
20	Norwegian, 6" x 3-⅙" x 12" (3.75 per S.F.) $500 per M	↓	4.83	8.52	13.35
21					
22					
23	Stone, veneer, fieldstone, 6" thick	S.F.	7.38	9.62	17
24	Marble, 2" thick		35.54	16.11	51.65
25	Limestone, 2" thick	↓	15.74	17.91	33.65
26					
27					
28					
29					
30					
31	Cut & patch to match existing construction, add, minimum		2%	3%	
32	Maximum		5%	9%	
33	Dust protection, add, minimum		1%	2%	
34	Maximum		4%	11%	
35	Equipment usage curtailment, add, minimum		1%	1%	
36	Maximum		3%	10%	
37	Material handling & storage limitation, add, minimum		1%	1%	
38	Maximum		6%	7%	
39	Protection of existing work, add, minimum		2%	2%	
40	Maximum		5%	7%	
41	Shift work requirements, add, minimum			5%	
42	Maximum			30%	

252

Figure 27.15

EXTERIOR CLOSURE	04.1-416	Wood Frame Exterior Wall

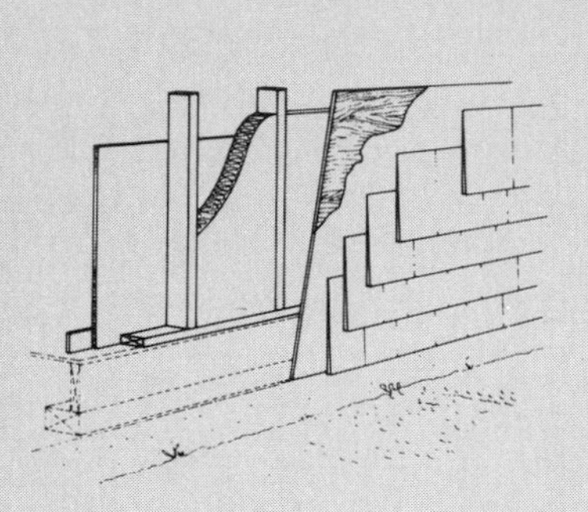

This page illustrates and describes a wood frame exterior wall system including wood studs, sheathing, felt, insulation, plasterboard, taped and finished, baseboard and painted interior. Lines 04.1-416-05 thru 13 give the unit price and total price per square foot for this system. Prices for alternate wood frame exterior wall systems are on Line Items 04.1-416-17 thru 27. Both material quantities and labor costs have been adjusted for the system listed.

Factors: To adjust for job conditions other than normal working situations use Lines 04.1-416-31 thru 42.

Example: You are to install the system with need for complete temporary bracing. Go to Line 04.1-416-42 and apply these percentages to the appropriate MAT. and INST. costs.

LINE NO.	DESCRIPTION	QUANTITY	COST PER S.F.		
			MAT.	INST.	TOTAL
01	**Wood stud wall, cedar shingle siding, building paper, plywood sheathing,**				
02	**Insulation, ⅝" drywall, taped, finished and painted, baseboard.**				
03					
04					
05	2" x 4" wood studs, 16" O.C.	.1 L.F.	.33	.57	.90
06	½" CDX sheathing	1 S.F.	.45	.40	.85
07	18" No. 1 red cedar shingles, 7-½" exposure	.008 C.S.F.	.75	.97	1.72
08	15# felt paper	.01 C.S.F.	.02	.09	.11
09	3-½" fiberglass insulation	1 S.F.	.30	.17	.47
10	⅝" drywall, taped and finished	1 S.F.	.32	.55	.87
11	Baseboard trim, stock pine, 9/16" x 3-½", painted	.1 L.F.	.08	.19	.27
12	Paint, 2 coats, interior	1 S.F.	.10	.27	.37
13	TOTAL	S.F.	2.35	3.21	5.56
14					
15					
16	For alternate exterior wall systems:				
17	Aluminum siding, horizontal clapboard	S.F.	2.70	3.32	6.02
18	Cedar bevel siding, ½" x 6", vertical , painted		3.39	3.34	6.73
19	Redwood siding 1" x 4" to 1" x 6" vertical, T & G		3.04	3.62	6.66
20	Board and batten		2.84	3.30	6.14
21	Ship lap siding		2.84	3.34	6.18
22	Plywood, grooved (T1-11) fir		2.14	3.08	5.22
23	Redwood		3.06	3.08	6.14
24	Southern yellow pine		2.21	3.08	5.29
25	Masonry on stud wall, stucco, wire and plaster		2.24	5.56	7.80
26	Stone veneer		6.44	6.75	13.19
27	Brick veneer, brick $250 per M	↓	3.88	8.01	11.89
28					
29					
30					
31	Cut & patch to match existing construction, add, minimum		2%	3%	
32	Maximum		5%	9%	
33	Dust protection, add, minimum		1%	2%	
34	Maximum		4%	11%	
35	Material handling & storage limitation, add, minimum		1%	1%	
36	Maximum		6%	7%	
37	Protection of existing work, add, minimum		2%	2%	
38	Maximum		5%	7%	
39	Shift work requirements, add, minimum			5%	
40	Maximum			30%	
41	Temporary shoring and bracing, add, minimum		2%	5%	
42	Maximum		5%	12%	

255

Figure 27.16

Division 6: Interior Construction

The most important thing to remember when performing an Assemblies estimate is to be sure to include all items. Since design data is usually limited, the estimator can easily overlook items that are assumed and must be anticipated. Each of the previous five divisions have included relatively few items. In commercial renovation, interior construction encompasses a great deal of work and must be carefully planned and estimated.

The major portion of the interior construction is wood stud partitions. Three types of partitions are required for the renovation: interior partitions within the apartments, one-hour firewalls between units and at the corridors, and furring and drywall at the existing stair enclosure. The appropriate system for the interior partitions is found in Figure 27.29. The insulation is included for soundproofing. The costs are entered on the estimate sheet for Division 6, in Figure 27.30. The firewalls are essentially the same system as the interior partitions. The drywall, however, must be 5/8'' fire resistant. Costs for the drywall are found in Figure 27.31 and incorporated into the system in Figure 27.32. Note that the costs in Figure 27.31 are per single square foot and must be doubled when substituted into the system price.

Assembly from Figures 27.15 and 27.16	Material	Installation	Total
Polystyrene Insulation (3/4")	$.45	$.34	$.79
Furring	.12	.56	.68
Drywall (1/2")	.22	.27	.49
Taping	.06	.27	.33
Paint	.10	.27	.37
Baseboard	.08	.19	.27
	$1.03	$1.90	$2.93
Factors 2% Material	.02	.06	.08
3% Installation			
Total per S.F., R-Value 4	$1.05	$1.96	$3.01
Wood Studs	$.33	$.57	$.90
Fiberglass Insulation (3-1/2")	.30	.17	.47
Drywall (Price for 1/2")	.22	.27	.49
Taping	.06	.27	.33
Paint	.10	.27	.37
Baseboard	.08	.19	.27
	$1.09	$1.74	$2.83
Factors 2% Material	.02	.05	.07
3% Installation			
Total per S.F., R-Value 4	$1.11	$1.79	$2.90

Figure 27.17

EXTERIOR CLOSURE	04.6-152	Doors, Sliding - Patio

This page illustrates and describes sliding door systems including a sliding door, frame, interior and exterior trim with exterior staining. Lines 04.6-152-04 thru 07 give the unit price and total price on a cost each basis for this system. Prices for alternate sliding door systems are on Line Items 04.6-152-11 thru 22. Both material quantities and labor costs have been adjusted for the system listed.

Factors: To adjust for job conditions other than normal working situations use Lines 04.6-152-27 thru 40.

Example: You are to install the system with temporary shoring and bracing. Go to Line 04.6-152-39 and apply these percentages to the appropriate MAT. and INST. costs.

LINE NO.	DESCRIPTION	QUANTITY	COST EACH		
			MAT.	INST.	TOTAL
01	**Sliding wood door, 6'-0" x 6'-8", with wood frame, interior and exterior**				
02	**Trim and exterior staining.**				
03					
04	Sliding wood door, standard, 6'-0" x 6'-8", insulated glass	1 Ea.	462	138	600
05	Interior & exterior trim	1 Set	12.32	22.88	35.20
06	Stain door & trim	1 Ea.	4.06	43.34	47.40
07	TOTAL	Ea.	478.38	204.22	682.60
08					
09					
10	For alternate sliding door systems:				
11	Wood, standard, 8'-0" x 6'-8", insulated glass	Ea.	798.74	254.02	1052.76
12	12'-0" x 6'-8"		1061.34	302.24	1363.58
13	Vinyl coated, 6'-0" x 6'-8"		739.08	203.52	942.60
14	8'-0" x 6'-8"		934.04	253.72	1187.76
15	12'-0" x 6'-8"		1396.84	316.74	1713.58
16					
17	Aluminum, standard, 6'-0" x 6'-8", insulated glass	Ea.	421.74	204.46	626.20
18	8'-0" x 6'-8"		490.31	255.01	745.32
19	12'-0" x 6'-8"		528.47	305.09	833.56
20	Anodized, 6'-0" x 6'-8"		509.74	201.46	711.20
21	8'-0" x 6'-8"		534.31	256.01	790.32
22	12'-0" x 6'-8"		632.97	305.59	938.56
23					
24	Deduct for single glazing	Ea.	52		52
25					
26					
27	Cut & patch to match existing construction, add, minimum		2%	3%	
28	Maximum		5%	9%	
29	Dust protection, add, minimum		1%	2%	
30	Maximum		4%	11%	
31	Equipment usage curtailment, add, minimum		1%	1%	
32	Maximum		3%	10%	
33	Material handling & storage limitation, add, minimum		1%	1%	
34	Maximum		6%	7%	
35	Protection of existing, work, add, minimum		2%	2%	
36	Maximum		5%	7%	
37	Shift work requirements, add, minimum			5%	
38	Maximum			30%	
39	Temporary shoring and bracing, add, minimum		2%	5%	
40	Maximum		5%	12%	
41					
42					

258

Figure 27.18

EXTERIOR CLOSURE	04.7-145	Windows - Wood

This page illustrates and describes a wood window system including double hung wood window, exterior and interior trim, hardware and insulating glass. Lines 04.7-145-04 thru 07 give the unit price and total price on a cost each basis for this system. Prices for alternate wood window systems are on Line Items 04.7-145-13 thru 23. Both material quantities and labor costs have been adjusted for the system listed.

Factors: To adjust for job conditions other than normal working situations use Lines 04.7-145-31 thru 40.

Example: You are to install the system during evening hours only. Go to Line 04.7-145-40 and apply this percentage to the appropriate INST. cost.

LINE NO.	DESCRIPTION	QUANTITY	COST EACH		
			MAT.	INST.	TOTAL
01	**Double hung wood window 2′-0″ x 3′-0″, exterior and interior trim,**				
02	**Hardware, glazed with insulating glass.**				
03					
04	2′-0″ x 3′-0″ double hung wood window, with insulating glass	1 Ea.	127.60	27.40	155
05	Exterior and interior trim	1 Set	11.55	21.45	33
06	Hardware	1 Set	1.65	11.45	13.10
07	TOTAL	Ea.	140.80	60.30	201.10
08					
09					
10					
11					
12	For alternate window systems:				
13	Double hung, 3′-0″ x 4′-0″	Ea.	165.55	71.55	237.10
14	4′-0″ x 4′-6″		201.85	94.25	296.10
15	Casement 2′-0″ x 3′-0″		164.45	38.65	203.10
16	2 leaf, 4′-0″ x 4′-0″ TYPE I		444.95	102.15	547.10
17	3 leaf, 6′-0″ x 6′-0″ TYPE II		931.15	154.95	1086.10
18	Awning, 2′-10″ x 1′-10″		173.80	62.30	236.10
19	3′-6″ x 2′-4″		211.75	70.35	282.10
20	4′-0″ x 3′-0″		232.65	93.45	326.10
21	Horizontal siding 3′-0″ x 2′-0″		150.70	60.40	211.10
22	4′-0″ x 3′-6″		213.95	68.15	282.10
23	6′-0″ x 5′-0″		320.65	47.45	368.10
24					
25					
26					
27					
28					
29					
30					
31	Cut & patch to match existing construction, add, minimum		2%	3%	
32	Maximum		5%	9%	
33	Dust protection, add, minimum		1%	2%	
34	Maximum		4%	11%	
35	Material handling & storage limitation, add, minimum		1%	1%	
36	Maximum		6%	7%	
37	Protection of existing work, add, minimum		2%	2%	
38	Maximum		5%	7%	
39	Shift work requirements, add, minimum			5%	
40	Maximum			30%	
41					
42					

262

Figure 27.19

EXTERIOR CLOSURE	04.6-142	Doors, Metal - Commercial

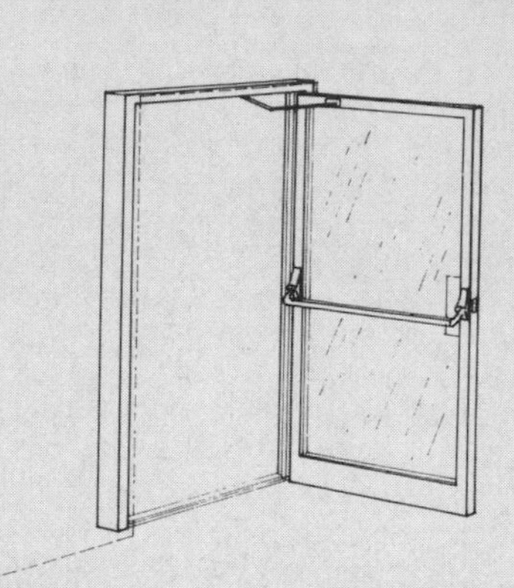

This page illustrates and describes a commercial metal door system, including a single aluminum and glass door, narrow stiles, jamb, hardware weatherstripping, panic hardware and closer. Lines 04.6-142-04 thru 10 give the unit price and total price on a cost each basis for this system. Prices for alternate commercial metal door systems are on Line Items 04.6-142-13 thru 25. Both material quantities and labor costs have been adjusted for the system listed.

Factors: To adjust for job conditions other than normal working situations, use Lines 04.6-142-31 thru 40.

Example: You are to install the system and cut and patch to match existing construction. Go to Line 04.6-142-32 and apply these percentages to the appropriate MAT. and INST. costs.

LINE NO.	DESCRIPTION	QUANTITY	COST EACH MAT.	COST EACH INST.	COST EACH TOTAL
01	**Single aluminum and glass door, 3′-0″x7′-0″, with narrow stiles, ext. jamb.**				
02	**Weatherstripping, ½″ tempered insul. glass, panic hardware, and closer.**				
03					
04	Aluminum door, 3′-0″ x 7′-0″ x 1-¾″, narrow stiles	1 Ea.	357.50	167.50	525
05	Exterior jamb and trim	1 Set	214.50	100.50	315
06	Hardware	1 Set	143	67	210
07	Tempered insulating glass, ½″ thick	20 S.F.	457.60	202.40	660
08	Panic hardware	1 Set	256.30	93.70	350
09	Automatic closer	1 Ea.	66	44	110
10	ENTRANCE	Ea.	1494.90	675.10	2170
11					
12	For alternate door systems:				
13	Single aluminum and glass with transom, 3′-0″ x 10′-0″	Ea.	1782.44	776.56	2559
14	Anodized aluminum and glass, 3′-0″ x 7′-0″		1745.15	792.35	2537.50
15	With transom, 3′-0″ x 10′-0″		2069.27	900.98	2970.25
16	Steel, deluxe, hollow metal 3′-0″ x 7′-0″ CORRIDOR		708.40	287.10	995.50
17	With transom 3′-0″ x 10′-0″		827.20	319.95	1147.15
18	Fire door, "A" label, 3′-0″ x 7′-0″		756.47	285.53	1042
19	Double, aluminum and glass, 6′-0″ x 7′-0″		2575.10	1119.90	3695
20	With transom, 6′-0″ x 10′-0″		3078.68	1369.32	4448
21	Anodized aluminum and glass 6′-0″ x 7′-0″		3008.23	1290.52	4298.75
22	With transom, 6′-0″ x 10′-0″		3559.93	1570.57	5130.50
23	Steel, deluxe, hollow metal, 6′-0″ x 7′-0″		1154.67	483.83	1638.50
24	With transom, 6′-0″ x 10′-0″		1392.27	549.53	1941.80
25	Fire door, "A" label, 6′-0″ x 7′-0″	↓	1218.80	486.20	1705
26					
27					
28					
29					
30					
31	Cut & patch to match existing construction, add, minimum		2%	3%	
32	Maximum		5%	9%	
33	Dust protection, add, minimum		1%	2%	
34	Maximum		4%	11%	
35	Material handling & storage limitation, add, minimum		1%	1%	
36	Maximum		6%	7%	
37	Protection of existing work, add, minimum		2%	2%	
38	Maximum		5%	7%	
39	Shift work requirements, add, minimum			5%	
40	Maximum			30%	
41					
42					

256

Figure 27.20

04.9-500	Hardware Selective Price Sheet			
LINE NO.	DESCRIPTION	COST EACH		
		MAT.	INST.	TOTAL
01	Door closer, rack and pinion	67.10	42.90	110
02	Backcheck and adjustable power	77	48	125
03	Regular, hinge face mount, all sizes, regular arm	71.50	43.50	115
04	Hold open arm	83.60	41.40	125
05	Top jamb mount, all sizes, regular arm	69.30	45.70	115
06	Hold open arm	79.20	45.80	125
07	Stop face mount, all sizes, regular arm	71.50	43.50	115
08	Hold open arm	82.50	42.50	125
09	Fusible link, hinge face mount, all sizes, regular arm	81.40	43.50	124.90
10	Hold open arm	93.50	41.40	134.90
11	Top jamb mount, all sizes, regular arm	79.20	45.70	124.90
12	Hold open arm	89.10	45.80	134.90
13	Stop face mount, all sizes, regular arm	81.40	43.50	124.90
14	Hold open arm	92.40	42.50	134.90
15				
16				
17	Door stops			
18				
19	Holder & bumper, floor or wall	3.30	11.45	14.75
20	Wall bumper	3.96	11.44	15.40
21	Floor bumper	1.54	11.46	13
22	Plunger type, door mounted	14.85	11.15	26
23	Hinges, full mortise, material only, per pair			
24	Low frequency, 4-½" x 4-½", steel base, USP	11.65		11.65
25	Brass base, US10	39		39
26	Stainless steel base, US32	55		55
27	Average frequency, 4-½" x 4-½", steel base, USP	21		21
28	Brass base, US10	44		44
29	Stainless steel base, US32	63		63
30	High frequency, 4-½" x 4-½", steel base, USP	55		55
31	Brass base, US10	83		83
32	Stainless steel base, US32	110		110
33	Kick plate			
34				
35	6" high, for 3'-0" door, aluminum	18.70	18.30	37
36	Bronze	38.50	18.50	57
37	Panic device			
38				
39	For rim locks, single door, exit	256.30	93.70	350
40	Outside key and pull	294.80	55.20	350
41	Bar and vertical rod, exit only	393.80	56.20	450
42	Outside key and pull	440	70	510
43	Lockset			
44				
45	Heavy duty, cylindrical, passage doors	95.70	24.30	120
46	Classroom	203.50	36.50	240
47	Bedroom, bathroom, and inner office doors	123.20	21.80	145
48	Apartment, office, and corridor doors	165	25	190
49	Standard duty, cylindrical, exit doors	63.80	27.20	91
50	Inner office doors	26.40	22.60	49
51	Passage doors	36.30	22.70	59
52	Public restroom, classroom, & office doors	75.90	34.10	110
53	Deadlock, mortise, heavy duty	102.30	32.70	135
54	Double cylinder	112.20	32.80	145
55	Entrance lock, cylinder, deadlocking latch	96.80	28.20	125
56	Deadbolt	115.50	34.50	150
57	Commercial, mortise, wrought knob, keyed, minimum	126.50	33.50	160
58	Maximum	225.50	39.50	265

267

Figure 27.21

There is no door schedule for the project. The estimator must be sure to establish the different door types, hardware sets, and quantities of each. When only one, or a "typical," floor plan is provided, it is easy to forget to multiply for repetitive floors. Crosschecking is necessary. The interior door system is found in Figure 27.33. The apartment entrance doors and stairway corridor doors must both be fire-rated, but different hardware sets are required from those as shown in Figure 27.34. The correct hardware is selected from Figure 27.35. The calculations are shown in Figure 27.36. Note that installation costs for the hinges are already included in the installation costs for the door. Costs for the remaining door types are obtained from Figures 27.37 and 27.38.

Throughout the estimate for Division 6, no factors have been added. This is because the interior construction for this project is similar to new construction, unimpeded by existing conditions. It is up to the estimator, using experience and discretion, to determine if and when factors are required for a particular project.

The estimator must be constantly aware of the project as a whole to be sure to include all requirements. The heating and cooling system is forced air. The ductwork will have to be concealed by a suspended ceiling. At this point, or in Division 8 of the Assemblies Section of *Means Repair and Remodeling Cost Data*, a rough duct layout should be made to identify areas that will require a suspended ceiling. The system to be used is shown in Figure 27.39. A factor is added because the installation is in small spaces, in the bathrooms and closets. This will entail a high waste factor for materials, as well as added labor expense.

Hardware Substitution

	Material	Installation	Total
Entrance Door 04.6-142-10	$1,494.90	$675.10	$2,170.00
Deduct Panic 04.6-142-08	(256.30)	(93.70)	(350.00)
Add Panic	393.80	56.20	450.00
	$1,632.40	$637.60	$2,270.00
Factors 2% Material 3% Installation	32.65	19.13	51.78
Total	$1,665.05	$656.73	$2,321.78

Figure 27.22

ROOFING | **05.1-492** | **Wood Frame Roof & Ceiling**

This page illustrates and describes a wood frame roof system including rafters, ceiling joists, sheathing, building paper, asphalt shingles, roof trim, furring, insulation, plaster and paint. Lines 05.1-492-04 thru 15 give the unit price and total price per square foot for this system. Prices for alternate wood frame roof systems are on Line Items 05.1-492-19 thru 27. Both material quantities and labor costs have been adjusted for the system listed.

Factors: To adjust for job conditions other than normal working situations use Lines 05.1-492-33 thru 40.

Example: You are to install the system while protecting existing work. Go to Line 05.1-492-38 and apply these percentages to the appropriate MAT. and INST. costs.

LINE NO.	DESCRIPTION	QUANTITY	COST PER S.F.		
			MAT.	INST.	TOTAL
01	**Wood frame roof system, 4 in 12 pitch, including rafters, sheathing,**				
02	**Shingles, insulation, drywall, thin coat plaster, and painting.**				
03					
04	Rafters, 2" x 6", 16" O.C., 4 in 12 pitch	1.08 L.F.	.46	.62	1.08
05	Ceiling joists, 2" x 6", 16 O.C.	1 L.F.	.43	.45	.88
06	Sheathing, ½" CDX	1.08 S.F.	.48	.44	.92
07	Building paper, 15# felt	.011 C.S.F.	.03	.09	.12
08	Asphalt shingles, 240#	.011 C.S.F.	.36	.61	.97
09	Roof trim	.1 L.F.	.08	.13	.21
10	Furring, 1" x 3", 16" O.C.	1 L.F.	.11	.79	.90
11	Fiberglass insulation, 6" batts	1 S.F.	.43	.20	.63
12	Gypsum board, ½" thick	1 S.F.	.22	.31	.53
13	Thin coat plaster	1 S.F.	.08	.36	.44
14	Paint, roller, 2 coats	1 S.F.	.10	.27	.37
15	TOTAL	S.F.	2.78	4.27	7.05
16					
17					
18	For alternate roof systems:				
19	Rafters 16" O.C., 2" x 8"	S.F.	2.99	4.31	7.30
20	2" x 10"		3.24	4.63	7.87
21	2" x 12"		3.40	4.71	8.11
22	Rafters 24" O.C., 2" x 6"		2.55	4	6.55
23	2" x 8"		2.71	4.02	6.73
24	2" x 10"		2.89	4.27	7.16
25	2" x 12"		3	4.33	7.33
26	Roof pitch, 6 in 12, add		3%	10%	
27	8 in 12, add	↓	5%	12%	
28					
29					
30					
31					
32					
33	Cut & patch to match existing construction, add, minimum		2%	3%	
34	Maximum		5%	9%	
35	Material handling & storage limitation, add, minimum		1%	1%	
36	Maximum		6%	7%	
37	Protection of existing work, add, minimum		2%	2%	
38	Maximum		5%	7%	
39	Shift work requirements, add, minimum			5%	
40	Maximum			30%	
41					
42					

270

Figure 27.23

05.9-300	Roofing & Ceiling Finish Selective Price Sheet			
LINE NO.	DESCRIPTION	COST PER S.F.		
		MAT.	INST.	TOTAL
01	Roofing, built-up, asphalt roll roof, 3 ply organic/mineral surface	.39	.71	1.10
02	3 plies glass fiber felt type iv, 1 ply mineral surface	.56	.79	1.35
03	Cold applied, 3 ply		.27	.27
04	Coal tar pitch, 4 ply asbestos felt	.64	.91	1.55
05	Mopped, 3 ply glass fiber	.56	.99	1.55
06	4 ply organic felt	.70	.90	1.60
07	Elastomeric, hypalon, neoprene unreinforced	1.60	1.68	3.28
08	Polyester reinforced	1.76	1.98	3.74
09	Neoprene, 5 coats 60 mils	4.79	5.86	10.65
10	Over 10,000 S.F.	4.46	3.04	7.50
11	PVC, traffic deck sprayed	1.27	3.05	4.32
12	With neoprene	1.38	1.23	2.61
13	Shingles, asbestos, strip, 14" x 30", 325#/sq.	1.41	.69	2.10
14	12' x 24", 167#/sq.	.74	.76	1.50
15	Shake, 9.35" x 16" 500#/sq.	2.37	1.23	3.60
16				
17	Asphalt, strip, 210-235#/sq.	.33	.50	.83
18	235-240#/sq.	.33	.55	.88
19	Class A laminated	.57	.63	1.20
20	Class C laminated	.52	.68	1.20
21	Slate, buckingham, 3/16" thick	3.69	1.56	5.25
22	Black, 1/4" thick	3.69	1.56	5.25
23	Wood, shingles, 16" no. 1, 5" exp.	1.21	1.09	2.30
24	Red cedar, 18" perfections	1.27	.98	2.25
25	Shakes, 24", 10" exposure	.94	1.11	2.05
26	18", 8-1/2" exposure	.88	1.37	2.25
27	Insulation, ceiling batts, fiberglass, 3-1/2" thick, R11	.24	.17	.41
28	6" thick, R19	.40	.20	.60
29	9" thick, R30	.55	.24	.79
30	12" thick, R38	.79	.28	1.07
31	Mineral, 3-1/2" thick, R13	.31	.17	.48
32	Fiber, 6" thick, R19	.48	.18	.66
33	Roof deck, fiberboard, 1" thick, R2.78	.30	.34	.64
34	Mineral, 2" thick, R5.26	.67	.34	1.01
35	Perlite boards, 3/4" thick, R2.08	.33	.34	.67
36	2" thick, R5.26	.80	.39	1.19
37	Polystyrene extruded, R5.26, 1" thick,	.36	.18	.54
38	2" thick	.70	.22	.92
39	Urethane paperbacked, 1" thick, R6.7	.53	.27	.80
40	3" thick, R25	1.03	.34	1.37
41	Foam glass sheets, 1-1/2" thick, R3.95	1.79	.34	2.13
42	2" thick, R5.26	2.27	.34	2.61
43	Ceiling, plaster, gypsum, 2 coats	.34	1.56	1.90
44	3 coats	.47	1.86	2.33
45	Perlite or vermiculite, 2 coats	.39	1.81	2.20
46	3 coats	.62	2.27	2.89
47	Gypsum lath, plain 3/8" thick	.39	.35	.74
48	1/2" thick	.42	.37	.79
49	Firestop, 1/2" thick	.42	.42	.84
50	5/8" thick	.44	.46	.90
51	Metal lath, rib, 2.75 lb.	.19	.40	.59
52	3.40 lb.	.22	.43	.65
53	Diamond, 2.50 lb.	.13	.41	.54
54	3.40 lb.	.14	.51	.65
55	Drywall, taped and finished standard, 1/2" thick	.28	.71	.99
56	5/8" thick	.32	.72	1.04
57	Fire resistant, 1/2" thick	.32	.72	1.04
58	5/8" thick	.34	.72	1.06

272

Figure 27.24

05.9-500	Roof Accessory Selective Price Sheet			
LINE NO.	DESCRIPTION	COST PER L.F.		
		MAT.	INST.	TOTAL
01	Downspouts per L.F., aluminum, enameled .024" thick, 2" x 3"	.62	1.57	2.19
02	3" x 4"	.99	2.14	3.13
03	Round .025" thick, 3" diam.	.74	1.57	2.31
04	4" diam.	1.08	2.13	3.21
05	Copper, round 16 oz. stock, 2" diam.	3.30	1.57	4.87
06	3" diam.	4.46	1.59	6.05
07	4" diam.	5.72	2.08	7.80
08	5" diam.	6.11	2.29	8.40
09	Rectangular, 2" x 3"	5.06	1.59	6.65
10	3" x 4"	6.27	2.08	8.35
11	Lead coated copper, round, 2" diam.	4.51	1.59	6.10
12	3" diam.	5.45	1.55	7
13	4" diam.	6.71	2.04	8.75
14	5" diam.	8.58	2.32	10.90
15	Rectangular, 2" x 3"	5.94	1.56	7.50
16	3" x 4"	7.70	2.05	9.75
17	Steel galvanized, round 28 gauge, 3" diam.	.52	1.57	2.09
18	4" diam.	.65	2.06	2.71
19	5" diam.	.79	2.30	3.09
20	6" diam.	1.21	2.85	4.06
21	Rectangular, 2" x 3"	.50	1.57	2.07
22	3" x 4"	1.43	2.06	3.49
23	Elbows, aluminum, round, 3" diam.	.85	2.99	3.84
24	4" diam.	1	2.99	3.99
25	Rectangular, 2" x 3"	.58	2.99	3.57
26	3" x 4"	.94	2.99	3.93
27	Copper, round 16 oz., 2" diam.	4.68	2.97	7.65
28	3" diam.	5.83	2.97	8.80
29	4" diam.	10.18	2.97	13.15
30	5" diam.	12.98	2.97	15.95
31	Rectangular, 2" x 3"	9.90	3	12.90
32	3" x 4"	13.97	2.98	16.95
33	Drip edge per L.F., aluminum, 5" girth	.17	.68	.85
34	8" girth	.33	.69	1.02
35	28" girth	1.32	2.75	4.07
36				
37	Steel galvanized, 5" girth	.24	.69	.93
38	8" girth	.29	.68	.97
39				
40				
41				
42				
43	Flashing 12" wide per S.F., aluminum, mill finish, .013" thick	.31	2.06	2.37
44	.019" thick	.73	2.06	2.79
45	.040" thick	1.49	2.06	3.55
46	.050" thick	1.79	2.07	3.86
47	Copper, mill finish, 16 oz.	2.86	2.59	5.45
48	20 oz.	3.58	2.72	6.30
49	24 oz.	4.29	2.86	7.15
50	32 oz.	5.72	2.98	8.70
51	Lead, 2.5 lb./S.F., 12" wide	3.47	2.03	5.50
52	Over 12" wide	3.47	2.03	5.50
53	Lead-coated copper, fabric backed, 2 oz.	1.43	.91	2.34
54	5 oz.	1.65	.91	2.56
55	Mastic backed, 2 oz.	.99	.91	1.90
56	5 oz.	1.60	.90	2.50
57	Paper backed, 2 oz.	.88	.91	1.79
58	3 oz.	1.10	.91	2.01

274

Figure 27.25

05.9-500	Roof Accessory Selective Price Sheet			
LINE NO.	DESCRIPTION	COST PER L.F.		
		MAT.	INST.	TOTAL
59	Polyvinyl chloride, black, .010" thick	.11	.96	1.07
60	.020" thick	.17	.95	1.12
61	.030" thick	.28	.95	1.23
62	.056" thick	.50	.95	1.45
63	Steel, galvanized, 20 gauge	.58	2.30	2.88
64	30 gauge	.22	1.87	2.09
65	Stainless, 32 gauge, .010" thick	2.04	1.92	3.96
66	28 gauge, .015" thick	2.53	1.93	4.46
67	26 gauge, .018" thick	3.14	1.91	5.05
68	24 gauge, .025" thick	3.80	1.90	5.70
69	Gutters per L.F., aluminum, 5" box, .027" thick	.91	2.50	3.41
70	.032" thick	1.10	2.49	3.59
71	Copper, half round, 4" wide	4.40	2.50	6.90
72	6" wide	5.01	2.59	7.60
73	Steel, 26 gauge galvanized, 5" wide	.83	2.49	3.32
74	6" wide	1.10	2.49	3.59
75	Wood, treated hem-fir, 3" x 4"	4.40	2.75	7.15
76	4" x 5"	5.50	2.75	8.25
77	Reglet per L.F., aluminum, .025" thick	1.10	1.22	2.32
78	Copper, 10 oz.	2.09	1.22	3.31
79	Steel, galvanized, 24 gauge	.69	1.22	1.91
80	Stainless, .020" thick	1.27	1.22	2.49
81	Counter flashing 12" wide per L.F., aluminum, .032" thick	1.43	1.99	3.42
82	Copper, 10 oz.	2.31	1.99	4.30
83	Steel, galvanized, 24 gauge	.58	2	2.58
84	Stainless, .020" thick	1.71	1.99	3.70

275

Figure 27.26

Painting is included in the systems for all the required interior construction except for the furring and drywall system, which was "assembled." The ceilings, in Division 3 of the Assemblies Section of *Means Repair and Remodeling Cost Data*, also included painting. The costs for painting, and for other interior finishes that may be substituted for those included in the systems, are found in Figure 27.40.

Without specifications, the estimator must also choose the type of carpeting to be used. This decision should be based on experience, or made with the advice of a floorcovering subcontractor. The costs are found in Figure 27.41.

Division 7: Conveying Systems

While no conveying systems are specified for the sample project, such work is often included in commercial renovation. Accessibility for handicapped persons has become a consideration in most building codes, and wheelchair lifts are becoming more common. The Assemblies pages of *Means Repair and Remodeling Cost Data* contain costs for hydraulic elevators. When other conveying systems are required, refer to the Unit Price pages of *Means Repair and Remodeling Cost Data*, as in Figure 27.42.

Division 8: Mechanical

The mechanical systems for the sample project include the bathrooms and HVAC systems.

A system similar to the bathrooms of the sample project is found in Figure 27.43. The water meter and controls are already in place and are not added to the costs. The calculations for the bathrooms are in Figure 27.44. The rough-in for the kitchen sink is not included in the systems price for kitchens (Division 11: Special) and is obtained from the Unit Price section (Division 15) of *Means Repair and Remodeling Cost Data*. Overhead and profit should be added to the bare material and labor costs. The plumbing costs are recorded on the estimate sheet in Figure 27.45.

Division 5: Calculations

	Material	Installation	Total
Shingles	$ .36	$ .61	$ .97
Roof Trim	.08	.13	.21
Subtotal	0.44	0.74	1.18
Factors	(5%) 0.02	(9%) 0.07	0.09
Total/S.F.	$0.46	$0.81	$1.27
Insulation (9")	.55	.24	0.79
Factors	(8%) .04	(10%) .02	.06
Total/S.F.	$ .59	$ .26	$ .85

Figure 27.27

Means® Forms

COST ANALYSIS

SHEET NO. 5 of 11

PROJECT Apartment Renovation

ESTIMATE NO. 89-2

ARCHITECT

DATE 1989

TAKE OFF BY: QUANTITIES BY: EBW PRICES BY: RSM EXTENSIONS BY: CHECKED BY:

DESCRIPTION	SOURCE/DIMENSIONS			QUANTITY	UNIT	MATERIAL		LABOR		EQ./TOTAL	
						UNIT COST	TOTAL	UNIT COST	TOTAL	UNIT COST	TOTAL
Division 5: Roofing											
Shingles & Roof Trim				4,216	SF	.46	1939	.81	3415		
Insulation				4,000	SF	.59	2360	.26	1040		
Downspouts				56	LF	.62	35	1.57	88		
Gutters				164	LF	.91	149	2.50	410		
Division 5: Totals							4483		4953		

Figure 27.28

INTERIOR CONSTR.	06.1-592	Partitions, Wood Stud

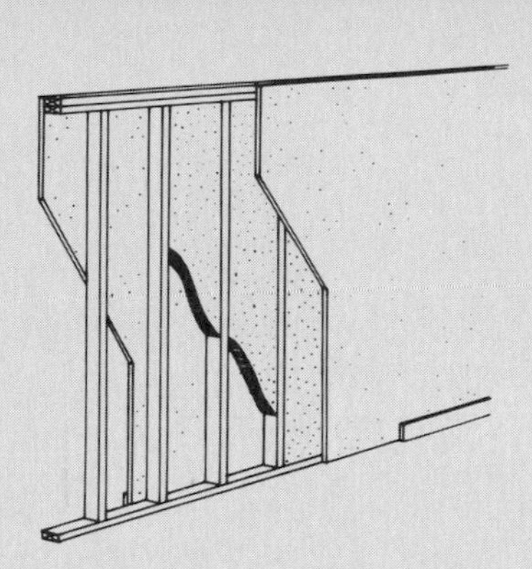

This page illustrates and describes a wood stud partition system including wood studs with plates, gypsum plasterboard - taped and finished, insulation, baseboard and painting. Lines 06.1-592-04 thru 10 give the unit price and total price per square foot for this system. Prices for alternate wood stud partition systems are on Line Items 06.1-592-13 thru 27. Both material quantities and labor costs have been adjusted for the system listed.

Factors: To adjust for job conditions other than normal working situations use Lines 06.1-592-29 thru 40.

Example: You are to install the system where material handling and storage present a serious problem. Go to Line 06.1-592-34 and apply these percentages to the appropriate MAT. and INST. costs.

LINE NO.	DESCRIPTION	QUANTITY	COST PER S.F.		
			MAT.	INST.	TOTAL
01	**Wood stud wall, 2"x4",16"O.C., dbl. top plate, sngl bot. plate, ⅝" dwl.**				
02	**Taped, finished and painted on 2 faces, insulation, baseboard, wall 8' high**				
03					
04	Wood studs, 2" x 4", 16" O.C., 8' high	1 S.F.	.35	.71	1.06
05	Gypsum drywall, ⅝" thick	2 S.F.	.53	.55	1.08
06	Taping and finishing	2 S.F.	.11	.55	.66
07	Insulation, 3-½" fiberglass batts	1 S.F.	.30	.17	.47
08	Baseboard, painted	.2 L.F.	.21	.43	.64
09	Painting, roller, 2 coats	2 S.F.	.20	.54	.74
10	TOTAL	S.F.	1.70	2.95	4.65
11					
12	For alternate wood stud systems:				
13	2" x 3" studs, 8' high, 16" O.C.	S.F.	1.68	2.91	4.59
14	24" O.C.		1.65	2.78	4.43
15	10' high, 16" O.C.		1.66	2.79	4.45
16	24" O.C.		1.61	2.68	4.29
17	2" x 4" studs, 8' high, 24" O.C.		1.67	2.81	4.48
18	10' high, 16" O.C.		1.68	2.81	4.49
19	24" O.C.		1.63	2.70	4.33
20	12' high, 16" O.C.		1.65	2.81	4.46
21	24" O.C.		1.60	2.70	4.30
22	2" x 6" studs, 8' high, 16" O.C.		1.88	3.03	4.91
23	24" O.C.		1.81	2.86	4.67
24	10' high, 16" O.C.		1.85	2.87	4.72
25	24" O.C.		1.77	2.74	4.51
26	12' high, 16" O.C.		1.80	2.90	4.70
27	24" O.C.		1.74	2.74	4.48
28					
29	Cut & patch to match existing construction, add, minimum		2%	3%	
30	Maximum		5%	9%	
31	Dust protection, add, minimum		1%	2%	
32	Maximum		4%	11%	
33	Material handling & storage limitation, add, minimum		1%	1%	
34	Maximum		6%	7%	
35	Protection of existing work, add, minimum		2%	2%	
36	Maximum		5%	7%	
37	Shift work requirements, add, minimum			5%	
38	Maximum			30%	
39	Temporary shoring and bracing, add, minimum		2%	5%	
40	Maximum		5%	12%	
41					
42					

277

Figure 27.29

Means® Forms

COST ANALYSIS

PROJECT Apartment Renovation | SHEET NO. 6 of 11
ESTIMATE NO. 89-2
ARCHITECT | DATE 1989
TAKE OFF BY: | QUANTITIES BY: EBN | PRICES BY: RSM | EXTENSIONS BY: | CHECKED BY:

DESCRIPTION	SOURCE/DIMENSIONS			QUANTITY	UNIT	MATERIAL		LABOR		EQ./TOTAL	
						UNIT COST	TOTAL	UNIT COST	TOTAL	UNIT COST	TOTAL
Division 6: Interior Construction											
Interior Partitions				7,208	SF	1.70	12254	2.95	21264		
Firewalls				2,584	SF	1.75	4522	2.95	7623		
Furring Wall				527	SF	.42	221	1.02	537		
Interior Doors				24	EA.	109.76	2634	141.04	3385		
Apartment & Stair Doors				10	EA.	540.74	5407	176.26	1763		
Bathroom Closet Doors				8	EA.	147.58	1181	135.91	1087		
Utility Room Doors				8	EA.	194.46	1556	142.34	1139		
Bi-Folds - Coat Closets				8	EA.	115.73	926	126.78	1014		
Bedroom Closets				16	EA.	19.60	3066	175	2800		
Suspended Ceiling				1,216	SF	.85	1034	.89	1082		
Painting				527	SF	.14	74	.34	179		
Corridor Carpet				304	SF	2.87	872	.57	173		
Apartment Carpet				5,824	SF	2.56	14909	.55	3203		
Bathroom & Kitchen Vinyl				1,280	SF	1.38	1766	1.14	1459		
Division 6: Totals							50442		46708		

Figure 27.30

06.9-300	Plaster & Drywall Selective Price Sheet			
LINE NO.	DESCRIPTION	COST PER S.F.		
		MAT.	INST.	TOTAL
01	Lath, gypsum perforated			
02				
03	Regular, ⅜" thick	.39	.35	.74
04	½" thick	.45	.43	.88
05	Fire resistant, ½" thick	.42	.42	.84
06	⅝" thick	.44	.46	.90
07	Moisture resistant, ½" thick	.46	.41	.87
08	⅝" thick	.48	.44	.92
09				
10	Metal lath			
11	Diamond painted, 2.5 lb.	.13	.36	.49
12	3.4 lb.	.14	.41	.55
13	Rib painted, 2.75 lb	.19	.40	.59
14	3.40 lb	.22	.43	.65
15				
16				
17	Plaster, gypsum, 2 coats	.34	1.36	1.70
18	3 coats	.47	1.65	2.12
19	Perlite/vermiculite, 2 coats	.39	1.55	1.94
20	3 coats	.62	1.93	2.55
21	Bondcrete, 1 coat	.62	.60	1.22
22				
23	Wood fiber, 2 coats	.24	1.98	2.22
24	3 coats	.30	2.48	2.78
25				
26				
27	Drywall, standard, ⅜" thick, no finish included	.21	.27	.48
28	½" thick, no finish included	.22	.27	.49
29	Taped and finished	.28	.56	.84
30	⅝" thick, no finish included	.26	.28	.54
31	Taped and finished	.32	.57	.89
32	Fire resistant, ½" thick, no finish included	.26	.28	.54
33	Taped and finished	.32	.57	.89
34	⅝" thick, no finish included	.29	.27	.56
35	Taped and finished	.34	.57	.91
36	Water resistant, ½" thick, no finish included	.33	.27	.60
37	Taped and finished	.39	.56	.95
38	⅝" thick, no finish included	.37	.28	.65
39	Taped and finished	.43	.57	1
40	Finish, instead of taping			
41	For thin coat plaster, add	.08	.36	.44
42	Finish, textured spray, add	.11	.38	.49

292

Figure 27.31

INTERIOR CONSTR.	06.1-592	Partitions, Wood Stud

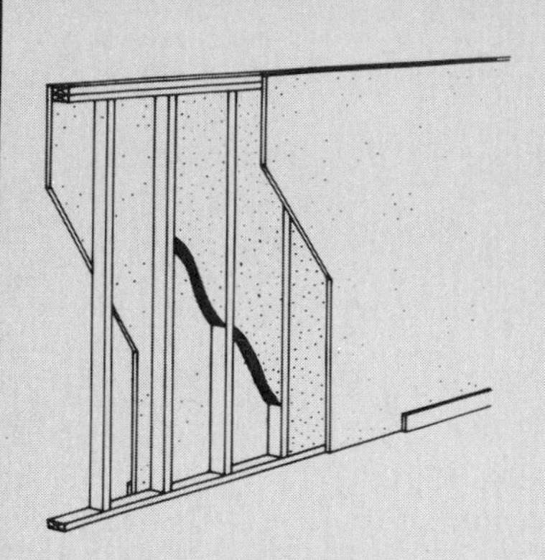

This page illustrates and describes a wood stud partition system including wood studs with plates, gypsum plasterboard - taped and finished, insulation, baseboard and painting. Lines 06.1-592-04 thru 10 give the unit price and total price per square foot for this system. Prices for alternate wood stud partition systems are on Line Items 06.1-592-13 thru 27. Both material quantities and labor costs have been adjusted for the system listed.

Factors: To adjust for job conditions other than normal working situations use Lines 06.1-592-29 thru 40.

Example: You are to install the system where material handling and storage present a serious problem. Go to Line 06.1-592-34 and apply these percentages to the appropriate MAT. and INST. costs.

LINE NO.	DESCRIPTION	QUANTITY	COST PER S.F.		
			MAT.	INST.	TOTAL
01	**Wood stud wall, 2"x4",16"O.C., dbl. top plate, sngl bot. plate, ⅝" dwl.**				
02	**Taped, finished and painted on 2 faces, insulation, baseboard, wall 8' high**				
03					
04	Wood studs, 2" x 4", 16" O.C., 8' high	1 S.F.	.35	.71	1.06
05	Gypsum drywall, ⅝" thick FIRE RESISTANT	2 S.F.	.58 ~~.53~~	✓.55	1.08
06	Taping and finishing	2 S.F.	.11	.55	.66
07	Insulation, 3-½" fiberglass batts	1 S.F.	.30	.17	.47
08	Baseboard, painted	.2 L.F.	.21	.43	.64
09	Painting, roller, 2 coats	2 S.F.	.20	.54	.74
10	TOTAL	S.F.	1.75 ~~1.70~~	✓2.95	4.65
11					
12	For alternate wood stud systems:				
13	2" x 3" studs, 8' high, 16" O.C.	S.F.	1.68	2.91	4.59
14	24" O.C.		1.65	2.78	4.43
15	10' high, 16" O.C.		1.66	2.79	4.45
16	24" O.C.		1.61	2.68	4.29
17	2" x 4" studs, 8' high, 24" O.C.		1.67	2.81	4.48
18	10' high, 16" O.C.		1.68	2.81	4.49
19	24" O.C.		1.63	2.70	4.33
20	12' high, 16" O.C.		1.65	2.81	4.46
21	24" O.C.		1.60	2.70	4.30
22	2" x 6" studs, 8' high, 16" O.C.		1.88	3.03	4.91
23	24" O.C.		1.81	2.86	4.67
24	10' high, 16" O.C.		1.85	2.87	4.72
25	24" O.C.		1.77	2.74	4.51
26	12' high, 16" O.C.		1.80	2.90	4.70
27	24" O.C.	↓	1.74	2.74	4.48
28					
29	Cut & patch to match existing construction, add, minimum		2%	3%	
30	Maximum		5%	9%	
31	Dust protection, add, minimum		1%	2%	
32	Maximum		4%	11%	
33	Material handling & storage limitation, add, minimum		1%	1%	
34	Maximum		6%	7%	
35	Protection of existing work, add, minimum		2%	2%	
36	Maximum		5%	7%	
37	Shift work requirements, add, minimum			5%	
38	Maximum			30%	
39	Temporary shoring and bracing, add, minimum		2%	5%	
40	Maximum		5%	12%	
41					
42					

277

Figure 27.32

INTERIOR CONSTR.	06.4-142	Doors, Interior Flush, Wood

This page illustrates and describes flush interior door systems including hollow core door, jamb, header and trim with hardware. Lines 06.4-142-04 thru 08 give the unit price and total price on a cost each basis for this system. Prices for alternate flush interior door systems are on Line items 06.4-142-11 thru 24. Both material quantities and labor costs have been adjusted for the system listed.

Factors: To adjust for job conditions other than normal working situations use Lines 06.1-142-27 thru 38.

Example: You are to install the system in an area where dust protection is vital. Go to Line 06.4-142-30 and apply these percentages to the appropriate MAT. and INST. costs.

LINE NO.	DESCRIPTION	QUANTITY	COST EACH		
			MAT.	INST.	TOTAL
01	**Single hollow core door, include jamb, header, trim and hardware, painted.**				
02					
03					
04	Hollow core Lauan, 1-3/8" thick, 2'-0" x 6'-8", painted	1 Ea.	21.44	59.56	81
05	Wood jamb, 4-9/16" deep	1 Set	35.38	24.14	59.52
06	Trim, casing	1 Set	19.71	36.61	56.32
07	Hardware, hinges, lockset	1 Set	11.65		11.65
08	TOTAL	Ea.	98.08	137.41	235.49
09					
10	For alternate door systems:				
11	Lauan (Mahogany) hollow core, 1-3/8" x 2'-6" x 6'-8"	Ea.	100.90	139.21	240.11
12	2'-8" x 6'-8"		103.16	140.64	243.80
13	3'-0" x 6'-8"		106.06	145.19	251.25
14					
15	Birch, hollow core, 1-3/8" x 2'-0" x 6'-8"	Ea.	102.48	138.01	240.49
16	2'-6" x 6'-8"		106.40	139.71	246.11
17	2'-8" x 6'-8"		109.76	141.04	250.80
18	3'-0" x 6'-8"		113.76	145.49	259.25
19	Solid core, pre-hung, 1-3/8" x 2'-6" x 6'-8"		207.60	136.51	344.11
20	2'-8" x 6'-8"		212.06	139.74	351.80
21	3'-0" x 6'-8"		216.06	140.19	356.25
22					
23					
24	For metal frame instead of wood, add	Ea.	50%	20%	
25					
26					
27	Cut & patch to match existing construction, add, minimum		2%	3%	
28	Maximum		5%	9%	
29	Dust protection, add, minimum		1%	2%	
30	Maximum		4%	11%	
31	Equipment usage curtailment, add, minimum		1%	1%	
32	Maximum		3%	10%	
33	Material handling & storage limitation, add, minimum		1%	1%	
34	Maximum		6%	7%	
35	Protection of existing work, add, minimum		2%	2%	
36	Maximum		5%	7%	
37	Shift work requirements, add, minimum			5%	
38	Maximum			30%	
39					
40					
41					
42					

283

Figure 27.33

INTERIOR CONSTR.	06.4-146	Doors, Interior Flush, Metal

This page illustrates and describes interior metal door systems including a metal door, metal frame and hardware. Lines 06.4-146-04 thru 07 give the unit price and total price on a cost each for this system. Prices for alternate interior metal door systems are on Line Items 06.4-146-11 thru 21. Both material quantities and labor costs have been adjusted for the system listed.

Factors: To adjust for job conditions other than normal working situations use Lines 06.5-146-29 thru 40.

Example: You are to install the system while protecting existing construction. Go to Line 06.4-146-37 and apply these percentages to the appropriate MAT. and INST. costs.

LINE NO.	DESCRIPTION	QUANTITY	COST EACH		
			MAT.	INST.	TOTAL
01	**Single metal door, including frame and hardware.**				
02					
03					
04	Hollow metal door, 1-3/8″ thick, 2′-6″ x 6′-8″, painted	1 Ea.	151.24	64.76	216
05	Metal frame, 5-3/4″ deep	1 Set	71.50	33.50	105
06	Hinges and passage lockset DEDUCT	1 Set	64.90	17.10	82
07	TOTAL	Ea.	287.64	115.36	403
08					
09					
10	For alternate systems:				
11	Hollow metal doors, 1-3/8″ thick, 2′-8″ x 6′-8″	Ea.	295.34	117.66	413
12	3′-0″ x 7′-0″	″	303.04	124.96	428
13					
14	Interior fire door, 1-3/8″ thick, 2′-6″ x 6′-8″	Ea.	329.44	118.56	448
15	2′-8″ x 6′-8″	↓	337.14	120.86	458
16	3′-0″ x 7′-0″	↓	350.34	122.66	473
17					
18	Add to fire doors:				
19	Baked enamel finish	Ea.	30%	90%	
20	Galvanizing	↓	9.50%		
21	Porcelain finish	↓	100%	150%	
22					
23					
24					
25					
26					
27					
28					
29	Cut & patch to match existing construction, add, minimum		2%	3%	
30	Maximum		5%	9%	
31	Dust protection, add, minimum		1%	2%	
32	Maximum		4%	11%	
33	Equipment usage curtailment, add, minimum		1%	1%	
34	Maximum		3%	10%	
35	Material handling & storage limitation, add, minimum		1%	1%	
36	Maximum		6%	7%	
37	Protection of existing work, add, minimum		2%	2%	
38	Maximum		5%	7%	
39	Shift work requirements, add, minimum			5%	
40	Maximum			30%	
41					
42					

285

Figure 27.34

06.9-500	Hardware Selective Price Sheet			
LINE NO.	DESCRIPTION	COST EACH		
		MAT.	INST.	TOTAL
01	Door closer, rack and pinion	67.10	42.90	110
02	Backcheck and adjustable power	77	48	125
03	Regular, hinge face mount, all sizes, regular arm	71.50	43.50	115
04	Hold open arm	83.60	41.40	125
05	Top jamb mount, all sizes, regular arm	69.30	45.70	115
06	Hold open arm	79.20	45.80	125
07	Stop face mount, all sizes, regular arm	71.50	43.50	115
08	Hold open arm	82.50	42.50	125
09	Fusible link, hinge face mount, all sizes, regular arm	81.40	43.50	124.90
10	Hold open arm	93.50	41.40	134.90
11	Top jamb mount, all sizes, regular arm	79.20	45.70	124.90
12	Hold open arm	89.10	45.80	134.90
13	Stop face mount, all sizes, regular arm	81.40	43.50	124.90
14	Hold open arm	92.40	42.50	134.90
15				
16				
17	Door stops			
18				
19	Holder & bumper, floor or wall	3.30	11.45	14.75
20	Wall bumper	3.96	11.44	15.40
21	Floor bumper	1.54	11.46	13
22	Plunger type, door mounted	14.85	11.15	26
23	Hinges, full mortise, material only, per pair			
24	Low frequency, 4-½″ x 4-½″, steel base, USP	11.65		11.65
25	Brass base, US10	39		39
26	Stainless steel base, US32	55		55
27	Average frequency, 4-½″ x 4-½″, steel base, USP	21		21
28	Brass base, US10	44		44
29	Stainless steel base, US32	63		63
30	High frequency, 4-½″ x 4-½″, steel base, USP	55		55
31	Brass base, US10	83		83
32	Stainless steel base, US32	110		110
33	Kick plate			
34				
35	6″ high, for 3′-0″ door, aluminum	18.70	18.30	37
36	Bronze	38.50	18.50	57
37	Panic device			
38				
39	For rim locks, single door, exit	256.30	93.70	350
40	Outside key and pull	294.80	55.20	350
41	Bar and vertical rod, exit only	393.80	56.20	450
42	For touch bar	440	70	510
43	Lockset			
44				
45	Heavy duty, cylindrical, passage doors	95.70	24.30	120
46	Classroom	203.50	36.50	240
47	Bedroom, bathroom, and inner office doors	123.20	21.80	145
48	Apartment, office, and corridor doors	165	25	190
49	Standard duty, cylindrical, exit doors	63.80	27.20	91
50	Inner office doors	26.40	22.60	49
51	Passage doors	36.30	22.70	59
52	Public restroom, classroom, & office doors	75.90	34.10	110
53	Deadlock, mortise, heavy duty	102.30	32.70	135
54	Double cylinder	112.20	32.80	145
55	Entrance lock, cylinder, deadlocking latch	96.80	28.20	125
56	Deadbolt	115.50	34.50	150
57	Commercial, mortise, wrought knob, keyed, minimum	126.50	33.50	160
58	Maximum	225.50	39.50	265

293

Figure 27.35

The owner has requested that the estimate include cooling, so a ducted forced air system is the only choice. The estimator has learned from the site visit that the gas service is of adequate capacity. The system chosen is shown in Figure 27.46. Note that the flue, in Figure 27.46, is connected to an existing chimney. There are no chimneys in the existing building, so metal flues must be included in the estimate. The costs for the flues are obtained from Division 15 of the Unit Price section of *Means Repair and Remodeling Cost Data*. Overhead and profit must be added to the costs in Figure 27.47.

The mechanical and electrical work would, most likely, be performed by a subcontractor. The material and installation costs in the Assemblies pages of *Means Repair and Remodeling Cost Data* include the overhead and profit of the installing contractor. When certain work is to be done by a subcontractor, the estimator should add a percentage for supervision by the general contractor. This percentage is usually 10% and can be added within the appropriate division costs rather than at the estimate summary.

Division 9: Electrical

Since there are no electrical plans, the estimator must use experience and sound judgment to determine the requirements and quantities of the electrical work.

The existing electrical service is old and inadequate and will be removed. The project calls for each apartment to be metered separately with a 100-Amp service. The appropriate system, as shown in Figure 27.48, must be modified. Obviously, the building will not have eight individual service entrances and grounding systems. The weathercap, service entrance cable, and grounding system are

Calculations for Apartment and Stairway Doors			
	Material	Installation	Total
Apartment Entrance and Stairway Doors 06.4-146-16	$350.34	$122.66	$473.00
Deduct Hardware	64.90	17.10	82.00
Subtotal	285.44	105.56	391.00
Add:			
Closer	69.30	45.70	115.00
Hinges	21.00		21.00
Lockset	165.00	25.00	190.00
Total	$540.74	$176.26	$717.00

Figure 27.36

INTERIOR CONSTR.	06.4-144	Doors, Interior Solid And Louvered

This page illustrates and describes interior, solid and louvered door systems including a pine panel door, wood jambs, header, and trim with hardware. Lines 06.4-144-04 thru 08 give the unit price and total price on a cost each basis for this system. Prices for alternate interior, solid and louvered systems are on Line Items 06.4-144-12 thru 24. Both material quantities and labor costs have been adjusted for the system listed.

Factors: To adjust for job conditions other than normal working situations use Lines 06.4-144-29 thru 40.

Example: You are to install the system during night hours only. Go to Line 06.4-144-40 and apply these percentages to the appropriate INST. costs.

LINE NO.	DESCRIPTION	QUANTITY	COST EACH		
			MAT.	INST.	TOTAL
01	**Single interior door, including jamb, header, trim and hardware, painted.**				
02					
03					
04	Solid pine panel door, painted 1-3/8" thick, 2'-0" x 6'-8"	1 Ea.	87.44	58.56	146
05	Wooden jamb, 4-5/8" deep	1 Set	35.38	24.14	59.52
06	Trim, casing	1 Set	19.71	36.61	56.32
07	Hardware, hinges, lockset	1 Set	21.55	17.10	38.65
08	TOTAL	Ea.	164.08	136.41	300.49
09					
10					
11	For alternate door systems:				
12	Solid pine, painted raised panel, 1-3/8" x 2'-6" x 6'-8"	Ea.	176.80	137.31	314.11
13	2'-8" x 6'-8"	↓	181.26	140.54	321.80
14	3'-0" x 6'-8"	↓	190.76	145.49	336.25
15					
16	Louvered pine, painted 1'-6" x 6'-8" BATH CLOSET	Ea.	147.58	135.91	283.49
17	2'-0" x 6'-8"	↓	170.75	138.36	309.11
18	2'-6" x 6'-8" UTILITY	↓	194.46	142.34	336.80
19	3'-0" x 6'-8"	↓	209.46	146.79	356.25
20					
21					
22	For prehung door, deduct	Ea.	5%	30%	
23					
24	For metal frame instead of wood, add	Ea.	50%	20%	
25					
26					
27					
28					
29	Cut & patch to match existing construction, add, minimum		2%	3%	
30	Maximum		5%	9%	
31	Dust protection, add, minimum		1%	2%	
32	Maximum		4%	11%	
33	Equipment usage curtailment, add, minimum		1%	1%	
34	Maximum		3%	10%	
35	Material handling & storage limitation, add, minimum		1%	1%	
36	Maximum		6%	7%	
37	Protection of existing work, add, minimum		2%	2%	
38	Maximum		5%	7%	
39	Shift work requirements, add, minimum			5%	
40	Maximum			30%	
41					
42					

284

Figure 27.37

INTERIOR CONSTR.	06.4-148	Doors, Closet

This page illustrates and describes an interior closet door system including an interior closet door, painted, with trim and hardware. Prices for alternate interior closet door system are on Line Items 06.4-148-05 thru 22. Both material quantities and labor costs have been adjusted for the system listed.

Factors: To adjust for job conditions other than normal working situations use Lines 0.64-148-29 thru 40.

Example: You are to install the system and match the existing construction. Go to Line 06.4-148-29 and apply these percentages to the appropriate MAT. and INST. costs.

LINE NO.	DESCRIPTION	QUANTITY	COST PER SET		
			MAT.	INST.	TOTAL
01	**Interior closet door painted, including frame, trim and hardware, prehung.**				
02					
03					
04	Bi-fold doors				
05	Pine paneled, 3'-0" x 6'-8"	Set	153.13	118.38	271.51
06	6'-0" x 6'-8"		278.50	164.10	442.60
07	Birch, hollow core, 3'-0" x 6'-8" COAT CL.		115.73	126.78	242.51
08	6'-0" x 6'-8" BEDROOM CL.		191.60	175	366.60
09	Lauan, hollow core, 3'-0" x 6'-8"		103.63	117.88	221.51
10	6'-0" x 6'-8"		168.50	164.10	332.60
11	Louvered pine, 3'-0" x 6'-8"		120.13	116.38	236.51
12	6'-0" x 6'-8"		207	165.60	372.60
13					
14	Sliding, bi-passing closet doors				
15	Pine paneled, 4'-0" x 6'-8"	Set	210.51	121.03	331.54
16	6'-0" x 6'-8"		265.30	167.30	432.60
17	Birch, hollow core, 4'-0" x 6'-8"		133.51	123.03	256.54
18	6'-0" x 6'-8"		160.80	166.80	327.60
19	Lauan, hollow core, 4'-0" x 6'-8"		122.51	124.03	246.54
20	6'-0" x 6'-8"		144.30	168.30	312.60
21	Louvered pine, 4'-0" x 6'-8"		147.81	123.73	271.54
22	6'-0" x 6'-8"		180.60	167	347.60
23					
24					
25					
26					
27					
28					
29	Cut & patch to match existing construction, add, minimum		2%	3%	
30	Maximum		5%	9%	
31	Dust protection, add, minimum		1%	2%	
32	Maximum		4%	11%	
33	Equipment usage curtailment, add, minimum		1%	1%	
34	Maximum		3%	10%	
35	Material handling & storage limitation, add, minimum		1%	1%	
36	Maximum		6%	7%	
37	Protection of existing work, add, minimum		2%	2%	
38	Maximum		5%	7%	
39	Shift work requirements, add, minimum			5%	
40	Maximum			30%	
41					
42					

286

Figure 27.38

INTERIOR CONSTR.	06.7-342	Ceiling, Suspended Acoustical

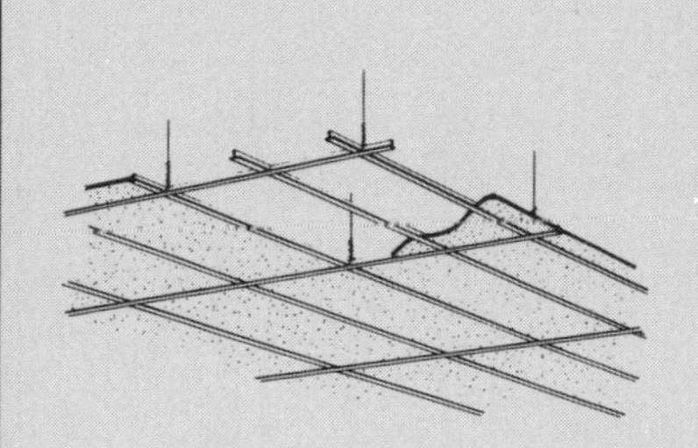

This page illustrates suspended acoustical board systems including acoustic ceiling board, hangers, and T bar suspension. Lines 06.7-342-04 thru 07 give the unit price and total price per square foot for this system. Prices for alternate suspended acoustical board systems are on Line Items 06.7-342-11 thru 24. Both material quantitites and labor costs have been adjusted for the system listed.

Factors: To adjust for job conditions other than normal working situations use Lines 06.7-342-29 thru 40.

Example: You are to install the system and protect existing construction. Go to Line 06.7-342-38 and apply these percentages to the appropriate MAT. and INST. costs.

LINE NO.	DESCRIPTION	QUANTITY	COST PER S.F.		
			MAT.	INST.	TOTAL
01	**Suspended acoustical ceiling board installed on exposed grid system.**				
02					
03					
04	Fiberglass boards, film faced, 2′ x 4′, ⅝″ thick	1 S.F.	.39	.43	.82
05	Hangers, #12 wire	1 S.F.		.05	.05
06	T bar suspension system, 2′ x 4′ grid	1 S.F.	.42	.34	.76
07	TOTAL	S.F.	.81	.82	1.63
08	FACTORS		.04	.07	
09			.85	.89	
10	For alternate suspended ceiling systems:				
11	2′ x 4′ grid, mineral fiber board, aluminum faced, ⅝″ thick	S.F.	1.58	.81	2.39
12	Standard faced		.97	.80	1.77
13	Plastic faced		1.29	1.08	2.37
14	Fiberglass, film faced, 3″ thick, R11		1.41	.94	2.35
15	Grass cloth faced, ¾″ thick		1.69	.93	2.62
16	1″ thick		1.85	.94	2.79
17	1-½″ thick, nubby face		2.20	.94	3.14
18	Wood fiber, reveal edge, painted, 1″ thick		1.25	.84	2.09
19	2″ thick		1.74	.89	2.63
20	2-½″ thick		2.18	.94	3.12
21	3″ thick		2.47	1	3.47
22					
23					
24	Add for 2′ x 2′ grid system	S.F.	.09	.09	.18
25					
26					
27					
28					
29	Cut & patch to match existing construction, add, minimum		2%	3%	
30	Maximum		5%	9%	
31	Dust protection, add, minimum		1%	2%	
32	Maximum		4%	11%	
33	Equipment usage curtailment, add, minimum		1%	1%	
34	Maximum		3%	10%	
35	Material handling & storage limitation, add, minimum		1%	1%	
36	Maximum		6%	7%	
37	Protection of existing work, add, minimum		2%	2%	
38	Maximum		5%	7%	
39	Shift work requirements, add, minimum			5%	
40	Maximum			30%	
41					
42					

288

Figure 27.39

06.9-600	Interior Wall Finish Selective Price Sheet			
LINE NO.	DESCRIPTION	COST PER S.F.		
		MAT.	INST.	TOTAL
01	Painting, on plaster or drywall, brushwork, primer and 1 ct.	.09	.26	.35
02	Primer and 2 ct.	.13	.40	.53
03	Rollerwork, primer and 1 ct.	.10	.22	.32
04	Primer and 2 ct.	.14	.34	.48
05	Woodwork incl. puttying, brushwork, primer and 1 ct.	.09	.38	.47
06	Primer and 2 ct.	.11	.51	.62
07	Wood trim to 6" wide, enamel, primer and 1 ct.	.07	.75	.82
08	Primer and 2 ct.	.09	1.07	1.16
09	Cabinets and casework, enamel, primer and 1 ct.	.08	1.10	1.18
10	Primer and 2 ct.	.12	1.49	1.61
11	On masonry or concrete, latex, brushwork, primer and 1 ct.	.13	.52	.65
12	Primer and 2 ct.	.15	.71	.86
13	For block filler, add	.09	.13	.22
14				
15	Varnish, wood trim, sealer 1 ct., sanding, puttying, quality work	.13	.52	.65
16	Meduim work	.10	.40	.50
17	Without sanding	.08	.31	.39
18				
19	Wall coverings, wall paper, at $9.70 per double roll, average workmanship	.18	.39	.57
20	At $20.00 per double roll, average workmanship	.36	.48	.84
21	At $44.00 per double roll, quality workmanship	.78	.59	1.37
22				
23	Grass cloths with lining paper, minimum	.63	.63	1.26
24	Maximum	1.83	.72	2.55
25	Vinyl, fabric backed, light weight	.51	.39	90
26	Medium weight	.68	.53	1.21
27	Heavy weight	1.02	.59	1.61
28				
29	Cork tiles, 12" x 12", 3/16" thick	2.09	1.06	3.15
30	5/16" thick	2.20	1.06	3.26
31	Granular surface, 12" x 36", ½" thick	.57	.66	1.23
32	1" thick	.80	.66	1.46
33	Aluminum foil	.63	.92	1.55
34				
35	Tile, ceramic, adhesive set, 4-¼" x 4-¼"	4.73	2.62	7.35
36	6" x 6"	4.40	2.35	6.75
37	Decorated, 4-¼" x 4-¼", minimum	2.48	.54	3.02
38	Maximum	15.40	.80	16.20
39	For epoxy grout, add	.66	.49	1.15
40	Pregrouted sheets	3.91	1.99	5.90
41	Plastic, 4-¼" x 4-¼", .050" thick	1.65	2.20	3.85
42	.110" thick	2.20	2.20	4.40
43	Metal, 4-¼" x 4-¼", thin set, aluminum, plain	2.04	3.41	5.45
44	Epoxy enameled	2.26	3.44	5.70
45	Leather on aluminum	17.60	4.40	22
46	Stainless steel	9.35	3.45	12.80
47	Brick, interior veneer, 4" face brick, running bond, minimum	1.73	5.92	7.65
48	Maximum	1.73	5.92	7.65
49	Simulated, urethane pieces, set in mastic	4.07	1.83	5.90
50	Fiberglass panels	1.87	1.37	3.24
51	Wall coating, on drywall, thin coat, plain	.08	.36	.44
52	Stipple	.08	.36	.44
53	Textured spray	.11	.38	.49
54				
55	Paneling not incl. furring or trim, hardboard, tempered, ⅛" thick	.29	1.13	1.42
56	¼" thick	.35	1.13	1.48
57	Plastic faced, ⅛" thick	.46	1.13	1.59
58	¼" thick	.63	1.13	1.76

295

Figure 27.40

06.9-800	Floor Finish Selective Price Sheet			
LINE NO.	DESCRIPTION	COST PER S.F.		
		MAT.	INST.	TOTAL
01	Flooring, carpet, acrylic, 26 oz. light traffic	1.89	.55	2.44
02	35 oz. heavy traffic CORRIDORS	2.87	.57	3.44
03	Nylon anti-static, 15 oz. light traffic	1.16	.52	1.68
04	22 oz. medium traffic	1.86	.47	2.33
05	26 oz. heavy traffic APARTMENTS	2.56	.55	3.11
06	28 oz. heavy traffic	3.11	.55	3.66
07	Tile, foamed back, needle punch	2.20	.41	2.61
08	Tufted loop	1.93	.40	2.33
09	Wool, 36 oz. medium traffic	3.20	.57	3.77
10	42 oz. heavy traffic	4.48	.63	5.11
11	Composition, epoxy, with colored chips, minimum	1.43	2.12	3.55
12	Maximum	2.40	2.90	5.30
13	Trowelled, minimum	1.93	2.55	4.48
14	Maximum	2.64	2.96	5.60
15	Terrazzo, ¼" thick, chemical resistant, minimum	3.47	3.83	7.30
16	Maximum	8.47	5.08	13.55
17	Resilient, asphalt tile, ⅛" thick	.77	.66	1.43
18	Conductive flooring, rubber, ⅛" thick	1.98	.84	2.82
19	Cork tile ⅛" thick, standard finish	1.27	.83	2.10
20	Urethane finish	1.71	.83	2.54
21	PVC sheet goods for gyms, ¼" thick	2.97	3.28	6.25
22	⅜" thick	3.30	4.40	7.70
23	Vinyl composition 12" x 12" tile, plain, 1/16" thick	.55	.84	1.39
24	⅛" thick	1.49	.83	2.32
25	Vinyl tile, 12" x 12" x ⅛" thick, minimum	1.65	.84	2.49
26	Maximum	7.98	.82	8.80
27	Vinyl sheet goods, backed, .093" thick BATH & KITCHEN	1.38	1.14	2.52
28	.250" thick	2.48	1.14	3.62
29	Slate, random rectangular, ¼" thick	3.66	3.29	6.95
30	½" thick	3.43	4.72	8.15
31	Natural cleft, irregular, ¾" thick	1.54	5.36	6.90
32	For sand rubbed finish, add	2.35		2.35
33	Terrazzo, cast in place, bonded 1-¾" thick, gray cement	2.05	7.75	9.80
34	White cement	2.64	7.76	10.40
35	Not bonded 3" thick, gray cement	2.61	9.69	12.30
36	White cement	3.39	9.71	13.10
37	Precast, 12" x 12" x 1" thick	6.60	17.40	24
38	1-¼" thick	8.25	16.75	25
39	16" x 16" x 1-¼" thick	7.70	21.30	29
40	1-½" thick	9.90	23.10	33
41	Marble travertine, standard, 12" x 12" x ¾" thick	4.84	7.41	12.25
42				
43	Tile, ceramic, natural clay, thin set	3.03	2.57	5.60
44	Porcelain, thin set	3.69	2.61	6.30
45	Specialty, decorator finish	5.28	2.57	7.85
46				
47	Quarry, red, mud set, 4" x 4" x ½" thick	3.08	3.97	7.05
48	6" x 6" x ½" thick	3.08	3.37	6.45
49	Brown, imported, 6" x 6" x ⅞" thick	5.28	3.97	9.25
50	9" x 9" x 1-¼" thick	6.60	4.30	10.90
51	Slate, vermont, thin set, 6" x 6" x ¼" thick	2.15	2.63	4.78
52				
53	Wood, maple strip, 25/32" x 2-¼", finished, select	2.67	1.92	4.59
54	2nd and better	1.90	1.92	3.82
55	Oak, 25/32" x 2-¼" finished, clear	1.90	1.92	3.82
56	No. 1 common	3.05	1.80	4.85
57	Parquet, standard, 5/16", finished, minimum	1.46	2.02	3.48
58	Maximum	5.03	3.06	8.09

299

Figure 27.41

141 | Dumbwaiters

141 200 | Electric Dumbwaiters

			CREW	DAILY OUTPUT	MAN-HOURS	UNIT	BARE COSTS MAT.	BARE COSTS LABOR	BARE COSTS EQUIP.	BARE COSTS TOTAL	TOTAL INCL O&P	
201	0010	DUMBWAITERS 2 stop, electric, minimum	2 Elev	.13	123	Ea.	3,575	2,975		6,550	8,500	201
	0100	Maximum		.11	145	"	8,500	3,525		12,025	14,700	
	0600	For each additional stop, add	↓	.54	29.630	Stop	850	715		1,565	2,025	

142 | Elevators

142 010 | Elevators

			CREW	DAILY OUTPUT	MAN-HOURS	UNIT	BARE COSTS MAT.	BARE COSTS LABOR	BARE COSTS EQUIP.	BARE COSTS TOTAL	TOTAL INCL O&P	
011	0012	ELEVATORS										011
	5000	Passenger, pre-engineered, 5 story, hydraulic, 2,500 lb. cap.	M-1	.04	800	Ea.	42,700	18,400	1,725	62,825	77,000	
	5100	For less than 5 stops, deduct	"	.29	110	Stop	5,780	2,525	240	8,545	10,500	
	5200	For 4,000 lb. capacity, general purpose, add				Ea.	7,400			7,400	8,150	
	5400	10 story, geared traction, 200 FPM, 2,500 lb. capacity	M-1	.02	600	"	41,400	36,700	3,450	81,550	105,500	
	5500	For less than 10 stops, deduct		.34	94.120	Stop	3,150	2,150	205	5,505	7,000	
	5600	For 4,500 lb. capacity, general purpose	↓	.02	600	Ea.	44,500	36,700	3,450	84,650	109,000	
	5800											
	7000	Residential, cab type, 1 floor, 2 stop, minimum	2 Elev	.20	80	Ea.	5,500	1,925		7,425	9,025	
	7100	Maximum		.10	160		9,800	3,875		13,675	16,700	
	7200	2 floor, 3 stop, minimum		.12	133		6,800	3,225		10,025	12,400	
	7300	Maximum		.06	267		16,500	6,450		22,950	28,000	
	7700	Stair climber (chair lift), single seat, minimum		1	16		3,550	385		3,935	4,500	
	7800	Maximum		.20	80		5,000	1,925		6,925	8,475	
	8000	Wheelchair, porch lift, minimum		1	16		3,200	385		3,585	4,125	
	8500	Maximum		.50	32		7,000	775		7,775	8,875	
	8700	Stair lift, minimum		1	16		5,800	385		6,185	6,975	
	8900	Maximum	↓	.20	80	↓	8,800	1,925		10,725	12,600	

144 | Lifts

144 010 | Lifts

			CREW	DAILY OUTPUT	MAN-HOURS	UNIT	BARE COSTS MAT.	BARE COSTS LABOR	BARE COSTS EQUIP.	BARE COSTS TOTAL	TOTAL INCL O&P	
011	0010	CORRESPONDENCE LIFT 1 floor 2 stop, 25 lb. capacity, electric	2 Elev	.20	80	Ea.	3,575	1,925		5,500	6,900	011
	0100	Hand, 5 lb. capacity	"	.20	80	"	1,475	1,925		3,400	4,600	

145 | Material Handling Systems

145 600 | Chutes

			CREW	DAILY OUTPUT	MAN-HOURS	UNIT	BARE COSTS MAT.	BARE COSTS LABOR	BARE COSTS EQUIP.	BARE COSTS TOTAL	TOTAL INCL O&P	
601	0010	CHUTES Linen or refuse, incl. sprinklers										601
	0020											
	0050	Aluminized steel, 16 ga., 18" diameter	2 Shee	3.50	4.570	Floor	525	110		635	750	
	0100	24" diameter	"	3.20	5	"	625	120		745	875	

176

Figure 27.42

MECHANICAL	08.1-931	Plumbing - Three Fixture Bathroom

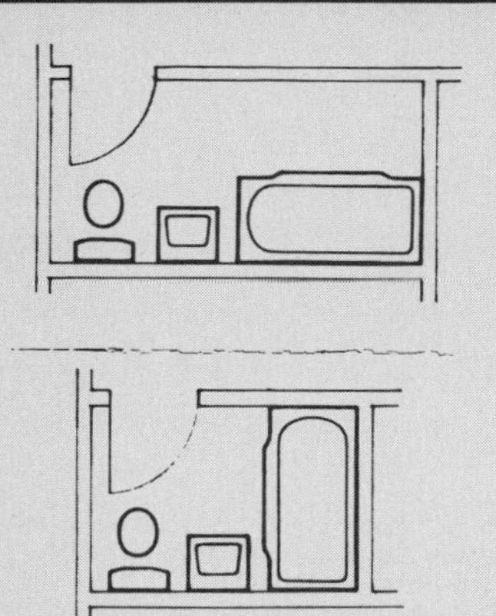

This page illustrates and describes a three fixture bathroom system including a water closet, tub, lavatory, accessories and service piping. Lines 08.1-931-04 thru 14 give the unit price and total price on a cost each basis for this system. Prices for an alternate three fixture bathroom system are on Line Item 08.1-931-17. Both material quantities and labor costs have been adjusted for the system listed.

Factors: To adjust for job conditions other than normal working situations use Lines 08.1-931-29 thru 40.

Example: You are to install the system and protect all existing work. Go to Line 08.1-931-38 and apply these percentages to the appropriate MAT. and INST. costs.

LINE NO.	DESCRIPTION	QUANTITY	COST EACH		
			MAT.	INST.	TOTAL
01	**Three fixture bathroom incl. water closet, bathtub, lavatory, accessories,**				
02	**And necessary service piping to install this system in 2 walls.**				
03					
04	Water closet, floor mounted, 2 piece, close coupled	1 Ea.	126.50	103.50	230
05	Rough-in waste & vent for water closet	1 Set	85.90	258.20	344.10
06	Bathtub, P.E. cast iron 5' long with accessories	1 Ea.	288.20	121.80	410
07	Rough-in waste & vent for bathtub	1 Set	73.73	296.77	370.50
08	Lavatory, 20" x 18" P.E. cast iron with accessories	1 Ea.	143	67	210
09	Rough-in waste & vent for lavatory	1 Set	122.46	322.54	445
10	Accessories				
11	Toilet tissue dispenser, chrome, single roll	1 Ea.	20.90	9.10	30
12	18" long stainless steel towel bar	2 Ea.	39.60	24.40	64
13	Medicine cabinet with mirror, 16" x 22", unlighted	1 Ea.	44	20	64
14	TOTAL	System	944.29	1223.31	2167.60
15					
16					
17	Above system installed in one wall with all necessary service piping	System	936.53	1192.07	2128.60
18					
19					
20					
21					
22					
23					
24	**NOTE: PLUMBING APPROXIMATIONS**				
25	WATER CONTROL: water meter, backflow preventer,				
26	Shock absorbers, vacuum breakers, mixer....10 to 15% of fixtures				
27	PIPE AND FITTINGS: .... 30 to 60% of fixtures		30%	30%	
28					
29	Cut & patch to match existing construction, add, minimum		2%	3%	
30	Maximum		5%	9%	
31	Dust protection, add, minimum		1%	2%	
32	Maximum		4%	11%	
33	Equipment usage curtailment, add, minimum		1%	1%	
34	Maximum		3%	10%	
35	Material handling & storage limitation, add, minimum		1%	1%	
36	Maximum		6%	7%	
37	Protection of existing work, add, minimum		2%	2%	
38	Maximum		5%	7%	
39	Shift work requirements, add, minimum			5%	
40	Maximum			30%	
41					
42					

305

Figure 27.43

deleted. Each apartment will have a meter socket, disconnect, and panelboard. To replace the deleted items, the estimator must determine costs for one large service entrance, meter trough, and the distribution feeders to each remote panel board. Since this is a unique installation, the estimator can get a budget price from an electrical subcontractor, usually with only a telephone call. The costs for one grounding rod, clamp, and cable must also be included. The calculations are shown in Figure 27.49.

The appropriate lighting fixtures are shown in Figures 27.50 and 27.51. The conduit is not required for this installation and is deleted from the cost.

Figure 27.52 gives wiring prices for various electrical devices using different types of wire. Note that wiring is included in costs for the lighting fixtures, so the costs for lighting wiring from Figure 27.52 are not included at this point. The cost analysis for Division 9 is shown in Figure 27.53. As with the mechanical costs, 10% is added for supervision of the electrical subcontractor.

Division 10: General Conditions

In the Systems Estimating format, General Conditions are priced in Division 10. For convenience in our example, we have calculated the General Conditions at the end of the estimate.

Division 11: Special

This division contains costs for various systems that are used in renovation but are not included in other divisions. The kitchen system, as shown in Figure 27.54, must be modified to meet the requirements of the sample project. A disposal unit, range and refrigerator must be added, the compactor deleted, and the cabinets changed. Costs for these items are found in Figures 27.55a and 27.55b.

Bathroom Calculations

	Material	Installation	Total
Bathroom 08.1-931-17	$936.53	$1,192.07	$2,128.60
Pipe and Fittings (30%)	280.96	357.62	638.58
Subtotal	$1,217.49	$1,549.69	$2,767.18
Factors 7% Material	85.22	170.47	255.69
11% Installation			
Total	$1,302.71	$1,720.16	$3,022.87

Figure 27.44

Means® Forms

COST ANALYSIS

SHEET NO. 7 of 11

PROJECT Apartment Renovation

ESTIMATE NO. 89-2

ARCHITECT

DATE 1989

TAKE OFF BY: QUANTITIES BY: EBW PRICES BY: RSM EXTENSIONS BY: CHECKED BY:

DESCRIPTION	SOURCE/DIMENSIONS	QUANTITY	UNIT	MATERIAL UNIT COST	MATERIAL TOTAL	LABOR UNIT COST	LABOR TOTAL	EQ./TOTAL UNIT COST	EQ./TOTAL TOTAL
Division 8: Mechanical									
Bathrooms		8	EA.	1302.71	10422	1720.16	13761		
Kitchen Sink Rough-In									
152-152-4980		8	EA.	65.85	527	190	1520		
O&P		8	EA.	6.59	53	100.51	804		
HVAC Systems		7,296	SF	2.78	20283	2.72	19845		
Vent Chimney									
155-680-0100		96	VLF	2.80	269	5.10	490		
O&P		96	VLF	.28	27	2.83	272		
Subtotal					31581		36692		
GC Supervision		10%			3158		3669		
Division 8: Totals					34739		40361		

Figure 27.45

MECHANICAL	08.3-320	Heating-Cooling, Gas, Forced Air

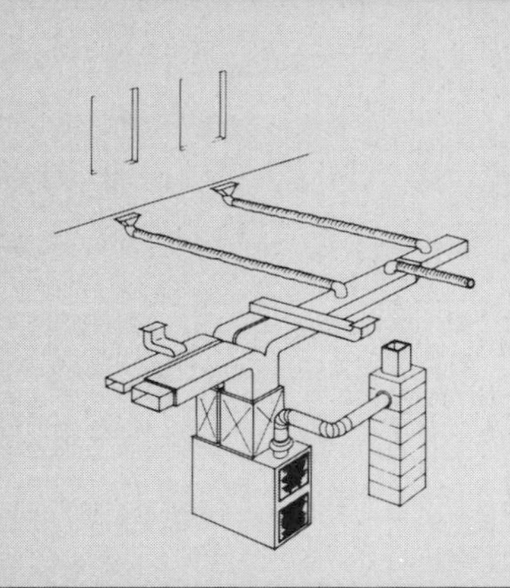

This page illustrates and describes a gas fired forced air system including a gas fired furnace, ductwork, registers and hookups. Lines 08.3-320-04 thru 13 give the unit price and total price per square foot for this system. Prices for alternate gas fired forced air systems are on Line Items 08.3-320-15 thru 26. Both material quantities and labor costs have been adjusted for the system listed.

Factors: To adjust for job conditions other than normal working situations use Lines 08.3-320-29 thru 40.

Example: You are to install the system with material handling and storage limitations. Go to Line 08.3-320-35 and apply these percentages to the appropriate MAT. and INST. costs.

LINE NO.	DESCRIPTION	QUANTITY	COST PER S.F. MAT.	INST.	TOTAL
01	**Gas fired hot air heating system including furnace, ductwork, registers**				
02	**And all necessary hookups.**				
03					
04	Area to 800 S.F., heat only				
05	Furnace, gas, AGA certified, direct drive, 44 MBH	1 Ea.	401.50	133.50	535
06	Gas piping	1 L.S.	100.38	33.37	133.75
07	Duct, galvanized steel	312 Lb.	442.73	1111.03	1553.76
08	Insulation, blanket type, ductwork	270 S.F.	124.74	417.96	542.70
09	Flexible duct, 6" diameter, insulated	100 L.F.	161.70	207.30	369
10	Registers, baseboard, gravity, 12" x 6"	8 Ea.	57.20	102.80	160
11	Return, damper, 36" x 18"	1 Ea.	73.21	31.79	105
12	TOTAL	System	1361.46	2037.75	3399.21
13		S.F.	1.70	2.55	4.25
14	For alternate heating systems:				
15	Gas fired, area to 1000 S.F.	S.F.	1.39	2.07	3.46
16	To 1200 S.F.		1.24	1.83	3.07
17	To 1600 S.F.		1.19	1.72	2.91
18	To 2000 S.F.		1.12	1.98	3.10
19	To 3000 S.F.		.92	1.58	2.50
20	For combined heating and cooling systems:				
21	Gas fired, heating and cooling, area to 800 S.F.	S.F.	3.21	3.05	6.26
22	To 1000 S.F.		2.65	2.50	5.15
23	To 1200 S.F.		2.28	2.19	4.47
24	To 1600 S.F.		2.07	2	4.07
25	To 2000 S.F.		1.89	2.22	4.11
26	To 3000 S.F.		1.60	1.76	3.36
27					
28					
29	Cut & patch to match existing construction, add, minimum		2%	3%	
30	Maximum		5%	9%	
31	Dust protection, add, minimum		1%	2%	
32	Maximum		4%	11%	
33	Equipment usage curtailment, add, minimum		1%	1%	
34	Maximum		3%	10%	
35	Material handling & storage limitation, add, minimum		1%	1%	
36	Maximum		6%	7%	
37	Protection of existing work, add, minimum		2%	2%	
38	Maximum		5%	7%	
39	Shift work requirements, add, minimum			5%	
40	Maximum			30%	
41					
42					

315

Figure 27.46

155 | Heating

		155 600 | Heating System Access.	CREW	DAILY OUTPUT	MAN-HOURS	UNIT	BARE COSTS MAT.	LABOR	EQUIP.	TOTAL	TOTAL INCL O&P	
651	8800	Urethane, with ASJ, -60°F to +225°F										651
	8960	1" wall, ½" iron pipe size	Q-14	240	.067	L.F.	1.05	1.43		2.48	3.43	
	9290	Urethane, with ultraviolet cover										
	9310	1" wall, ½" pipe size	Q-14	216	.074	L.F.	1.16	1.59		2.75	3.80	
	9320	¾" pipe size	"	207	.077	"	1.16	1.66		2.82	3.91	
	9600	Minimum labor/equipment charge	1 Plum	4	2	Job		49		49	75	
671	0010	TANKS										671
	0020	Fiberglass, underground, U.L. listed, not including										
	0030	manway or hold-down strap										
	0100	1000 gallon capacity	Q-6	2	12	Ea.	1,370	275		1,645	1,925	
	0140	2000 gallon capacity	Q-7	2	16		2,000	375		2,375	2,775	
	0160	4000 gallon capacity		1.30	24.620		2,580	575		3,155	3,725	
	0180	6000 gallon capacity		1	32		3,270	750		4,020	4,750	
	0500	For manway, fittings and hold-downs, add					20%	15%				
	2000	Steel, liquid expansion, ASME, painted, 15 gallon capacity	Q-5	17	.941		200	21		221	250	
	2040	30 gallon capacity		12	1.330		220	29		249	285	
	2080	60 gallon capacity		8	2		310	44		354	410	
	2120	100 gallon capacity		6	2.670		450	59		509	585	
	3000	Steel ASME expansion, rubber diaphragm, 19 gal. cap. accept.		12	1.330		710	29		739	825	
	3020	31 gallon capacity		8	2		925	44		969	1,075	
	3040	61 gallon capacity		6	2.670		1,300	59		1,359	1,525	
	3080	119 gallon capacity		4	4		1,550	88		1,638	1,850	
	4000	Steel, storage, above ground, including supports, coating,										
	4020	fittings, not including mat, pumps or piping										
	4040	275 gallon capacity	Q-5	5	3.200	Ea.	200	71		271	330	
	4060	550 gallon capacity	"	4	4		900	88		988	1,125	
	4080	1000 gallon capacity	Q-7	4	8		1,200	190		1,390	1,600	
	4120	2000 gallon capacity	"	3	10.670		1,500	250		1,750	2,025	
	5000	Steel underground, sti-P3, set in place, incl. hold-down bars.										
	5500	Excavation, pad, pumps and piping not included										
	5520	1000 gallon capacity, 7 gauge shell	Q-7	4	8	Ea.	1,500	190		1,690	1,925	
	5540	5000 gallon capacity, ¼" thick shell		1	32		4,500	750		5,250	6,100	
	5560	10,000 gallon capacity, ¼" thick shell		.70	45.710		5,800	1,075		6,875	8,025	
	5600	20,000 gallon capacity, 5/16" thick shell		.30	107		12,000	2,500		14,500	17,000	
	9000	Minimum labor/equipment charge	Q-5	4	4	Job		88		88	135	
680	0010	VENT CHIMNEY Prefab metal, U.L. listed										680
	0020	Gas, double wall, galvanized steel										
	0080	3" diameter	Q-9	72	.222	V.L.F.	2.30	4.81		7.11	10	
	0100	4" diameter		68	.235		2.80	5.10		7.90	11	
	0120	5" diameter		64	.250		3.32	5.40		8.72	12.05	
	0140	6" diameter		60	.267		3.88	5.75		9.63	13.25	
	0160	7" diameter		56	.286		5.32	6.20		11.52	15.45	
	0180	8" diameter		52	.308		5.92	6.65		12.57	16.85	
	0200	10" diameter		48	.333		12.59	7.20		19.79	25	
	0220	12" diameter		44	.364		16.83	7.85		24.68	31	
	0260	16" diameter		40	.400		37.90	8.65		46.55	55	
	0300	20" diameter	Q-10	36	.667		57.10	14.95		72.05	86	
	0340	24" diameter	"	32	.750		88.20	16.85		105.05	125	
	5000	Vent damper bi-metal 6" flue	Q-9	16	1	Ea.	57	22		79	96	
	5100	Gas, auto., electric		8	2		110	43		153	190	
	5120	Oil, auto., electric		8	2		132	43		175	215	
	7800	All fuel, double wall, stainless steel, 6" diameter		60	.267	V.L.F.	15.20	5.75		20.95	26	
	7802	7" diameter		56	.286		19.05	6.20		25.25	31	
	7804	8" diameter		52	.308		22.75	6.65		29.40	35	
	7806	10" diameter		48	.333		31.90	7.20		39.10	46	
	7808	12" diameter		44	.364		42.90	7.85		50.75	59	
	7810	14" diameter		42	.381		56.20	8.25		64.45	75	

For expanded coverage of these items see *Means Mechanical Cost Data 1989*

205

Figure 27.47

ELECTRICAL	09.1-230	Residential Service - Single Phase

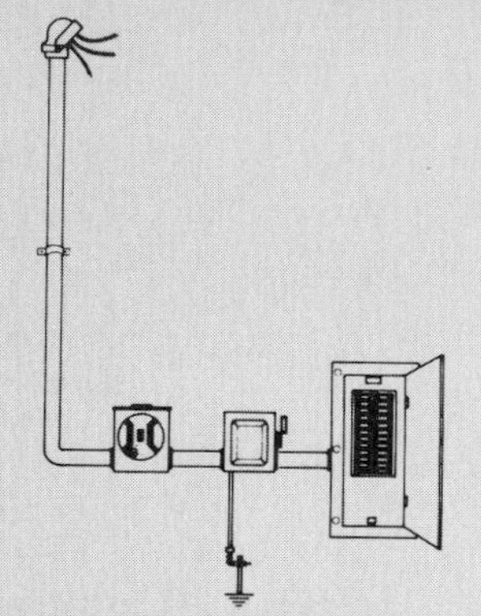

This page illustrates and describes a residential, single phase system including a weather cap, service entrance cable, meter socket, entrance switch, ground rod, ground cable, EMT, and panelboard. Lines 09.1-230-03 thru 10 give the unit price and total price on a cost each basis for this system. Prices for an alternate residential, single phase system are on Line Item 09.1-230-25. Both material quantities and labor costs have been adjusted for the system listed.

Factors: To adjust for job conditions other than normal working situations use Lines 09.1-230-29 thru 40.

Example: You are to install the system with a minimum equipment usage curtailment. Go to Line 09.1-230-33 and apply these percentages to the appropriate MAT. and INST. costs.

LINE NO.	DESCRIPTION	QUANTITY	COST EACH MAT.	COST EACH INST.	COST EACH TOTAL
01	**100 Amp Service, single phase**				
02					
03	Weathercap	1 Ea.	4.18	24.82	29
04	Service entrance cable	.2 C.L.F.	19.36	58.64	78
05	Meter socket	1 Ea.	23.10	91.90	115
06	Entrance disconnect switch	1 Ea.	121	154	275
07	Ground rod, with clamp	1 Ea.	12.10	60.90	73
08	Ground cable	.1 C.L.F.	13.20	18.30	31.50
09	Panelboard, 12 circuit	1 Ea.	106.70	243.30	350
10	TOTAL	Ea.	299.64	651.86	951.50
11					
12					
13			COST EACH		
14			MAT.	INST.	TOTAL
15	200 Amp Service , single phase				
16					
17	Weathercap	1 Ea.	12.27	36.73	49
18	Service entrance cable	.2 C.L.F.	42.90	84.10	127
19	Meter socket	1 Ea.	40.70	154.30	195
20	Entrance disconnect switch	1 Ea.	245.30	224.70	470
21	Ground rod, with clamp	1 Ea.	24.20	66.80	91
22	Ground cable	.1 C.L.F.	10.07	9.93	20
23	¾" EMT	10 L.F.	4.84	22.66	27.50
24	Panelboard, 24 circuit	1 Ea.	260.70	454.30	715
25	TOTAL	Ea.	640.98	1053.52	1694.50
26					
27					
28					
29	Cut & patch to match existing construction, add, minimum		2%	3%	
30	Maximum		5%	9%	
31	Dust protection, add, minimum		1%	2%	
32	Maximum		4%	11%	
33	Equipment usage curtailment, add, minimum		1%	1%	
34	Maximum		3%	10%	
35	Material handling & storage limitation, add, minimum		1%	1%	
36	Maximum		6%	7%	
37	Protection of existing work, add, minimum		2%	2%	
38	Maximum		5%	7%	
39	Shift work requirements, add, minimum			5%	
40	Maximum			30%	
41					
42					

317

Figure 27.48

There are 19 L.F. of cabinets in each kitchen of the sample project. Since the types and sizes of the cabinets are not specified, costs for 20 L.F. in Figure 27.54 are adjusted. The calculations for the kitchen are found in Figure 27.56.

Mailboxes are to be installed. Costs for mailboxes are not found in the Systems pages so Unit Prices are used from Figure 27.57. The bare material and labor costs must be changed to include the installing contractor's overhead and profit. Figure 27.58 is the cost analysis for Division 11.

Division 12: Site Work

The amount of site work in renovation will vary greatly from project to project. In the sample project, the site work is relatively extensive. The scope of work is developed through discussions with the owner. Excavation is required for the retaining wall and footing in Division 1 of the Assemblies Section of *Means Repair and Remodeling Cost Data*. The existing parking lot is to be resurfaced and new loam and seeding is specified for the entire lawn. Demolition costs are also included in Division 12 of the Assemblies Section of *Means Repair and Remodeling Cost Data*.

The estimator must be very careful when evaluating the site of an existing building for excavation work. Utility locations must be identified at the site. A water line break can be a very expensive mistake. After thorough examination, the estimator has determined that there should be no utilities at the area to be excavated. The

Electrical Service Calculations

	Material	Installation	Total
Service			
Meter Socket	$ 23.10	$ 91.90	$ 115.00
Disconnect	121.00	154.00	275.00
Panel Board	106.70	243.30	350.00
Subtotal	$ 250.80	$ 489.20	$ 740.00
Factors 5% Material	12.54	44.03	56.57
9% Installation			
Total	$ 263.34	$ 533.23	$ 796.57
Entrance, Trough and Feeders	$2,400.00	$3,360.00	$5,760.00
Ground Rod	17.43	37.00	54.43
Ground Wire	8.40	16.15	24.55
Total	$2,425.83	$3,413.15	$5,838.98

Figure 27.49

ELECTRICAL	09.2-900	Lighting, Fluorescent

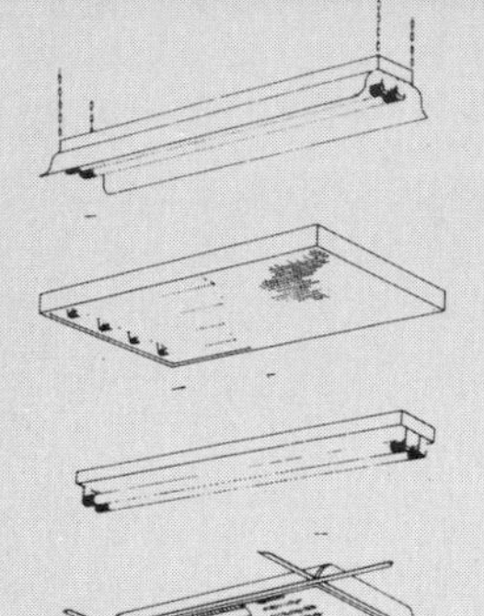

This page illustrates and describes fluorescent lighting systems including a fixture, lamp, outlet box and wiring. Lines 09.2-900-04 thru 08 give the unit price and total price on a cost each basis for this system. Prices for alternate fluorescent lighting systems are on Line Items 09.2-900-13 thru 15. Both material quantities and labor costs have been adjusted for the system listed.

Factors: To adjust for job conditions other than normal working situations use Lines 09.2-900-29 thru 40.

Example: You are to install the system during evening hours. Go to Line 09.2-900-39 and apply this percentage to the appropriate INST. cost.

LINE NO.	DESCRIPTION	QUANTITY	COST EACH MAT.	INST.	TOTAL
01	**Fluorescent lighting, including fixture, lamp, outlet box and wiring.**				
02					
03					
04	Recessed lighting fixture, on suspended system	1 Ea.	63.80	61.20	125
05	Outlet box	1 Ea.	1.75	21	22.75
06	#12 wire	.66 C.L.F.	3.59	17.53	21.12
07	Conduit, EMT, ½" conduit DEDUCT	20 L.F.	(6.82)	(34.58)	41.40
07	Fixture whip	Ea.			
08	TOTAL	Ea.	75.96	134.31	210.27
09					
11					
12	For alternate lighting fixtures:				
13	Surface mounted, 2' x 4', acrylic prismatic diffuser	Ea.	(98.77)	(137.38)	236.15
14	Strip fixture, 8' long, two 8' lamps	↓	62.47	129.68	192.15
15	Pendant mounted, industrial, 8' long, with reflectors		108.67	147.48	256.15
16					
17			98.77	137.38	
18			− 6.82	− 34.58	
19			91.95	102.80	
20					
21					
22					
23					
24					
25					
26					
27					
28					
29	Cut & patch to match existing construction, add, minimum		2%	3%	
30	Maximum		5%	9%	
31	Dust protection, add, minimum		1%	2%	
32	Maximum		4%	11%	
33	Equipment usage curtailment, add, minimum		1%	1%	
34	Maximum		3%	10%	
35	Material handling & storage limitation, add, minimum		1%	1%	
36	Maximum		6%	7%	
37	Protection of existing work, add, minimum		2%	2%	
38	Maximum		5%	7%	
39	Shift work requirements, add, minimum			5%	
40	Maximum			30%	
41					
42					

318

Figure 27.50

ELECTRICAL	09.2-910	Lighting, Incandescent

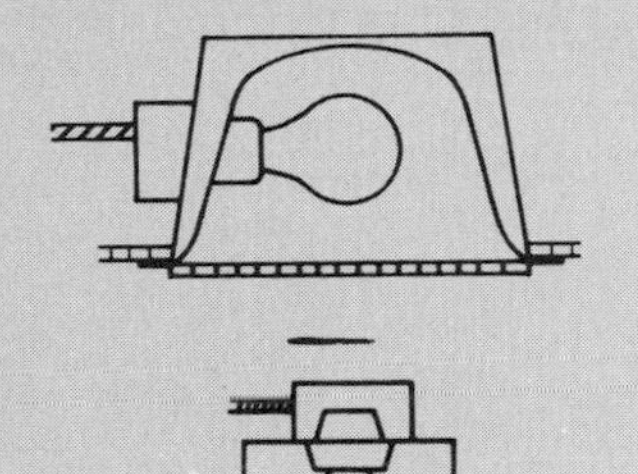

This page illustrates and describes incandescent lighting systems including a fixture, lamp, outlet box, conduit and wiring. Lines 09.2-910-04 thru 08 give the unit price and total price on a cost each basis for this system. Prices for an alternate incandescent lighting system are on Line Item 09.2-910-18. Both material quantities and labor costs have been adjusted for the system listed.

Factors: To adjust for conditions other than normal working situations use Lines 09.2-910-29 thru 40.

Example: You are to install the system and cut and match existing construction. Go to Line 09.2-910-30 and apply this percentage to the appropriate INST. costs.

LINE NO.	DESCRIPTION	QUANTITY	COST EACH MAT.	COST EACH INST.	COST EACH TOTAL
01	**Incandescent light fixture, including lamp, outlet box, conduit and wiring.**				
02					
03					
04	Recessed wide reflector with flat glass lens	1 Ea.	35.20	43.80	79
05	Outlet box	1 Ea.	1.75	21	22.75
06	Armored cable, 3 wire DEDUCT	.2 C.L.F.	6.60	29.40	36
07					
08	TOTAL	Ea.	43.55	94.20	137.75
09			6.60	29.40	
10			36.95	64.80	
11					
12					
13	Recessed, R-40 flood lamp with reflector skirt	1 Ea.	50.60	36.40	87
14	150 watt R-40 flood lamp	.01 Ea.	4.49	2.26	6.75
15	Outlet box	1 Ea.	1.75	21	22.75
16	Romex, 12-2 with ground	.2 C.L.F.	5.50	26.50	32
17	Conduit, ½" EMT	20 L.F.	6.82	34.58	41.40
18	TOTAL	Ea.	69.16	120.74	189.90
19					
20					
21					
22					
23					
24					
25					
26					
27					
28					
29	Cut & patch to match existing construction, add, minimum		2%	3%	
30	Maximum		5%	9%	
31	Dust protection, add, minimum		1%	2%	
32	Maximum		4%	11%	
33	Equipment usage curtailment, add, minimum		1%	1%	
34	Maximum		3%	10%	
35	Material handling & storage limitation, add, minimum		1%	1%	
36	Maximum		6%	7%	
37	Protection of existing work, add, minimum		2%	2%	
38	Maximum		5%	7%	
39	Shift work requirements, add, minimum			5%	
40	Maximum			30%	
41					
42					

319

Figure 27.51

09.9-500	Wiring Devices Selective Price Sheet			
LINE NO.	DESCRIPTION	COST EACH		
		MAT.	INST.	TOTAL
01	Using non-metallic sheathed, cable, air conditioning receptacle	9.15	29.85	39
02	Disposal wiring	8.01	32.99	41
03	Dryer circuit	22.88	53.12	76
04	Duplex receptacle	9.15	22.85	32
05	Fire alarm or smoke detector	44	29	73
06	Furnace circuit & switch	12.58	49.42	62
07	Ground fault receptacle	40.70	37.30	78
08	Heater circuit	9.15	36.85	46
09	Lighting wiring	9.15	18.85	28
10	Range circuit	34.10	75.90	110
11	Switches single pole	9.15	18.85	28
12	3-way	11.44	24.56	36
13	Water heater circuit	11.44	58.56	70
14	Weatherproof receptacle	66	49	115
15	Using BX cable, air conditioning receptacle	13.73	35.27	49
16	Disposal wiring	11.44	39.56	51
17	Dryer circuit	28.60	64.40	93
18	Duplex receptacle	13.73	27.27	41
19	Fire alarm or smoke detector	44	29	73
20	Furnace circuit & switch	16.02	58.98	75
21	Ground fault receptacle	46.20	44.80	91
22	Heater circuit	12.58	44.42	57
23	Lighting wiring	13.73	22.27	36
24	Range circuit	47.30	87.70	135
25	Switches, single pole	13.73	22.27	36
26	3-way	17.16	29.84	47
27	Water heater circuit	16.02	69.98	86
28	Weatherproof receptacle	70.40	59.60	130
29	Using EMT conduit, air conditioning receptacle	16.02	43.98	60
30	Disposal wiring	14.87	49.13	64
31	Dryer circuit	31.90	78.10	110
32	Duplex receptacle	16.02	33.98	50
33	Fire alarm or smoke detector	49.50	43.50	93
34	Furnace circuit & switch	20.59	73.41	94
35	Ground fault receptacle	49.50	55.50	105
36	Heater circuit	16.02	54.98	71
37	Lighting wiring	17.16	27.84	45
38	Range circuit	53.90	111.10	165
39	Switches, single pole	16.02	27.98	44
40	3-way	19.45	36.55	56
41	Water heater circuit	19.45	85.55	105
42	Weatherproof receptacle	74.80	75.20	150
43	Using aluminum conduit, air conditioning receptacle	19.69	58.95	87
44	Disposal wiring	18.15	65.50	84
45	Dryer circuit	34.10	105.30	140
46	Duplex receptacle	19.25	45.35	65
47	Fire alarm or smoke detector	58.30	58.95	115
48	Furnace circuit & switch	26.40	98.25	125
49	Ground fault receptacle	53.90	73.70	130
50	Heater circuit	20.40	73.70	94
51	Lighting wiring	21.50	36.85	58
52	Range circuit	61.60	147.40	210
53	Switches, single pole	25.30	36.85	62
54	3-way	27.50	49.15	77
55	Water heater circuit	27.50	117.90	145
56	Weatherproof receptacle	83.60	98.25	180
57	Using galvanized steel conduit	18.81	62.70	82
58	Disposal wiring	17.00	70.20	87

322

Figure 27.52

Means® Forms

COST ANALYSIS

SHEET NO. 8 of 11

ESTIMATE NO. 89-2

DATE 1989

PROJECT Apartment Renovation

ARCHITECT

TAKE OFF BY: QUANTITIES BY: EBW PRICES BY: RSM EXTENSIONS BY: CHECKED BY:

DESCRIPTION	SOURCE/DIMENSIONS	QUANTITY	UNIT	MATERIAL UNIT COST	MATERIAL TOTAL	LABOR UNIT COST	LABOR TOTAL	EQ./TOTAL UNIT COST	EQ./TOTAL TOTAL
Division 9: Electrical									
Service: Meter Socket, Disconnect, Panel Board		8	EA.	263.34	2107	533.23	4266		
Entrance & Feeders, Ground		1	EA.	2426.	2426	3413	3413		
Fixtures: Fluorescent		8	EA.	91.95	736	102.80	822		
Incandescent		24	EA.	36.95	887	64.80	1555		
Devices: Disposal		8	EA.	8.01	64	32.99	264		
Receptacles		156	EA.	9.15	1427	22.85	3565		
Smoke Detectors		10	EA.	44	440	29.	290		
Furnace		8	EA.	12.58	101	49.42	395		
GFI		8	EA.	40.70	326	37.30	298		
Range		8	EA.	34.10	273	75.90	607		
Switches		40	EA.	9.15	366	18.85	754		
Subtotal					9153		16229		
GC Supervision		10%			915		1623		
Division 9: Totals					10068		17852		

Figure 27.53

SPECIAL	11.1-242	Kitchens

This page illustrates and describes kitchen systems including top and bottom cabinets, custom laminated plastic top, single bowl sink, and appliances. Lines 11.1-242-04 thru 11 give the unit price and total price on a cost each basis for this system. Prices for alternate kitchen systems are on Line Items 11.1-242-15 and 16. Both material quantities and labor costs have been adjusted for the system listed.

Factors: To adjust for job conditions other than normal working situations use Lines 11.1-242-29 thru 40.

Example: You are to install the system and protect the work area from dust. Go to Line 11.1-242-32 and apply these percentages to the appropriate MAT. and INST. costs.

LINE NO.	DESCRIPTION	QUANTITY	COST EACH		
			MAT.	INST.	TOTAL
01	**Kitchen cabinets including wall and base cabinets, custom laminated**				
02	**Plastic top,sink & appliances,no plumbing or electrical rough-in included.**				
03					
04	Prefinished wood cabinets, average quality, wall and base *19 LF*	20 L.F.	1320	360	1680
05	Custom laminated plastic counter top *19 LF*	20 L.F.	110	183	293
06	Stainless steel sink, 22" x 25"	1 Ea.	311.30	168.70	480
07	Faucet, top mount	1 Ea.	31.90	30.10	62
08	Dishwasher, built-in	1 Ea.	253	72	325
09	Compactor, built-in *DELETE*	1 Ea.	385	35	420
10	Range hood, 30", ductless	1 Ea.	88	86	174
11	TOTAL	Ea.	2499.20	934.80	3434
12					
13					
14	For alternate kitchen systems:				
15	Prefinished wood cabinets, high quality	Ea.	4155.80	1065.20	5221
16	Custom cabinets, built in place, high quality	"	5137	1244	6381
17					
18					
19					
20					
21					
22	NOTE: No plumbing or electric rough-ins are included in the above				
23	Prices, for plumbing see Division A15, for electric see Division A16.				
24					
25					
26					
27					
28					
29	Cut & patch to match existing construction, add, minimum		2%	3%	
30	Maximum		5%	9%	
31	Dust protection, add, minimum		1%	2%	
32	Maximum		4%	11%	
33	Equipment usage curtailment, add, minimum		1%	1%	
34	Maximum		3%	10%	
35	Material handling & storage limitation, add, minimum		1%	1%	
36	Maximum		6%	7%	
37	Protection of existing work, add, minimum		2%	2%	
38	Maximum		5%	7%	
39	Shift work requirements, add, minimum			5%	
40	Maximum			30%	
41					
42					

324

Figure 27.54

11.9-100	Kitchen Selective Price Sheet			
LINE NO.	DESCRIPTION	COST EACH		
		MAT.	INST.	TOTAL
01	Cabinets standard wood, base, one drawer one door, 12" wide	104.50	20.50	125
02	15" wide	110	25	135
03	18" wide	115.50	24.50	140
04	21" wide	126.50	23.50	150
05	24" wide	137.50	22.50	160
06				
07	Two drawers two doors, 27" wide	170.50	24.50	195
08	30" wide	176	24	200
09	33" wide	187	28	215
10	36" wide	192.50	27.50	220
11	42" wide	209	26	235
12	48" wide	220	30	250
13	Drawer base (4 drawers), 12" wide	110	20	130
14	15" wide	121	24	145
15	18" wide	126.50	23.50	150
16	24" wide	148.50	26.50	175
17	Sink or range base, 30" wide	110	25	135
18	33" wide	132	28	160
19	36" wide	143	27	170
20	42" wide	154	26	180
21	Corner base, 36" wide	143	32	175
22	Lazy susan with revolving door	198	32	230
23	Cabinets standard wood, wall two doors, 12" high, 30" wide	88	22	110
24	36" wide	93.50	21.50	115
25	15" high, 30" high	93.50	21.50	115
26	36" wide	104.50	25.50	130
27	24" high, 30" wide	110	25	135
28	36" wide	126.50	23.50	150
29	30" high, 30" wide	126.50	28.50	155
30	36" wide	137.50	27.50	165
31	42" wide	148.50	31.50	180
32	48" wide	165	30	195
33	One door, 30" high, 12" wide	82.50	22.50	105
34	15" wide	88	27	115
35	18" wide	99	26	125
36	24" wide	104.50	25.50	130
37	Corner, 30" high, 24" wide	132	33	165
38	36" wide	170.50	34.50	205
39	Broom, 84" high, 24" deep, 18" wide	154	56	210
40	Oven, 84" high, 24" deep, 27" wide	176	69	245
41	Valance board, 4' long	26.40	5.60	32
42	6" long	39.60	8.40	48
43	Counter tops, laminated plastic, stock 25" wide w/backsplash, min.	5.50	9.15	14.65
44	Maximum	16.83	11.17	28
45	Custom, 7/8" thick, no splasn	16.34	9.66	26
46	Cove splash	21.23	8.77	30
47	1-1/4" thick, no splash	19.25	9.75	29
48	Square splash	24.20	9.80	34
49	Post formed	8.58	9.17	17.75
50				
51	Maple laminated 1-1/2" thick, no splash	33	10	43
52	Square splash	37.40	9.60	47
53				
54				
55	Appliances, range, free standing, minimum	286	24	310
56	Maximum	1320	55	1375
57	Built-in, minimum	385	35	420
58	Maximum	880	70	950

327

Figure 27.55a

11.9-100	Kitchen Selective Price Sheet			
LINE NO.	DESCRIPTION	COST EACH		
		MAT.	INST.	TOTAL
59	Counter top range 4 burner. maximum	220	35	255
60	Maximum	473	72	545
61	Compactor, built-in, minimum	385	35	420
62	Maximum	539	41	580
63	Dishwasher, built-in, minimum	253	72	325
64	Maximum	429	76	505
65	Gargage disposer, sink-pipe, minimum	60.50	59.50	120
66	Maximum	214.50	60.50	275
67	Range hood, 30" wide, 2 speed, minimum	33	49	82
68	Maximum	264	61	325
69	Refrigerator, no frost, 12 cu. ft.	330	45	375
70	20 cu. ft.	649	61	710
71	Plumb. not incl. rough-ins, sinks porc. C.I., single bowl, 21" x 24"	150.70	169.30	320
72	21" x 30"	173.80	166.20	340
73	Double bowl, 20" x 32"	211.20	208.80	420
74				
75	Stainless steel, single bowl, 19" x 18"	282.70	167.30	450
76	22" x 25"	311.30	168.70	480
77				
78				
79				
80				
81				
82				
83				
84				

328

Figure 27.55b

costs are shown in Figure 27.59. Note that the costs as shown are for 360 CY (a full day's cost) and per cubic yard. A half-day rental is usually a minimum for heavy equipment. Even though the amount to be excavated is 91 CY (less than one half of 360 CY), the costs for a half day are included in the estimate.

The concrete slab at the patios was not included in Division 2, Substructure, because the individual components required more closely resemble a sidewalk than an interior floor slab. The costs are derived from Figure 27.60. The edge forms are deducted because only a very small amount of forming will be necessary compared to a normal sidewalk. The factors are added because access for the truck is impossible and the concrete will be placed by hand. Calculations are shown in Figure 27.61.

The parking lot is to receive only a new topping coat. The existing base and binder are adequate. Therefore, only those costs that apply are taken from Figure 27.62. The loam and seed costs are shown in Figure 27.63. It is important that the estimator determine those areas that are to receive loam and seed to avoid any possible confusion. A quick sketch of the site will help to define the limits of the work.

Depending upon the precision necessary for the budget Assemblies Estimate, the estimator may elect to use a lump sum cost for demolition based on past experience and the thorough evaluation of the site. If greater accuracy is required, Figure 27.64 provides costs for selective demolition. In this example, a lump sum figure is used.

The costs for Division 12, Site Work, are summarized on the estimate sheet in Figure 27.65.

Calculations for Kitchen

	Material	Installation	Total
Cabinets (adj. for 19 L.F.)	$1,254.00	$ 342.00	$1,596.00
Counter top	104.50	173.85	278.35
Sink	311.30	168.70	480.00
Faucet	31.90	30.10	62.00
Dishwasher	253.00	72.00	325.00
Range Hood	88.00	86.00	174.00
Range	286.00	24.00	310.00
Disposer	60.50	59.50	120.00
Refrigerator	330.00	45.00	375.00
Total	$2,719.20	$1,001.15	$3,720.35

Figure 27.56

105 | Lockers, Protective Covers and Postal Specialties

	105 520 | Mail Boxes		CREW	DAILY OUTPUT	MAN-HOURS	UNIT	BARE COSTS MAT.	LABOR	EQUIP.	TOTAL	TOTAL INCL O&P	
521	0010	**MAIL BOXES** Horiz., key lock, 5"H x 6"W x 15"D, alum., rear load	1 Carp	34	.235	Ea.	35	5.05		40.05	47	521
	0100	Front loading		34	.235		38	5.05		43.05	50	
	0200	Double, 5"H x 12"W x 15"D, rear loading		26	.308		58	6.60		64.60	74	
	0300	Front loading		26	.308		63	6.60		69.60	80	
	0500	Quadruple, 10"H x 12"W x 15"D, rear loading		20	.400		100	8.55		108.55	125	
	0600	Front loading		20	.400		107	8.55		115.55	130	
	1600	Vault type, horizontal, for apartments, 4" x 5"		34	.235		30	5.05		35.05	41	
	1700	Alphabetical directories, 35 names		10	.800		140	17.10		157.10	180	
	1900	Letter slot, residential		20	.400		26	8.55		34.55	42	
	2000	Post office type		8	1		107	21		128	150	
	9000	Minimum labor/equipment charge		5	1.600	Job		34		34	55	

106 | Partitions and Storage Shelving

	106 150 | Demountable Partitions		CREW	DAILY OUTPUT	MAN-HOURS	UNIT	BARE COSTS MAT.	LABOR	EQUIP.	TOTAL	TOTAL INCL O&P	
152	0010	**PARTITIONS, MOVABLE OFFICE** Demountable, add for doors										152
	0100	Do not deduct door openings from total L.F.										
	0900	Asbestos cement, 1-3/4" thick, prefinished, low walls	2 Carp	80	.200	L.F.	35	4.28		39.28	45	
	1000	Full height	"	40	.400		43	8.55		51.55	61	
	1200	Economy grade, 1-3/4" thick, deduct					20%					
	1300	High quality, 4" thick, add					100%					
	3400	Metal, to 9'-6" high, enameled steel, no glass	2 Carp	30	.533		75	11.40		86.40	100	
	3500	Steel frame, all glass		30	.533		90	11.40		101.40	115	
	3700	Vinyl covered, no glass		30	.533		115	11.40		126.40	145	
	3800	Steel frame with 52% glass		30	.533		125	11.40		136.40	155	
	4000	Free standing, 4'-6" high, steel with glass		100	.160		110	3.42		113.42	125	
	4100	Acoustical		100	.160		63	3.42		66.42	75	
	4300	Low rails, 3'-3" high, enameled steel		100	.160		48	3.42		51.42	58	
	4400	Vinyl covered		100	.160		70	3.42		73.42	83	
	5500	For acoustical partitions, add, minimum				S.F.	1			1	1.10	
	5550	Maximum				"	3.50			3.50	3.85	
	5700	For doors, not incl. hardware, hollow metal door, add	2 Carp	4.30	3.720	Ea.	160	80		240	305	
	5800	Hardwood door, add		3.40	4.710		108	100		208	280	
	6000	Hardware for doors, not incl. closers, keyed		29	.552		76	11.80		87.80	105	
	6100	Non-keyed		29	.552		68	11.80		79.80	94	
	9000	Minimum labor/equipment charge		3	5.330	Job		115		115	185	

	106 300 | Portable Partitions		CREW	DAILY OUTPUT	MAN-HOURS	UNIT	BARE COSTS MAT.	LABOR	EQUIP.	TOTAL	TOTAL INCL O&P	
304	0010	**PARTITIONS, PORTABLE** Divider panels, free standing, fiber core										304
	0020	Fabric face straight										
	0100	2'-6" long, 4'-0" high	2 Carp	100	.160	L.F.	70	3.42		73.42	83	
	0200	5'-0" high		90	.178		74	3.80		77.80	88	
	0500	6'-0" high		75	.213		88	4.57		92.57	105	
	0900	5'-0" long, 4'-0" high		175	.091		47	1.96		48.96	55	
	1000	5'-0" high		150	.107		49	2.28		51.28	58	
	1500	6'-0" high		125	.128		56	2.74		58.74	66	
	1600	6'-0" long, 5'-0" high		162	.099		50	2.11		52.11	58	
	3100	Curved, 2'-6" long, 5'-0" high		90	.178		93	3.80		96.80	110	
	3150	6'-0" high		75	.213		112	4.57		116.57	130	
	3200	Economical panels, fabric face, 4'-0" long, 5'-0" high		132	.121		31	2.59		33.59	38	

164

Figure 27.57

Means® Forms

COST ANALYSIS

SHEET NO. 9 of 11

PROJECT Apartment Renovation

ESTIMATE NO. 89-2

ARCHITECT

DATE 1989

TAKE OFF BY: QUANTITIES BY: EBW PRICES BY: RSM EXTENSIONS BY: CHECKED BY:

DESCRIPTION	SOURCE/DIMENSIONS	QUANTITY	UNIT	MATERIAL UNIT COST	MATERIAL TOTAL	LABOR UNIT COST	LABOR TOTAL	EQ./TOTAL UNIT COST	EQ./TOTAL TOTAL
Division 11: Special									
Kitchen									
(From Table III.56)		8	EA.	2719	21752	1001	8008		
Mailboxes									
105-521-0600		2	EA.	107.	214	8.55	17		
O&P		2	EA.	11.	22	5.18	10		
Division 11: Totals					21988		8035		

Figure 27.58

SITE WORK	12.1-464	Excavation, Foundation

This page illustrates and describes foundation excavation systems including a backhoe-loader, operator, equipment rental, fuel, oil, mobilization, hauling material, no back filling. Lines 12.1-464-06 thru 12 give the unit price and total price per cubic yard for this system. Prices for alternate foundation excavation systems are on Line Items 12.1-464-16 thru 22. Both material quantities and labor costs have been adjusted for the system listed.

Factors: To adjust for job conditions other than normal working situations use Lines 12.1-464-29 thru 40.

Example: You are to install the system with the use of temporary shoring and bracing. Go to Line 12.1-464-39 and apply these percentages to the appropriate TOTAL costs.

LINE NO.	DESCRIPTION	QUANTITY	COST PER C.Y.		
			EQUIP.	LABOR	TOTAL
01	**Foundation excav w/ ¾ C.Y. backhoe-loader, incl. operator, equip**				
02	**Rental, fuel, oil, and mobilization. Hauling of excavated material is**				
03	**Included. Prices based on one day production of 360 C.Y. in medium soil**				
04	**Without backfilling.**				
05					
06	Equipment operator	8 Hr.		270	270
07	Backhoe-loader, ¾ C.Y.	1 Day	134		134
08	Operating expense (fuel, oil)	8 Hr.	52.40		52.40
09	Hauling, 12 C.Y. trucks, 1 mile round trip	360 C.Y.		820.80	820.80
10	Mobilization	1 Ea.		255	255
11	1 DAY TOTAL	360 C.Y.	186.40	1345.80	1532.20
12		C.Y.	.52	3.74	4.26
13					
14	For alternate size excavations: ½ DAY		93.20	672.90	
15					
16	100 C.Y.	C.Y.	.15	6.35	6.50
17	200 C.Y.		.52	4.32	4.84
18	300 C.Y.		.54	3.88	4.42
19	400 C.Y.		.51	3.66	4.17
20	500 C.Y.		.52	3.55	4.07
21	600 C.Y.		.50	3.45	3.95
22	700 C.Y.		.51	3.38	3.89
23					
24					
25					
26					
27					
28					
29	Dust protection, add, minimum		1%	2%	
30	Maximum		4%	11%	
31	Equipment usage curtailment, add, minimum		1%	1%	
32	Maximum		3%	10%	
33	Material handling & storage limitation, add, minimum		1%	1%	
34	Maximum		6%	7%	
35	Protection of existing work, add, minimum		2%	2%	
36	Maximum		5%	7%	
37	Shift work requirements, add, minimum			5%	
38	Maximum			30%	
39	Temporary shoring and bracing, add, minimum		2%	5%	
40	Maximum		5%	12%	
41					
42					

331

Figure 27.59

SITE WORK	12.7-104	Sidewalks

This page illustrates and describes sidewalk systems including concrete, welded wire and broom finish. Lines 12.7-104-05 thru 13 give unit price and total price per square foot for this system. Prices for alternate sidewalk systems are on Line Items 12.7-104-17 thru 19. Both material quantities and labor costs have been adjusted for the system listed.

Factors: To adjust for job conditions other than normal working situations use Lines 12.7-104-29 thru 40.

Example: You are to install the system and match existing construction. Go to Line 12.7-104-29 and apply these percentages to the appropriate MAT. and INST. costs.

LINE NO.	DESCRIPTION	QUANTITY	COST PER S.F.		
			MAT.	INST.	TOTAL
01	**4" thick concrete sidewalk with welded wire fabric**				
02	**3000 psi air entrained concrete, broom finish.**				
03					
04					
05	Gravel fill, 4" deep	.012 C.Y.	.04	.06	.10
06	Compact fill	.012 C.Y.		.02	.02
07	Hand grade	1 S.F.		.11	.11
08	Edge form DEDUCT	.25 L.F.	.06	.44	.50
09	Welded wire fabric	.011 S.F.	.09	.21	.30
10	Concrete, 3000 psi air entrained	.012 C.Y.	.73		.73
11	Place concrete	.012 C.Y.		.16	.16
12	Broom finish	1 S.F.		.43	.43
13	TOTAL	S.F.	.92	1.43	2.35
14			− .06	− .44	
15					
16	For alternate sidewalk systems:		.86	.99	
17	Asphalt (bituminous), 2" thick	S.F.	.37	.42	.79
18	Brick, on sand, bed, 4.5 brick per S.F.		2.50	4.50	7
19	Flagstone, slate, 1" thick, rectangular		2.40	5.40	7.80
20					
21					
22					
23					
24					
25					
26					
27					
28					
29	Cut & patch to match existing construction, add, minimum		2%	3%	
30	Maximum		5%	9%	
31	Dust protection, add, minimum		1%	2%	
32	Maximum		4%	11%	
33	Equipment usage curtailment, add, minimum		1%	1%	
34	Maximum		3%	10%	
35	Material handling & storage limitation, add, minimum		1%	1%	
36	Maximum		6%	7%	
37	Protection of existing work, add, minimum		2%	2%	
38	Maximum		5%	7%	
39	Shift work requirements, add, minimum			5%	
40	Maximum			30%	
41			9%	17%	
42					

336

Figure 27.60

Division 10: General Conditions

In the Assemblies Estimating format, the General Conditions Division is placed in the tenth position. This part of the estimate, however, is typically done after all other portions of the building project have been priced out. The prices shown in the Assemblies section of *Means Repair and Remodeling Cost Data* include the installing contractor's overhead and profit. Chapter 25 of this book provides a complete explanation of how labor rates, overhead, and profit are included and used.

1. **Project Overhead.** It is necessary to identify those project overhead items that have not been previously included in the estimate. Such items include field supervision, tools and minor equipment, field office, sheds, and photos. All have costs that must be included. Division 1 in the Unit Price section of *Means Repair and Remodeling Cost Data*, lists many items that fall in the Project Overhead category. Depending on the estimating precision necessary, the estimator can be as specific or as general as the job dictates. Figure 27.66 may prove beneficial for general percentage markups only.
2. **Office Overhead.** There are certain indirect expense items that are incurred by the requirement to keep the shop doors open and to attract business and bid work. The percentage of main office overhead expense declines with increased annual volume of the contractor. Overhead is not appreciably increased when there is an increase in the volume of work. Typical main office expenses range from 2% to 20%, with the median being about 7% of the total volume (gross billings). Figure 27.67 shows approximate percentages for some of the different items usually included in a general contractor's main office overhead. These percentages may vary with different accounting procedures.

Calculations for Retaining Wall

	Material	Installation	Total
Sidewalk Slab	$ .92	$1.43	$2.35
Deduct Forms	(.06)	(0.44)	(0.50)
Subtotal	.86	0.99	1.85
Factors 9% Material	.08	0.17	0.25
11% Installation			
Total	$ .94	$1.16	$2.10

Figure 27.61

SITE WORK	12.5-514	Parking Lots, Asphalt

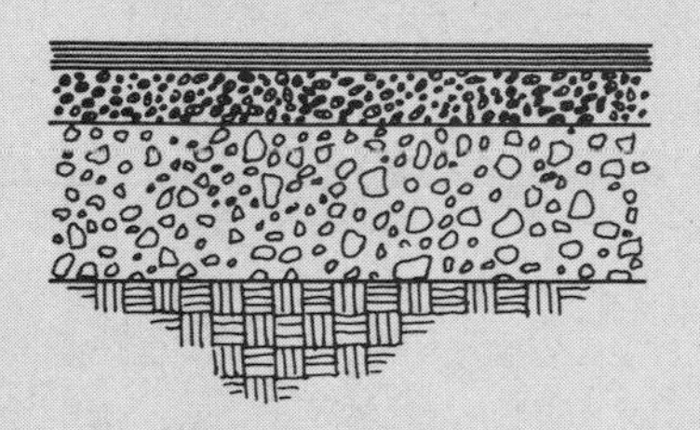

This page illustrates and describes asphalt parking lot systems including asphalt binder, topping, crushed stone base, painted parking stripes and concrete parking blocks. Lines 12.5-514-04 thru 11 give the unit price and total price per square yard for this system. Prices for alternate asphalt parking lot systems are on Line Items 12.5-514-15 thru 25. Both material quantities and labor costs have been adjusted for the system listed.

Factors: To adjust for job conditions other than normal working situations use Lines 12.5-514-29 thru 40.

Example: You are to install the system and match existing construction. Go to Line 12.5-514-29 and apply these percentages to the appropriate MAT. and INST. costs.

LINE NO.	DESCRIPTION	QUANTITY	COST PER S.Y.		
			MAT.	INST.	TOTAL
01	**Parking lot consisting of 2″ asphalt binder and 1″ topping on 6″**				
02	**Crushed stone base with painted parking stripes and concrete parking blocks**				
03					
04	Fine grade and compact subgrade	1 S.Y.		1.10	1.10
05	6″ crushed stone base, stone	.32 Ton	4.33	.58	4.91
06	2″ asphalt binder	1 S.Y.	3.28	1.20	4.48
07	1″ asphalt topping	1 S.Y.	1.75	.43	2.18
08	Paint parking stripes	.5 L.F.	.01	.05	.06
09	6″ x 10″ x 6′ precast concrete parking blocks	.02 Ea.	.46	.18	.64
10	Mobilization of equipment	.005 EA		1.28	1.28
11	TOTAL	S.Y.	9.83	4.82	14.65
12					
13					
14	For alternate parking lot systems:				
15	Above system on 9″ crushed stone	S.Y.	12.02	5.02	17.04
16	12″ crushed stone		14.16	5.18	19.34
17	On bank run gravel, 6″ deep		7.08	4.67	11.75
18	9″ deep		7.87	4.30	12.17
19	12″ deep		8.65	4.95	13.60
20	3″ binder plus 1″ topping on 6″ crushed stone		11.47	4.45	15.92
21	9″ deep crushed stone		13.66	4.65	18.31
22	12″ deep crushed stone		15.80	4.81	20.61
23	On bank run gravel, 6″ deep		8.72	4.30	13.02
24	9″ deep		9.51	3.93	13.44
25	12″ deep	↓	10.29	4.58	14.87
26					
27					
28					
29	Cut & patch to match existing construction, add, minimum		2%	3%	
30	Maximum		5%	9%	
31	Dust protection, add, minimum		1%	2%	
32	Maximum		4%	11%	
33	Equipment usage curtailment, add, minimum		1%	1%	
34	Maximum		3%	10%	
35	Material handling & storage limitation, add, minimum		1%	1%	
36	Maximum		6%	7%	
37	Protection of existing work, add, minimum		2%	2%	
38	Maximum		5%	7%	
39	Shift work requirements, add, minimum			5%	
40	Maximum			30%	
41					
42					

334

Figure 27.62

SITE WORK | 12.7-604 | Landscaping - Lawn Establishment

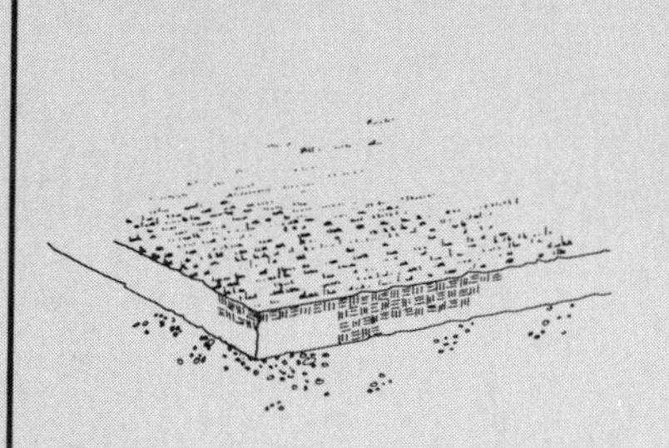

This page describes landscaping — lawn establishment systems including loam, lime, fertilizer, side and top mulching. Lines 12.7-604-04 thru 08 give the unit price and total price per square yard for this system. Prices for alternate landscaping — lawn establishment systems are on Line Items 12.7-604-11 and 12. Both material quantities and labor costs have been adjusted for the system listed.

Factors: To adjust for job conditions other than normal working situations use Lines 12.7-604-29 thru 40.

Example: You are to install the system and provide dust protection. Go to Line 12.7-604-32 and apply these percentages to the appropriate MAT. and INST. costs.

LINE NO.	DESCRIPTION	QUANTITY	COST PER S.Y.		
			MAT.	INST.	TOTAL
01	**Establishing lawns with loam, lime, fertilizer, seed and top mulching**				
02	**On rough graded areas.**				
03					
04	Furnish and place loam 4" deep	.11 C.Y.	1.78	.53	2.31
05	Fine grade, lime, fertilize and seed	1 S.Y.	.19	1.56	1.75
06	Hay mulch, 1 bale/M.S.F.	1 S.Y.	.03	.27	.30
07	Rolling with hand roller	1 S.Y.		.50	.50
08	TOTAL	S.Y.	2	2.86	4.86
09					
10	For alternate lawn systems:				
11	Above system with jute mesh in place of hay mulch	S.Y.	2.63	2.86	5.49
12	Above system with sod in place of seed	"	3.49	2.14	5.63
13					
14					
15					
16					
17					
18					
19					
20					
21					
22					
23					
24					
25					
26					
27					
28					
29	Cut & patch to match existing construction, add, minimum		2%	3%	
30	Maximum		5%	9%	
31	Dust protection, add, minimum		1%	2%	
32	Maximum		4%	11%	
33	Equipment usage curtailment, add, minimum		1%	1%	
34	Maximum		3%	10%	
35	Material handling & storage limitation, add, minimum		1%	1%	
36	Maximum		6%	7%	
37	Protection of existing work, add, minimum		2%	2%	
38	Maximum		5%	7%	
39	Shift work requirements, add, minimum			5%	
40	Maximum			30%	
41					
42					

337

Figure 27.63

12.9-300	Demolition Selective Price Sheet		
LINE NO.	DESCRIPTION	UNIT	TOTAL COST
01	Cabinets, base	L.F.	5.40
02	Wall	"	5.40
03	Carpet, bonded	S.F.	.22
04	Tackless		.05
05	Ceiling, tile, adhesive bonded		.48
06	On suspension system		.57
07	Sheetrock, on furring		.54
08	On suspension system		.60
09	Plaster, on wire lath		.76
10	On suspension system		.76
11	Chimney, brick, 16" x 16"	V.L.F.	15.15
12	20" x 20"	"	27
13	Concrete, footing, 1' thick, 2' wide	L.F.	9.10
14	2' thick, 3' wide	"	15.65
15	Slab, 6" thick, plain	S.F.	3.30
16	Mesh reinforced		3.64
17	Wall, interior, 6" thick		7.75
18	12" thick		12.35
19	Door and frame, wood	EA.	33.05
20	Hollow metal	"	47.55
21	Ducts, small size, 4" x 8"	L.F.	1.08
22	Large size, 30" x 72"		4.33
23	Fascia, to 6" wide		.43
24	To 10" wide		.54
25	Flooring, brick	S.F.	.91
26	Ceramic		.78
27	Linoleum		.31
28	Resilient tile		.43
29	Terrazzo, cast in place		1.56
30			
31	Wood, strip	S.F.	.84
32	Block		.68
33	Subflooring, tongue and groove boards		.84
34	Plywood		.45
35	Framing, steel girders, 10" x 12"	L.F.	9.45
36			
37	Wood, studs, 2" x 4"	L.F.	.22
38	Rafters, 2" x 8"	"	.60
39			
40			
41			
42			
43	Gutters, attached	L.F.	1.08
44	Built in	"	2.16
45	Masonry, veneer, by hand, brick to 4" thick	S.F.	1.95
46	Marble to 2" thick		1.52
47	Granite to 4" thick		1.61
48	Stone to 8" thick		1.56
49	Walls, brick, 4" thick		3.47
50	8" thick		4.59
51	12" thick		5.60
52	16" thick		6.83
53	Block, 4" thick		.77
54	6" thick		1.02
55	8" thick		1.53
56	12" thick		1.87
57	Paneling, plywood		.43
58	Woodboards, tongue and groove		.62

338

Figure 27.64

Means® Forms

COST ANALYSIS

PROJECT: Apartment Renovation	SHEET NO. 10 of 11
ARCHITECT:	ESTIMATE NO. 89-2
	DATE 1989

TAKE OFF BY: QUANTITIES BY: EBW PRICES BY: RSM EXTENSIONS BY: CHECKED BY:

DESCRIPTION	SOURCE/DIMENSIONS			QUANTITY	UNIT	MATERIAL		LABOR		EQ./TOTAL	
						UNIT COST	TOTAL	UNIT COST	TOTAL	UNIT COST	TOTAL
Division 12: SiteWork											
Excavation (1/2 Day)				91	CY		93		673		
Sidewalk				356	SF	.94	335	1.16	413		
Asphalt				467	SY	2.22	1037	1.94	906		
Landscaping				890	SY	2.00	1780	2.86	2545		
Demolition				1	LS				12000		
Division 12: Totals							3245		16537		

Figure 27.65

3. **Profit.** The profit assumed in *Means Repair and Remodeling Cost Data* is 10% on material and equipment, and 15% on labor. Since this is profit margin for the installing contractor, a percentage must be added to cover the profit of the general or prime contractor when subcontractors perform the work. A figure of 10% is used in the appropriate divisions of the sample estimate.

An allowance for the general or prime contractor's project and office overhead, project management, supervision, and markup should be

General Contractor's Overhead

Items of General Contractor's Indirect Costs	% of Direct Costs: As a Markup of Labor Only	% of Direct Costs: As a Markup of Both Material and Labor
Field Supervision	6.0%	2.4%
Main Office Expense	9.2	7.7
Tools and Minor Equipment	1.0	0.4
Field Office, Sheds, Photos, etc.	2.0	0.8
Performance and Payment Bond (0.5 to 1.2% Average)	0.7	0.7

Figure 27.66

General Contractor's Main Office Overhead

Item	Typical Range	Average
Managers, clerical and estimators' salaries	40% to 55%	48%
Profit sharing, pension and bonus plans	2% to 20%	12%
Insurance	5% to 8%	6%
Estimating and project management (not including salaries)	5% to 9%	7%
Legal, accounting and data processing	0.5% to 5%	3%
Automobile and light truck expenses	2% to 8%	5%
Depreciation of overhead capital expenditures	2% to 6%	4%
Maintenance of office equipment	0.1% to 2%	1%
Office rental	3% to 5%	4%
Utilities including phone and light	1% to 3%	2%
Miscellaneous	5% to 15%	8%

Figure 27.67

added to the prices in *Means Repair and Remodeling Cost Data*. This markup ranges from approximately 10% to 20%. A figure of 15% is used for the sample project for general conditions in Figure 27.68.

Estimate Summary

It is common practice to include an allowance for contingencies to cover unforeseen construction difficulties or for oversights during the estimating process. Different factors should be used for the various stages of design completion. The following serve as guides to determine contingency factors.

Conceptual Stage: add 15% to 20%

Preliminary Drawings: add 10% to 15%

Working Drawings, 60% Design Complete: add 7% to 10%

Final Working Drawings, 100% Checked Finals: add 2% to 7%

Field Contingencies: add 0% to 3%

As far as the construction contract is concerned, changes in the project can and often will be covered by extras or change orders. The contractor should consider inflationary price trends and possible material shortages during the course of the job. Escalation factors depend on both economic conditions and the anticipated time between the estimate and actual construction. In the final summary, contingencies are a matter of estimating judgment. Once the estimate is complete, an analysis is required of items such as sales tax on materials and rental equipment, as well as wheel and use taxes for the city, county, and/or state where the project will be constructed. There are some locations that tax construction labor in addition to the above items. It is crucial that the estimator check the local regulations for the area of the project to find the necessary tax information. Percentages for contingencies and sales tax are included in the Cost Analysis sheet in Figure 27.69.

A 10% figure for architectural and engineering fees is added to give the owner a better idea of the total project costs. The actual costs would be negotiated by the owner and architect.

The Assemblies Estimate for the sample project could be completed in less than one day, after initial discussions and a site evaluation. When performed with sound judgment and proper estimating practice, the Assemblies Estimate is an invaluable budgetary tool.

This book has stressed the importance of the site visit and evaluation. Every existing structure is different and must be treated accordingly. Only with a thorough understanding of how the existing conditions will affect the work will the estimator be properly prepared to analyze and estimate remodeling and renovation projects.

Means® Forms

COST ANALYSIS

PROJECT Apartment Renovation	SHEET NO. 11 of 11
	ESTIMATE NO. 89-2
ARCHITECT	DATE 1989
TAKE OFF BY: QUANTITIES BY: EBW PRICES BY: RSM EXTENSIONS BY:	CHECKED BY:

DESCRIPTION	SOURCE/DIMENSIONS			QUANTITY	UNIT	MATERIAL		LABOR		EQ./TOTAL	
						UNIT COST	TOTAL	UNIT COST	TOTAL	UNIT COST	TOTAL
Estimate Summary											
Foundations							1753		3754		5507
Substructure							174		208		382
Superstructure							4131		6029		10160
Exterior Closure							30945		21622		52567
Roofing							4483		4953		9436
Interior Construction							50422		46708		97130
Conveying systems							—		—		—
Mechanical							34739		40361		75100
Electrical							10068		17852		27920
Special							21988		8035		30023
Site Work							3245		16537		19782
Subtotal							161948		166059		328007
General Conditions				15%			24292		24909		49201

Figure 27.68

Means® Forms

COST ANALYSIS

PROJECT Apartment Renovation

ARCHITECT

SHEET NO. 11 of 11

ESTIMATE NO. 89-2

DATE 1989

TAKE OFF BY:

QUANTITIES BY: EBN

PRICES BY: RSM

EXTENSIONS BY:

CHECKED BY:

DESCRIPTION	SOURCE/DIMENSIONS	QUANTITY	UNIT	MATERIAL UNIT COST	MATERIAL TOTAL	LABOR UNIT COST	LABOR TOTAL	EQ./TOTAL UNIT COST	EQ./TOTAL TOTAL
Estimate Summary									
Foundations					1753		3754		5507
Substructure					174		208		382
Superstructure					4131		6029		10160
Exterior Closure					30945		21622		52567
Roofing					4483		4953		9436
Interior Construction					50422		46708		97130
Conveying Systems					—		—		—
Mechanical					34739		40361		75100
Electrical					10068		17852		27920
Special					21988		8035		30023
SiteWork					3245		16537		19782
Subtotal					161948		166059		328007
General Conditions		15%			24292		24909		49201
Subtotal					186240		190968		377208
Contingencies		10%			18624		19097		37721
Sales Tax			6%		12292				12292
Total Construction Costs					217156		210065		427221
Arch. & Eng. Fees		10%							42722
Total Budget									469943

Figure 27.69

INDEX

A

Access panels, metal, 170
Access problems, 17-18, 20
Acoustical tile, 192
Adjacent areas, protecting, 46
Adjacent materials, protecting, 51
Air conditioning, 234-38
Alkyds, 202
Architectural equipment, 211-12
  unit price estimating for, 349
Architectural wood doors, 167
Ashlar stone, 128
Assemblies estimating, 8-11, 375-444
  conveying systems, 403
  electrical work, 412-20
  estimate summary, 444
  exterior closure, 381-83
  foundations, 377
  general conditions, 420, 438-44
  interior construction, 383-403
  mechanical work, 403-12
  project description, 375-77
  roofing, 383
  site work, 425-33
  special, 420-25
  substructure, 377
  superstructure, 377

B

"Bare" unit costs, 63
Beams, decorative, 147
Bid bond, 101
Bids
  pre-bid scheduling, 91-93
  subcontractor, 63
  subcontractors and, 74
Blast doors, 170
Blinds, 217
Bonds, 77, 101-3
Booths, 215-16
Bracing, temporary, 46
Brick, 117-19
Builder's Risk Insurance, 262
Building stone, 126-28

C

Cabinets, 147-51, 217
Carpentry
  finish, 147-51
  rough, 145-47
Carpeting, 195-98
Carpet tiles, 197-98
"Catalogue cuts," 68
Catalyzed epoxies, 202
Caulking, 156
Ceilings, 191-95
  inspecting, 21-22
Ceramic tile, 189-91
Chlorinated rubber, 202
Clean-up, as project overhead, 101
Coat racks, 217
Cold-storage doors, 170-73
Columns, decorative, 147
Commercial wood doors, 167
Completion bond, 101
Composition flooring, 198
Computer rooms, air conditioning of, 235-38
Concrete block, 119
Concrete floors on slab form, 114
Concrete topping, 114
Concrete work, 111-14
  unit price estimating for, 302
Construction, special, 219
Construction equipment, 44
Construction estimates, types of costs in, 66
Construction materials. *See* Materials
Construction methods, old versus modern, 18
Construction Specifications Institute (CSI), MASTERFORMAT Divisions, 8, 254
Construction standards, variation in, 31
Contingencies, 81-83
Control joints, 111-12
Conveying systems, 221
  in assemblies estimating, 403
  evaluating, 32-35
  unit price estimating for, 353
Cooling. *See* Heating and cooling systems
Cost. *See also* Direct costs; Indirect costs
  allocating equipment costs, 73-74
  of drywall, 189

in-house historical cost data, 63
insulation costs, 157
millwork costs, 151
as presented in *Means Repair & Remodeling Cost Data*, 254-56
"soft" costs, 42
subcontract costs, 85
of tile, 191
types of costs, 64-66
Cost information, sources of, 63-64
Costs for units, 5
Counters, 147-51
Countertops, 217
Crew, for typical construction elements, 259-61
Critical Path Method (CPM), 91
CPM scheduling, 91-93
Cubic foot estimates, 6, 13
Curtain walls, 178
Cutting and patching, 42-44

D

Damage assessment, 28
Dampproofing, 155
Debris removal, 51
Decking, metal, 137
Decorator wood doors, 167
Deluge sprinkler systems, 231
Demolition, 47-51, 105-9
unit price estimating for, 287-302
Direct costs, 64, 67-77
bonds, 77
equipment, 73-74
labor, 68-69
material, 67-68
project overhead, 74-77
sales tax, 77
subcontractors, 74
Disposal, provisions for, 34
Distribution systems, 239-43
Dividers, 217
Door and Frame Schedule, 59
Door frames
features of, 165
metal, 169-70
Doors, 165-170
evaluating, 25
metal, 169-70
special, 170-73
unit price estimating for, 331-41
wood, 167-69
Door trim, 147
Dormitory furniture, 215
Downspouts, inspecting, 28
Drainage, inspecting, 20
Dressing compartments, 208
Drilling waste, disposal of, 34
Dry pipe sprinkler systems, 231
Drywall, 185-89
types of, 186
Drywall suspension systems, 191
Dust protection, 44

E

Electrical systems, 239-50
distribution systems, 239-43
evaluating, 38-42
lighting and power, 243-50
Electrical work
in assemblies estimating, 412-20
unit price estimating for, 363
Elevators, 32-34, 221
Entrances, 174
Environmental Protection Agency (EPA), 49
Epoxy esters, 202
Equipment
as a direct cost, 73-74
costs of, 44, 261
determining appropriate, 109
Estimates
final adjustments to, 86-89
organizing data on, 85-86
pricing of, 63-66
types of, 5-13
unit price, 7-8
Estimate summary, 85-89
in assemblies estimating, 444
in unit price estimating, 363-71
Estimating. *See also* Assemblies estimating; Unit price estimating
approaches to, 97-98
by MASTERFORMAT division, 99-104
process of, 3
Estimator, 3
Evaluation. *See* Site visits
Excavation, 47-49
problems with, 17-18
Expansion joints, 112-14
Exterior closure
in assemblies estimating, 381-83
evaluating, 24-27
Exteriors, cleaning, 24-25

F

Fasteners, for drywall, 185
Fieldstone, 126
Files and filing systems, 216-17
Finish carpentry, 147-51
Finish coats, 203
Finished work, protection of, 101
Finishes, 183-206
carpeting, 195-198
ceilings, 191-95
composition flooring, 198
drywall, 185-89
lath and plaster, 183-85
painting and finishing, 202-5
resilient floors, 198-200
terrazzo, 200-201
tile, 189-91
unit price estimating for, 341-49
wall coverings, 205-6
wood floors, 201-2
Finish hardware, 178
Fire doors, 167-69

Fire protection, 230-34
Fire resistance, for ceilings, 193
Firewalls, 32, 187
Flashing, inspection of, 28
Flooring
  composition, 198
  inspecting, 21-22
  resilient floors, 198-99
  stone, 128
  terrazzo, 200-201
  wood, 201-2
Floor mats, 217
Floor slabs, 111-14
Fluorescent lighting, 243-47
Flush doors, 167
Folding chairs, 215
Folding tables, 216
Forms, preprinted, 55-59
Foundations
  in assemblies estimating, 377
  evaluating, 16-19
Framing, 185
Framing members, 146-47
Fully adhered roofing, 162
Furnishings, 213-18
  furnishing items, 215-18
  furniture, 213-15
  moving of, 218
  unit price estimating for, 349
Furring, 146

G

Generic specification, 68
Glass, 173-78
  unit price estimating for, 331-41
Glass block, 128-31
Glazing, 173-78
Grounds, 146
Gutters, inspecting, 28
Gypsum board ceilings, 192

H

"Hard" costs. *See* Direct costs
Hardware
  for doors, 165
  finish, 178
Hazardous materials, removal of, 49
Heating, 234-38
Heating and cooling systems,
  installation of, 37-38
High-intensity discharge lighting, 248
Historic preservation guidelines, 25
Hotel furniture, 214
Hydraulic elevators, 32-34

I

Incandescent lighting, 247-48
Indirect costs, 64, 79-83
Information
  obtaining and recording, 64
  sources of cost information, 63-64
In-place units, 5
Installed unit, 5
Insulation, 156-57
  evaluating, 28
Insurance, 79, 103, 262
Interior construction
  in assemblies estimating, 383-403
  evaluation of, 30-32

J

Joints
  for concrete slabs, 111-14
  reinforcing, 119-26

L

Labor, 68-69
  costs of, 261
  for interior reconstruction, 31
Ladder reinforcing, 119
Laminated construction, 147
Latex binders, 202
Lath, 183-85
Lease costs, 73
Lecture hall seating, 216
Library furniture, 215
Lighting, 243-50
  estimates for temporary, 42
  estimating, 249-50
Loose-laid and ballasted roofing, 161

M

Masonry, 117-31
  brick, 117-19
  building stone, 126-28
  concrete block, 119
  glass block, 128-31
  joint reinforcing, 119-26
  stone floors, 128
  structural facing tile, 126
  tips for estimating, 131
  unit price estimating for, 302-10
MASTERFORMAT Divisions, 8
Material only units, 5
Materials
  analyzing material quotations, 67-68
  costs of, 261
  in demolition, 49
  of exteriors, 24
  of foundations, 16-17
  handling and storage of, 44-46
  protecting adjacent, 51
  roofing, 27
Means Repair & Remodeling Cost Data,
    253-66
  city cost indexes, 263-66
  format and data, 253-56
  reference section, 263
  unit price section, 256-63
Mechanical installations, 20
Mechanical systems, evaluating, 35-36
Mechanical work, 227-38
  in assemblies estimating, 403-12
  unit price estimating for, 353-63
Metal doors, 169-70
Metals, 135-42

miscellaneous and ornamental, 137-42
steel joists and deck, 137
structural steel, 135-37
Metal work, unit price estimating for, 310-17
Millwork, 147-51
Miscellaneous expenses, 80
Modular carpet tiles, 197
Moisture problems, in foundations, 17
Moisture protection, 155-56
unit price estimating for, 323-31
Moldings, 147
Multiple seating, 216

O
Office, costs of, 79-81
Office furniture, 213-14
Office manager, 80
Operating costs, 73
Operating overhead, 79-81
Order of magnitude estimates, 6, 13
Ornamental metals, 137-42
Overhead, 262. *See also* Project overhead
Owner's salary, 80
Ownership costs, 73

P
Painting, 202-4
Paneling, 147, 217
Parquet floors, 202
Partially adhered roofing, 161
Partitions, 208
Patching, 42-44
Performance bond, 101-3
Permits, 103
Personnel
as project overhead, 100
temporary nonconstruction, 46
Physical plant expenses, 80
Piping, inspecting, 36
Plaster ceilings, 192
Plaster work, 183-85
Plastics, 145-51
unit price estimating for, 317-23
Plumbing, 227-30
Plumbing installations, 36
Power, estimates for temporary, 42
Pre-action sprinkler systems, 231
Pre-bid scheduling, 91-93
Precedence chart, 91
Prefinished manufactured items, 207-8
Preprinted forms, 55-59
Pricing, of estimates, 63-66
Primers, 203
Productivity, 69
Professional services, 80
Profit, 81, 262-63
Project description, 273-86
in assemblies estimating, 375
in unit price estimating, 273
Project manager, 80
Project overhead, 74-77, 99-104
bonds, 101-3
general requirements, 103-4
job clean-up, 101
miscellaneous, 103
personnel, 100
temporary construction, 100-101
temporary services, 100
Project Overhead Summary, 363-71
Proprietary specification, 68
Public liability, 262
Published cost data, 64

Q
Quantity takeoff, 5, 55-62
preprinted forms for, 55-59
takeoff guidelines, 62

R
Receivables, uncollected, 80-81
Rental costs, 73
Residential wood doors, 167
Resilient floors, 198-99
Restaurant furniture, 214
Restoration, of interiors, 30
Rolling platforms, 103
Roof accessories, 163
Roof area, determining, 157
Roofing
in assemblies estimating, 383
evaluating, 27-30
inspecting the roof surface, 28
sheet metal, 162
single-ply, 161-62
Roofing materials, 161
Roofing squares, 160
Room Finish Schedule, 59
Rough carpentry, 145-47
"R" value, 157

S
Safety, 100-101
in constructing elevator shafts, 35
Sales tax, 77
Scaffolding, 103
Scheduling, pre-bid, 91-93
School furniture, 215
Screens, for privacy, 208
Sealants, 156, 203
Security, 100
Settling, of foundations, 17
Shaftwall, 187
Sheet-metal roofing, 162
Shelves, 147-51
Shift work, costs and, 46
Shingles, 157-61
Shoring, temporary, 46
Shortcuts, for quantity takeoff, 59
Siding, 161
Single-ply roofing, 161-62
Site visits, 15-53
evaluating conveying systems, 32-35
evaluating electrical systems, 38-42

evaluating exterior closure, 24-27
evaluating foundations, 16-18
evaluating general conditions, 42-47
evaluating interior construction, 30-32
evaluating mechanical systems, 35-38
evaluating roofing, 27-30
evaluating special items, 47
evaluating substructure, 19-21
evaluating superstructure, 21-24
Site work, 47-51, 105
in assemblies estimating, 425-33
unit price estimating for, 287-302
Skyroofs, 163
Social Security (FICA), 262
"Soft" costs, 42
Sound insulation, 157
Soundproofing, 187
Special construction, 219
unit price estimating for, 353
Special items, locations for, 47
Specialties, 207-8
unit price estimating for, 349
Sprinkler systems, 230-34
installation of, 37
Square foot estimates, 6, 13
Stairs, inspecting, 22
Steel joists, 137
Stone, building, 126-28
Stone floors, 128
Storage, 100
Storefronts, 174
Strip flooring, 201-2
Structural facing tile, 126
Structural steel, 135-37
Studs, 145-46
Subcontract costs, 85
Subcontractor bids, 63
Subcontractors
as a direct cost, 74
coordination of, 47
Subfloors, 146
Substructure
in assemblies estimating, 377
evaluating, 19-21
Superstructure
in assemblies estimating, 377
evaluating, 21-24
Surety bond, 103
Suspension systems, 191
Systems estimates, 6, 8-11
Systems estimating. *See* Assemblies estimating
Systems estimating divisions, 11

T

Takeoff guidelines, 62
Taxes, 79
Temporary construction, as project overhead, 100-101
Temporary personnel, 46
Temporary services, as project overhead, 100
Temporary structures, 46
Terrazzo flooring, 200-201
Thermal protection, 156-63
unit price estimating for, 323-31
Tile, 189-91
acoustical, 192
carpet tiles, 197-98
structural facing tile, 126
Time-saving measures, for quantity takeoff, 59
Toilet partitions, 208
Tongue-and-groove decks, 146
Tools, allowances for, 103
Topcoats, 203
Trash receivers, 218
Truss reinforcing, 119

U

"Unburdened" unit costs, 63
Uncollected receivables, 80-81
Undercarpet distribution systems, 239-43
Unemployment insurance, 262
Unit price estimates, 6, 7-8
Unit price estimating, 273-371
architectural equipment, 349
concrete, 302
conveying systems, 353
doors, windows, and glass, 331-41
electrical work, 363
estimate summary, 363-71
finishes, 341-49
furnishings, 349
general requirements, 287
masonry, 302-10
mechanical work, 353-63
metals, 310-17
project description, 273-86
site work and demolition, 287-302
special construction, 353
specialties, 349
thermal and moisture protection, 323-31
wood and plastics, 317-23
Units of measure, 5
Upholstery, 216
Utilities, inspecting, 36
Utility installations, 20

V

Vendor quotes, 63
Ventilating, 234-38
Venting, evaluating, 28

W

Wage rates, 68-69
Wall coverings, 205-6
Walls
drywall, 185-89
firewalls, 187
shaftwall, 187

sheathing on, 146
soundproofing, 187
Wardrobes, 217
Waste, in roofing, 160
Waterproofing, 155-56
Wet pipe sprinkler systems, 231
Windows, 173
evaluating, 25
unit price estimating for, 331-41
Wiring, 39-41
Wood, 145-51. *See also* Carpentry
unit price estimating for, 317-23
Wood doors, 167-69
Wood floors, 201-2
Workers' Compensation Insurance, 262
Worker protection, 100-101